Green Issues
and Debates

Green Issues and Debates

An A-to-Z Guide

The SAGE Reference Series on
Green Society
Toward a Sustainable Future

HOWARD S. SCHIFFMAN, GENERAL EDITOR
New York University

PAUL ROBBINS, SERIES EDITOR
University of Arizona

Los Angeles | London | New Delhi
Singapore | Washington DC

Los Angeles | London | New Delhi
Singapore | Washington DC

FOR INFORMATION:

SAGE Publications, Inc.
2455 Teller Road
Thousand Oaks, California 91320
E-mail: order@sagepub.com

SAGE Publications Ltd.
1 Oliver's Yard
55 City Road
London EC1Y 1SP
United Kingdom

SAGE Publications India Pvt. Ltd.
B 1/I 1 Mohan Cooperative Industrial Area
Mathura Road, New Delhi 110 044
India

SAGE Publications Asia-Pacific Pte. Ltd.
33 Pekin Street #02-01
Far East Square
Singapore 048763

Publisher: Rolf A. Janke
Assistant to the Publisher: Michele Thompson
Senior Editor: Jim Brace-Thompson
Production Editors: Kate Schroeder, Tracy Buyan
Reference Systems Manager: Leticia Gutierrez
Reference Systems Coordinator: Laura Notton
Typesetter: C&M Digitals (P) Ltd.
Proofreader: Sandy Zilka
Indexer: Julie Sherman Grayson
Cover Designer: Gail Buschman
Marketing Manager: Kristi Ward

Golson Media
President and Editor: J. Geoffrey Golson
Author Manager: Ellen Ingber
Editors: Mary Jo Scibetta, Kenneth Heller
Copy Editors: Tricia Lawrence, Holli Fort, Barbara Paris

Copyright © 2011 by SAGE Publications, Inc.

Printed in the United States of America

Library of Congress Cataloging-in-Publication Data

Green issues and debates : an A-to-Z guide / Howard S. Schiffman, editor.

p. cm. — (The Sage reference series on green society: toward a sustainable future)
Includes bibliographical references and index.

ISBN 978-1-4129-9694-5 (cloth) — ISBN 978-1-4129-7572-8 (ebk)

1. Environmentalism. 2. Sustainable living. I. Schiffman, Howard, (Howard S.)

GE195.G736 2011 333.703—dc22 2011007301

11 12 13 14 15 10 9 8 7 6 5 4 3 2 1

Contents

About the Editors

Green Series Editor: Paul Robbins

Paul Robbins is a professor and the director of the University of Arizona School of Geography and Development. He earned his Ph.D. in Geography in 1996 from Clark University. He is the general editor of the *Encyclopedia of Environment and Society* (2007) and author of several books, including *Environment and Society: A Critical Introduction* (2010); *Lawn People: How Grasses, Weeds, and Chemicals Make Us Who We Are* (2007); and *Political Ecology: A Critical Introduction* (2004).

Robbins's research focuses on the relationships between individuals (homeowners, hunters, professional foresters), environmental actors (lawns, elk, mesquite trees), and the institutions that connect them. He and his students seek to explain human environmental practices and knowledge, the influence nonhumans have on human behavior and organization, and the implications these interactions hold for ecosystem health, the local community, and social justice. Past projects have examined chemical use in the suburban United States, elk management in Montana, forest product collection in New England, and wolf conservation in India.

Green Issues and Debates General Editor: Howard S. Schiffman

Howard S. Schiffman is visiting associate professor of Environmental Conservation Education at New York University, Steinhardt School of Culture, Education and Human Development, where he teaches courses in Contemporary Environmental Debates and Environmental Governance. He holds a B.A. from Boston University, a J.D. from Suffolk University Law School, an LL.M. (Master of Laws) from George Washington University Law School, and a Ph.D. from Cardiff University (Wales, UK). He began his career as a criminal defense lawyer, working for several years as a staff attorney with the Legal Aid Society of New York. Afterward, Schiffman served as the founding academic director of the NYU graduate program in Global Affairs. His book, *Marine Conservation Agreements: The Law and Policy of Reservations and Vetoes,* was published by Martinus Nijhoff-Brill and was featured as part of the Publications in Ocean Development Series. In 2010, Dr. Schiffman served as a Senior Fulbright Scholar at Waikato University Law School in New Zealand, where he researched the development of the South Pacific Regional Fisheries Management Organization. Schiffman has authored numerous scholarly articles addressing marine conservation law and policy and is a corresponding editor with the *Journal of International Wildlife Law and Policy*. He is committed to environmental education as a tool to address conservation challenges.

Introduction

Debates over how best to provide for our needs while conserving the environment are not new. Even so, with the growing urgency of environmental problems and the promise of new technologies, these debates come into sharper focus. Biotechnology has the potential to transform modern living but at what cost? How long can we base our energy needs on fossil fuels? Why is it so hard to transition to renewable energy? Will we see a resumption of commercial whaling? Is organic farming truly a better option? Is carbon sequestration really a viable solution to ever-growing carbon dioxide buildup in the atmosphere? Should we value nature intrinsically as opposed to its use value to human beings? Any discussion of "Green Issues and Debates" requires the input of multiple disciplines. This final volume in the series addresses some of the more contentious environmental issues of our time, offering a variety of perspectives.

The natural sciences, philosophy, political science, law, economics, international affairs, education, and other disciplines all have something to add in resolving the vital environmental questions of our age. To understand debates over energy, for example, one must understand not only certain technical aspects but also the economic and political interests present in the debate. To understand questions of animal ethics, it is not enough to understand a legal and regulatory framework, but also the value many assign to animals as a component of nature. In the many debates flowing from climate change, one must be able to navigate the perspectives of developed and developing countries and appreciate the genuine differences of opinion that exist in how best to address this key challenge to our environment.

The friction between the needs of developing states to address poverty and lack of opportunity on the one hand, and sustainable policies promoting conservation of resources on the other, became apparent at the United Nations Conference on the Human Environment in Stockholm in 1972. At Stockholm, the Group of 77—the original number of developing countries that formed a negotiating bloc to represent the interests of those states—made a clear statement on the priority of developmental needs. This message would resonate for decades and would be addressed in virtually every major environmental initiative since Stockholm.

Many environmentalists, admittedly many from developed countries, are uncomfortable compromising environmental agendas for the sake of economic developmental needs in the developing world. The uneasy fusion of environmental concerns with developmental needs labeled "sustainable development," enshrined at the 1992 United Nations Conference on Environmental and Development in Rio de Janeiro, has hardly resolved the issue. On the contrary, the goals of environmental conservation and economic development are often difficult, if not seemingly impossible, to reconcile. This friction runs through many debates today, including forestry, fisheries and agricultural policy, and of course, climate change.

As everyone knows, not all environmental debates occur at the global level. Debates happen over whether or not to build a power plant in a neighborhood. Should water be diverted for irrigation? Should we tolerate higher levels of pollution from a factory, or the risk of an oil spill, if such commercial activity economically benefits the community? These debates, as well as global ones, reflect disagreements about particular values that must be understood and respected if workable solutions are to be found. In some cases, new, even yet-to-be discovered, technologies will overtake the debates discussed herein. In many more cases, however, it is important for observers of green debates to realize that technology by itself is not the sole solution. On the contrary, the history of scientific discovery teaches us that new technology brings new challenges and ultimately more debate. More importantly, science and technology will never release us from the commonsense realization that actions, and inactions, have consequences. Responsibility for the stewardship of Planet Earth remains with imperfect human beings.

Environmental debates, global and local alike, will be with us far into the future, and there is no better resource to ensure a constructive process than a well-educated citizenship. This includes an understanding of the various interested parties to these debates, including environmental nongovernmental organizations and other nonprofits, corporations, government agencies, intergovernmental organizations, community groups, professional associations, private foundations, lobbyists, and individuals. Suspicions often run deep between these actors, and one must be vigilant to recognize the difference between zealous, yet scrupulous, advocacy and falsehood. One need only look at the evolution of the discourse on climate change to see that this is so.

The many actors participating in environmental debates, as well as observers of those debates, must be well informed and appreciate the dizzying range of interests connected with modern environmental issues. Participants and observers must have the tools available to sift through competing claims, arguments, and counterarguments. This volume is dedicated to that ideal. Drawing on multiple themes and concepts introduced in the earlier volumes, *Green Issues and Debates* relates to many articles appearing throughout the series.

To the extent urgency is expressed in the debates addressed in this volume, it is because nothing less is at stake than the continued viability of life on Earth. The need for drinkable water, breathable air, sufficient energy, robust ecosystems, and livable coastlines leaves little room for error as we seek solutions to environmental problems. Open, honest, educated, and fair-minded debate is a vital asset in the pursuit of those solutions.

Howard S. Schiffman
General Editor

Reader's Guide

(Volume numbers refer to previous volumes in the Green Society series. The articles listed within each of the preceding 11 volumes appear in this volume on *Green Issues and Debates* and are also discussed further in the corresponding volumes.)

Volume I. Green Energy

Biomass
Carbon Sequestration
Carbon Tax
Carbon Trading/Emissions Trading
Ethanol
Green Pricing
Nuclear Power (Energy)
Oil
Oil Sands
Solar Energy (Cities)
Wind Power
Yucca Mountain

Volume II. Green Politics

Anarchism
Capitalism
Citizen Juries
Commodification
Common Property Theory
Ecocapitalism
Ecological Imperialism
Ecosocialism
Ecotax
Environmental Justice (Business)
International Whaling Commission (IWC)
North American Free Trade
    Agreement (NAFTA)
North–South Debate
Not in My Backyard (NIMBY)

Precautionary Principle
    (Uncertainty)
Resource Curse
Skeptical Environmentalism
Sustainable Development (Business)
Tragedy of the Commons
Utilitarianism

Volume III. Green Food

Agribusiness
Animal Welfare
Aquaculture
Bacillus Thuringiensis (Bt)
Ecolabeling
Fisheries
Genetically Modified
    Organisms (GMOs)
Hunting
Irradiation
Nanotechnology and Food
Organic Farming
Pesticides
Roundup Ready Crops

Volume IV. Green Cities

Adaptation
Asthma (Cities)
Biofuels (Cities)
Carbon Trading (Cities)
Green Jobs (Cities)

List of Articles

List of Contributors

Arnold, Jeffrey R.
*U.S. Army Engineer Institute for
Water Resources*

Bandyopadhyay, Kaushik Ranjan
*Asian Institute of Transport
Development*

Beck, Diana L.
Knox College

Becker, Chad A.
Indiana State University

Beder, Sharon
University of Wollongong

Bellestri, Tani E.
Independent Scholar

Boslaugh, Sarah
*Washington University
in St. Louis*

Bremer, Leah L.
*San Diego State University/University
of California, Santa Barbara*

Burns, William C. G.
*Journal of International Wildlife
Law & Policy*

Cameron-Glickenhaus, Jesse
Independent Scholar

Dedekorkut, Aysin
Griffith University

Denault, Jean-Francois
Independent Scholar

Duffy, Lawrence K.
University of Alaska, Fairbanks

Farley, Kathleen A.
San Diego State University

Feldpausch-Parker, Andrea M.
Texas A&M University

Fenner, Charles R., Jr.
State University of New York, Canton

Giessen, Lukas
University of Göttingen

Griffin, Mary Ruth
Independent Scholar

Gunter, Michael M., Jr.
Independent Scholar

Helfer, Jason A.
Knox College

Hiner, Colleen C.
University of California, Davis

Hurst, Kent L.
Independent Scholar

Imran, Muhammad
Massey University

Inkoom, Daniel Kweku Baah
*Kwame Nkrumah University of Science
and Technology*

Jarvie, Michelle E.
Independent Scholar

Kaur, Meera
University of Manitoba

Khetrapal, Neha
Indian Institute of Information Technology

Kirchhoff, Christine
University of Colorado at Boulder

Kte'pi, Bill
Independent Scholar

Lanfair, Jordan K.
Knox College

Lepsoe, Stephanie
Independent Scholar

Liss, Jodi
Independent Scholar

Menard, Nicole
Independent Scholar

Mukhopadhyay, Kausiki
University of Denver

Mullaney, Emma Gaalaas
Pennsylvania State University

Nash, Hazel
Cardiff University

Neo, Harvey
National University of Singapore

Ogale, Swati
Independent Scholar

Pajewski, Amy
Independent Scholar

Paul, Pallab
University of Denver

Pearce, Joshua M.
Queen's University

Reed, Matt
Countryside and Community Research Institute

Richerson, Kate
University of California, Santa Cruz

Rowe, Briony MacPhee
Independent Scholar

Schneider, Jen
Colorado School of Mines

Schroth, Stephen T.
Knox College

Sekerka, Leslie E.
Menlo College

Silva, Carlos Nunes
University of Lisbon

Smith, Dyanna Innes
Antioch University New England

Stancil, John L.
Florida Southern College

Steffens, Ron
Green Mountain College

Stimel, Derek
Menlo College

Swanson, Barry L.
Knox College

Trevino, Marcella Bush
Barry University

Tucker, Corrina
Independent Scholar

Turrell, Sophie
Independent Scholar

Tyman, Shannon
University of Washington

Vedwan, Neeraj
Montclair State University

Vynne, Stacy Johna
Independent Scholar

Wang, Yiwei
University of California, Santa Cruz

Yovovich, Veronica
University of California, Santa Cruz

Green Issues and Debates Chronology

12,000–6,000 B.C.E.: During the Neolithic Revolution, early humans learn to domesticate plants and animals, developing agriculture and the beginnings of settlements in the Fertile Crescent and Indus River Valley.

1,000 B.C.E.: The first known consumption of fossil fuels occurs in China. Coal is unearthed and likely used to smelt copper in rudimentary blast furnaces.

c. 500 B.C.E.: Athens, Greece, establishes what may have been the first city dump in the Western world, accompanied by a ban against throwing garbage into the streets.

c. 1530: Commercial whaling begins as the Basques begin the pursuit of right whales in the North Atlantic, killing an estimated 25,000 to 40,000 whales over the next 80 years.

1833: English chemist and meteorologist Luke Howard describes the "urban heat island" effect in *The Climate of London*, noting that the city "partakes much of an artificial warmth, induced by its structure, by a crowded population, and the consumption of great quantities of fuel."

1839: French physicist Becquerel discovers the photovoltaic effect, the process by which electromagnetic radiation is absorbed by matter, which in turn emits electrons, creating electricity.

1862: With much of the agricultural south in the United States not voting because of the Civil War, the U.S. creates the Department of Agriculture, which assumes the responsibility of promoting agriculture production and the land grant university system.

1863: The world's first subway opens in London, England.

1867: Karl Marx writes of commodification in *Das Kapital*.

1872: U.S. President Ulysses Grant signs a bill into law designating the area of Yellowstone as the world's first national park.

1873: The first U.S. federal law against animal cruelty is passed. Called the "Twenty-Eight Hour Law," it requires that "livestock transported across country be provided with water and rest at least once every 28 hours."

1878: C.V. Riley, an entomologist for the U.S. Department of Agriculture, advocates the use of biological pest controls for farming, a method used today in organic farming.

1882: The world's first hydroelectric power station, the Vulcan Street Power Plant, is built in Appleton, Wisconsin.

1886: Swedish scientist Arrhenius speculates on the greenhouse effect and suggests that higher levels of carbon dioxide in the atmosphere trap solar radiation in the form of heat.

1888: Charles F. Brush adapts the first large windmill to generate electricity in Cleveland, Ohio. Electricity-generating mills are coined "wind turbines." General Electric later acquires Brush's company, Brush Electric.

1890: The Sempervirens Club, California's oldest land conservation organization, is founded.

1891: Baltimore inventor Clarence Kemp patents the first commercial solar water heater.

1892: British reformer Henry S. Salt, a socialist, pacifist, and vegetarian, publishes a landmark work on animal welfare, *Animal Rights Considered in Relation to Social Progress*.

1892: The Sierra Club is founded in the city of San Francisco by preservationist John Muir.

1905: Upton Sinclair publishes his novel *The Jungle* in serial format in the socialist magazine *Appeal to Reason*. Public outcry over the filthy conditions of the meatpacking industry portrayed in this novel led to passage of the Pure Food and Drug Act in 1906.

1906: U.S. President Theodore Roosevelt signs the Pure Food and Drug Act, effectively creating the Food and Drug Administration.

1917: Article 27 of the Mexican constitution changes its land tenure forms, nationalizing many lands and subdividing many into a communal system of *ejidos*.

1918: The Save the Redwoods League forms in the United States for the purpose of purchasing the remaining redwood forests, which have been extensively harvested for lumber.

1920: In response to the perceived failures of existing federal laws to deal with mining of coal and oil resources, the United States passes the Mineral Leasing Act to regulate mining on public lands. The law governs deposits of coal, oil, gas, oil shale, phosphate, potash, sodium, and sulfur.

1934: The Taylor Grazing Act establishes grazing districts on public lands that have formerly been unregulated and grants leases to individuals to use them. The fees are deliberately set low and in general remain lower today than the fees charged for comparable private grazing land.

1939: Swiss chemist Paul Müller discovers that the synthetic chemical dichlorodiphenyltrichloroethane (DDT) is effective at killing insects. The chemical is widely used in the armed forces to control mosquitoes and lice and becomes popular in agriculture as well after World War II.

1942: Jerome Irving Rodale begins publication of *Organic Farming and Gardening*, popularizing the concept of organic food production as advocated by the British writers Sir Albert Howard and Lord Northbourne.

1946: The International Convention for the Regulation of Whaling holds its first meeting in Washington, D.C., and sets quotas for whaling that are intended to allow the whaling industry to continue at reduced levels so that whales are not hunted to extinction.

1949: The International Convention for the Regulation of Whaling establishes the International Whaling Commission, which is intended to regulate whaling but has been beset by conflicts between nations with traditional whaling industries (e.g., Japan, Iceland, and Norway) and those that wish to impose a moratorium on all whaling.

1952: London, England, experiences a thermal inversion in December, which combined with air pollution from automobiles, factories, and coal-burning furnaces, blankets the city in smog. The smog is believed to have caused about 3,000 deaths.

1954: The introduction of highly industrialized fishing vessels into the codfish industry in the United States and Canada greatly increases the annual catch, causing a steady decline in the cod population to the point where, in 1977, it was determined that the number of spawning codfish off the coast of Newfoundland had decreased by 92 percent since 1962.

1956: The first victims of Minamata disease, caused by alkyl mercury poisoning from eating contaminated fish, are identified in Japan. Victims suffer from neurological impairments, including retardation, contorted limbs, and sensory disturbances. The source of the contamination is traced to the Chisso Company, which had been dumping industrial pollutants into Minamata Bay on the west coast of Kyushu Island.

1965: Scientist James Lovelock proposes the Gaia hypothesis.

1968: Bowling Green State University's Center for Environmental Programs is established.

1968: Garret Hardin publishes the *Tragedy of the Commons* in *Science*, which is widely cited by many environmentalists, but also criticized for its insistence that markets are the solution to all environmental problems.

1970: The University of Colorado becomes the first college to establish an Environmental Center.

1970: Norman Borlaug, father of the "green revolution," which is credited with substantially increasing crop yield in the third world, wins the Nobel Peace Prize. Although few question that the green revolution saved millions of people from starvation, many particularly in more recent years have criticized Borlaug's reforms because they rely heavily on chemical fertilizers, irrigation, and seeds that must be purchased annually from multinational corporations or first-world research institutions, thus increasing corporate control of third-world agriculture and dependency on modern science.

1970: The National Environmental Policy Act (NEPA) is passed, requiring all federal agencies to compile and submit an Environmental Impact Statement.

1971: Frances Moore Lappé publishes *Diet for a Small Planet.* Lappé advocates for the adoption of a vegetarian diet, both for reasons of health and because of the much greater resources required to produce meat rather than vegetables and grains.

1972: The Environmental Protection Agency bans the use of DDT in the United States.

1973: The Arab Oil Embargo begins as a response to the U.S. decision to supply the Israeli military during the Yom Kippur War.

1973: Ernst Friedrich Schumacher's *Small Is Beautiful: Economics as if People Mattered* criticizes the assumption that economic development requires adoption of large-scale Western technologies and a lifestyle based on acquisition of consumer goods.

1973: Philosopher Arne Naess coins the term *deep ecology* to describe a philosophical orientation that focuses not only on reducing pollution but also recognizes that nature has a significant intrinsic value.

1974: The U.S. Department of Energy forms a branch dedicated to national research and development of solar energy—the Solar Energy Research Institute.

1975: The Convention on International Trade in Endangered Species of Wild Fauna and Flora (CITES), signed in 1973, comes into force. The convention's purpose is to end international trade in endangered or threatened animal and plant species or products made from them and requires signatories to implement domestic laws (such as the Endangered Species Act in the United States) to carry out the convention's principles.

1975: Jim Hightower coins the term *McDonaldization* in his book *Eat Your Heart Out* and warns of the danger of international corporations such as McDonald's destroying local cuisines as well as driving small farms and restaurants out of business.

1976: Wes and Dana Jackson found the Land Institute in Salina, Kansas, to develop sustainable agriculture practices. The institute includes both a school and an agricultural research station that investigate alternatives to industrial agricultural practices, such as monoculture and extensive use of pesticides and herbicides.

1976: Bioethics professor Peter Singer publishes *Animal Liberation: A New Ethics for Our Treatment of Animals.*

1977: The First Intergovernmental Conference on Environmental Education is held in Tbilisi, Georgia.

1978: Supertanker *Amoco Cadiz* runs aground off the coast of France, emptying the entire cargo of 1.6 million barrels of oil into the water. The largest oil spill in history, it is estimated that $250 million in damages occurred.

1979: A partial core meltdown occurs at the Three Mile Island nuclear generating station, releasing radioactive gases into the Pennsylvania air. This near-catastrophe draws attention to the dangers involved in nuclear power generation and halts construction of new nuclear power facilities in the United States.

1980: Alex Pacheco and Ingrid Newkirk found People for the Ethical Treatment of Animals (PETA) to protect animal rights. Philosophers such as Tom Regan and Peter Singer heavily influence the PETA founders, in particular that their belief that "speciesism," or the belief that man is superior to all other species, is incorrect.

1983: The World Commission on Environment and Development, commonly known as the Brundtland Commission, is held.

1984: An accident at a Union Carbide pesticide-producing plan in Bhopal, India, releases a toxic cloud of methyl isocyanate, causing more than 2,000 deaths immediately and at least 2,000 more in ensuing years.

1985: A group of activists found the Rainforest Action Network in San Francisco, California, with the purpose of protecting the world's rainforests and the people who live in them from environmental destruction. Their first major action is a boycott of the American fast food chain Burger King, which at that time imported much of its beef from Central and South America, where rainforest destruction was hastened by the economic incentive of clearing the forest and turning it into grazing land for cattle.

1985: Robyn Van En coins the term *community-supported agriculture* and establishes the first collective in the United States in Massachusetts. In this system, community members buy shares of a farmer's crops in advance and receive their portion of the crops, which are delivered as they are harvested. This shifts some of the risk of crop loss, as well as the benefits if a growing season is particularly productive, from the farmer to members of the collective.

1986: Chernobyl, Ukraine, then part of the Soviet Union, becomes the site of the worst nuclear disaster in history.

1986: The world's largest solar thermal facility is commissioned in Kramer Junction, California.

1986: Arcigola Slow Food, the forerunner of the international Slow Food movement, is founded in Italy as a protest against standardized food produced by international corporations such as McDonald's as well as a celebration of local products and traditions.

1986: Historian Alfred Cosby writes about ecological imperialism, describing how ecosystems were transformed through the colonization of the Americas.

1987: The United Church of Christ Commission for Racial Justice issues a study demonstrating that the location of toxic waste sites is more closely related to the race of neighborhood residents than to either income or social class.

1987: Burger King announces that it will no longer import beef from rainforest areas.

1987: American activist Dave Foreman publishes *Ecodefense: A Field Guide to Monkeywrenching*, which advocates sabotage to prevent environmentally destructive development and other commercial activities. Many of Foreman's suggested tactics are illegal, including driving metal spikes into trees to prevent their being logged, sabotaging earthmoving equipment such as bulldozers, removing surveyor's stakes, and pulling down power lines. The book's title refers to *The Monkey Wrench Gang*, a 1975 novel by Edward Abbey, which called for individuals to take direct action to halt the destruction of wilderness.

1988: Negative publicity about polluted beaches in New York and New Jersey leads to passage of the Ocean Dumping Ban Act, which takes force in 1991. The act prohibits dumping industrial waste and sewage sludge into the ocean and is more stringent than previous laws, which allowed some dumping as long as it did not seriously degrade the marine environment.

1988: The Dutch development agency Solidaridad introduces fair-trade coffee in the Netherlands under the label "Max Havelaar" (referring to a fictional character who opposed exploitation of coffee pickers).

1988: Ron Arnold and Alan Gottlieb publish *The Wise Use Agenda*, which states that the opposition of the Wise Use Movement in the western United States led to increased government regulation of commercial uses of public lands.

1989: The worst oil spill in American history at the time occurs when the supertanker ship *Exxon Valdez* grounds on a reef and spills more than 11 million gallons of oil into Prince William Sound near Valdez, Alaska.

1989: Kalle Lasn founds the Canadian advocacy group Adbusters, which popularizes "Buy Nothing Day," corresponding to the biggest day of the U.S. holiday shopping season.

1990: The U.S. Food and Drug Administration approves the Pfizer product Chy-Max chymosin, a genetically modified version of rennet used in cheese. Currently, 60 percent of U.S. cheeses are produced using genetically modified chymosins.

1990: The Intergovernmental Panel on Climate Change (IPCC) releases its first report on the status of climate science.

1990: The Organic Foods Production Act, which sets national standards for the production, handling, and labeling of organic food, is passed.

1991: The Flavr Savr tomato, the world's first genetically engineered food crop, is developed by Calgene and approved by the U.S. Food and Drug Adminstration.

1991: U.S. President George H. W. Bush incorporates the Solar Energy Research Institute into the new National Renewable Energy Laboratory. Its mission is to develop renewable energy and energy-efficient technologies and practices, advance related science and engineering, and transfer knowledge and innovations to addressing the nation's energy and environmental goals by using scientific discoveries to create market-viable alternative energy solutions.

1992: Under President Clinton, the Energy Policy Act of 1992 is passed by U.S. Congress. It is organized under several titles enacting legislation on such subjects as energy efficiency, conservation, and management; electric motor vehicles; coal power and clean coal; renewable energy; alternative fuels; natural gas imports and exports; and various others. Among the new directives is a section that designates Yucca Mountain in Nevada as a permanent disposal site for radioactive materials from nuclear power plants. It also reforms the Public Utility Holding Company Act to help prevent an oligopoly and provide further tax credits for using renewable energy.

1992: The Earth Summit held in Rio de Janeiro results in the document "Agenda 21," which calls for national governments to adapt strategies for sustainable development and to cooperate with nongovernmental organizations and other countries in implementing them.

1992: The Environmental Protection Agency introduces the voluntary Energy Star program to help consumers evaluate the energy efficiency of common products. Specific standards vary by product, but most Energy Star products represent an improvement of at least 20–30 percent in energy efficiency over the traditional version of the same product.

1992: The European Union introduces the European Ecolabel to help consumers identify products—from paper to televisions to paint—that have relatively low impact on the environment over the life cycle of the product.

1993: The U.S. Green Building Council is founded as a nonprofit trade organization that promotes self-sustaining building design, construction, and operation. The council develops the Leadership in Energy and Environmental Design (LEED) rating system and organizes Greenbuild, a conference promoting environmentally responsible materials and sustainable architecture techniques.

1993: U.S. President Bill Clinton's administration suggests a tax on energy consumption per British thermal unit (btu), which is widely criticized.

1994: U.S. President Bill Clinton signs Executive Order 12898, requiring federal agencies to determine the impact that environmental degradation has on low-income communities.

1994: The North American Free Trade Agreement (NAFTA) takes effect, and rebels in Chiapas, Mexico, choose the moment to launch their counterinsurgency to protest the impacts of NAFTA on rural and indigenous Mexico.

1996: Monsanto plants the first commercial fields with the Roundup Ready soybean, the first widely adopted commercial genetically modified crop in the United States. The beans are engineered to resist the common herbicide glyphosate, which can therefore be sprayed on the fields without damaging the soybean crop. By 2009, more than 90 percent of soybeans grown in the United States are engineered to be herbicide resistant.

1996: William Rees and Mathis Wackernagel develop the concept of the "ecological footprint," which signifies all the resources used by a particular species, in their book *Our Ecological Footprint: Reducing Human Impact on the Earth.*

1997: The term *affluenza* is introduced in a documentary of the same name to refer to a cycle the filmmakers see in contemporary Western society: the unbounded pursuit of material possessions that provide ever-decreasing satisfaction but produce the desire to accumulate still more goods.

1997: The Fair-Trade Labeling Organizations International is founded in Germany with the goals of bringing together disparate fair-trade organizations and harmonizing standards for fair-trade certification.

1997: The U.S. Department of Agriculture publishes draft rules for organic food production that include genetically modified organisms (GMOs), sewage sledge, and irradiation, drawing a significant amount of protest from consumers and food groups.

1997: The Kyoto Protocol, an international agreement linked to the United Nations Framework Convention on Climate Change that aims to reduce or prevent global warming, is adopted. Under the protocol, which goes into effect in 2005, most industrialized countries agree to reduce their emissions of greenhouse gases. Some were given specific targets, and some were given the goal of reducing their emissions to 1990 levels, while others were allowed to reduce their levels. The protocol also allows countries to trade carbon emissions in order to meet their goals.

1998: Great Britain announces that by 2008, more than 60 percent of new housing starts will be on brownfield grounds; that is, they will reuse previously developed lands, such as abandoned commercial developments, that are underutilized or neglected.

1999: A study by Cone Millennial Cause finds that most (89 percent) of Americans age 13–25 would be willing to switch to a brand associated with a "good cause" and that many would prefer to work for a socially responsible company.

2000: New York City introduces Green Building Tax Credits, which offer tax breaks to developers whose buildings meet energy efficiency standards.

2000: The Biomass Research and Development Board is created as part of a U.S. Congress act attempting to coordinate federal research and development of bio-based fuels obtained by living (as opposed to long-dead fossil fuels) biological material, such as wood or vegetable oils. Biofuel industries begin to expand in Europe, Asia, and the Americas.

2001: The U.S. Green Building Council founds the Green Building Certification Institute to certify Leadership in Energy and Environmental Design (LEED) professionals who are qualified to evaluate the sustainability of buildings.

2002: The Center for Food Safety publishes *Fatal Harvest: The Tragedy of Industrial Agriculture*, a collection of essays criticizing the effects of industrial agriculture, which is the norm in the United States, on the environment and on human health. Specific criticisms include overuse of pesticides, loss of topsoil, use of genetically engineered species, loss of agricultural diversity, and loss of family farms.

2002: William McDonough and Michael Braungart popularize the term *cradle to cradle,* which was introduced by Walter Stahel in the 1970s. Cradle to cradle refers to the principle that companies should be responsible for recycling materials from their products after they are discarded.

2004: Bjørn Lomborg writes *The Skeptical Environmentalist.*

2005: NanoJury UK, a citizen jury debating the risks and merits of nanotechnology, offers a verdict on the technology's future.

2005: The European Union Emission Trading Scheme, a carbon-trading scheme involving 25 of the then-27 European countries, officially begins.

2005: In *Lost Child in the Woods*, Richard Louv coins the term *nature deficit disorder* to describe how children are developing behavioral problems as a result of spending less time outdoors than previous generations.

2005: The Energy Policy Act is passed by U.S. Congress and signed into law by George W. Bush, making sweeping reforms in energy legislation, mostly in the way of tax deductions and subsidies. Loans are guaranteed for innovative technologies that avoid greenhouse gases, and alternative energy resources such as wind, solar, and clean coal production are given multimillion-dollar subsidies.

2006: The United Nations releases a study showing that the U.S. meat industry has a larger impact on global warming than any other industry.

2006: The *New Oxford American Dictionary* selects "carbon neutral" as its word of the year.

2006: Walmart launches an initiative requiring its suppliers to reduce packaging to the lowest possible levels.

2007: San Francisco, California, bans polystyrene foam (Styrofoam) containers for takeout food, requiring that containers used for that purpose be compostable or recyclable.

2007: The College of the Atlantic announces that it will become the first "carbon-neutral" university.

2007: The Energy Independence and Security Act of 2007 is passed by U.S. Congress and signed by George W. Bush. Its stated purposes are "to move the United States toward greater energy independence and security, to increase the production of clean, renewable fuels, to protect consumers, to increase the efficiency of products, buildings and vehicles, to promote research on and deploy greenhouse gas capture and storage options, and to improve the energy performance of the federal government," as well as various other goals. Included in the new provisions is a requirement of government and public institutions to lower fossil fuel use 80 percent by 2020.

2007: In the U.S. Supreme Court case *Massachusetts v. Environmental Protection Agency*, the court rules 5–4 that the EPA has the legal authority to regulate the emission of heat-trapping gases in automobile emissions.

2007: The state of California enacts laws that spur the creation of its Green Chemistry Initiative.

2007: California Governor Arnold Schwarzenegger establishes the low-carbon fuel standard, which aims to reduce the carbon intensity of fuels by 10 percent by 2020.

2008: President George W. Bush signs the Higher Education Sustainability Act (HESA) into law.

2009: Amid a global recession, the American Recovery and Reinvestment Act of 2009 is one of the inaugural acts signed by President Barack Obama. It makes provisions for job creation, tax relief, and infrastructure investment, but is also heavily focused on energy efficiency and science. Multibillion-dollar funding is appropriated toward energy-efficient building practices, green jobs, electric and hybrid vehicles, and modernizing the nation's electric grid.

2009: President Barack Obama halts 20-year-old plans to make Yucca Mountain the national repository of nuclear waste.

2010: Three administrative Nuclear Regulatory Commission judges ruled that President Barack Obama did not have the power to close the Yucca Mountain repository unilaterally. It ruled that doing so would require another act of Congress.

2010: The largest offshore oil spill in U.S. history occurs as an explosion rocks the Deepwater Horizon oil rig, spilling approximately 35,000 to 60,000 barrels of oil per day into the Gulf of Mexico and severely harming the region's environment.

March 2011: The nuclear debate is reignited around the world as a result of the Japanese nuclear emergency at the Fukushima Daiichi power plant, which is located about 150 miles north of heavily populated Tokyo. The leak of radiation, brought on as a result of Japan's 9.0 earthquake and subsequent tsunami, prompted safety reviews of nuclear power plants in China and the European Union. In the United States, President Barack Obama asked the Nuclear Regulatory Commission to do a "comprehensive review" of domestic plants in light of the events in Japan.

Dustin Mulvaney
University of California, Berkeley

Adaptation

Adaptation to the extreme weather caused by climate change includes creating higher dams and building larger reservoirs like Hoover Reservoir, which holds 20 billion gallons of water and provides drinking water to more than 500,000 people.

Source: http://Columbus.gov

The Intergovernmental Panel on Climate Change (IPCC) defines adaptation to climate risks as adjustments to reduce vulnerability or enhance resilience—the ability to withstand change or shocks—in response to observed or expected changes in climate. Adaptation includes management, planning, policy, or other actions taken by individuals, communities, governments, or other entities to reduce the risks or impacts posed by climatic variability or climate change. Humans have a long history of adapting to climate variability such as irrigating crops instead of relying solely on precipitation, constructing reservoirs to store water during water short periods, and storing food and feed. In addition, humans have adapted to climatic extremes such as armoring coastlines and strengthening building codes to withstand storm surges and high winds, constructing levees to hold back high stream flows to lessen flood risks, or planting drought-resistant crops.

While humans have historically adapted to climate risks, climate change may make adaptation more challenging because the range of past experiences may not adequately bound future climate-related risks. The uncertainty around the magnitude of future climate risks has made it difficult to achieve consensus on the appropriate course of action on climate change. The debate centers around whether global resources should be directed toward aggressive mitigation—reducing greenhouse gases that contribute to anthropogenic climate change—or aggressive adaptation efforts, or whether resources should be directed at a combination of adaptation and mitigation efforts.

Background

To understand the debate requires first understanding a little about the climate of the Earth system. Earth's climate is affected by the incoming energy from the sun, the energy radiated from the Earth, and the exchange of energy between the atmosphere, land, oceans, ice, and living things. The composition of the atmosphere is important because it affects the flow of incoming solar radiation and outgoing infrared radiation. Carbon dioxide, methane, ozone, nitrous oxide, and water vapor are all greenhouse gases naturally occurring in the atmosphere. These naturally occurring greenhouse gases warm Earth's surface by blocking infrared radiation from Earth and redirecting a portion of that heat back toward the planet surface. This natural greenhouse effect is essential for life to exist on the planet.

A million years before the Industrial Revolution, the carbon dioxide (CO_2) concentration in the atmosphere ranged between 170 and 280 parts per million (ppm). Since the Industrial Revolution, the concentration of CO_2 has increased to 387 ppm in large part from the burning of carbon-based fossil fuels. Concentrations of methane and nitrous oxide have also increased as a result of the burning of fossil fuels as well as from farming practices, industrial activities, and land use changes. The increased concentrations of these greenhouse gases trap infrared radiation, further warming the planet. According to the Fourth Assessment Report of the IPCC, warming attributed to anthropogenic climate change has increased the global average temperature by 0.8 degrees Celsius since the beginning of the industrial period. Furthermore, because greenhouse gases like carbon dioxide persist in the atmosphere for centuries, the current elevated greenhouse gas concentrations likely commit the world to warming of nearly 2 degrees Celsius.

Most of the attention from research and policymakers has been to increase understanding of the climate system to better attribute the causes and consequences of climate change in order to inform mitigation policies. Over the past 20 years, advancements in the science of climate change have greatly improved our understanding of the climate system. Unfortunately, while understanding of the climate system has increased, there has been little progress to date in reducing greenhouse gas emissions.

Global efforts to mitigate greenhouse gases can be traced to the United Nations Framework Convention on Climate Change (UNFCCC), which entered into force in 1994. The UNFCCC established a framework for intergovernmental efforts to address greenhouse gas emissions and to cooperate for adapting to the impacts of climate change. The UNFCCC does not bind signatories to specific emissions reductions targets. Rather, emissions targets were established in the 1997 Kyoto Protocol. Unfortunately, existing international agreements have not resulted in significant reductions in greenhouse gas emissions or in reducing the risks posed by climate change. In fact, global consumption of fossil fuels continues to add about 6 billion metric tons of carbon to the atmosphere each year. The concentration of carbon dioxide is now about 30 percent higher than it was at the start of the Industrial Revolution.

The inability to slow greenhouse gas emissions means that the planet is already committed to some anthropogenic climate change. And, in fact, climate change impacts are already manifesting in higher average air and ocean temperatures, shifting precipitation patterns, melting snow and ice, and rising sea levels. Cold days, nights, and frost events are less frequent, while heat waves are becoming more common. More frequent heavy precipitation events on one hand and more frequent and intense droughts on the other are also attributed to the impacts of climate change. Furthermore, the impacts of climate change are not distributed evenly across the globe. Temperature changes are greatest at the poles, while near the equator, increasing water vapor leads to more intense rain and storm events.

The Adaptation Versus Mitigation Debate

Some argue that because (1) the global political process has thus far failed to reduce or limit greenhouse gas emissions and (2) climate change is already evident, more attention should be paid to adaptation. That is, because global efforts have failed to arrest emissions to limit the impacts of climate change and the likelihood of the success of future global efforts to limit greenhouse gases is small, resources should be focused on preparing for climate change impacts and building resilience to climate change. Urbanization can increase resilience to climate-related risks through increased access to public infrastructure and services and early warning systems. Urban areas can increase resilience by investing in no-regrets options that provide benefits with or without climate change and by investing in increasing margins of safety such as building higher dikes and dams that can withstand larger-magnitude events or that provide more reservoir capacity with only incremental increases in cost. A third strategy involves investing in flexible and reversible decision making and long-term planning. This strategy emphasizes investment in robust, forward-looking decisions and planning that reduces path dependency and enhances resilience. While it makes sense to invest in more localized adaptation efforts to build resilience, especially given the present failure of the global political process to limit greenhouse gas emissions, there is a countervailing argument that suggests adaptation without mitigation is risky.

While it is true that climate change is already evident and that existing greenhouse gas emissions present in the atmosphere commit the planet to further climatic changes, abandoning mitigation efforts risks undermining adaptation. Though it appears "easier" to adapt to climate change than to "fight" or mitigate climate change, it may be more difficult to adapt if mitigation efforts fail. For example, if greenhouse gas concentrations increase the global average temperature above 2 degrees Celsius—the level beyond which experts predict the world will experience disruptive climate change impacts—the physical impacts of future climate change may include ecosystem collapse, species extinction, and greater sea-level rise. If temperatures rise even higher, more dire consequences are predicted, including the extinction of some 50 percent of species worldwide, inundation of coastal wetlands and other low-lying areas resulting from a sea-level rise of several meters, die-off of coral reefs, salinization of coastal aquifers, and other mostly irreversible impacts. At present, some 600 million people globally and 15 of the world's 20 megacities are located in low-elevation coastal areas and are at risk from rising sea levels and storm surges. The risk of unabated climate change may render adaptation in these and other areas moot if rising sea levels swamp coastline defenses, make water supplies too salty to drink, and inundate low-lying urban areas. On the other hand, stabilizing emissions may limit the risk of catastrophic events and bound future climatic conditions, supporting more effective adaptation efforts.

The debate over how much mitigation to pursue or how much adaptation to pursue is important because the discussion brings into sharp focus the consequences of this false either/or proposition. Resources directed toward global mitigation efforts, if successful, can keep more localized options on the table, avoiding irreversible consequences of climate change. Similarly, investments in adaptation strategies can help cope with the otherwise unavoidable impacts of climate change. In the end, pursuing both mitigation and adaptation efforts is the prudent course.

See Also: Carbon Emissions (Personal Carbon Footprint); Mitigation; Sustainable Development (Politics).

Further Readings

Adger, W. N., S. Dessai, M. Goulden, M. Hulme, et al. "Are There Social Limits to Adaptation to Climate Change?" *Climatic Change*, 93:3–4 (2009).
Bierbaum, R., M. Fay, J. Bucknall, S. Frankhouser, et al., eds. "World Development Report 2010: Development and Climate Change." Washington, DC: World Bank, 2010.
Intergovernmental Panel on Climate Change (IPCC). *Climate Change 2007: Impacts, Adaptation and Vulnerability.* Contribution of Working Group II to the Fourth Assessment Report of the IPCC, M. L. Parry, O. E. Canziani, J. P. Palutikof, P. J. van der Linden et al., eds. Cambridge, UK: Cambridge University Press, 2007.
Intergovernmental Panel on Climate Change (IPCC). *Climate Change 2007: Mitigation.* Contribution of Working Group III to the Fourth Assessment Report of the IPCC, B. Metz, O. R. Davidson, P. R. Bosch, R. Dave, et al., eds. Cambridge, UK: Cambridge University Press, 2007.
Intergovernmental Panel on Climate Change (IPCC). *Climate Change 2007: The Physical Science Basis.* Contribution of Working Group I to the Fourth Assessment Report of the IPCC, S. Solomon, D. Qin, M. Manning, Z. Chen, et al., eds. Cambridge, UK: Cambridge University Press, 2007.
United Nations Framework Convention on Climate Change (UNFCCC). "Essential Background." http://unfccc.int/essential_background/items/2877.php (Accessed September 2010).

Christine Kirchhoff
University of Colorado at Boulder

AGRIBUSINESS

Agribusiness refers to the farms and firms directly involved with food production. It includes the food corporations that supply seed, fertilizers, and pesticides (e.g., ConAgra, Monsanto, Cargill, Archer Daniels Midland [ADM], and Syngenta). Due to extensive vertical integration and contract farming, agribusiness extends to the businesses of food processing, distribution, marketing, sales, and research (e.g., Tyson Foods and McDonalds). The term is used to describe large-scale corporate food businesses at the confluence of commodity production and processing. It refers to a type of agriculture associated with modernity and practiced most notably in the United States. Other terms that allude to agribusiness are *factory farms*, *corporate agriculture*, and *agro-industrialization*.

The mid-20th century witnessed a dramatic change in agricultural production. Industry-wide, food production has become enlarged, intensified, and incorporated. Cheap fuel and cheap land made this quick expansion possible. Technological breakthroughs such as the internal combustion engine, rail transport, and refrigeration cars have advanced agricultural productivity. The result of these changes has been a shift toward fewer and larger farms. Food production is now characterized as highly mechanized and governed by efficiency, profit, and share price, following an industrial model. Agribusiness thus mirrors the capitalist practices and values of industry.

The debates surrounding agribusiness extend to larger questions regarding the current state and the future of the world food system. There are myriad ecological and social problems associated with agribusiness. The ecological problems include a break in the soil-nutrient cycle, pollution of land and water, soil erosion, lower water tables, and a threat to ecological diversity. Social concerns include the concentration of power and wealth, increasingly centralized decision making, threats to human health, and the destruction of rural economies and communities. The benefits inherent to agribusiness lie in increased productivity and lower economic costs of food. Alternative forms of agriculture, including agroecology, provide sustainable solutions to the problems posed by agribusiness.

Environmental Threats

Agribusinesses engage in farming practices that negatively affect the environment through soil erosion, water use, chemical use, biodiversity and habitat loss, effluents and pollution, energy requirements, and genetically modified organisms (GMOs). Following a capitalist model of efficiency of scale, the agribusiness model of food growing is highly specialized. In other words, it usually involves growing one crop of genetically similar plants, or mono-cropping. Such farming represents a threat to biodiversity and habitat. It is also at greater risk of disease and pesticide attack. Ironically, though the industrial model of agribusiness is founded on the concept of efficiency, it has not had an efficient energy return. Each calorie of food grown in the United States takes approximately 10 calories of fuel to produce.

Twentieth-century agriculture has been highly dependent on fossil fuel–based inputs in the form of pesticides, fertilizers, and fungicides. In addition, increasing dependency on mechanization has increased the amount of fossil fuels that are used in farming operations. The result, according to a 2004 estimate by the United Nations' Intergovernmental Panel on Climate Change (IPCC), is that agriculture accounts for 13.5 percent of global greenhouse gases (GHGs). Livestock accounts for another 18 percent, and Greenpeace estimates that the contribution might be as high as 32 percent if the GHGs are calculated throughout the food production chain, including processing, packaging, and distribution. Production of synthetic fertilizers and pesticides also requires significant energy input. Land deforested for agro-industrial production directly contributes to planetary warming by decreasing carbon sequestration. Because of its scale, agribusiness is a leading cause of these impacts. A detailed report published in spring 2010 by the Agribusiness Action Initiative, "A Harvest of Heat: Agribusiness and Climate Change," calculates the climate impacts of ADM, Cargill, Dean Foods, Dow, Tyson, and Monsanto. Environmentalists argue that sustainable forms of small-scale agriculture that grow for their local community would greatly reduce the energy use of agriculture. Other agriculturalists believe that a higher number of smaller farms would result in a higher energy expenditure.

Food in the United States travels an average of 1,500 miles from farm to plate. This accounts for 80 percent of the energy associated with the life cycle of each food item. Locally grown food mitigates this huge environmental cost. Factory farms, however, are designed to efficiently produce a vast quantity of food that exceeds local demand and thus gets shipped to supermarkets around the world. The increasing prevalence of supermarkets globally has led to the critique of agribusiness for co-opting local food markets. Such multinational relationships to food, critics argue, lead to a lack of responsibility to communities, and accountability for social, economic, and environmental costs is made more complex.

The global versus local debate touches on social justice issues as well. Though most agribusiness profits are collected by inhabitants of the global north, workers, farmers, and consumers in the global south often pay social and environmental costs. International trade liberalization policies in agriculture have benefited transnational agribusiness and encouraged export-oriented monocrop agriculture. The industry argues that the consumer benefits with more food choices at a lower cost.

Animal Production

Concentrated animal feeding operations (CAFOs), also referred to as confined animal feeding operations, are a common means by which agribusiness raises animals for meat, dairy, and egg production. Thousands of animals can be held in one CAFO facility. There are approximately 450,000 CAFOs in the United States alone. Critics of CAFOs and industrial agriculture argue that factory farming has a negative impact on animal welfare, rural quality of life, human health and well-being, and the environment. A 2008 report by the Pew Commission on Industrial Farm Animal Production (PCIFAP), "Putting Meat on the Table: Industrial Farm Animal Production in America," detailed these concerns.

CAFOs and other industrial models of animal production raise ethical questions regarding the treatment of farm animals as machines. Steps have been taken by some companies to mitigate the stress to the animals, but critics argue that the conditions are unacceptable for raising animals and that it is not possible to humanely raise animals in such concentrated numbers. For this reason, People for the Ethical Treatment of Animals (PETA) supports the development of technology such as in vitro meat that would alleviate the need to raise live animals for slaughter. Other environmental and animal rights groups see genetic engineering as an environmental and social threat and instead promote vegetarianism and veganism and urge people to eat less meat.

The large-scale use of antibiotics in CAFOs prevents disease outbreaks but also increases the number of antibiotic-resistant bacteria. In the United States, 70 percent of all antibiotics sold are fed to industrially raised cattle, swine, and poultry for purposes other than treating disease. There is concern that the widespread use of these antibiotics could lead to antimicrobial resistance (AMR) in foodborne pathogens. Such pathogens may have a negative impact on human health because they cannot be treated with commonly used medications. In June 2010, the U.S. Food and Drug Administration (FDA) released a draft statement titled "The Judicious Use of Medically Important Antimicrobial Drugs in Food-Producing Animals," suggesting guidelines for the use of antibiotics. There has been increasing public pressure in the United States to create legally binding regulations to protect public health from the negative impacts of extraneous antibiotic use.

Animal waste is another serious environmental concern. In cases where there are thousands of animals in a small area, the waste is generally dumped into open-air lagoons that can contaminate local water systems with runoff that contains high nitrogen content. This runoff in turn causes algal blooms that lead to eutrophication, or the depletion of oxygen from the water, due to excess nutrient buildup. The result is massive die-offs and blooms of toxic algae that kill fish and can afflict humans.

Technology and Agribusiness

The model of agricultural production encouraged by agribusiness is heavily dependent on technological innovation. The technocratic assumption is that new technology will solve the environmental problems associated with large-scale, single-crop production. Among the technologies most promoted by agribusiness is genetic engineering (GE). In the face of tremendous population growth and unpredictable weather related to climate change, proponents of agribusiness argue that large-scale crop and animal production as well as biotechnology are necessary to feed the world.

Scholars and activists critical of agribusiness see this convergence of science, technology, and business strategy as a threat to small farmers, indigenous farming practices, and land ownership. Biotechnology research is expensive and has been concentrated in the hands of a few transnational companies. Monsanto, a multinational agrochemical company, for example, produces about 70 percent of all GE crops grown worldwide. The consolidation of GE technology mirrors the concentration of the rest of the agricultural industry. The industry sees genetically modified crops as a solution to the unpredictable impacts of climate change. These climate-ready seeds, according to the industry, would be drought and disease resistant. Others have observed that genetically modified organisms (GMOs) require constant environmentally costly inputs such as herbicides like Roundup, which in turn contribute to the buildup of carbon dioxide. A report published by the Union of Concerned Scientists, "No Sure Fix: Prospects for Reducing Nitrogen Fertilizer Pollution Through Genetic Engineering," encourages systems-based approaches and diverse technologies to alleviate the problem of nitrogen optimization. The report also encourages more publicly supported research so that there is greater democratic control of research agendas.

A final critique of agribusiness-owned GE technology is the plight of small farmers around the world. GE seeds are self-reproducing, so farmers must buy stock from seed companies each year. In order to receive a high yield from their crops, they must also buy accompanying fertilizers, pesticides, and herbicides, often from the same company that sold them the seeds. In this way, farmers remain stuck in the GE technology treadmill. As a result, farmers around the world have found themselves in tremendous debt, and a rash of farmer suicides has resulted. At the 2003 World Trade Organization (WTO) meeting in Cancun, Mexico, Korean farmer Lee Kyung Hae publicly committed suicide to draw attention to the plight of small farmers around the world. The free-market logic that currently governs global agricultural policies, he and others have argued, benefits agribusiness at the expense of small farmers, who cannot make a living wage.

Social Problems

Critics of agribusiness argue that the corporate profits engendered by large-scale businesses threaten individual rights by concentrating decision-making power in the hands of a few. This concentration of wealth and power is thus a threat to a democratic food system whereby individuals have control over their food system and are a part of decisions affecting the way their food is grown. Due to efficiency of scale and vertical integration, these businesses are able to provide food at a lower price than small independent farmers, putting many out of business. It is argued that the cost of growing/raising food does not decrease with scale, but that the environmental and social costs are externalized to the rest of society. In this sense, agribusiness is a threat to both food sovereignty and food security.

Increasingly, power in the food industry has shifted from food producers to food processors and from publicly to privately funded research initiatives. In addition, there is a trend toward increasing concentration of ownership, whereby a few companies control vast shares of world food markets. For example, ADM and Cargill control 65 percent of the global grain market. Decision making is thus also concentrated. Current international policies promoted by the World Bank (WB) and the International Monetary Fund (IMF) support export-oriented crop production. This vision of agricultural production supports the agribusiness model of large-scale, single-crop production vertically integrated with processing, distribution, and marketing. There is no place in such an economy for subsistence farming.

As communities have less say in land use and food choices, rural communities and small farmers are at risk. Seed ownership is of particular concern as companies patent traits. Monsanto, for example, owns 90 percent of the genetically modified soybeans planted in the United States, and 90 percent of the soybeans planted in the United States are genetically modified. This has led to the observation by groups such as the Center for Food Safety that there is extraordinarily high market concentration in biotechnology.

The current food system provides significant health risks, including nutrition-related disease, foodborne pathogens, and food insecurity. Food safety is a growing concern, especially in light of the 2006 *E. coli* scare, which was eventually traced to California spinach, and the 2010 egg recall, the largest in history. In addition, there are growing concerns about the nutrition and quality of foods grown and processed by agribusinesses. Value is increasingly added to food post-production in the stages of processing, packaging, and advertising. The proliferation of prepackaged, processed foods, then, has been correlated with the growth of vertically integrated companies that control all stages from field to shelf. Increasing rates of obesity and diabetes, diseases whose origins and onset lie heavily with diet, have been correlated with the fast food industry and agribusiness, most notably in *Fast Food Nation* by Eric Schlosser.

The scale of agricultural production as dictated by agribusiness is such that workers labor in the fields much like factory lines. Unlike family farms where the family provides a majority of the labor, larger farms do not have a good track record of responding to worker rights. Outrage at the treatment of farmworkers made public by the protests of the United Farm Workers and César Chávez in the mid-1960s led to union organizing and some accountability for the chemicals used in the fields, but farmworkers continue to struggle for better working conditions.

Solutions

Consumer pressure has in some cases resulted in the adoption of environmentally friendly policies on the part of agribusiness. Activists and scholars critical of agribusiness continue to cite concerns about the industry, including the increasing size of farming operations, lack of labor standards, and corporate ownership of the organics industry. An important part of their platform is that government policies in the United States and Canada encourage corporate large-scale ownership, thus driving family farms—defined as farms where the family provides most of the equity and labor—out of business and threatening human health and the environment. In exchange, they seek socially just, economically viable, and ecologically friendly agricultural practices over those that are governed only by higher profits and efficiency. Organizations such as Robert Kennedy Jr.'s brainchild, Waterkeepers, have advocated stronger regulation of agribusiness. Other groups such as the New England Small Farm Institute advocate for small farms and sustainable farming practices. Other outspoken advocates of alternative agriculture include Indian physicist Vandana Shiva and environmentalist David Suzuki. Industry representatives see agribusiness as the future direction of the food industry in the world market.

Organic agriculture has been presented as a solution to the problems presented by industrial agriculture. Though organic food is produced without using most conventional pesticides, fertilizers made with synthetic ingredients, sewage sludge, bioengineering, or ionizing radiation, organic production is not immune to the capitalist logic of agribusiness. In fact, since the U.S. Department of Agriculture (USDA) mandated regulations for Certified Organic agriculture in 2001, the $4-billion-a-year industry has reproduced the patterns of agricultural modernization. Many organic farms produce thousands of acres of genetically similar monocultures. In addition, ownership is highly concentrated. According to research done at the University of California, Davis, 2 percent of California's organic farms accounted for half of the state's organic sales in the mid- to late 1990s.

The alternative to the industrial model of agriculture, whether or not organic, is small-scale, mixed farming with closed nutrient cycles. *Agroecological farming*, the umbrella term for system-based agricultural practices, emphasizes context rather than promoting a single type of agricultural practice. Critics of industrial agriculture promote an agricultural system that pursues values beyond maximum efficiency such as human rights, democracy, culture, and geography. The most common critique of such an agricultural system is that it will not be able to produce enough food to feed the world's population. This is disputed by estimates compiled by the Food and Agricultural Organization (FAO) that 50 percent of the world's food is wasted after being grown.

See Also: Agriculture; Animal Welfare; Antibiotic/Antibiotic Resistance; Genetically Engineered Crops; Organic Farming; Pesticides; Roundup Ready Crops.

Further Readings

Agribusiness Action Initiatives. http://www.agribusinessaction.org (Accessed September 2010).

Allen, Gary and Ken Abala. *The Business of Food: Encyclopedia of the Food and Drink Industries*. Westport, CT: Greenwood Press, 2007.

Jansen, Kees and Sietze Vellema. *Agribusiness and Society: Corporate Responses to Environmentalism, Market Opportunities and Public Regulation*. London: Zed Books, 2004.

Magdoff, Fred, John Bellamy Foster, and Frederick H. Buttel, eds. *Hungry for Profit: The Agribusiness Threat to Farmers, Food, and the Environment*. New York: Monthly Review Press, 2000.

Norberg-Hodge, Helena, Todd Merrifield, and Steven Gorelick. *Bringing the Food Economy Home: Local Alternatives to Global Agribusiness*. West Hartford, CT: Kumarian Press, 2002.

Shannon Tyman
University of Washington

AGRICULTURE

The green revolution and advances in technology, such as this combine, allowed more food to be produced by far fewer people and created widespread environmental and societal changes.

Source: iStockphoto

In any debate about sustainability, agriculture features very highly because it is the primary way that humans physically sustain themselves and relate with nature, however indirectly. Agriculture is the deliberate manipulation of plants, animals, and ecosystems to provide food, fiber, and fuel for human ends. To that extent, only people who live in cultures of gathering and hunting from wild ecosystems do not engage in agriculture.

As an important relationship between humans and natural systems and due to its impact in modifying ecosystems, agriculture is involved in a wide range of debates about the appropriate use of the technology and the social goals of agriculture. Debates about agriculture can be solely about food production, or they can be proxies or symbols of larger debates about the goals of a community. Given the reliance of agriculture on natural resources, as planetary processes such as climate change or resource exhaustion become more prominent, the role of agriculture becomes more important in societal debates.

History and Development

The present configuration of agriculture as a set of technologies, animals, and plants came into place during the middle part of the 20th century, largely driven by the use of oil and its derivatives. Gasoline and diesel became the force behind farm machinery, replacing draft animals such as horses, mules, camels, and oxen. Fertilizers, pesticides, and other chemicals that increase the productivity of plants and animals are also broadly derived from oil.

These technologies are most effective when used in combination with varieties of plants and, later, animals that are specifically bred to respond well to them. This process of the intensification of agricultural production is generally known as the "green revolution." This intensification allows far more food to be produced by far fewer people and leads to a concentration of farms into larger units with an increasing focus on producing food commodities for trade in the global market. Fewer people are needed to work the land, meaning that more labor is available for use in other industries. It started the demographic trends that led to humans moving from being a predominantly rural species to an urban one by the beginning of the 21st century. Scientists such as Norman Borlaug saw the green revolution as a way of escaping mass starvation in the post–World War II period, and it was seen by others as a way of ensuring that the West won the Cold War.

Impact of Agricultural Intensification

This intensification of agriculture came at a price, not just in profound changes to farms and rural communities due to the spread of new technologies, but also to the farmed environment and the ecosystems surrounding it. As the Dust Bowl in the midwestern states of the United States demonstrated in the 1930s, the new technologies could wreak havoc on fragile ecosystems, which in this case caused soil degradation.

The introduction of systemic pesticides after World War II saw these novel chemicals deployed on a wide scale and also had unexpected impacts on wildlife. These chemicals accumulated in insects and in animals higher up in the food chain. As they ate pesticide-treated plants, insects and other animals ingested small amounts of the chemicals. The effects were unexpected; bird populations in some areas dropped significantly after their eggshells were weakened as a result of the contamination, and some mammals, like otters, proved to be very sensitive, and their numbers declined sharply. It was also found that humans metabolized these chemicals into their body fat, and in turn, this could be transmitted via breast milk to infants. The exposure of these interconnections by Rachel Carson in her 1962 book *Silent Spring* caused many of the existing critiques to coalesce and provoke an environment critique of the use of these chemicals.

The intensification of production was not confined to plants but was taking place in animals as well. It was found that confining animals and feeding them rations with high protein content could raise productivity. Furthermore, their health could be maintained with the routine use of antibiotics and, later, with artificial growth hormones. These methods transformed the production of chickens for both eggs and meat, giving rise to the term *battery farming,* although it went on to influence the production of pork and North American beef. As with the intensification of cereal crops, this led to a concentration of production on larger units but also to a greater absolute availability of the produce and often a considerable drop in retail prices.

According to many critics, another consequence was the forming of a new relationship between humans and animals that was based on cruel treatment. Although farmed animals were always intended for slaughter, animals in these new systems were unable to express any of their natural behaviors and lived in conditions of extreme stress. To stop hens from pecking one another, their beaks were removed. Likewise, pigs had their teeth removed in order to prevent them from biting one another. The environmental consequences of this intensive form of production were twofold. In the immediate area, there were problems of disposing of the manure from these units. This waste, in such volume and concentration, is highly damaging to ecosystems, particularly if it enters the watercourse or rivers. As these units became increasingly divorced from farms, material that could potentially be used as fertilizer becomes a pollutant. The second effect was that in feeding animals concentrated feeds, animal production became an important user of grains that could otherwise be fed directly to humans. Rather than grazing or foraging as in other farming systems, animals were fed the crops from other increasingly distant farms. For example, nations such as Brazil and Argentina began to grow soy to feed animals in Europe and the United States. Critics of this form of animal production accuse it of making food less available for the world's poor in order to satiate the world's rich.

Globalization and Agriculture

Food products have always been traded, as evidenced by the ancient amphora found on the bed of the Mediterranean Sea or the distribution of the chicken (a native of southeast

Asia but found around the globe even before the modern era) testify. Contemporary agriculture has focused on growing a relatively narrow range of crops and varieties that can be easily traded globally. This has been boosted by innovations that allow humans to control the ripening of produce so that they can be transported using refrigerated logistics chains across the planet. Fruit grown in the summer in the southern hemisphere can be shipped and sold during the winter in the north. Remote locations such as New Zealand and Chile can become important players in fresh produce production and global distribution, rather than solely processed goods such as wine. Consumers can become familiar with produce that is exotic to their area—not just fruit in the winter, but fruits from different cultures. This has increasingly brought the cost of oil and the creation of carbon dioxide from transporting food across such great distances to the forefront of public discussions, with the environmental footprint of such foods being questioned. At the same time, others have questioned the impact upon local food cultures by the constant availability of a homogenized range of foodstuffs.

Debate and Controversy

In the past decade, there has been a renewed interest in agriculture beyond what for many years had been seen as a narrow and increasingly technical area; this reflects in many ways the environmental constraints now facing agriculture. This has created a number of controversies that speak to past debates and discussions about the intensification of agriculture, and to problems that are unique to the present.

The introduction of genetically modified (GM) crops in North America in the early 1990s was largely uncontroversial. The large corporations behind these new technologies argued that they had simply increased the speed and accuracy of traditional plant technologies. Plants would now be resistant to insect attack after genes from a *bacillus* (BT) was introduced into the plant's genome, or they would be resistant to herbicides so that only weeds would be killed. The proponents of these new technologies argued that these new crops would boost productivity and allow agrichemicals to be applied more sparingly, thus lessening the impact of agriculture on the wider environment. Major agricultural producers, such as Brazil, Argentina, and Australia, appeared to be favorably disposed to these innovations.

The attempt to introduce these technologies into Europe led to them being questioned and debated. The counterarguments to these new technologies were that they consolidated the hold of a few multinational companies over the food chain, as they owned the patents on the genes and the chemicals, and thus consolidated a form of agribusiness that endangered people and the planet. Test plots of the plants were destroyed as protestors courted arrest and, in some cases, were sent to jail. Naked protestors invaded meetings about the crops, and consumers boycotted products with even traces of the crops. This led to protests in North America and to many other nations questioning the technology, with arguments that it had no place in the lucrative European market. The future of agriculture became divided between those who saw it as based in some way on GM technology and those who opposed it.

Exacerbating these debates was a decision by the administration of U.S. President George W. Bush to boost the energy security of the United States by turning some agriculture crops into fuels. Using sugars derived from plants to fuel engines has a long history and has even been part of Brazil's national fuel strategy for several decades due to the availability of sugarcane. According to its advocates, fuel derived from plants (also called biofuel) offers a way of literally growing fuel that will help lower both carbon footprints and dependency on fossil fuels. The policy was an immediate success, with millions of tons of U.S. corn being refined into fuel. This was closely followed, as some would argue, by a global rise in the price of food. Grains globally traded, and on which many of the poor in large cities of the developing world rely on, became more expensive as they were being turned into fuel. Other commentators, such as the British environmentalist George Monbiot, demonstrated that the energy used to grow and refine the corn into fuel exceeded that produced by it. Some proponents of biofuels acknowledged the problems and argued that attention should be shifted to second-generation biofuels—those made from waste products—with the eventual aim of algae being used to produce ethanol to be used as fuel. This opened the possibility that a future growth area of agriculture may be growing algae as a source of biofuels.

For many years, hunger plagued human societies, particularly for the poor and marginalized. Subsistence agriculture, growing enough for a family's needs from a plot of land controlled and possibly owned by that family, was a basic survival strategy and the cornerstone of many societies. In contemporary societies, where the majority of people live in towns and cities and agriculture is conducted on large farms, agriculture has become a more distant part of life. For many environmentalists, this distance from the land and the natural world is of profound importance. Some, such as E. O. Wilson, have gone as far as to say that humans have innate feelings toward living things, what he terms *biophilia*, which informs agriculture. Some writers and thinkers, such as Wendell Berry, have pointed to the virtues of small-scale farming. Others, such as Vandana Shiva, have argued that the technological advances of the green revolution have led to a greater distance of people from nature and each other, leading to increased inequality and violence. Popular social movements such as the organic food and farming movement or, more recently, a global alliance of small farmers and peasants called Via Campesina, have contested the trajectory of contemporary agriculture. These environmentalist arguments are premised on people having a more direct relationship with agriculture and, in turn, a better connection to the ecosystems sustaining life on the planet.

The specter of hunger in part informs the discussion about the resilience of agriculture and how it can provide food security. Food security is a multifaceted topic, but in this context it can be best explored through considering the connection between agriculture and man-made climate change. In William Ruddiman's 2005 book *Plows, Plagues and Petroleum*, he states that agriculture had an impact on changing the climate before the Industrial Revolution, although at a much lower level than contemporary trends. Agriculture has been implicated in the production of greenhouse gases like methane, which is actually more damaging than carbon dioxide, and is produced from rice production and expelled from ruminant animals, particularly cattle. Agricultural systems that are highly reliant on fossil fuels have a particularly high carbon footprint, especially through the use of nitrogen-based fertilizers. Production is only part of the picture when it comes to contributions toward climate change, as the distribution of food can also have a considerable impact. The scale and scope of these contributions is disputed as methodologies are devised and revised. However essential agriculture is, it is becoming clear that it can cause a range of effects in producing greenhouse gases depending on the technologies used and the crops or animals being raised.

Agriculture can also play a role in lowering emissions that contribute to global warming through a variety of routes. First, there is an increasing interest in discovering ways through which agricultural soils can sequester carbon, thereby acting as a carbon sink and preventing carbon dioxide from entering the atmosphere. Agricultural soils often have high levels of carbon in them because of vegetative materials, as well as permanent covers such as grasses either for grazing or for use as biofuels. It has been suggested that this could be augmented by the introduction of biochar charcoal, which would be buried and in turn become incorporated into the soil. Proponents argue that through correct management, agriculture could play an important role in stabilizing climate change. This optimistic perspective on agriculture under conditions of climate change is particularly welcome as weather and climatic conditions continue to be a central factor in the success of agriculture.

Conclusion

The International Assessment of Agricultural Knowledge, Science and Technology for Development (IAASTD), a 2008 intergovernmental session in Johannesburg, South Africa, considered the role of agriculture in providing food, health, and ecological and social sustainability. The report they issued was sympathetic to many of the goals of environmental sustainability but also emphasized the interaction between poverty and environmental degradation that had not been a mainstream part of many of the environmental critiques of agriculture. The topic of agriculture will likely remain a focus of social debate.

See Also: Agribusiness; Animal Ethics; Animal Welfare; Biofuels (Cities); Biological Control of Pests; Carbon Sequestration; Ethanol; Genetically Modified Organisms (GMOs); Irradiation; Nanotechnology and Food; Organic Farming; Pesticides and Fertilizers (Home); Roundup Ready Crops.

Further Readings

Allen, Gary and Ken Abala. *The Business of Food: Encyclopedia of the Food and Drink Industries.* Westport, CT: Greenwood Press, 2007.
International Assessment of Agricultural Knowledge, Science and Technology for Development (IAASTD). http://www.agassessment.org (Accessed January 2011).
Magdoff, Fred, John Bellamy Foster, and Frederick H. Buttel, eds. *Hungry for Profit: the Agribusiness Threat to Farmers, Food, and the Environment.* New York: Monthly Review Press, 2000.

Matt Reed
Countryside and Community Research Institute

ALTERNATIVE ENERGY

There is not a single, universally accepted definition of what constitutes *alternative energy* but in general the term refers to energy sources not derived from fossil fuels. Often, there is an additional stipulation that alternative energy sources be renewable and cause less harm to the environment (particularly that they are less likely to promote global warming) than traditional sources of energy such as petroleum or coal. Another feature often included in the definition of alternative energy is that to be "alternative," it must be less commonly used than "traditional" energy sources, so that hydroelectric and nuclear power would not qualify as "alternative." New alternative energy sources are constantly being developed, but this article will concentrate on several in which there is sufficient research and experience of actual use to make meaningful statements about issues and controversies regarding them: solar, wind, and biomass.

Energy is necessary for human society: even the simplest societies use energy sources such as wood or dung for cooking. However, energy use varies widely across societies: the World Resources Institute estimates that the global average energy consumption per capita in 2005 was 1,778 kgoe (kilograms of oil equivalent) per person, while the average for North America was 7,942.9 kgoe, and for South America 1,152.2 kgoe. Per capita energy consumption tends to be greater in high-income (5,523.6 kgoe) than in low-income (491.8 kgoe) countries and greater in developed (4,720 kgoe) than developing (975.9 kgoe) countries. However, in most countries energy use is increasing, with some of the most rapid increases coming in developing regions of the world. For instance, per capita energy use in Asia increased more than 35 percent from 1990 (775.8 kgoe) to 2005 (1,051.5 kgoe). Although per capita consumption is higher in more developed regions of the world, the search for ways to conserve energy and find alternative sources has led to a decrease in some regions: for instance, in Europe per capita consumption fell about 7.5 percent in the same time period, from 4,080.4 kgoe in 1990 to 3,773.4 kgoe in 2005.

The global demand for energy is expected to increase in the future, making the search for alternative energy crucial to developing countries as well as to those already at a high level of development. Already, increased use of traditional sources of energy such as wood has led to depletion of those resources in some regions, while increased use of energy sources such as petroleum and coal has produced levels of air pollution that endanger human health. Each type of alternative energy offers a different set of potential advantages and disadvantages, but salient concerns include the feasibility of replacing traditional energy sources with alternatives, the cost of energy produced from alternative sources, and the specific requirements and environmental costs associated with each type of alternative energy.

Solar Power

Since ancient times, humans have used the energy of the sun to do useful work such as heating and cooking, and these uses continue in modern times. For instance, many modern buildings are designed to use the sun's rays for passive solar heating, and in many parts of the world, solar cookers and water heaters are in common use. The sun's energy is virtually limitless, and these applications present almost no environmental cost other than that of manufacturing and transporting the necessary equipment, but the direct application of such simple technologies is limited. However, even in developed countries, this type of application can be usefully combined with other technologies: for instance, designing buildings to take advantage of the sun's rays can significantly lower energy consumption and costs for heating and lighting, even if those buildings also use traditional energy sources such as electricity.

Most discussions about solar power today are concerned with methods to convert solar energy to electricity, a process discovered in the 1950s and in practical use since the 1970s. Photovoltaic systems can be self-sufficient, making them useful for isolated locations, but their usefulness is somewhat limited by their inefficiency: large arrays of solar cells are necessary to generate practical amounts of electricity because the amount of electricity generated is proportional to the area of cells exposed to the sun. In addition, the cells can only generate electricity when the sun is shining, meaning that they are most useful in areas with few obstructions casting shade such as trees or tall buildings (hence solar panels are typically installed on the roofs of homes) and in which overcast days are rare. A storage system or backup source of electricity is required for the nighttime and for days when the sun is obscured by clouds. Often, homeowners remain attached to their local electrical grid after installing a solar power system so they can sell excess electricity generated during the sunlight hours to the local power system (although because photovoltaic systems generate direct current, it must be converted to alternating current).

More efficient generation of energy is possible with concentrated solar power systems that use lenses or mirrors to focus the sun's rays on photovoltaic cells or to generate heat to power a turbine (similar to a conventional power plant). This type of installation is designed to generate large amounts of electricity for a large area, similar to a conventionally fueled municipal power plant. These systems are more efficient in terms of converting the sun's energy to electricity but also require a dedicated location, often in a desert area (for instance, the Solar Energy Generating Systems in the Mojave Desert of the United States) that may be ecologically fragile. Infrastructure similar to those of conventional power systems is required to distribute the energy, and roads and other infrastructure are necessary to build and maintain a concentrated solar power plant. However, the greatest barrier remains that of cost: electricity generated from solar power is generally more expensive (15 to 17 cents per kilowatt hour according to the U.S. Energy Information Administration) than electricity generated by conventional means (generally below 10 cents per kilowatt hour in the United States).

Wind Power

Man has harnessed the energy of the wind for centuries to do useful work. For instance, windmills have been used since at least the 9th century to pump water and grind grain and are still used for those purposes today. However, modern discussions of wind power generally center on the use of wind turbines to operate generators that produce electricity. Like solar energy, wind energy can be harnessed on a small scale, making it possible for an isolated location to generate power for its own use, or it can be used to operate large-scale wind farms connected to a grid that distributes the generated power to homes, businesses, and so on.

A study by the European Commission found that wind power has the lowest external costs of several different types of energy when taking into consideration global warming, building and crop damage, human health, and ecological impact; the energy types used for comparison were solar photovoltaic, hydropower, nuclear energy, biomass, gas, oil, and coal. Wind power creates virtually no pollution beyond that which is necessary to manufacture and transport the equipment and electrical grid plus roads and other infrastructure necessary to allow building and servicing the equipment. Placement of wind turbines is important not only because they need to be widely spaced in order to not compete with each other (i.e., so one turbine does not limit the wind available to another) but also because of noise pollution. Wind turbines produce three types of noise: the clearly audible sound of the blades plus low-frequency and ultra-low-frequency sound (20 Hz, below the normal range of human hearing) from the turbine's motors. Although studies have not identified risks to human health from low- and ultra-low-frequency sound, people living near wind turbines have reported a variety of medical conditions, including nausea, headaches, sleep disturbance, and irregular heartbeat, which some researchers believe are associated with exposure to the noise produced by the turbines.

To limit the noise problem, wind farms are usually placed some distance from cities, for instance, on rural wind farms or offshore. Wind farms require large areas of land, but most of it is not used for the turbines (they are generally tall towers with a rotor raised well above ground level) and can therefore be used for grazing, agriculture, and forestry. One exception is that the area around the base of the turbines may need to be cleared during installation, particularly in a forested area. Another environmental consideration is the risk wind turbines create for bird populations, both from the hazard of birds flying into the turbines and the potential disturbance of habitat and nesting areas. However, several studies have determined that the risk to birds from wind farms is relatively small compared to other human activities such as automobiles and aircraft, and environmental regulations about the siting of wind farms can minimize habitat destruction. A final issue in siting wind turbines is that they have been subjected to NIMBY (not in my backyard) objections, meaning that individuals are not opposed to the construction of wind farms per se but do not wish them to be located near their property (or to be visible from their property). A famous example is the opposition expressed to Cape Wind, a proposed offshore wind farm in Nantucket Sound (Massachusetts) that has been opposed by some residents on the grounds that it would obstruct views of the ocean, thus decreasing property values and also potentially hurting the area's tourist trade.

As with the sun, the energy of the wind is renewable and virtually inexhaustible, but harnessing wind energy for large-scale power generation is most feasible in certain regions of the world and not in others. In addition, turbines only generate power when the wind is blowing, so alternative means of power may be required. Wind farms are not practical near cities because tall buildings interfere with access to the wind; also, the large quantities of land required favor installations in less expensive geographic regions. Near-shore and off-shore locations offer the benefit of higher average wind speeds while removing the noise of the turbines from populated areas but have been contested in some areas because they were felt to spoil the natural beauty of the region. The cost of electricity generated by wind has dropped rapidly since the early 1990s and in some regions of the United States is competitive (less than 5 cents per kilowatt hour) with that produced by conventional energy plants.

Biomass

Biomass refers to organic materials, often from plants, that can be used to generate heat or electricity. Using organic materials as an energy source is an ancient practice (e.g., burning wood for cooking and heating), but in the modern sense, biomass refers most often to using such materials to generate electricity. There are many potential sources of biomass: some of the most common are wood (including forest residue left behind by logging operations), biodegradable wastes, and plants such as hemp, corn, and sugarcane and switchgrass (both crops grown to create biofuel and the wastes of crops grown for other purposes). Biomass is a relatively new energy source but one with great potential: it is not only renewable but also creates a practical use for wastes that need to be disposed of anyway (e.g., garbage that would otherwise be left in a municipal landfill).

When biomass is burned, it creates air pollution, but studies by the U.S. Energy Information Administration suggest that operating coal-fired power plants with biomass would result in substantial reductions in emissions of carbon dioxide, sulfur dioxide, and nitrogen oxide. Closed-loop processes, in which power is generated using purpose-grown feedstocks (material grown specifically to be burned, e.g., switchgrass), can achieve almost net zero carbon dioxide emissions, even when including in the calculations the emissions required for harvesting, transportation, and feed preparation. Gasification of biomass can also be used to operate power plants: for instance, the McNeil Generating Station in Burlington, Vermont, operates primarily through wood combustion but also includes a wood gasifier that extracts gas from wood chips that are also used to augment the energy supplied to the boiler.

Biomass currently is used on a small scale for electricity generation, and several questions remain about the feasibility of large-scale use of biomass to produce electricity. One is the supply of raw material that also has competing uses and whose price can fluctuate accordingly. In addition, the use of waste materials from agriculture and forestry may have unexpected ecological effects and may play an important part in the ecosystem of the soil and the forest. In addition, the relatively high start-up costs of biomass power facilities, as opposed to those that use more traditional fuels, are also expected to limit expansion of this type of energy production.

As world demand for energy is expected to increase while stocks of traditional fuels such as petroleum decline further, development and utilization of alternate energy sources is expected in the future. The practicability of these and other forms of energy production varies by location and is affected by many factors, including the availability of government or other funds for initial construction, the relative value placed on the costs of the energy thus generated, and the environmental damage associated with more traditional forms of energy.

See Also: Biomass; Hydroelectric Power; Hydropower; Solar Energy (Cities); Wind Power.

Further Readings

American Wind Energy Association. "Wind Energy Costs." http://www.awea.org/faq/wwt_costs.html (Accessed September 2010).

European Commission. "External Costs: Research Results on Socio-Environmental Damages Due to Electricity and Transport" (2003). http://www.externe.info/externpr.pdf (Accessed September 2010).

Science Daily. "Renewable Energy News." http://www.sciencedaily.com/news/earth_climate/renewable_energy (Accessed September 2010).

U.S. Energy Information Administration. "Biomass for Electricity Generation." http://www.eia.doe.gov/oiaf/analysispaper/biomass (Accessed September 2010).

U.S. Energy Information Administration. "Renewable and Alternative Fuels." http://www.eia.doe.gov/fuelrenewable.html (Accessed September 2010).

World Resources Institute. "Earth Trends: Energy Consumption." http://earthtrends.wri.org/text/energy-resources/variable-351.html (Accessed September 2010).

Sarah Boslaugh
Washington University in St. Louis

ANARCHISM

Anarchism is a political ideology based on the idea that the best way to organize society is as a free association of all members, with no overarching authority such as government or police. It should not be confused with "anarchy," which is a state of affairs evidenced by a complete

Pierre-Joseph Proudhon, the first person to proclaim himself an anarchist, believed in society as a free association of members and considered central government unnecessary and harmful.

Source: Wikimedia

breakdown of organized society resulting in an "every man and woman for himself/herself" condition; instead, anarchism is the political philosophy that argues that a centralized, coercive government is both unnecessary and harmful. In general, most anarchists have as their goal a society that organizes its social, political, and economic structures as voluntary federations of decentralized, directly democratic, policymaking bodies. Within this broad framework, many different strains of anarchism have emerged, with some that go beyond the broad rejection of the state and argue for the abolishment of all social hierarchies, while other anarchists are focused more specifically on the abolishment of capitalism or of violence, for example. Some anarchists define themselves as collectivists and others as individualistic; some embrace a revolutionary ideology that advocates a violent reordering of society, and others seek a more pacifistic route to anarchy. Other disparate strains of anarchism include anarcho-capitalism, anarcho-socialism, and green anarchism. Thus, anarchism is a widely disputed label, encompassing a vast assortment of ideologies, though the rejection of all hierarchical structures and relationships is common to all who identify as "anarchist."

Anarchism: A Brief History

The use of the term *anarchia*, meaning "without rulers," can be traced back to ancient Greek writings, such as Homer's *Iliad*, though many frame the entire period of time before recorded human history as anarchist in nature because this era lacked established authority or formal political institutions. Still, anarchism as a school of thought did not develop until the rise of hierarchical societies; at that point, anarchist ideology arose as a response to what was seen as coercive political institutions and hierarchical social relationships. During the Middle Ages in Europe, a variety of anarchist religious movements developed, including the Anabaptists, who repudiated all law, believing that the Holy Spirit would guide good people toward benevolent behavior. Religious anarchist movements were predicated on the notion that the Earth's blessings should be shared by all, that God is the ultimate authority, and that no person had the right to wield authority over another.

Evidence of anarchism can be found in the North American British colonies, particularly in religious dissenters such as Anne Hutchinson, who, in 1638, founded the town of Pocasset, site of modern-day Portsmouth, Rhode Island. Hutchinson, believing in absolute religious liberty for the individual, declared that the government did not have the right to rule the individual at all. Throughout the 18th and 19th centuries, in the United States and elsewhere, anarchist thought continued to develop, as evidenced by the 1793 publication of William Godwin's *An Enquiry Concerning Political Justice*, which is considered by many to be the first anarchist treatise, though Godwin did not use the term *anarchism*; Godwin advocated the abolition of government through a gradual process of reform and enlightenment. The Frenchman Pierre-Joseph Proudhon, who in 1840 published *What Is Property?* is credited as being the first person to proclaim himself an anarchist. Proudhon developed the theory of spontaneous order in society, believing that organization emerges best without a central coordinator imposing its ideas of order against the wills of individuals acting in their own self-interest. Proudhon was opposed to the state as well as to organized religion and certain aspects of capitalism, though he also opposed communism. He advocated an economic system known as "mutualism," marked by workers who organized themselves as small, democratic societies with equal treatment for all; he believed that capitalism resulted in exploitation of workers, akin to slavery. Proudhon's ideas were welcomed and embraced by many among the French working class and manifested in the French Revolution of 1848.

Other activists and writers associated with anarchism in the 19th century include Russians Mikhail Bakunin and Peter Kropotkin, who are associated with collectivist anarchism and anarcho-communism, respectively. In the United States during this time, a significant anarchist movement emerged, mostly focused on workers' rights, which nurtured the growth of unions, particularly the Industrial Workers of the World. The late 19th and early 20th centuries saw anarchists participating in labor movements and various uprisings, and many immigrants to the United States during this time period were anarchists, though the Red Scare of 1919 led to anarchists in the United States going "underground." The rise of fascism in Europe along with the Spanish Civil War, World War II, and the obliteration of anarchists by Joseph Stalin in the Soviet Union led to a realigning of the wider political dichotomy as a conflict between capitalism and communism, with anarchism becoming not much more than a latent notion. Anarchism reemerged during the 1960s and 1970s, and by the 1980s, many anarchists were declaring that civilization as we know it—not just the state—needed to be abolished in order to create true liberty and a just social order. The 1990s saw a further renaissance of anarchism, as the antiglobalization movement took root, fueled in large part by anarchists.

Anarchism: Liberty for All

Murray Bookchin is one of the most influential anarchist philosophers and writers of the 20th century, and he advocated the complete overthrow of capitalism. Though Bookchin later declared himself a "communalist," having grown disillusioned with the way anarchist thought had evolved, most contemporary anarchists still find in Bookchin a guiding anarchist light. Contemporary anarchists believe in the following principles:

- The essence of anarchism is free cooperation between equals to maximize their liberty and individuality.
- While knowledgeable, skilled, wise experts are valued, they should not have the authority to force people to follow their recommendations.
- No human being should dominate another.
- Society should be structured in such a way as to nurture human interactions that enhance the liberty of all.

Further, in relation to their critiques of capitalism and of organized authority and hierarchical relationships, anarchists reject racism, sexism, and homophobia; anarchists are also extremely critical of the exploitation of nature as well as the ways that technology is used to reinforce domination and hierarchy. One of the most common criticisms of anarchy is that it is rooted in chaos and a complete lack of structure, but anarchists flatly deny this claim and instead insist that liberty cannot exist without society and organization. They envision a society anchored by the following:

- Free agreement, direct democracy, and massive decentralization of power
- Organizations with accountability that are marked by power sharing, skill sharing, and task rotation
- Workers organized into unions, people into communes, communes into regions, regions into nations, and nations into one great federation

Anarchism: Impractical Utopianism

Critics of anarchism today come from both the right and left wings of the political spectrum; further, various anarchist strains are critical of each other. Infighting among anarchists revolves around issues such as collectivism versus individualism and revolutionary action versus pacifism. Further, anarchists disagree over whether or not modified versions of capitalism can exist within anarchist frameworks and also find conflict in defining degrees of change; that is, anarchists disagree, for example, in regard to their level of opposition to technology.

Capitalist critiques of anarchism include the following ideas:

- Capitalism is not inherently evil and is the only viable method for structuring the world's economic system, meeting the needs of the world's population, and thwarting problems such as ecological devastation.
- Anarchism is coercive and limits personal freedoms, while capitalism is a contractual system based on rationality and freedom.
- Anarchism is based on irrational notions of human behavior that do not acknowledge the inherent selfishness of people; a coercive government force is required in order to protect individual rights.

Socialist critiques of anarchism argue that the following is true:

- Anarchism does not offer a coherent strategy for fighting oppression.
- Anarchism lacks class-consciousness, fails to understand the relationship between exploitation and oppression, and neglects to address class-based subjugation.
- Anarchism has become more focused on protesting authority than on debating the roots of oppression—this emphasis on practice over theory represents a retreat from a goal-based, long-term strategy for reordering society.
- Contemporary anarchism ignores the power of the state and instead focuses on creating horizontal networks of self-governing institutions; believing that the state will simply dissolve in the face of the multiplication of these "autonomous zones," anarchists have no plan for what will replace the state.

As contemporary society faces profound challenges related to globalization, ecological devastation, and intense income and wealth stratification, anarchism offers for some a framework for building the world anew, while others see anarchism as nothing more than an unrealizable utopian vision, rooted in distorted notions of human nature.

See Also: Antiglobalization; Capitalism; Green Anarchism.

Further Readings

Ernesto Aguilar: Grassroots Media, Politics, Culture, Technology, and More. "Critiques of Anarchism: Resources." http://ernestoaguilar.org/critiques-of-anarchism-resources (Accessed September 2010).

Guerin, Daniel, ed. *No Gods No Masters: An Anthology of Anarchism,* Paul Sharkey, trans. Oakland, CA: AK Press, 2005.

Milstein, Cindy. *Anarchism and Its Aspirations.* Oakland, CA: AK Press, 2010.

Tani E. Bellestri
Independent Scholar

ANIMAL ETHICS

Environmental philosophers have long questioned the moral relationship between humans and animals, but animal ethics did not rise to true prominence until the late 20th century. The two most prominent schools of thought in animal ethics are animal welfare, based on ensuring the humane treatment of animals, and animal rights, based on ensuring that the inherent rights of animals as sentient beings were respected. Outcomes of the increased philosophical interest in the moral status of animals and human obligations to animals have included raised awareness and criticism of animal handling and mistreatment in the agricultural, medical, and consumer industries, leading to improvements in those areas. Philosophy has been transformed into action through social activist organizations such as People for the Ethical Treatment of Animals (PETA) and the Humane Society.

The late 20th century brought a rising philosophical interest regarding the moral status of animals, a subject that philosophers had only briefly considered in the past. Modern environmental philosophers' challenge to older anthrocentric philosophies that placed humans at the apex of evolution has necessarily led to a reevaluation of animal ethics and man's relationship with the animal kingdom. They have also challenged older views stating that humans have no moral obligations to animals. These beliefs were based in part on the philosophical view that animals were machines or automata, put forth by such philosophers as Rene Descartes.

There are two central philosophical schools of thought regarding the moral status of animals: animal welfare (liberation) and animal rights. Both animal welfare and animal rights proponents agree that animals are sentient beings able to feel pleasure and pain and with intrinsic value of their own outside their usefulness to humans, both of which should be considered by human moral agents. Many scientists and philosophers also accept the fact that animals have other mental capacities as well. Both also agree that any unnecessary animal suffering in their use by humans should be avoided when practical. Utilitarian philosopher Jeremy Bentham was one of the first to challenge older views and inspire anticruelty advocates.

Both animal welfare and animal liberation activists believe that humans are not inherently superior to other species based on their place on the evolutionary ladder, technological or cultural developments, ability to reason, or any other measure. Environmental philosopher Peter Singer claimed that such an anthrocentric view was tantamount to speciesism, which was as morally wrong as racism or sexism. Differences between animal welfare and animal rights include debates as to whether it is ethically acceptable to place human use value above animals' intrinsic values when done in a humane manner, whether animals deserve moral consideration, and whether animals have intrinsic rights that humans have an obligation to respect.

Older strains of animal welfare were primarily concerned with the humane treatment of animals in their relationships with the human world. The philosophical basis for this form of animal welfare bases the humane treatment of animals on such treatment's human benefits, an indirect duty to be kind to animals out of an obligation to maximize human good rather than a moral consideration of the animal's intrinsic value. For example, those people who practice cruelty to animals have been shown to be more likely to treat other people in the same manner, such as the statistical links between animal abuse and later serial killing. Proponents of this theory have included Immanuel Kant and St. Thomas Aquinas.

Peter Singer and Animal Liberation

Peter Singer was a key founder of the modern environmental philosophy of animal welfare or, as he termed it in his book of the same name, *animal liberation*. Modern utilitarian environmental philosophy posited that the ultimate goal of a moral agent was to base his rules and actions on the goal of accomplishing the greatest good (pleasure) and not increasing or inflicting evil (pain or suffering). Singer applied this ethic to animals as sentient beings capable of feeling both pleasure and pain, meaning that they were entitled to human moral consideration.

This second philosophical basis for animal welfare bases the humane treatment of animals on the moral importance of an animal's sentient ability to experience pleasure and pain, a direct duty to the animals themselves regardless of human benefits. Some advocates of this approach have stated that animals in the modern agribusiness, medical and scientific research, and other human enterprises suffer, often unnecessarily. Other advocates argue that such a blanket rule does not take into account the overall good that could be accomplished. For example, an animal's pain through the course of medical research could lead to a breakthrough that reduces or eliminates the pain of possibly countless humans. Thus, complex calculations of the many factors involved should be made before a determination as to the ethical (best) course of action.

Animal welfare or liberation proponents allow for the human use of animals as food sources and within the context of biomedical research as long as uses are humanely conducted. While Singer personally advocated vegetarianism due to the mistreatment of animals within modern agribusiness, animal welfare theoretically viewed meat eating as permissible if the animal was raised and slaughtered humanely. Disagreements may arise among animal welfare philosophers over what constitutes "humane" treatment.

Animal ethicists also debate the value of human–animal interactions when judging what is or is not considered humane or a human use value that justifies any negative consequences for the animal. Specific considerations have included the morality of hunting for sport versus food, animal experimentation in the interest of medical research versus consumer product testing, and the use of animals in sports such as bull, cock, and dog fighting.

These different human use values of animals are not judged to have the same value and are not judged in the same way by everyone.

Tom Regan and Animal Rights

Tom Regan was a key founder of the environmental philosophy of animal rights, which arose in the 1970s as a response to perceived failings within animal welfare or liberation that still allowed for the humane use and killing of animals. He believed that animals possessed not just a sentient ability to feel pleasure and pain, but also a sense of self that includes other mental capabilities such as desiring and aversion. This sense of self imparts an intrinsic value. Regan stated that human use of animals for food or medical research could not be condoned when done humanely because animals have their own intrinsic value that humans are bound to respect. Regan has also been involved in efforts to develop legal arguments recognizing the rights of animals.

The theory of animal rights posits that humans may not use animals for human uses such as food or biomedical research regardless of whether such uses respect the welfare of the animals involved. Other more extreme animal rights theorists have stated that man should not make any human use of animals, calling, for example, for the liberation of all domesticated animals and household pets, including dogs and cats. Disagreements among animal rights proponents have also arisen based on the determination of whether all animals or only certain animals possess rights. Regan limited such rights to animals that consciously experienced their own lives or had a sense of self.

Animal rights proponents have also clashed with members of the larger environmental community as a whole. Many environmentalists feel that the animal rights approach does not give moral consideration to other components of the natural world, such as whole or endangered species and ecosystems and inanimate objects such as rivers and trees because of their focus on the intrinsic values and rights of individual animals. They also state that focusing on individual animal rights does not take into account the pressing need to defend, preserve, and restore the larger natural world of which they are a part. Finally, they note that a blanket insistence on animal rights may hinder opportunities at humane reform within the industries they wish to end.

Animal rights proponents like Regan counter that environmentalists sacrifice the intrinsic values and rights of individual animals to the collective intrinsic values and rights of species or ecosystems, a process Regan labeled environmental fascism (sometimes called ecofascism). Such beliefs include the idea that the collective intrinsic value of ecosystems can be greater than the sum of the individual intrinsic values of their component parts. They also note the difficulties that would be encountered in seeking to assign rights to plants or other nonsentient beings or inanimate objects and to protect those rights.

What the Differences Mean

The differences between the philosophies of animal welfare and animal rights have important practical implications for human–animal interactions and animal rights activism. The philosophy of animal welfare advocates a reevaluation of the human moral understanding and treatment of animals. The philosophy of animal rights makes the case for a revolutionary new approach to the human moral understanding and treatment of animals that forbids the use of animals for purely human use value, regardless of the goodness of that value in utilitarian terms, such as the development of model organ donation or animal-based medicines such as insulin. Animal ethics has also extended beyond philosophy to related fields such as medicine, history, anthropology, and law.

A number of prominent groups and individuals have transformed these two main philosophies of animal ethics and their variants into social action, which has also led to debates over what animal ethics should entail and how best to achieve it. All of these activists agree that there are moral similarities between animals and humans and that humans have a duty at least to consider animals' intrinsic value during the course of human–animal interactions. They differ, however, on whether that extends to the recognition of animal rights that humans are morally bound to respect. They also differ as to the best tactics to use when promoting their goals.

Organizations such as the Humane Society follow the strand of animal welfare as they seek to ensure the humane treatment of domestic animals such as pets, a movement with a longer history dating back to the 19th century. These groups seek the reevaluation and modification of human–animal interactions. More recent organizations, such as People for the Ethical Treatment of Animals (PETA), have placed greater emphasis on animal rights in the agricultural, scientific, medical, and consumer industries, calling for a completely new approach to human–animal relations. Some activists have also criticized the extreme tactics used by animal rights organizations, such as PETA members throwing ketchup on those wearing fur coats to symbolize the blood of the animals used to make the garment and the ongoing dangerous wars between Greenpeace boats and whaling industry boats.

See Also: Agribusiness; Animal Welfare; Commodification; Ecofascism; Hunting.

Further Readings

Bekoff, Marc. *Animal Passions and Beastly Virtues: Reflections on Redecorating Nature.* Philadelphia, PA: Temple University Press, 2006.

Carruthers, Peter. *The Animals Issue.* Cambridge, UK: Cambridge University Press, 1992.

Clark, S. S. L. *The Moral Status of Animals.* Oxford, UK: Oxford University Press, 1977.

Cohen, C. and T. Regan. *The Animal Rights Debate.* Lanham, MD: Rowman & Littlefield, 2001.

Elliot, Robert. *Environmental Ethics.* New York: Oxford University Press, 1995.

Favre, David. *Animals: Welfare, Interests, and Rights.* East Lansing, MI: Animal Legal & Historical Center, 2003.

Frey, R. G. *Interests and Rights: The Case Against Animals.* Oxford, UK: Oxford University Press, 1980.

Groves, Julian McAllister. *Hearts and Minds: The Controversy Over Laboratory Animals.* Philadelphia, PA: Temple University Press, 1997.

Hargrove, Eugene C. *The Animal Rights, Environmental Ethics Debate: The Environmental Perspective.* Albany: State University of New York Press, 1992.

Leahy, M. *Against Liberation: Putting Animals in Perspective.* Lanham, MD: Rowman & Littlefield, 1991.

Machan, T. *Putting People First: Why Humans Are Favored by Nature.* Lanham, MD: Rowman & Littlefield, 2004.

Pluhar, Evelyn. *Beyond Prejudice: The Moral Significance of Human and Nonhuman Animals.* Durham, NC: Duke University Press, 1995.

Post, Stephen Garrard. *Bioethics for Students: How Do We Know What's Right? Issues in Medicine, Animal Rights, and the Environment*. New York: Macmillan Reference USA, 1999.

Regan, Tom. *Animal Rights, Human Wrongs: An Introduction to Moral Philosophy*. Lanham, MD: Rowman & Littlefield, 2004.

Regan, Tom. *The Case for Animal Rights*, 2nd ed. Berkeley: University of California Press, 2004.

Regan, Tom and Peter Singer, eds. *Animal Rights and Human Obligations*, 2nd ed. Englewood Cliffs, NJ: Prentice Hall, 1991.

Rowland, Mark. *Animal Rights: A Philosophical Defense*. New York: Macmillan, 1998.

Singer, Peter. *Animal Liberation*, 2nd ed. New York: New York Review of Books, 1990.

Singer, Peter, ed. *In Defense of Animals*. New York: Blackwell, 1985.

Squatriti, Paolo. *Nature's Past: The Environment and Human History*. Ann Arbor: University of Michigan Press, 2007.

Wenz, Peter S. *Environmental Ethics Today*. New York: Oxford University Press, 2001.

Marcella Bush Trevino
Barry University

Animal Welfare

Although the human–animal relationship is steeped in antiquity (e.g., the goat was domesticated by humans about 12,000 to 15,000 years ago), it is only fairly recently in this long history that animal welfare emerged as an organized concern in society. The earliest animal welfare group, the Royal Society for the Prevention of Cruelty to Animals, was only founded in 1824 in London. Although animals have for ages been seen as unthinking, unfeeling objects that deserve little or no moral consideration from humans, few now would disagree that many animals, especially mammals, are beings that are able to feel pain and suffer from being inflicted with pain. The need for animal welfare is commonly justified on this sentient ground.

Animal rights and animal welfare reemerged in the public sphere in the 1970s with the publication of philosopher Peter Singer's iconic book *Animal Liberation*. In it, Singer argues from a utilitarian perspective that the maximization of societal good and bad should also take into account nonhumans. A failure to do so is akin to practicing a form of racism called "speciesism." On the other hand, philosophers like Tom Regan have argued that the key rationale for ensuring the welfare of animals is not to maximize social good but rather to acknowledge that animals have certain rights. Such moral rights include the right to be treated with respect and the right not to be harmed.

In reality, considerations for and acceptance of animal welfare policies among the general population are often predicated on a complex combination of utilitarian concern and the affirmation of animal rights. The success and extent of animal welfare regulations are also dependent on the specific space in which humans and animal interact. In other words, the degree to which people support measures to increase or ensure animal welfare depends on what the animals are and how they have been "used." In this article, we will consider four different spaces of interaction between humans and animals. They are the home (animals as household pets); farms and slaughterhouses (animals being reared as food and clothing); laboratories (animal testing and genetically modified animals); and zoos and other recreation spaces (animals used for human entertainment).

Spaces of Human–Animal Interaction

The case for animal welfare is perhaps the strongest for household pets. According to a survey conducted by the Humane Society of the United States in 2009, it is estimated that 39 percent of American households own at least one dog. In general, animals as pets evoke the strongest emotional ties with humans. It is not surprising, then, that incidences of pet abuse or abandonment provoke strong condemnation from many people, particularly pet owners. The majority of pets, however, are treated well by their owners, and it is fair to say that animals as pets have the highest level of welfare compared to their counterparts in other spaces of interaction. One explanation for this is that pets are not treated as mere instrumental objects, compared with other types of animal use.

In recent years, there has been increased attention shone on the intensified production methods of the modern meat industry. Large animal factories intent on producing more meat in a shorter time allegedly compromise on the well-being of the animals. Animal rights groups have surreptitiously exposed how livestock are born, reared, and slaughtered in appalling conditions in the modern meat production system. Responding to such criticisms, the United Kingdom has adopted five key principles of "freedom" that are considered to be a global standard in ensuring the welfare of livestock. These principles are freedom from malnutrition, thermal and physical discomfort, injury and disease, and fear and stress, and freedom to express most normal patterns of behavior. The welfare of livestock animals has gained comparatively less attention among the general population because of the way such animals are seen as having a singular functional purpose (to feed consumers) and the way the meat industry commodifies them. In many cases, because of the relatively cheaper prices of animals farmed intensively, consumers choose to continue to support such methods of farming even when they are cognizant of the unsatisfactory welfare standards of factory-farmed animals. The low costs of intensively farmed chickens coupled with the lack of widespread knowledge of the welfare standards of animals in such farms are obstacles toward achieving and maintaining higher animal welfare standards in the livestock industry.

Animal testing in laboratories represents another significant space for human–animal interaction and from which the question of animal welfare arises. It is estimated that more than 3 million animal testing procedures were performed in the United Kingdom (UK) in 2007, of which 4,000 were conducted on nonhuman primates. Critics have often argued that most animal testing for medical and pharmaceutical purposes has little scientific merit (e.g., in that the results cannot be easily extrapolated to humans) and inflict unnecessary suffering on the animals. Counterarguments by pharmaceutical conglomerates have rested on the fact that because human safety is of paramount importance, animal testing of new pharmaceutical products is a necessary evil. Since 1966, the United States has had the Laboratory Animal Welfare Act, which seeks to minimize suffering by animals and to approve animal testing only when it is absolutely essential. On the other hand, animal testing for cosmetics has seen increasing legislative opposition in recent years, with the European Union imposing a near-total ban in 2009 on cosmetic products that involve animal testing. The contrasting outcomes for the pharmaceutical and cosmetic industries in their use of animal testing demonstrate that the instrumental value of the product in question can influence the level of acceptance of animal welfare and animal rights (with medicine arguably seen as more critical for human survival than cosmetics).

Finally, welfare issues also emerge in regard to animals that are used, broadly speaking, for entertainment purposes. These include animals in zoos, circuses, hunting, and sports. While "modern" zoos should logically be places where animals in captivity have considerable welfare (since the zoo needs the animals to attract visitors), a significant minority of the hundreds of zoos around the world have been accused of bad management and appalling infrastructure. This results in highly stressed animals that are kept in poor, confined spaces that curtail their natural instincts to forage or roam. Performing animals in circuses and other places have also been a focus of attention. Sporadic accidents that see performing animals suddenly turn aggressive and attack their trainers have supported the claim that subjecting animals to performances is unnatural, cruel, and dangerous. Legal hunting is predicated on conservation (e.g., to control animal population) and/or recreational grounds. In the UK, critics have argued that, despite the justification on cultural grounds or on the need for "pest" control, fox hunting is a cruel activity that brings unnecessary suffering to foxes and to the hounds that help the hunters. Indeed, the Hunting Act passed in 2004 outlawed the use of dogs to hunt for other mammals. Using "culture" to justify activities that negatively affect animal welfare is also common in "animal sports." Perhaps the most well-known example is bullfighting or tauromachy. Spain, probably the bastion of bullfighting, has seen increasing efforts by animal rights activists to curtail what they call a "blood sport." There are many other "animal sports" popular around the world, including dog racing and cock fighting, although bullfighting remains one of the most organized of all "animal sports."

The most recent turning point for the advancement of animal welfare came in 2006 when the Dutch political party Partij voor de Dieren (Party for the Animals) won seats in the Dutch house of representatives as well as its senate. It is the first and only political party in the world whose main agenda includes advancing animal rights and welfare. Such formalized politics represent the next frontier in the progress and evolution of animal welfare.

See Also: Agribusiness; Animal Ethics; Commodification; Hunting.

Further Readings

Fraser, D. *Animal Welfare and the Intensification of Animal Production.* FAO Readings in Ethics 2. Rome: Food and Agricultural Organization of the United Nations, 2005.

Haynes, R. P. *Animal Welfare: Competing Conceptions and Their Ethical Implications.* New York: Springer, 2008.

Kalof, L. and A. Fitzgerald, eds. *The Animals Reader: The Essential Classic and Contemporary Writings.* New York: Berg, 2007.

Regan, T. *The Case for Animal Rights.* Berkeley: University of California Press, 2004.

Singer, P. *Animal Liberation.* New York: Ecco, 2002 [1975].

Woods, M. "Fantastic Mr. Fox? Representing Animals in the Hunting Debate." In *Animal Spaces, Beastly Places—New Geographies of Animal-Human Relations*, C. Philo and C. Wilbert, eds. London: Routledge, 2000.

Harvey Neo
National University of Singapore

ANTHROPOCENTRISM VERSUS BIOCENTRISM

Opposing the anthropocentric view that human well-being is the central consideration, Ecuador added the Rights of Nature to its constitution to protect its diverse ecosystems like the Amazon rainforest and the Galapagos Islands (pictured).

Source: Rosalind Cohen, NODC, NOAA

This article evaluates anthropocentrism and biocentrism, identifying some of the pros and cons of both approaches to environmental conservation and protection. In order to fully understand this debate, it is first important to review their definitions. Anthropocentrism is a perspective that regards humans as the most important entity on the planet. An anthrocentric—literally "human-centered"—approach to environmental protection translates into a conviction that human well-being is the central consideration. In contrast, biocentrism deems humans to be merely one of many biological species, assigning inherent value to nonhuman organisms as well. With this perspective in mind, biocentrism holds that environmental conservation should not focus solely (or even in large part) on humans, but rather hold all living things equally valuable. Although there are advocates and opponents of both approaches, each with valid arguments, the discourse is often reduced to a contentious debate—one approach "versus" the other.

The debate becomes more complicated when theory becomes practice. In application, the two approaches do not in fact embody a neat dichotomy but demonstrate considerable overlap. In this regard, this entry then assesses recent legal and policy formulations: the September 2008 passage of the Ecuadorian constitution and the potential articulation of another crime under the jurisdiction of the International Criminal Court (ICC)—the crime of ecocide. These developments pinpoint some of the advantages and disadvantages of these theories in practical application. The entry concludes with a consideration of the implications of such developments.

An Evaluation of Anthropocentrism: Pros and Cons

Anthropocentrism is fundamentally dependent upon the belief that humans alone possess intrinsic value (or, at the very least, that their intrinsic value is much greater than that of any other living thing). "Extrinsic (instrumental or use) value" is the value of something as a means to achieving a further end. Conversely, "intrinsic value" constitutes an end in and of itself—the entity has its own inherent value. Anthropocentrism posits that with respect to the environment, its use or even exploitation may be justified if it is for the benefit of humans—clearing the rainforest for agricultural purposes or the utilization of lumber, harvesting the deep-sea bed for mineral resources, or fishing extensively in certain waters to feed the world's population. The benefit of this maximization of environmental resources is an increased quality of life and economic and social progress for all humans. The environment should be protected because its conservation promotes human health and well-being. Anthropocentrism does recognize that there is a fine line between use and exploitation of the environment, but it deems humans the primary focus, and all issues are evaluated with that focus in mind. What this means is that humans are given an elevated status above all other living species, which in turn means that the converse is automatically true—other species are accorded an inherently inferior status.

Anthropocentrism believes that humans are unique, considering that they are the most evolved, sentient, and intellectually superior—and therefore deserving—of the planet's species. It is therefore axiomatic that they should have the right to use the resources available to them for their enjoyment and survival. Anthropocentrism is egalitarian in the sense that it accords all humans the same elevated status over nonhumans. This is especially relevant for developing countries and is another argument in its favor, since it provides all peoples with the opportunity to use the resources available to them to advance their way of life.

The primary argument against adopting an anthrocentric approach may be found in a fundamental limitation of the theory—in order to establish unacceptable damage to the environment, it is necessary to prove detriment to its associated humans, and that may be very difficult to achieve. It is possible that negative human impact may not be immediately discernible and, in fact, damage to the environment today may not adversely affect the human population for generations. Therefore, application of an anthrocentric approach to the environment may well mean that years of environmental degradation may occur before any remedial action is taken.

The "precautionary principle" has been developed to address this concern. It holds that the absence of evidence of harm to the environment does not necessarily mean that steps should not be taken to protect it in advance of such harm. It places the burden of proof on those wishing to conduct the questionable practices to show that there will be no detrimental effects. It takes the cautious stance that humans need a sustained environment in order to survive and that the world's ecosystems are extremely delicate and interconnected; thus, if the equilibrium of the ecosystem is disturbed, the results may not be known or understood for generations.

Another negative consequence of an anthrocentric approach is the fact that environmental degradation does not always adversely impact humans alone—species in the underlying ecosystem may also be affected. Therefore, while there may be no damage to humans, there may be irreversible destruction of the environment, and this factor may be discounted. Furthermore, it is recognized that unrestrained resource use may be shortsighted: sacrificing long-term judicious and sustainable use of a resource in favor of an immediate benefit to humans is not necessarily or unarguably a viable trade-off.

The solution posited is the promotion of sustainable development. However, developing nations argue that sustainable development can be quite expensive and that they do not or may not have the resources or financial ability to achieve it. These countries argue that the developed nations had the opportunity to advance in a way that was cheap and efficient, without any regard to the manner in which they were using the resources available to them. They consider that pushing sustainable development on them could be unfair. This gives rise to another related debate outside the scope of this article—whether the responsibility for sustainable development of resources is local or global in light of the interconnectedness and corresponding reciprocal effects of activities around the world today.

An Evaluation of Biocentrism: Pros and Cons

Environmental ethics is a subdiscipline of philosophy that questions, examines, and evaluates the moral relationship of humans to their natural environment. Until it emerged in the 1970s, Western moral and ethical perspectives of the status of humans in their natural environment were dominated by anthropocentrism. This new inquiry into the assumed superiority of humans led to the development, in part, of biocentrism, which challenged the anthrocentric status quo. Biocentrism takes the position that all living things possess equal and independent value—that is, intrinsic, not merely extrinsic value—and have an inherent right to protection. Consequently, humans do not enjoy a special or an elevated status, with all other living entities existing for the sole purpose of enhancing human existence.

Another term that is often discussed in relation to biocentrism is *ecocentrism*, but it is not necessarily synonymous with biocentrism, and therefore the terms are not interchangeable. Ecocentrism incorporates a biocentric approach but takes the perspective one step further, assigning inherent value to Earth's nonliving systems or processes as well. This is a more holistic approach that emphasizes the interdependent and interconnected nature of Earth, stressing that humans are reliant upon its entire ecosystem in order to sustain themselves. For the purposes of this article, ecocentrism is considered a tangential issue, but the objective of ecocentrism has implications not just for biocentrism but for anthropocentrism as well.

The primary advantage of the biocentric approach with respect to environmental conservation is that visible degradation of the environment is sufficient to justify action to halt the destructive processes. Thus, it obviates the need to prove damage to humans. This is considered useful because damage to humans may not occur or be discernible for generations, if at all. Lowering the threshold for proof makes establishing the case for environmental destruction that much easier. There is also no need to prove a direct causal link between a specific environmental impact and a specific human harm. Even though such a link may in fact exist, the proof may simply not be discoverable, and biocentrism removes the requirement of establishing human harm.

Another advantage of adopting a biocentric approach is that the value judgments underlying the discourse are framed differently. When humans stop being the only measure of value, the scope of the debate is expanded. If all living entities are afforded the same elevated status as that of humans, then more extensive considerations for their promotion and protection are possible and/or available. Biocentrism does away with the necessity for humans to have a full and comprehensive understanding of the ramifications of their actions vis-á-vis the environment. There may well be a delicate balance in nature that humans have not fully defined and do not completely understand. Reducing every debate to the benefits, or for that matter detriment, to humans narrows the focus in a manner that fails to consider the overarching and interconnected environmental relationships within which humans are only a tangential component.

A major argument that has been posited against biocentrism is that equating all living species to humans reduces the status of humans unacceptably, since there is no clear demarcation line as to where this equivalency should end. Does protecting the rights of every species of plant and animal have the same intrinsic value as protecting those of humans? Those who oppose biocentrism would answer this question in the negative. They would argue that every species demonstrably does not have the same intrinsic value—that there is a recognizable and definable hierarchy, with "lower" life forms at the bottom of the scale and "higher" life forms toward the top, and of course, humans at the very top.

Biocentrism is also often charged with promoting an antihuman perspective that risks sacrificing human welfare and advancement in the name of and for the sake of an ill-defined "greater good." Treating the environment as though it has its own inherent rights prevents humans from maximizing the benefits they may glean from its use. The contrary argument is that the quality of human life is in fact directly reliant upon a healthy environment, and preservation of that environment ultimately benefits humans. This debate is thus reduced to a philosophical argument as to whether the world is made better or worse—and humans benefited or harmed—as a result of careless stewardship of the environment. At this point in the debate, the demarcation lines between anthropocentrism and biocentrism begin to blur.

Highlighting the Pros, the Cons, and the Blurring of the Dichotomy: Recent Developments

As shown, there are both pros and cons on both sides of the anthrocentric versus biocentric debate. These positions are not necessarily absolutes; there is a spectrum with respect to the fervency of their advocacy, and some considerations may even fall on both sides of the debate. The approach adopted becomes even more relevant in practical application, and more often than not may well be a combination of the two theories. The debate appears to be moving toward recognition of the inescapable interconnectedness of the two theoretical structures and the danger of being too narrowly in favor of one or the other approach. If the overall objective is to protect biodiversity and the environment for its own sake, then humans must also be protected as an integral component of that diversity. On the other hand, it is being recognized that in order to truly change the discourses surrounding environmental protection, there needs to be a shift in viewing the environment as primarily a resource for human development; the international community needs to begin thinking of the environment independently of its human connection, if only because of the factors that may be having an unknown or unknowable impact on humans.

This recognition of the need for common ground may be slowly becoming a reality, and this is where the Constitution of Ecuador and the ICC come into play. As the next two examples demonstrate, a biocentric approach is being adopted or considered at both national and international levels to define the inherent rights that accrue to, and crimes perpetrated against, the environment.

Ecuador and the Rights of Nature

In September 2008, Ecuador ratified its new constitution, which recognized the environment as an independent bearer of rights. The text states that "nature . . . has the right to exist, persist, maintain and regenerate its vital cycles, structure, functions and its processes in evolution." Although certain states in the United States such as Maine and Pennsylvania have adopted biocentric laws, Ecuador is the first country in the world to elevate the rights of the environment to this status, effectively creating a purely biocentric right. This is an example of biocentrism at work; the rights of nature are included in the same chapter as human rights, effectively placing them on par.

In this instance, a biocentric approach has several advantages. Ecuador has some of the world's richest and most diverse natural ecosystems, including the Galapagos Islands and the Amazon rainforest. Codifying into law that these ecosystems have the "right to exist" signifies that they have become the primary object of the right—a rights-bearing entity. This guarantees its protection without the need to demonstrate an adverse effect on humans. Moreover, harm to the environment does not have to occur before action can be taken, which stresses proactive rather than reactive care in protecting the environment. Finally, the text allows for individuals or communities, regardless of their nationality, to bring a case before Ecuador's public bodies if these rights are violated. This provides for a form of recourse and a mechanism by which nationals (and non-nationals) can bring a case.

Anthrocentrics may consider it a disadvantage that the text of Ecuador's constitution takes a biocentric approach by placing the rights of the environment and the rights of humans on par. This preference may appear to have the potential to dilute the importance of the rights of humans. Biocentrics may contend that the exact opposite may be true: protection of the environment without reference to human involvement may well be the surest way to protect that environment.

Ecocide and the International Criminal Court

The codification of the rights of nature in the Ecuadorian constitution focuses on the environment's inherent rights. There have also been several developments at the international level designed to identify and prosecute environmental degradation deemed to rise to the level of criminality. Although there is no comprehensive treaty that discusses violations of environmental rights, there has been an ongoing evolution with respect to environmental protection and environmental crimes and, in particular, a growing recognition that the environment can be the sole object of a war crime. This is a biocentric approach to crimes against the environment.

Until the mid-1970s, destruction in wartime was almost the norm, and it was not a concern of the international community unless it harmed human beings. During the aftermath of the Vietnam War, military attacks by states that damaged the nonhuman environment were criminalized through the adoption of Article 55 (Protection of the Natural Environment) of Protocol I to the Geneva Conventions prohibiting means of warfare resulting in "widespread, long-term and severe" damage to the environment and having a detrimental effect on the human population. While Protocol I was an important step in addressing destruction of the environment, countering unnecessary harm caused to the human population was still the objective, and thus it still maintained a primarily anthrocentric focus.

The text of the Rome Statute of the ICC assigns direct individual criminal responsibility (rather than state responsibility) for crimes against the environment. Similar to Protocol I of the Geneva Conventions, the language of Article 8(2)(b)(iv) of the Rome Statute prohibits excessive attacks in international armed conflict that would cause damage to civilians, civilian objects, or the environment. The "or" in the text is important to note since it criminalizes direct harm to "either" the environment "or" humans. This provision is the first example of a juridical tool guaranteeing what might be termed biocentric environmental protection. However, it established a high standard of proof in order to convict a defendant: that, with scientific certainty, the attack would cause long-term negative effects, and that the harm was inflicted intentionally. Furthermore, because the definition only refers to international armed conflict, it does not apply to intrastate conflict or civil war, which constitutes the majority of conflicts today.

There is now a movement to push the International Criminal Court toward articulation of a fifth crime under its jurisdiction (in addition to war crimes, crimes against humanity, genocide, and the crime of aggression)—the crime of ecocide. As it is currently being defined, ecocide is excessive damage to or destruction of ecosystems by anthropogenic or natural causes that drastically reduces enjoyment of the environment by the local population(s). The argument for this crime is that there is a link between environmental destruction and the onset of conflict. What is interesting here is that the language of the crime is biocentric, but there is still a reference to human harm, to justify why ecocide should be deemed a crime that may be brought before the ICC, making this movement a hybrid of anthropocentrism and biocentrism. It remains to be seen whether it will gain any traction.

Unlike the human rights field, the environmental legal field does not have established mechanisms in the form of courts, commissions, and other adjudicating bodies to enforce environmental rights. An advantage of identifying ecocide as a crime under the ICC's jurisdiction is that it may provide this forum. A possible disadvantage, however, is that it does place crimes against the environment on par with heinous crimes such as genocide, war crimes, and crimes against humanity. This has the potential of diluting the severity of these atrocities.

Conclusion—How Will the Debate Proceed?

As demonstrated, anthrocentric or biocentric approaches may become blurred in their application. Historically, anthropocentrism dominated the debate for years, but there may now be a burgeoning movement toward defining a biocentric approach to dealing with environmental issues. An interesting component of the debate is the intersection of the two approaches, overt or implied. Is the shift toward biocentrism the culmination of an evolutionary process in which the environment will now begin to be considered a universally acknowledged rights-holder? Or is this new focus merely an innovative way to protect the environment for humans but without the added burden of proving harm?

What do these developments indicate? The paradigm shift in thinking about the environment that has occurred may have much larger repercussions in other national contexts than originally anticipated, but this remains to be seen. There is still the overarching issue of enforcement—it remains unclear which individuals or entities will be able to seek redress or what the remedies for environmental destruction will be. In tandem with the movement to define ecocide as a crime that may be brought before the ICC, a Universal Declaration of Planetary Rights, modeled after the Declaration of Human Rights, has been proposed. It will be voted on at the General Assembly of the United Nations at the Earth Summit of 2012 and, if adopted, will provide the impetus for an enforcement mechanism to deal with acts of environmental degradation.

Elevation of the status of the environment was achieved initially by tying environmental damage to human harm, thus providing a justification for protecting the environment. That approach may have served its purpose, and the rising ascendancy of biocentrism may be an indication that the time may have come to abandon the difficult necessity of proving a causal nexus between environmental and human harm, albeit with the protection of human interests still, perhaps, in mind.

See Also: Intrinsic Value Versus Use Value; Precautionary Principle (Uncertainty); Utilitarianism Versus Anthropocentrism.

Further Readings

Edmund A. Walsh School of Foreign Service: Center for Latin American Studies, Georgetown University. Political Database of the Americas. "Republic of Ecuador, Constitution of 2008." http://pdba.georgetown.edu/Constitutions/Ecuador/ecuador08.html (Accessed September 2010).

Flores, Alejandro and Timothy W. Clark. "Finding Common Ground in Biological Conservation: Beyond the Anthropocentric vs. Biocentric Controversy." *Yale School of Forestry and Environmental Studies Bulletin*, 105 (2001).

Jowit, Juliette. "British Campaigner Urges UN to Accept 'Ecocide' as International Crime: Proposal to Declare Mass Destruction of Ecosystems a Crime on a Par With Genocide Launched by Lawyer." *The Guardian* (April 9, 2010).

Lawrence, Jessica C. and Kevin Jon Heller. "The First Ecocentric Environmental War Crime: The Limits of Article 8(2)(b)(iv) of the Rome Statute." *Georgetown International Environmental Law Review*, 20 (2007).

Stanford Encyclopedia of Philosophy. "Environmental Ethics." http://plato.stanford.edu/entries/ethics-environmental (Accessed September 2010).

Tuhus-Dubrow, Rebecca. "Sued by the Forest: Should Nature Be Able to Take You to Court?" *The Boston Globe* (July 19, 2009).

United Nations Office of the High Commissioner for Human Rights. "Protocol Additional to the Geneva Conventions of 12 August 1949, and relating to the Protection of Victims of International Armed Conflicts (Protocol 1)," adopted on June 8, 1977, by the Diplomatic Conference on the Reaffirmation and Development of International Humanitarian Law applicable in Armed Conflicts; entered into force December 7, 1979. http://www.unhchr.ch/html/menu3/b/93.htm (Accessed September 2010).

Briony MacPhee Rowe
Independent Scholar

ANTIBIOTIC/ANTIBIOTIC RESISTANCE

Antibiotics are a type of antimicrobial agent, a general term for substances such as drugs or chemicals that kill or slow the growth of bacteria or other microbes (organisms so small they can only be seen under a microscope) such as viruses, fungi, and parasites. The term *antibiotics* technically refers only to antimicrobial agents made from molds or bacteria, such as penicillin and streptomycin, although this distinction is not always observed in everyday language. The development of effective antibiotics is one of the great success stories of modern medicine: before the widespread use of antibiotics, infectious diseases were a common cause of death, while today such deaths are a relative rarity in developed nations (with the exception of acquired immune deficiency syndrome [AIDS], for which no cure has been discovered). Noncontagious chronic diseases such as cancer, stroke, and cardiovascular disease have largely replaced infectious diseases as the major threats to life. A similar trend exists in many developing countries, where chronic diseases are fast displacing infectious diseases as the leading causes of death. Antibiotic use is also widespread in livestock production, where it has enabled farmers and ranchers to produce greater quantities of food (e.g., beef and eggs) at lower cost than would be possible if antibiotics did not exist.

However, widespread use of antibiotics has created a new problem: antibiotic resistance. The reason antibiotic resistance develops is that the principle of natural selection applies to bacteria as well as to other living organisms. There is a great variety among bacteria, even within a single subspecies, and any particular antibiotic that kills most of a particular type of bacteria may have no effect on a few individual bacteria. These bacteria are said to be resistant to that particular antibiotic, and while this is not a problem if only a few antibiotic-resistant bacteria exist, repeated exposure to the same antibiotic creates ideal conditions for breeding an antibiotic-resistant strain of bacteria whose numbers may become large enough to cause serious infection. This happens because the few surviving bacteria pass on their traits to the next generation, which is also resistant to the antibiotics and will thrive while the nonresistant bacteria will die off. As this process of natural selection (only antibiotic-resistant bacteria live to reproduce) continues, each new generation will contain more of the antibiotic-resistant bacteria. Eventually, their numbers may be large enough to cause serious illness, and if no other antibiotics are available to which the bacteria are not resistant, then medical science may have little to offer the patient thus infected. The process of breeding antibiotic-resistant bacteria is also aided by individuals who do not follow a full course of treatment for diseases such as tuberculosis, exposing the bacteria to an antibiotic, then interrupting the treatment before all the bacteria are killed.

Fortunately, antibiotic resistance is acquired separately for specific drugs or classes of drugs, so for diseases in which several treatment options exist, it may still be possible to cure the infection. However, this only forestalls the ultimate problem because with exposure upon exposure to novel drugs, the natural selection process that created the antibiotic-resistant bacteria in the first place just begins again. In addition, the need to test the effectiveness of several drugs can delay effective treatment and also increase treatment cost and place additional burdens on healthcare systems. In addition, pharmaceutical companies had little to offer when drug-resistant strains of bacteria started to become a problem, having shifted most of their research and development programs toward drugs to treat what had become more common diseases such as cancer and stroke.

Infectious diseases are difficult to contain due to the mobility provided individuals in the modern world. This means that drug-resistant strains of different diseases can easily be spread from one country to another, and few safeguards are in place to prevent this. In a much-publicized 2007 case, for instance, an American lawyer infected with multiple-drug-resistant tuberculosis traveled from the United States to France to Prague to Canada and back to the United States on several airlines as well as ground transportation, exposing hundreds of people (at least) to potential infection.

Issues regarding antibiotic resistance generally fall into two categories: those relating to antibiotic use in humans and those relating to antibiotic use in animals, particularly in the livestock industry. In both cases, the questions that must be considered include the risks and benefits of the use of antibiotics, what constitutes overuse, who determines this, and what kind of enforcement mechanisms should be used.

Antibiotic Use in Humans

No one would seriously contest the fact that the development and use of effective antibiotics has made major improvements in human life. They are among the most common drugs prescribed today and offer potentially lifesaving treatment of many diseases caused by bacteria, from strep throat to syphilis. However, some people regard antibiotics as a sort of "miracle drug" that can cure anything rather than a specific agent that is best suited to killing particular types of microbes. The recent trend of adding antibiotic ingredients to household soaps trades on this belief: although there is no evidence that such soaps promote hygiene better than ordinary soaps, the word *antibiotic* or *antibacterial* on the label is a selling point that clearly attracts some purchasers. It is not clear whether such soaps are harmful: in vitro (test tube) studies have demonstrated that they can promote the growth of antibiotic-resistant bacteria, but no consequences to human health have yet been demonstrated.

The average person cannot distinguish between an earache or sore throat caused by bacteria (in which case an antibiotic might be effective treatment) and one caused by a virus (in which case an antibiotic would be useless). Misunderstandings about the distinction and the relative effectiveness of treatment with antibiotics sometimes lead people to demand prescriptions for antibiotics even when such treatment is not medically advisable. A physician may also feel pressured to write such a prescription or may do so on the theory that it cannot hurt and might help. In countries where antibiotics are available over the counter (i.e., without requiring a prescription), overuse of antibiotics to treat viral illnesses is common. The problem with this behavior is that while it may not be dangerous for the individual taking useless antibiotics, when considered at a population level, this approach to medicine creates an environment ideal for the breeding of drug-resistant strains of bacteria. This raises an issue of social responsibility: Should one person be expected to give up something he/she wants (antibiotics) for the sake of the greater good (not breeding antibiotic-resistant bacteria) when the gain to the individual may be minimal or nonexistent and the effect is also very diffuse (i.e., a single act of antibiotic overuse has very little effect—it takes many such acts to create a public health problem)? Should a physician be expected to refrain from writing a prescription that will please one of his patients, who will probably be able to find another physician to write the prescription if he refuses? Should there be some kind of legislation to prevent this overuse, inserting the law into the physician–patient relationship? Who will make the judgments about appropriate and inappropriate use (which both physicians and patients might see as infringing on their rights), and how could such legislation be enforced?

The World Health Organization (WHO) in 2001 called for creation of a global strategy to combat the spread of antimicrobial resistance, recognizing that national regulations may be insufficient given the ease with which drug-resistant bacteria can cross national boundaries. A number of interventions were recommended, aimed at all levels from the individual patient to national governments, but they are only recommendations because WHO has no policing power. There are disadvantages as well as advantages to legislation reducing access to antibiotics. For instance, a requirement that antibiotics only be dispensed to individuals who hold a prescription would raise the cost of accessing an antibiotic when needed (since the price would probably be higher due to increased paperwork, plus a physician visit would be required to get a prescription), which might place them out of reach of many of the world's poor.

A second common cause of antibiotic resistance is the failure of patients taking prescribed antibiotics to continue the treatment for the full period of time directed. This can happen for several reasons: one obvious reason is if the supply of medication runs out. A second is that patients sometimes stop taking their medication as soon as they start feeling better, not realizing that they still harbor harmful bacteria. This is a broadly observed phenomenon but in the case of diseases like tuberculosis (TB) has severe consequences for society, although not necessarily for the patient, because it can result in strains of the disease that are resistant to the most common drugs or all known drugs. Because TB is contagious (an estimated 20–30 percent of people exposed to TB bacilli become infected), this means that the drug-resistant strain can spread to other people.

TB provides an excellent case study for the development of antibiotic-resistant bacteria because it has run a cycle from being a greatly feared disease responsible for many deaths to being a rare disease (at least in industrialized countries) to being once again a deadly disease not necessarily curable by modern medicine. Development of streptomycin (in 1946) and other drugs that killed the TB bacilli nearly eliminated TB as a threat in the industrialized world: with antibiotic treatment, the disease could be cured, and new cases became rare because there were few infected people to spread the disease. However, the disease made a recrudescence in the 1980s due in part to the emergence of drug-resistant strains of the bacilli: the WHO estimates that in 2010, in some parts of the world, one-quarter of the TB cases are caused by strains that cannot be treated with standard drug regimens. Treating these cases is extremely expensive: for instance, a standard six-month course of TB drugs costs about $20, while drugs necessary to treat multi-drug-resistant or extensively drug-resistant TB can cost $5,000 or more. In addition, the course of treatment for drug-resistant TB can last for two years, and the drugs are more toxic (i.e., more harmful to the individual) than the standard drugs used to treat ordinary TB.

It is possible to bring drug-resistant TB under control as has largely been achieved in the United States through aggressive diagnosis and treatment programs. Directly observed therapy, a method of treatment in which patients are observed taking their medication, was developed in order to increase the probability that patients would take their medicine correctly and on schedule. This method is recommended worldwide by the WHO because it requires no particular technology to execute. Modern technology has also created a novel way to prevent the spread of TB—the negative pressure room. TB patients are sometimes isolated in negative pressure rooms in hospitals because air will flow into the room but not out of it, preventing airborne bacteria from escaping when the door is open. However, construction of negative pressure rooms is expensive and requires technology not available in all countries, and the practice of isolating patients raises questions of civil liberties because if an infected patient leaves the hospital room, he/she may spread the disease to others. Among the questions raised are these: Should patients who are ill with serious contagious diseases be kept under what would essentially constitute "house arrest" if they are prevented from leaving their room? Should they be prevented from traveling? The latter question was raised in 2007 in the case of Andrew Speaker, an American lawyer who traveled abroad (including flying on several airlines) while infected with multi-drug-resistant TB that was believed at the time to be the more serious extensively drug-resistant TB. One revelation from this incident was that in the United States, there was no federal authority to detain or isolate an American citizen infected with a serious disease, although a foreign national could be ordered into isolation or quarantine.

Antibiotic Use in Animals

Antibiotics are heavily used on some types of farm animals, including cattle, hogs, and chickens. Because humans and animals are vulnerable to some of the same microbes, the creation of antibiotic-resistant strains of bacteria in animals also poses a health risk to humans. The use of antibiotics to treat sick animals is not questioned, but the mass use of antibiotics to prevent disease and promote growth (often by adding small amounts continuously to their feed) has become a regular aspect of intense animal husbandry, also known as factory farming or confined animal feeding operations. Some growth promoters are in classes of microbials, such as streptogramins, which are essential drugs in human medicine and, therefore, the creation of strains of bacteria resistant to these drugs can pose a serious human health concern.

There are no accurate statistics regarding the amount of antimicrobials used in food animals, but one estimate is that globally, about half the antimicrobials produced are used for this purpose. In the United States, the Union of Concerned Scientists estimates that about 70 percent of antibiotics (about 13 million pounds annually) produced are used for nontherapeutic purposes (such as growth promotion) in animals. In Europe, it has been estimated that production of one kilogram of meat for human consumption is associated with the use of 100 milligrams of microbials. Prophylactic and growth-promotion use of antimicrobials in animals is expected to rise in the future because increased meat production in developing countries is usually tied to adoption of the factory farm model, which depends on the regular administration of antibiotics.

Studies in the United Kingdom, the United States, and elsewhere have demonstrated that use of microbials in food animals can result in antimicrobial-resistant strains of bacteria that can infect humans. For instance, after fluoroquinolone began to be used in poultry, fluoroquinolone-resistant *Salmonella* and campylobacter were isolated in both animals and humans. The U.S. Food and Drug Administration estimates that about 1.4 million *Salmonella* and 2.4 million campylobacter infections occur annually in humans in the United States, with about one in five *Salmonella* infections and about half of the campylobacter infections being drug resistant. A significant proportion of these infections in humans are resistant to multiple drugs: in the United States, about 100,000 *Salmonella* infections annually are resistant to at least five antibiotics, and about 326,000 cases of campylobacter infection annually are resistant to at least two antibiotics. To take another example, the introduction of vancomycin as a growth promoter in food animals was followed by the isolation of vancomycin-resistant enterococci in animals, food, and humans.

Some countries have outlawed the use of antibiotics in animal feed in order to try to stem this trend, but the United States, one of the largest meat producers in the world, continues to permit their use. Given the critical role antibiotics play in maintaining human health, many have argued that greater steps should be taken to curb their widespread use in agriculture. However, a major reason they remain in use is economic: antibiotics allow a greater amount of meat to be produced more cheaply because it allows farmers to maintain more animals in a smaller space and to prevent illnesses that would otherwise result. Use of antibiotics also accelerates growth, and in the case of cattle, allows them to be fed on corn rather than grass, promoting more rapid growth and higher milk production. Using antibiotics nontherapeutically therefore gives food producers an advantage over other producers who do not use them. In addition, if the use of nontherapeutic antibiotics in farm animals were outlawed, the likely outcome would be a rise in the cost of meat. Some argue that this would place an undue burden on poor families and that therefore no restrictions should be placed on the use of antibiotics on farms.

Antibiotics are one of the great medical advances of the 20th century and have greatly enhanced human life and increased the average life span. Antibiotics are also an integral part of modern meat production in many countries and allow meat to be produced more cheaply than would otherwise be possible. However, widespread use of antibiotics has led to the development of antibiotic-resistant strains of many common types of bacteria, and some of these pose a significant threat to human life. Decisions about the use of antibiotics require weighing considerations of freedom of choice, the rights of individuals versus the rights of the entire public, and the values of cheap food and medicines versus preventing growth of more antibiotic-resistant strains of bacteria.

See Also: Agribusiness; Animal Ethics; Antiseptics; Healthcare Delivery.

Further Readings

Centers for Disease Control and Prevention. "Antibiotic/Antimicrobial Resistance." http://www.cdc.gov/drugresistance/index.html (Accessed September 2010).
Mlot, Christine. "Antibiotic Resistance: The Agricultural Connection." In *Controversies in Science and Technology: From Maize to Menopause*, Daniel Lee Kleinman, Abby J. Kinchy, and Jo Handelsman, eds. Madison: University of Wisconsin Press, 2005.
Union of Concerned Scientists. "Antibiotic Resistance and Food Safety." http://www.ucsusa.org/food_and_agriculture/solutions/wise_antibiotics/food-safety-antibiotics.html (Accessed September 2010).
Walters, Mark Jerome. "The Travels of Antibiotic Resistance: *Salmonella* DT104." In *Six Modern Plagues and How We Are Causing Them*. Washington, DC: Island Press, 2003.
World Health Organization. "Antimicrobial Resistance." Fact Sheet No. 194 (rev. January 2002). http://www.who.int/mediacentre/factsheets/fs194/en (Accessed September 2010).
World Health Organization. "Drug Resistant Tuberculosis Now at Record Levels." News release (March 18, 2010). http://www.who.int/mediacentre/news/releases/2010/drug_resistant_tb_20100318/en/index.html (Accessed September 2010).
World Health Organization. "Use of Microbials Outside Human Medicine and Resultant Antimicrobial Resistance in Humans." Fact Sheet No. 268 (January 2002). http://www.who.int/mediacentre/factsheets/fs268/en/index.html (Accessed September 2010).

Sarah Boslaugh
Washington University in St. Louis

ANTIGLOBALIZATION

Antiglobalization cannot be understood without first considering what is meant by "globalization." While globalization is not a new phenomenon, the most recent incarnation of it does stand out from previous globalization "waves" that (while the exact timing is debated) date back to when explorers and traders began canvassing the world for new lands and resources in the 15th and 16th centuries. The most recent wave of globalization began circa 1980 and continues to the present. It is characterized by increasing interconnectivity and interdependence of people and activities around the globe and by rapid technological development. While various antiglobalization concerns have been in existence for decades, it was not until the late 1990s that a movement was created against globalization: the antiglobalization movement.

It is the current manifestation of globalization that has been met with what we can term "antiglobalization" sentiment. While antiglobalization is a commonly understood and relatively widely used term, it is somewhat misleading. The term *antiglobalization* gives the impression that this position seeks a complete rejection of globalization, and it also suggests that it is a unified stance. There are a number of phrases that are used as alternatives to antiglobalization, including *alterglobalization, antiglobal capitalism, anticorporatization,* or *anticapitalism.* These terms come closer to the kinds of concerns that are broadly shared when people talk about antiglobalization. To discuss the range of concerns that comprise antiglobalization requires first outlining the key features of contemporary globalization.

Globalization

Modern globalization is difficult to define in detail because it has so many different processes and implications associated with it. Core implications are seen in the economy, the social world, in politics, and in the environment. The implications of globalization have been compelled most noticeably by developments in the increased efficiency of global travel, the rapid development of communications technology, and the spread of capitalism and of liberal democratic political systems. These developments have enabled people, countries, governments, and organizations around the world to connect in ways not previously possible.

Economic globalization refers to a multitude of activities that involve the greater connection around the planet of people as consumers and of the production of goods and services. Global trade, free trade agreements (FTAs), and the development of international organizations and multinational corporations are all part of economic globalization. Capitalism—an economic system based on the premise of private ownership and the rights of the individual—is the dominant system in wealthy, developed nations and is expanding globally. This economic system is the underpinning factor involved in economic globalization.

The increased importing and exporting of goods around the world has been assisted by faster, cheaper, and easier travel and communications. Many countries rely heavily on their exports to increase their gross domestic product (GDP), while others have come to rely on imported goods to sustain their way of life. High-profile meetings have taken place in different parts of the world with the aim of reaching agreement over free trade deals between multiple countries.

It is not unusual for countries to have multiple FTAs. These agreements include details of specific commodities to be traded with a particular country or countries. They involve removing barriers to trade such as tariffs, taxes, or other kinds of fees and quotas. The idea is that by removing trade barriers, each country will have more opportunity to trade easily, which should be beneficial to all parties involved in the agreement.

International organizations have played a pivotal role in economic globalization, including determining how global trade should function. Some of the main organizations include the World Trade Organization (WTO), International Monetary Fund (IMF), the World Bank, World Economic Forum (WEF), Asia-Pacific Economic Cooperation (APEC), and the European Union (EU).

The WTO is an international organization that makes decisions about how trade should operate between countries. It has developed a series of WTO agreements that are designed to assist producers of goods and services, importers, and exporters. These agreements are signed by countries involved in trade and then ratified within their own country's parliamentary system.

The IMF is an organization that involves 187 countries, working together toward a multitude of financial outcomes. These outcomes include global monetary cooperation, financial stability and security, global trade, high employment levels, sustainable economic growth, and reduction of poverty. The IMF strives for these outcomes by monitoring the global economy, lending money to economically struggling countries, and providing practical support to such countries.

The WEF is a nonprofit organization based in Geneva, Switzerland. Its motto is "entrepreneurship in the global public interest." It states that economic progress must coexist with social development, and its aim is the facilitation of both. To achieve this vision, it meets regularly with world leaders to discuss concerns, initiatives, and ideas for moving forward.

APEC has representation from 21 countries, with its prime objectives being threefold. It seeks liberalization of trade and investment, business facilitation, and technical and economic cooperation.

The EU has been in existence since 1957. It basically works toward making it easier for people in its member nations to trade by working through legislative changes in this area. At present, its member nations number 27.

Another key factor in economic globalization is multinational corporations (MNCs). These are companies that operate in a number of different countries but are managed from within one. They play a big role in the movement of goods and services around the world. MNCs are becoming bigger as they take over or merge with other companies, oftentimes increasing the scope of their interests and eliminating smaller-scale competitors from the market along the way. The top 100 companies according to asset value are presently dominated by huge organizations, including MNCs in the banking and financial and the oil and gas industries, followed by telecommunication services, and pharmaceuticals and biotechnology.

The social aspects of modern globalization incorporate a wide area that is concerned with people. Specifically, it involves how people live and work as individuals, as families, and as societies. These areas are furthermore equated with matters related to security, culture, identity, and belonging. The effects of globalization in these areas can be viewed as twofold: in some respects, it brings people closer together through the increased efficiency of global travel and through telecommunications. On the other hand, not being able to access these features of globalization can mean that globally we become more divided.

Political globalization is currently mainly associated with the United Nations and multilateralism, whereby countries work together primarily on economic, foreign policy, and defense policies internationally. This trend involves a move toward increased state cooperation as well as increased international nongovernmental organizations (NGOs), which act as watchdog organizations that keep an eye on government activity.

The United Nations, as a main player in political globalization, emerged in 1945, when it was established by 51 countries. Its main commitments are toward world security, peace, social progress, and increasing the standard of living for people around the world. In more recent years, the UN has taken an increasing interest in environmental sustainability, increasing food production, disaster relief, and many more social issues. The UN has several main bodies associated with it, including the International Court of Justice, the Economic and Social Council, and the Security Council. At this writing, it has 193 member states.

Environmental globalization is an aspect of globalization that recognizes how environmental problems are becoming increasingly global problems that require global solutions. Climate change is one of the most widely recognized environmental problems that has involved the meeting of international bodies in order to find global resolution for addressing this problem. With the increased wealth realized through globalization, many countries have more resources, including technology, available to combat environmental degradation at both local and global levels.

While these four areas of globalization—economic, social, political, and environmental—can be discussed separately, they are all very much interlinked with each other. The problems that these areas raise are also interlinked; hence, antiglobalization also incorporates a broad scope.

Critique of Antiglobalization

Antiglobalization refers to a body of critique that is related to the current manifestation of globalization. Much of the critique is aimed at capitalism as the essential underlying feature of economic globalization. Economic globalization is in turn critically related to other areas of globalization, and hence of antiglobalization sentiment.

The antiglobalization movement that comprises a diverse array of activists is demonstrative of the range of concerns that fall within the bounds of antiglobalization. This movement is recognized as beginning on June 18, 1999. This date—alternately known as J18—saw antiglobalization protests organized around the world, including one in the state of Oregon, where a small-scale riot erupted. Later in the same year, on November 30, another mass protest occurred. This time Seattle, Washington, was the venue, chosen because the WTO was meeting there. Large protests have continued to occur around the world, particularly when and where there is a meeting of the G8 (Group of 8), G20 (Group of 20), or the aforementioned APEC or WTO.

Antiglobalization critiques tend to coalesce around these main issues: the advocating of an alternative economic system to capitalism (usually socialism or a radical, direct-participatory-style democracy), human rights or humanitarianism concerns (especially as they relate to indigenous peoples), and environmental sustainability.

Economic globalization is especially problematic. In particular, the problematic elements include FTAs, international organizations that promote free trade and general liberalization of the economy, and MNCs. FTAs are supposed to bring about benefits for each party to the agreement. But benefits are not equally distributed. Job losses, worker exploitation, and environmental degradation are argued to emerge due to FTAs, and it is the poorer nations that are argued to suffer more. MNCs have a role to play here, as these are the companies that employ hundreds of thousands of workers and that are exploiting resources for profit. The international bodies G8, G20, APEC, and WTO (among others) are focal points for antiglobalization protests, as these organizations each contribute toward these problems of economic globalization.

G8 refers to the "group of eight," which involves leaders from Canada, France, Germany, Italy, Japan, Russia, the United Kingdom, and the United States. The G20 is a group of 20 finance ministers and bank governors who represent 19 countries and the European Union. The countries represented by the G20 include those involved in the G8 along with 11 others. G8, G20, APEC, and the WTO all have agendas that involve economic development and liberalization, of which FTAs are deemed a central aspect. Slogans used by the antiglobalization movement include "IMF? Shut it down! G20? Shut it down!" and so on. These companies are criticized for putting profit and greed before people and planet, as captured by the antiglobalization slogan "the Earth is not for sale." These organizations are also argued to be taking decision making, and hence power, out of the hands of local peoples.

Humanitarian or social concerns are closely aligned with economic critiques of globalization. The antiglobalization movement includes groups that represent indigenous peoples from all around the world, feminist groups, worker unions, and farmers. These are groups that are poor or marginalized, and often both. Globalization is seen by these people as responsible for increasing poverty in already poor nations, which most often affects women and indigenous peoples. FTAs are seen as responsible for pushing workers to their limits to produce as many goods as possible in poor conditions for poor pay.

Meanwhile, farmers around the world are being challenged by the introduction of genetically engineered seed, which is patented. Farmers are not allowed to save seed that has been produced from patented plant seed because they need to pay royalties, and on top of this, they need to pay more for the seed. Given widespread reports about genetically engineered plants spreading, particularly in North America, lawsuits have been launched by biotechnology MNCs to seek money from farmers who are growing their seed without paying for it. This has occurred even where farmers have said that they did not plant the seed and that it is a result of their property becoming "contaminated" by genetically engineered seed. The livelihoods of farmers, among others, have come under threat by many aspects of globalization.

Environmental sustainability concerns are again linked closely to economic globalization. To be able to produce goods at competitive prices involves sourcing the raw materials as cheaply and quickly as possible. This can mean that the environment suffers as a result, as nonrenewable resources are taken with little thought as to what will take their place in the future. In addition, renewable sources are not being replaced at a fast enough rate. Pollution—of the air, water, and land—is a by-product of industry also. Pollutants threaten plant, animal, and human life. Biodiversity is in turn threatened by pollutants, and also by intellectual property as the push to patent or own the genetic sequences of plant life (primarily) is increased. The larger the amount of plant life that becomes "owned," the less there is available for people to use in ways that they may have done for millennia (particularly concerning medicinal uses). Every year, an increasing number of our staple crops, including maize, wheat, soy, cotton, and canola, are owned by MNCs (for example, Monsanto's Roundup Ready soybeans). Monocultures of patented crops are argued to produce problems for not only the environment but for farmers who cannot save the seed and who struggle to afford the patented seed.

What Do Proponents of Antiglobalization Want to See?

This question is not easily answered. Just as there are a variety of critiques regarding globalization, there are a variety of responses relating to what is needed to address the problems. Essentially, the approach taken by those involved with antiglobalization politics depends on what and how many aspects of globalization are deemed problematic. The range of responses varies from those that advocate a complete jettison of capitalist economy inasmuch as is possible by being as self-reliant and sustainable as possible. Anarchists and eco-anarchists tend to favor this approach. Others seek a fair political system, whether that is socialism, green democracy, social democracy, or a more direct, participatory-style system. Those of this latter persuasion argue that with better governance and more incorporation of the values and will of the people, the negative aspects of globalization can be better addressed. Looking after the people of this planet and the environment is argued as needing to be brought to the forefront.

Regardless of the multiplicity of antiglobalization interests and concerns, a world that is not centered on capitalism enterprise is shared across the movement. Capitalism and corporatism recur as common features of antiglobalization critique and are interwoven with humanitarian, cultural, and environmental concerns.

See Also: Capitalism; Organic Farming; Sustainable Development (Business).

Further Readings

Ainger, Katharine, Graeme Chesters, Tony Credland, John Jordan, et al. *We Are Everywhere: The Irresistible Rise of Global Anticapitalism.* London: Verso, 2003.

Brecher, Jeremy, Tim Costello, and Brendan Smith. *Globalization From Below: The Power of Solidarity.* Boston, MA: South End Press, 2000.

Klein, Naomi. *No Logo.* London: Flamingo, 2000.

Corrina Tucker
Independent Scholar

ANTISEPTICS

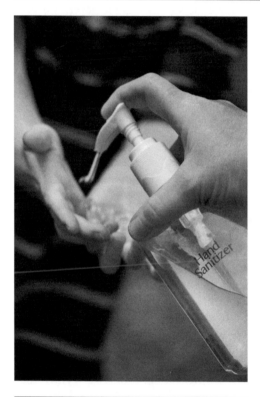

Antiseptics, such as this hand sanitizer, can be applied to skin to kill bacteria and are made from different chemicals based on their intended use as a healthcare, consumer, or food-industry product.

Source: iStockphoto

The term *antiseptics* refers to a class of drugs that can be applied to skin surfaces or mucous membranes for their anti-infective effects, and they can be either bactericidal or bacteriostatic. Antiseptics prevent infections by killing or hindering the growth of microorganisms through the use of biocides, also known as antimicrobial chemicals. Antiseptic products are divided into three categories based on their supposed usage: healthcare antiseptics, consumer antiseptics, and food-industry antiseptics. With the broad spectrum of use, many benefits such as reducing the number of bacteria on the skin and providing a residual effect have been discovered. However, there are many concerns about the use of antiseptics, including an increase of resistant microbials, cross-resistance to antibiotics, and their effect on the environment.

Healthcare antiseptics are products specifically intended for use by healthcare professionals and consist of personnel hand washes, surgical hand scrubs, and patient preoperative skin preparation. They are said to reduce bacteria on the skin prior to any patient care or surgery as well as to reduce or eliminate bacteria that could potentially cause an infection. These are different from consumer antiseptics in that they are also called antiseptic hand washes and are usually marketed as antibacterial soaps that can be either liquid or solid and are intended to be used for hand washing and general body cleansing. They are often marketed to reduce bacteria on the hands, lessen body odor, and prevent infection. Lastly, food handler antiseptics are intended to be used for hand washing and general body cleansing as well as to reduce the risk of foodborne disease and illness.

Antiseptics are made from different chemicals depending on their class. For example, antiseptics used for skin cleansing include benzalkonium chloride, chlorhexidine, hexachlorophene, iodine compounds, mercury compounds, alcohol, and hydrogen peroxide. The ingredient chlorhexidine has been used safely when applied to mucous membranes and is also prevalent in oral rinses and total body washes, while benzalkonium chloride is primarily used for hand scrubs and face washes. It should also be noted that it can be used as a disinfecting agent for instruments and helps act as a preservative for certain drugs, especially ophthalmic solutions.

The iodine compounds include tincture of iodine and povidone iodine, which are used as the broadest spectrum of all the topical anti-infectives found in antiseptics. They are highly effective in protection against bacteria, fungi, viruses, spores, protozoa, and yeasts. While tincture of iodine also contains alcohol, it has been found abrasive to the skin, while povidone iodine is an organic compound and therefore less irritating. However, it proves to be less effective in treating the broad spectrum. Povidone is most commonly used in hand scrubs and disinfection of surgical sites.

The mercury compound thimerosal helps protect against bacteria and yeasts but has also been linked with causing mercury toxicity. The final ingredient, hydrogen peroxide, acts through the liberation of oxygen gas when applied and is found at a usual concentration of 3 percent in a consumer antiseptic product. The bubbles produce an effervescent action that is thought to aid in wound cleansing by removing old tissue and debris.

Benefits of Consumer Antiseptics

Consumer antiseptics are a class of antimicrobial (biocide) drugs marketed for the general public to serve in a variety of settings. These can be marketed as antibacterial soaps, hand sanitizers, and antibacterial wipes and should be fast acting, broad spectrum, and persistent (remaining on the skin). There are many different uses such as using the antiseptic in place of regular hand washing after such activities as changing a diaper, assisting sick persons, or before contacting another person under medical care. Since antiseptics' purpose is to prevent infection, they provide a convenient alternative to hand washing on a daily basis. Similarly, antiseptic body washes are intended to kill germs that cause body odor and help treat certain skin infections and diseases.

Like healthcare antiseptics, consumer hand washes are meant to reduce the number of transient microorganisms on intact skin after use. In order to fully evaluate the antiseptics, there are two different kinds of studies performed: in vitro and in vivo tests. In vitro studies determine the product's spectrum and kinetics of antimicrobial activity and study the potential risk of resistance associated with the use of the antiseptic.

In vivo tests study the criteria based on the idea that bacterial reductions lead to a reduced potential of infection and that bacterial reduction can actually be adequately shown by using tests that simulate actual use. These tests are most important because they essentially show the overall effectiveness of a product in the presence of bacteria. For example, a subject is used and bacteria are placed on his or her hand. Then, the amount of bacteria is evaluated after each use of the test product or antiseptic up to 10 times. By the 10th wash, the product must achieve a specified reduction after the first wash.

Because of these tests, antiseptics can be found to be broad spectrum, fast acting, and persistent. In vivo tests find that the element of persistence, the ability of the antimicrobial ingredient to remain on the skin after one application, is found to be more prevalent in antiseptics as compared to nonantimicrobial soap, meaning that the overall effectiveness in removing bacteria from the skin and remaining on the skin to prevent future bacteria is greater than regular soap-and-water washing. It should also be noted that antiseptic body wash has been found to target resident organisms that are likely to cause body odor and skin infections. Thus adults, children, the elderly, daycare workers, food preparers, hospital patrons, and persons caring for the ill are all protected from infection and bacteria by the use of antiseptics.

Potential Concerns of Antiseptics

A major concern of antiseptic use is the potential development of antimicrobial resistance. Because of the growth of the antiseptic consumer industry, many concerns have been raised that overuse or misuse of the antiseptic products may lead to decreased bacterial susceptibility to biocides as seen in antibiotics. The American Medical Association (AMA) discovered that when significant bacteria resistance is present, the product is to be discontinued for consumer use unless studies can be done demonstrating the resistance has no effect on overall public health.

However, it is difficult to test for resistance when biocide action is not fully understood. They are believed to have a nonspecific action that may act on multiple targets in the bacteria cell itself. Many biocides destroy the bacterial cytoplasmic membrane in some way, resulting in damage inflicted on the bacterium. It is suggested that it is easy for bacteria to become less susceptible to biocide after it is applied at less than the necessary dose to fully kill it.

This is also similar to what has been seen in regard to antibiotic resistance. In theory, the use of biocides/antiseptics in consumer products can target bacterial strains that are resistant to antibiotics, which could exacerbate the antibiotic resistance problem. In studies, bacterial strains that showed reduced susceptibility to antiseptics also demonstrated the same in antibiotics, but there is no clear evidence associating the two. Nonetheless, it is a major concern that if a biocide does not kill the bacteria, resistance will grow.

This resistance is also of concern when considering the effects antiseptics have on the environment. Disinfectants and antiseptics can be released into the environment when they are disposed of in residential drains where local water treatment centers cannot remove many chemicals in the biocide, resulting in the environmental release to surface waters such as rivers and lakes. The U.S. Food and Drug Administration (FDA) considers harm to the environment to include toxicity to organisms as well as environmental effects such as lasting effects on the ecological community. However, there are some ways to remove triclosan from the water, such as sedimentation; it has also been found to be degraded by sunlight. The adverse effects of this toxin to the environment suggest that it inhibits the growth of algae and reduces algal species because the biocide directly affects the organisms at the bottom of the food chain hierarchy.

It is therefore important to consider the use and need for antiseptics. While they can provide the ability to sanitize a work environment or hands after taking care of an ill person, they can also harm the environment and other people through bacterial mutation. There is a need for encouragement of consumers to use the product as directed because misuse and overuse can directly affect the environment and could aversely create resistant strains that cannot be killed by antiseptics or antibiotics.

See Also: Antibiotic/Antibiotic Resistance; Healthcare Delivery; Health Insurance Reform.

Further Readings

Levy, Stuart B. "Antibiotic and Antiseptic Resistance: Impact on Public Health." *Pediatric Infectious Diseases*, 19 (2000).
McDonnell, Gerald E. *Antisepsis, Disinfection, and Sterilization: Types, Action, and Resistance*. Washington, DC: ASM Press, 2007.
Sheldon, Albert T., Jr. "Antiseptic Resistance: What Do We Know and What Does It Mean?" *Clinical Laboratory Science* (Summer 2005).

Amy Pajewski
Independent Scholar

AQUACULTURE

Aquaculture inevitably produces exchange with the environment. For example, farmed salmon like these can have sea lice that transfer to wild salmon as they migrate to sea.

Source: NOAA

Aquaculture, also called fish farming, is the culture of aquatic organisms for food and other uses. A broad spectrum of species is produced through aquaculture methods, including finfish, shellfish, invertebrates, and algae. By providing alternative sources of food and relieving consumptive pressure on native fisheries, aquaculture holds promise as a long-term solution to one of the world's biggest sustainability questions. However, there is broad debate over its negative effects on the environment, local economies, and native wildlife populations. Culture systems that rely on component exchange with the natural environment, like net pens suspended in coastal waters, contribute excess nutrients from uneaten food, fish waste, and overpopulation that can cause pollution. Fish escape, disease, and parasites affect the health of native species near the culture site, and accidental release damages native populations. Although aquaculture provides economic benefits through new jobs and increased revenue in communities, it can also have a negative impact on the livelihood of communities built for generations on the economy of their local, wild fishery. Both the science and technology of aquaculture face the challenge of successfully creating high-yield production that is at once environmentally and economically sustainable. Advances in technology, systems design, location choice, feed management, and genetics are all being developed to improve the future of aquaculture.

The Status of Aquaculture

Aquaculture exists throughout the world, artificially producing commercial aquatic species in fresh, marine, and brackish water systems. Examples of fish raised in aquaculture systems are salmon, tilapia, carp, catfish, hybrid striped bass, walleye, perch, sturgeon, tuna, clown fish, and other ornamental aquarium varieties. Mussels, oysters, clams, abalone, Pacific white shrimp, and tiger shrimp are among the types of aquacultured mollusks and crustaceans. Sea urchins, sea cucumbers, Spirulina algae, kelp, and even alligators are successfully produced using aquaculture methods.

According to the Food and Agriculture Organization of the United Nations (FAO), aquaculture accounted for more than half of the 142 million metric tons of fish, crustaceans, and mollusks caught or cultured commercially in 2008. China is the world leader in aquaculture production by weight, contributing nearly 33 million metric tons toward this total, followed distantly by India (3.5 million), Vietnam (2.5 million), Indonesia (1.7 million), and Thailand (1.4 million). The United States contributes 0.5 million. Statistics of aquaculture production based on value, however, show Japan as the leader.

The majority of aquaculture production occurs in freshwater. The U.S. Department of Agriculture (USDA) estimates the value of U.S. aquaculture production at $1 billion, with catfish as the largest sector.

Environmental Impacts of Aquaculture

Debates over aquaculture largely revolve around its negative environmental impacts. Most aquaculture takes place in open systems that have some level of exchange into the natural environment, whether intentional or accidental. Typical systems use created ponds, raceways, or floating net pens to contain a monoculture, or single species. In order to create the high yields necessary for production, aquaculture systems are densely packed. Because of this, their cumulative environmental impact on a natural system such as the body of water hosting the enclosure can be harmful, throwing off the natural balance in multiple ways.

Open systems rely on the area in which they are sited to duplicate certain biotic requirements of the species being raised, such as water and air temperature. Other elements of survival, like food, must be artificially introduced. Food is dispersed into the water above the pen, and whatever is not consumed before it passes through the culture enclosure becomes nutrient waste. The amount of food needed to sustain dense aquaculture populations, combined with fish waste and dead fish, can lead to increased nutrient loads. Excess nutrients increase phytoplankton and algae growth, leading to lower oxygen levels in the water. The resulting decrease in water quality can damage native wildlife living near the aquaculture system.

Typically, food is released into an open system in one large load. Any food not consumed either falls through the bottom of the pen or is flushed out of the system into the surrounding habitat. Food is also highly concentrated since it is not efficiently digested, amplifying the nutrient loads from uneaten surplus. To address these concerns, new technologies and feeding mechanisms allow for less waste, such as switching from sinking pellet food to floating pellets, giving the fish more time to find and consume the food.

Feeding schedules can be used to more closely coincide with natural fish behaviors, rather than bulk feedings or on-demand feed mechanisms that encourage overfeeding and waste. New feed products are being manufactured to be more digestible and biodegradable.

Siting enclosures in areas of tidal action or other flow of water can reduce nutrient impacts through flushing. Co-siting a fed species with a mollusk species, which requires nutrient-rich water for successful culture, can increase water quality through filter-feeding activities. This is an example of Integrated Multi-Trophic Aquaculture (IMTA), which uses one species to biomitigate the effects of another. This practice can also lead to increased economic stabilization by culturing more than one species in a commercial enterprise. Other efforts to recapture effluent waste include improved structural designs in systems and facility siting near estuary and wetland buffers that offer increased natural filtration.

Culture densities have other effects on native wildlife. Fish in cultures stressed by their enclosure conditions are more susceptible to the natural diseases and parasites found in native populations. Once these enter a cultured population, they can spread rampantly, amplifying the unhealthy situation back into the ecosystem. Sea lice are one parasite commonly found in farmed fish that are thought to be the cause of dramatic increases in infection rates of native species found in areas near fish farms. Wild salmon traveling from hatch sites upstream from salmon pens pick up sea lice where it is "incubating" in the cultured setting, leading to a decline in numbers of wild juveniles surviving their migration to sea. Complicating the situation, fish farms treat the outbreak of disease and parasites with antibiotics and other medicines and chemicals, which flush out of the pens, accumulating in natural systems. This has led to concerns over pests like sea lice developing resistance to treatment. Integrated pest management solutions may provide natural alternatives to chemical and medicinal treatment.

The accidental introduction of cultured species into the natural environment during storm events and by other means is a growing cause for concern. Escapees can interact with wild populations, competing for food and habitat and introducing weakened genetic lines through crossbreeding with wild stock. It is estimated that a quarter of all salmon in the wild either escaped from cultured systems or are offspring of farmed salmon. Another example comes from certain areas of the Everglades National Park, where non-native blue tilapia thought to have escaped from Florida aquaculture facilities are forcing local fish populations out of their native habitat and nesting sites. While a hotly debated topic in itself, genetic modification of aquaculture stock is thought by some to be an answer to a number of these issues. Selective breeding for disease- and stress-resistant strains, improved vaccines, and use of recirculating systems are all examples of strategies being employed to combat issues from accidental release.

Food pellets used for carnivorous species aquaculture, primarily salmon and shrimp aquaculture, are another source of controversy in the industry. The pellets are made from fishmeal and fish oil, ingredients obtained from sardines and other wild fish stock. Depending on the ratio of ingredients, the use of these pellets arguably creates a fishery that consumes more fish than it produces. Some estimates conclude that it takes four pounds of wild fish to make one pound of cultured fish. Since shrimp and salmon species make up more than 20 percent of all aquaculture production, by value, this adds up to a considerable requirement in pellet feed. Suppliers are experimenting with alternative protein sources for the pellets, such as soy, but others argue this solution does not necessarily lessen the impact but simply shifts it to impacts from another form of agriculture.

Environmental justice concerns have also been raised around the use of prey species as pellet food, pointing out that sardines and other prey species are typically caught from waters offshore from developing countries, which rely on the fish as a primary food source.

By creating pellets, the industry reduces subsistence fisheries to feed a luxury food source in developed countries, such as the United States.

Economic Effects

Regardless of the evidence of native fishery decline, fishing communities reliant on the market for wild-caught fish express increasing concern for the markets generated by aquaculture. Aquaculture enclosures sited in traditional fishing territories cause conflict in communities that have relied on the caught fishery as their trade and local economic structure for generations. In response, the National Oceanic and Atmospheric Administration (NOAA) is among the groups offering training to fishermen who are interested in learning the basics of cod aquaculture. The "Cod Academy" is offered as one solution to the economic tension between wild and cultured fisheries.

Aquaculture as an independent venture can be unstable to start up due to its high initial expense and lengthy return on investment. Because of this, larger facilities tend to receive more funding from banks and other sources. Smaller community-based operations with the potential for higher sustainability and lower environmental impact are not supported. There is promise, however, in the growing use of closed-loop systems. These recirculating aquaculture systems (RAS) confine the entire operation to a series of tanks with internal filtration and water-quality treatment. Smaller-sized entrepreneurial aquaculture businesses are using RAS to culture high-value species like ornamental fish for the aquarium trade. This serves to reduce the pressures on wild harvesting of protected reef species, for example, in a more sustainable form of aquaculture with less environmental impact and better return on investment.

Groups like the Marine Stewardship Council and the Aquaculture Stewardship Council are working with aquaculture producers and researchers from around the world to create and market certification labels, identifying fish cultured with best sustainability practices.

The Monterey Bay Aquarium publishes and updates its Seafood Watch, a list of fish available on the market rated based on green-ness to help inform consumer choices for aquaculture and caught fish.

See Also: Ecolabeling; Fisheries; Organic Foods.

Further Readings

Asche, F. and F. Khatun. "Aquaculture: Issues and Opportunities for Sustainable Production and Trade." ICTSD Natural Resources, International Trade and Sustainable Development Series Issue Paper No. 5. Geneva: International Centre for Trade and Sustainable Development, 2006.

Chopin, Thierry, et al. "Integrating Seaweeds Into Marine Aquaculture Systems: A Key Toward Sustainability." *Journal of Phycology*, 37:975 (2001).

Krkosek, Martin, et al. "Declining Wild Salmon Populations in Relation to Parasites From Farm Salmon." *Science*, 318:1772 (2007).

Monterey Bay Aquarium. "Seafood Watch." http://www.montereybayaquarium.org/cr/cr_seafoodwatch/sfw_aboutsfw.aspx (Accessed September 2010).

Dyanna Innes Smith
Antioch University New England

Asthma (Cities)

Scientists and environmentalists have noted increased rates of asthma in the late 20th and early 21st centuries, especially among children. Although asthma is a complex medical condition that can involve multiple causal factors, environmental triggers such as exposure to allergens and pollutants can play a key role in precipitating attacks. Scientific evidence as to whether environmental pollutants can induce asthma in otherwise healthy individuals is less clear. Proponents of green cities advocated an ecological approach to urban planning in part to improve human health issues such as asthma that have some causal links to the urban environment.

Asthma is a complex medical condition that can develop in childhood or adulthood. Its symptoms include unpredictable periods of extreme shortness of breath related to changes in the flow of air within the lungs' air passages. Asthma can be induced by a variety of causal factors, both biological and environmental. Causes of asthma may include exercise-induced asthma, a viral infection, allergy, indoor dust mites, or the ingestion of certain foods or pharmaceuticals. There may also be a genetic predisposition, based on its occurrence in families. This link is not as yet scientifically understood. Possible psychological factors are even less fully understood. Asthma caused by allergens or other external factors is called extrinsic asthma; otherwise, it is known as intrinsic asthma.

Some scientific studies have revealed increased rates of asthma in the developed world over time, especially among children. They also note the larger percentage of urban populations among developed countries. By the 21st century, urban residents accounted for more than half of the global population but close to 80 percent of the populations of developed countries. Current levels of outdoor air pollution have also been epidemiologically linked to the increased frequency and severity of symptoms in those already suffering from the disease.

Other theories for the modern rise of asthma rates among urban children in developed countries include the hygiene hypothesis and the belief that cockroach allergies can trigger severe asthmatic reactions. Proponents of the hygiene hypothesis point to the fact that most developing nations and rural areas have low asthma and allergy rates while such rates are on the rise in developed countries and urban areas. The hygiene hypothesis posits that one reason for the disparity may be the cleanliness of the modern urban household, in which infants and young children are not exposed to sufficient germs to allow the development of a healthy, properly functioning immune system.

The result is an improper immune defense response that can contribute to the onset of asthma. In particular, scientists testing the hygiene hypothesis have studied the role of endotoxin, also known as bacterial lipopolysaccharide (LPS), in the development of immune response. Proponents of the cockroach allergy theory note the increased exposure to cockroaches in developed, crowded urban environments. Scientific and medical tests have shown the sensitivity of some people to cockroach allergens and the correlation between cockroach allergies and asthma attacks. They also note the likelihood that urban children spend more time indoors due to safety concerns.

While some researchers and environmentalists have credited the rise in asthma rates to environmental or medical factors, others debate whether the statistics truly reflect rising rates at all. Critics counter that the increase in asthma rates may be wholly or mostly related to better awareness of the disease and its symptoms and improved diagnosis among medical personnel rather than an actual rise in the disease itself. They also note the scientific difficulty of determining the likelihood that environmental pollutants can actually induce asthma in otherwise healthy individuals, arguing that they merely induce attacks among those who already have a genetic predisposition or have been diagnosed with the condition. They also note that factors not related to environmental pollutants, such as secondhand smoke, diet, and the increased prevalence of indoor dust mites, may instead account for rising rates.

Many environmentalists feel that asthma has an important environmental component that may bring on or worsen the condition when an individual is exposed to environmental pollutants such as cigarette smoke, automobile exhaust fumes, or smog that are especially notable in many urban areas. For example, scientific studies have linked exposure to sulfur dioxide with narrowing of the lungs' air passages. Although possible natural environmental triggers such as pollen may be difficult to impossible to control, environmental pollutants have become the target of environmental regulations and the green cities movement.

Urban areas can be particularly vulnerable to pollutant-induced asthma attacks when atmospheric conditions hold the pollution close to ground level, often in the form of thick smog. Historical examples of smog so severe that multiple people died of asthma and related health issues include well-known tragedies in Donora, Pennsylvania, and London, England. Although instances of killer smog are rare, they have spurred calls for urban environmental awareness and reform. Many urban governments or media sources now routinely issue air quality forecasts or high pollution, smog, or ozone warnings in recognition of the connection between environmental pollutants and medical conditions such as asthma.

The international port city of Yokkaichi, Japan, became a leading symbol for environmentalists concerned with the links between urban industrial development and declining human health. Yokkaichi's spectacular industrialization and population growth in the post–World War II period was accompanied by a concomitant rise in pollution and subsequent human health problems such as asthma. Scientific research documented the relationship between air pollution caused by sulfur dioxide and the development of bronchial asthma and chronic bronchitis. The rate of victims increased, and eventually victims won a lawsuit against the industrial owners in 1972, which spurred the Japanese government's passage of antipollution legislation. As amounts of air pollution gradually decreased, so did the instances of asthma and chronic bronchitis.

The result of rising rates of asthma and other health-related issues in urban environments such as Yokkaichi, Japan, prompted some environmentalists and scientists to focus on improving urban environmental conditions. Urban ecosystems are largely anthropogenic, or man-made, and are dependent on their surrounding natural ecosystems for resources such as energy and food supplies because they are not self-sufficient. Environmentalists are also concerned with urban sprawl, also known as the urban fringe, as cities continually expand along their borders with the natural environment.

The health of cities affects not just city residents themselves but also affects the natural world both along urban borders and in totality. Air pollution, the exhaust fumes from automobiles, and other toxic contaminants are part of the by-products of a larger urban ecosystem that urban ecologists seek to understand. They argue that environmental actions cannot overlook the fact that humans are a key determinant in the future of both urban and natural ecosystems. While some positive air quality changes have occurred, including reduced emissions of pollutants that are the by-products of burning coal, air pollution from automobiles and other sources have increased, as has damage to the ozone layer.

Canadian researcher William Rees developed the concept of the carbon footprint to measure dependence on natural resources based on the land required to produce the necessary resources to support a particular person, family, or city. The movement for green cities seeks to reduce this carbon footprint in part due to its potential effects on human health, as evidenced through such measures as increased rates of asthma. Continued urban growth without ecological planning could result in more of the pollutants linked to asthma, both within and outside urban areas. Measures adopted by the green cities movement include green buildings that utilize renewable and efficient energy sources, electric and hybrid vehicles that reduce exhaust and reliance on fossil fuels, and bike-friendly and mass transit initiatives that reduce the number of vehicles.

Movements for sustainability, green living, and green cities have arisen from this understanding of the interconnectedness of the urban and natural worlds. A key component of the green cities movement is to utilize urban planning, government regulation, and scientific and technological tools in order to reduce the air pollutants that may be associated with the rise of asthma and related bronchial diseases and conditions. Urban planning and governance based on green city initiatives are designed to promote better health and quality of life for both urban and rural residents, as well as the sustainability of the urban ecosystem.

See Also: Asthma (Health); Environmental Justice (Ethics and Philosophy); Healthcare Delivery.

Further Readings

Cook, Allan R. *Environmentally Induced Disorders Sourcebook*. Detroit, MI: Omnigraphics, 1997.

Guest, Greg. *Globalization, Health, and the Environment: An Integrated Perspective*. Lanham, MD: AltaMira Press, 2005.

Kessel, A. *Air, the Environment, and Public Health*. New York: Cambridge University Press, 2006.

Lappe, Marc. *Evolutionary Medicine: Rethinking the Origins of Disease*. San Francisco, CA: Sierra Club Books, 1994.

Lomborg, Bjorn. *The Skeptical Environmentalist: Measuring the Real State of the World*. New York: Cambridge University Press, 2001.

Mitman, Gregg. *Breathing Space: How Allergies Shape Our Lives and Landscapes*. New Haven, CT: Yale University Press, 2007.

Shrader-Frechette, K. S. *Taking Action, Saving Lives: Our Duties to Protect Environmental and Public Health*. New York: Oxford University Press, 2007.

Targ, Nicholas. "The Health Politics of Asthma: Environmental Justice and Collective Illness Experience." In *Power, Justice, and the Environment: A Critical Appraisal of the Environmental Justice Movement*, David Pellow, N. Pellow, and Robert J. Brulle, eds. Cambridge, MA: MIT Press, 2005.

Marcella Bush Trevino
Barry University

ASTHMA (HEALTH)

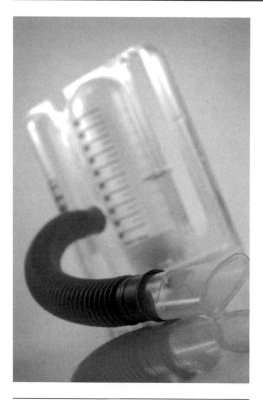

Devices like this spirometer, which measures the lungs' breathing capacity, can help diagnose asthma, which is the most common chronic disease among children.

Source: iStockphoto

Asthma is a chronic medical condition characterized by inflamed and swollen airways (the air passages leading to and from the lungs), increased secretion of mucus, and episodic attacks in which a person may suffer from coughing, wheezing, and shortness of breath. Repeated attacks can scar the airways, leading to more frequent and more severe attacks in the future, and severe asthma attacks can be fatal. Incidence of asthma seems to be on the increase, both in the United States and worldwide, imposing a serious burden on the population and on the healthcare system. Although no one would dispute that asthma is a serious disease, there are several controversies regarding it. One is whether the observed rise in the incidence (number of new cases) and prevalence (number of total cases) of asthma is due to an actual increase in the number of cases of the disease or whether it is due to increased monitoring and reporting and also perhaps to greater willingness on the part of physicians to give a diagnosis of asthma. A second question is whether exposure to pollutants or other human-generated factors are responsible for the increased number of cases, assuming that the observed increase is real and not merely due to differences in diagnosis or reporting.

The causes of asthma are not entirely understood, and there is no known cure for the disease: instead, medical practice focuses on trying to prevent attacks and limit their severity through avoidance of "triggers" (substances that can set off an attack in a particular individual) and the use of medications. Despite being a fairly common disease, asthma is also complex, with much variability in symptoms and risk factors among cases, both of which complicate efforts both to diagnose individual cases and to do epidemiological research regarding the disease. For instance, it is difficult to calculate the rates of occurrence in a single population and even more complex to compare disease rates across different populations because diagnostic criteria differ in different countries, and many cases of asthma remain undiagnosed. Even within an industrialized country such as the United States, significant changes have been made in the past 50 years or so in the process for diagnosing asthma.

There seems to be a genetic component involved in the disease; that is, certain people are born with a predisposition to develop asthma. Clinical studies suggest that a child is at increased risk of asthma if one or both parents have the disease: for one parent, the risk is about 30 percent (versus less than 10 percent in the general population), while a child whose parents both have asthma has a risk of about 50 percent of developing the disease. However, environmental factors, including exposure to certain substances, also appear to play a role, and the influence of genetic and environmental factors can be difficult to separate because in most cases, children and parents share a home environment. However, studies of identical twins raised apart have produced evidence that environmental exposures play a role even among individuals with identical genetic makeup. The most important environmental risk factors for asthma are exposure to allergens (substances that provoke an allergic reaction) and other airway irritants such as smoke and dust. The reaction to a particular environmental exposure can vary broadly among individuals, so a substance that might provoke a fatal asthma attack in one person may cause no reaction at all in another. The range of substances that have been identified as asthma triggers is quite broad and includes animal dander (materials shed from the body of animals); many different types of foods, including peanuts, eggs, wheat, fish and shellfish, and food preservatives such as sulfites; exercise; medications such as aspirin; dust, smoke, and other irritants; cold, humidity, and temperature changes; and strong emotions and stress.

"Cockroach dust," meaning droppings and body parts of cockroaches, has also been identified as an asthma trigger and one that is particularly significant because roaches are found in many inner-city areas where asthma rates are also high. Despite the scientific evidence that cockroach dust can act as an asthma trigger, assertion of this fact has courted controversy because some feel it suggests that poor or inner-city parents are poor housekeepers and are therefore responsible for their children's illness. This assertion is easily countered, however, by the observation that large apartment buildings that are typical in many cities are often infested so that independent of the state of any particular apartment, those living in the building will inevitably be exposed to cockroach dust.

Asthma imposes a heavy burden of disease on the world's population, and substantial resources are devoted to its treatment and prevention. According to the World Health Organization (WHO), asthma is the most common chronic disease among children, and worldwide about 300 million people (children and adults) suffer from asthma, with about 255,000 deaths caused by asthma in 2005. Although sometimes characterized as a disease of industrialized countries, in fact asthma occurs in all countries regardless of their level of development, and more than 80 percent of asthma deaths occur in low- and middle-income countries. In the United States, the Centers for Disease Control and Prevention (CDC) estimates the prevalence of asthma among children at 9.4 percent (about 7 million children) and 7.3 percent among adults (about 16.4 million adults), with almost 3,500 deaths annually due to asthma. In the United States, asthma is also more common among African Americans and people of Puerto Rican descent as opposed to whites and people of Mexican descent and in people living below the poverty level as opposed to those above the poverty level.

Diagnosis

Increases in asthma incidence and prevalence have been observed both globally and in individual countries. However, this statistical information requires careful interpretation because asthma diagnosis is a complex process best made in a clinical setting and with several sources of information, including a medical history, physical examination, and specialized tests to assess breathing function. Any attempts to compare rates of asthma, whether in different time periods or in different geographical areas, must consider how cases of asthma were defined because differences in diagnostic methods may account for large observed differences in apparent rates of asthma. Comparison of other statistics such as physician or hospital visits due to asthma must also take into consideration social and economic factors: for instance, in a country where everyone has guaranteed healthcare, more visits for asthma might be observed simply because there is no economic barrier to care.

Many symptoms have been associated with asthma, including uncomfortable breathing, chest discomfort or tightness, persistent coughing in the absence of an infection, excessive mucus production, and wheezing during exercise. Often, self-reporting questionnaires are used to assess whether the patient has a history of symptoms that would indicate asthma; in the case of a child, a parent would provide this information. The problem is that answers to such questions can be highly subjective, and one patient might report a symptom (for instance, wheezing during exercise) while another might consider the same behavior to be normal (for instance, as an indication of their own poor conditioning) and therefore not report it. Greater awareness of asthma among patients, parents, and physicians in recent years could influence people to report symptoms they would previously have ignored, thus leading to higher rates of diagnosis without necessarily higher rates of the disease. Greater awareness could also lead to less severe cases of asthma being diagnosed, when in past years they might have passed unnoticed. Changes in coding systems may also play a role in apparent changes in prevalence because epidemiologic studies are often based on hospital and clinical records where diagnoses are recorded using numerical codes. For instance, changes in the 10th revision of the International Classification of Diseases (ICD-10) in 1999 are generally assumed to be partly responsible for a decline in asthma mortality rates in the United States.

Several specialized medical tests are available to help in the diagnosis of asthma and to differentiate between asthma and other diseases that may present with similar symptoms. The most common are spirometric measures of lung function, tests that were developed as specialized procedures only available in research centers but that have become more commonly available in the United States and other industrialized countries. Spirometry requires a patient to take a deep breath of air and exhale it forcefully into the spirometer or peak flow meter, which measures lung volume and air flow, producing information that can be compared to norms based on gender, age, and height. For these measurements to be accurate, the subject must cooperate by taking in all the air he/she can and exhaling as hard as possible. Even more important to bear in mind when comparing rates of asthma is that when some data are based merely on reports of symptoms while others are based on spirometrics, it is likely to lead to a misleading result because of the different methods of diagnosis used.

Risk Factors for Asthma

Many different environmental factors have been associated with increased risk of asthma, and some scholars have suggested that an increase in environmental pollution may lie behind the observed increase in the number of asthma cases. Among the risk factors already identified for asthma are air pollution (in particular, diesel exhaust, nitrogen dioxide, and sulfur dioxide), tobacco smoke (parental smoking is highly associated with childhood asthma, and smoking is associated with asthma in teenagers and adults), indoor air pollutants (including cleaning agents, dust mites, fungi, and wood dust), and ozone. However, causality is often difficult to establish because of the complex nature of asthma, the varying methods of diagnosis, and the fact that different people respond differently to the same risk factors.

Some of the strongest evidence that environmental exposures can increase the risk of asthma comes from occupational studies that compare the incidence of asthma in people working in particular professions or with particular exposures (to dust, vapors, etc.) with that of the general population. A literature review by Kjell Toren and Paul Blanc based on studies from 19 countries and six continents and numerous types of occupational exposure concluded that 16.3 percent of all adult-onset asthma (i.e., the disease was not present in the individual in childhood) could be attributed to occupational exposure. The authors argue that preventive action is necessary to reduce exposure to asthma-causing agents.

Parents sometimes try to provide their children with the cleanest possible environment in hope of keeping them healthy. Ironically, in the case of asthma, this may be counterproductive as the hygiene hypothesis suggests that exposure to potential allergens and endotoxins while the immune system is still developing may have a protective effect, conferring an immunologic tolerance rather than promoting an asthmatic response. This hypothesis was first put forward in 1989 in an article by David Strachan that argued that British children from large families suffered from a lower incidence of allergic diseases such as eczema and hay fever because their large number of siblings gave them greater exposure to infectious agents. This hypothesis has been invoked to explain why asthma rates are persistently lower in rural areas and less developed countries: children growing up in the countryside or in the developing world are more likely to be exposed to environmental allergens and endotoxins early in their lives, while children growing up in the indoor, hygienic environment typical of developed countries have less exposure and therefore are more prone to developing allergic and asthmatic reactions.

Specifically, the hygiene hypothesis states that during the critical period for the development of immune response, the extremely clean environment of many homes in the industrialized world fails to provide the immune system with sufficient exposure to pathogens so it may develop properly. Although there is both epidemiological and laboratory evidence to support this hypothesis, many questions remain unanswered, and other factors such as differences in diagnostic practices may also cloud the picture. Increased rates of asthma reported in many settings, including inner-city environments in the United States and in African towns, suggest that there is more at work than simply exposure to a clean or dirty environment. In addition, it raises the question of what, if anything, could be done in practical terms should the hygiene hypothesis become generally accepted: parents are unlikely to favor exposing their children to less sanitary conditions, and medical advice is unlikely to stop recommending universal immunization against common childhood diseases (measles, whooping cough, etc.) because the consequences of those disease can be deadly.

Asthma is a serious disorder that places a heavy burden on the world's population, both in terms of the loss of health and possibly loss of life and also in terms of the economic cost of providing treatment for the disease. However, several important controversies remain, including whether rates of asthma are increasing, and if so, where and by how much, and how great a role environmental exposure plays in the development of asthma. Research into these questions is ongoing and may produce answers about how best to prevent and control this disease.

See Also: Asthma (Cities); Healthcare Delivery; Sustainable Development (Cities).

Further Readings

Centers for Disease Control and Prevention. "Asthma." http://www.cdc.gov/asthma (Accessed September 2010).

Gautrin, D., A. J. Newman-Taylor, H. Nordman, and J. L. Malo. "Controversies in Epidemiology of Occupational Asthma." *European Respiratory Journal*, 22:3 (2003).

Moorman, Jeanne E., Rose Anne Rudd, Carol A. Johnson, Michael King, et al. "National Surveillance for Asthma—United States, 1980–2004." *Morbidity and Mortality Weekly Report*, 56:SS08 (2007).

Strachan, David P. "Hay Fever, Hygiene and Household Size." *BMJ*, 299 (1989).

Toren, Kjell and Paul D. Blanc. "Asthma Caused by Occupational Exposures Is Common—A Systematic Analysis of Estimates of the Population-Attributable Fraction." *BMC Pulmonary Medicine*, 9:7 (January 2009).

U.S. Food and Drug Administration. "Asthma: The Hygiene Hypothesis" (2009). http://www .fda.gov/biologicsbloodvaccines/resourcesforyou/consumers/ucm167471.htm (Accessed September 2010).

World Health Organization. "Asthma: Fact Sheet No. 307" (May 2008). http://www.who.int/ mediacentre/factsheets/fs307/en/index.html (Accessed September 2010).

Sarah Boslaugh
Washington University in St. Louis

B

BACILLUS THURINGIENSIS (BT)

Pesticides used to control the European corn borer (pictured) have dropped by about one-third since Bt crops, created with the Bt genetic material that allows plants to create their own toxin, were introduced.

Source: Agricultural Research Service/U.S. Department of Agriculture

Public awareness of the unintended consequences of biotechnology has escalated in response to the use of *Bacillus thuringiensis* (Bt) in genetically engineered crops. Bt, a bacterium that lives in soils around the world, naturally produces a toxin that is fatal to certain insects and has been used since the 1920s as an insecticide spray. Since 1995, when the U.S. Environmental Protection Agency (EPA) first approved the technology, crops such as potatoes, corn, cotton, and rice have been genetically engineered to produce Bt toxin. The insertion of Bt genetic material, cloned from the host bacterium and engineered for expression, allows plants to produce toxic proteins and thereby protect themselves from certain insects. This genetic modification makes Bt crops resistant to some of the most pervasive pests threatening commercial crops. This technological innovation was argued to reduce the need for intensive chemical pesticide use in commercial agriculture. However, there is also widespread concern for the environmental impacts and risk posed to consumers from genetically modified crop production.

Uses

A solution of Bt toxin crystals was discovered, in 1901, to be highly effective against certain crop pests, including the corn borer, corn rootworm, tobacco budworm, bollworm, and pink bollworm; it was first used commercially as an insecticide spray in 1958. According to industry reports, these insecticidal properties constituted more than 95 percent of the biopesticide market as of 2010. Different strains of this bacterium are currently used to control for several insects and their larva that feed on fruits, vegetables, and other cash crops, including potatoes, corn, and cotton.

Bt crop varieties are engineered to produce a protein toxic to specific insects. This technique is used most effectively in areas with high levels of infestations of targeted pests. Since 1995, when the EPA first approved use of the technology, commercial production of Bt crops has increased dramatically. Bt cotton acreage, for example, grew to more than 75 percent in China and 73 percent in the United States. Plantings of Bt corn expanded to 63 percent of U.S. corn acreage in 2010, according to the U.S. Department of Agriculture (USDA), though plantings have fluctuated significantly, rising and falling depending on pest infestation levels. The use of Bt crops has been approved recently and is increasing in some countries, including Mexico, India, and China. In other countries, the technology has received greater degrees of opposition. The European Union, for example, imposed a de facto moratorium on the cultivation of certain genetically modified crops until the World Trade Organization ruled this illegal in 2006, and many countries continue to ban or heavily regulate the use of Bt crops. New research continues on introducing Bt technology to new crops such as eggplant.

Advantages

The use of insect-resistant Bt plants can potentially reduce use of chemical insecticide sprays, which are extremely toxic and expensive. Applications of conventional pesticides recommended for control of the European corn borer, for example, have dropped by about one-third since Bt corn was introduced, according to EPA estimates. Though more research is needed on the environmental impact of Bt crops, the protein is believed to target a narrow range of pests with little impact on nontarget organisms.

In instances where pest populations are high enough to cause severe damage to crops, the use of Bt plants offers a clear economic advantage to farmers. Though Bt technology does not directly lead to increased yields, it does aid in preventing crop loss to damage from certain pests.

Bt proteins are considered safe for human consumption. Although highly toxic to certain insect species, Bt is thought to be nontoxic to humans because a human body lacks the digestive enzymes needed to dissolve Bt protein crystals into their active form. However, any introduction of new genetic material is potentially a source for allergens, and for this reason, certain strains of Bt are not approved for human consumption.

Disadvantages

There is a great degree of uncertainty about the risk that Bt toxin may pose to human health and the impact of Bt proteins on the environment, and existing regulations and safety protections are not always enforced successfully. According to the EPA, more research is needed to study Bt protein accumulation in soils, the risks posed to nontarget organisms, and the likelihood of gene flow from Bt crops, particularly Bt cotton, to wild relatives.

Pest Resistance

Of great concern in the widespread cultivation of Bt crops is the potential for insects to develop a resistance to the toxin as a result of repeated exposure, which would render useless one of the most environmentally benign insecticides in use today. Already, certain moth and cotton pest populations are reported to have acquired resistance. EPA-mandated risk management strategies include the planting of refuges, such as a plot of non-Bt corn near a field planted in Bt corn, in order to maintain a local population of insects that remain susceptible to Bt toxin.

Lepidopteron Toxicity

Though the EPA initially considered Bt a negligible risk to nontarget organisms, entomologist John Losey observed that half of monarch larvae died within days of eating milkweed dusted with Bt corn pollen. Given that corn pollen can be dispersed over 60 meters by wind, and Bt corn is proliferating in the United States, the findings of Losey's study suggest that Bt corn could present a threat to the monarch butterfly. Losey emphasized that his lab-based results were preliminary, but his paper, published in *Nature* in 1999, nonetheless generated intense public debate about the potential risks of Bt corn and genetically engineered crops more broadly.

Loss of Farmer Control

Wealthy countries like the United States dominate the biotech industry, for the research and development necessary to pursue genetic engineering is prohibitively expensive and requires resources unavailable to many developing countries. Critics of the technology point out that ownership and control of Bt varieties, among other genetically modified crops, is consolidated in the hands of a small number of powerful corporations that impose restrictive conditions on farmers buying their seed. This trend has been accelerated by the widespread patenting of plant genes and germplasm in the United States and some other countries. Instead of having the choice to save and replant seed from their harvests, farmers are legally obligated to purchase new seed each year. Though Bt crops tend to reduce the need for chemical pesticides, they, like many commercial varieties, may require greater amounts of fertilizer, water, and other inputs than locally adapted varieties, thus putting increased financial strain on farmers.

Threats to Biodiversity

Critics of genetically modified crops argue that they will replace native varieties and result in an irreversible loss of biodiversity. Though such risk is not limited to Bt or other genetically engineered plants, the rapid expansion of these and other commercial crop varieties does increase the likelihood of genetic erosion among crop populations. As farmers increasingly plant Bt instead of locally adapted varieties, they narrow the gene pool of a particular crop, thereby reducing the genetic diversity necessary for resilience and resistance to diseases and pests. Strategies to conserve the biodiversity of crops include ex situ conservation (storing native varieties in seed banks) and in situ conservation (cultivating local varieties in their native environment).

See Also: Agribusiness; Biotechnology; Genetically Modified Organisms (GMOs); Pesticides.

Further Readings

Glare, Travis R. and Maureen O'Callaghan. *Bacillus Thuringiensis: Biology, Ecology and Safety*. Hoboken, NJ: Wiley, 2000.

Losey, John E., Linda S. Rayor, and Maureen E. Carter. "Transgenic Pollen Harms Monarch Larvae." *Nature*, 399:214 (1999).

Tabashnik, B. E., A. J. Gassmann, D. W. Crowder, and Y. Carriére. "Insect Resistance to Bt Crops: Evidence Versus Theory." *Nature Biotechnology*, 26:2 (2008).

Thompson, Jennifer. *GM Crops: The Impact and The Potential*. Collingwood, Australia: CSIRO Publishing, 2006.

U.S. Department of Agriculture, Economic Research Service. "Adoption of Genetically Engineered Crops in the U.S." http://www.ers.usda.gov/Data/biotechcrops (Accessed August 2010).

U.S. Environmental Protection Agency. "EPA's Regulation of Bacillus thuringiensis (Bt) Crops." http://www.epa.gov/oppbppd1/biopesticides/pips/regofbtcrops.htm (Accessed August 2010).

Emma Gaalaas Mullaney
Pennsylvania State University

BIOFUELS (BUSINESS)

Driven by concerns over climate change, energy supply security, and spikes in oil prices, interest in liquid biofuels is at an all-time high. Though the industry has grown several fold in the past few decades—production of ethanol and biodiesel, the most common biofuels, grew by 26 percent and 172 percent, respectively, between 2004 and 2006 alone—biofuels remain a small source of the world's energy, accounting for just over 1.5 percent of total road transport fuel consumption in 2008 and for 4 percent (ethanol alone) of the 1.3 trillion liter global gasoline consumption in 2007. Only one-tenth of total biofuel production volume is traded internationally. As several countries have committed to meeting national biofuel consumption targets and fuel blending mandates, the future is likely to see additional growth in both production and trade in biofuels, though this growth may be slowed by the global financial crisis, a drop in oil prices, and a reconsideration of possible adverse social and environmental impacts. This entry will examine the pros and cons of the expansion of the biofuels industry from a business perspective, with attention to the role of government support and private investment, the industry's potential to transform the energy sector, who stands to gain or lose in a larger biofuel market, and its effects on the environment and world food prices.

The State of the Business: Are Biofuels Economically Competitive?

With the exception of sugarcane ethanol in Brazil, ethanol and biodiesel fuels are typically more expensive than petroleum oil. Given their cost of production, which ranges from around $0.30 to $1.00 per liter or more (depending on the feedstock and country of origin), most biofuels are not economically competitive if oil prices are below $60 to $70 per barrel ($0.38–0.44 per liter). Large increases in oil prices make biofuels more attractive—as in the 1970s and again in the first decade of the 2000s—but absent these advantageous conditions, the industry has been able to survive mainly due to governmental and policy support.

To date, 25 countries have established biofuel targets (which set a desired level of national biofuel consumption) and/or blending mandates (which require a certain percentage of ethanol or biodiesel to be mixed with traditional fossil fuel) for transport and other uses. The U.S. Renewable Fuel Standard Program, for example, mandates the use of 36 billion gallons of biofuel by 2022, while the European Union's Biofuels Directive sets the ambitious goal of achieving a 10 percent share for biofuels in transport energy by 2020. By 2012, Thailand aims to replace all gasoline with E10, a blend of gasoline containing 10 percent ethanol, and Brazil currently requires all gasoline to be blended with 20 to 25 percent ethanol. Tax credits, price guarantees, public loans, high import tariffs on oil, and other subsidies represent additional strategies that are or have been used by various countries to support producers and increase the supply of biofuel.

These policies have protected fledgling industries and provided security to attract investors and corporations to the business. But in many countries, the growth in biofuel production and use will be limited by the nature of domestic transportation infrastructure as well as the amount of land available for the cultivation of energy crops. If a greater use of biofuel is desirable (a topic to be discussed below), the transition will necessitate a wider economic restructuring as well as significant advances in biofuels technology.

Is the Transportation Sector Ready for Increased Biofuel Use?

In the United States, E10 is currently the highest blend of ethanol and gasoline that automakers are willing to cover under vehicle warranties, fearing damage to engine components not designed for higher blends. With gasoline consumption at 137 billion gallons per year, this means that the domestic market for ethanol—the most widely used biofuel—will reach saturation at around 13 billion gallons (10 percent of gasoline consumption). Accommodating greater volumes in the U.S. economy (in order to meet the 36 billion gallon renewable fuel target in 2022, for example) requires not just an increase in biofuel production or importation, but a more fundamental change in the automotive industry: automakers must produce vehicles compatible with higher biofuel blends, and filling stations must be outfitted with the additional equipment needed to store the biofuel. It will take several years and more exacting government standards to replace a significant number of American cars with biofuel-compatible alternatives.

This kind of restructuring, while ambitious, is not impossible. Brazil, the second largest biofuel producer behind the United States, has been building its transportation sector around sugarcane ethanol for a number of years. Supported by a ban on personal diesel-powered vehicles, high blending mandates, and favorable biofuel prices, flexible-fuel vehicles able to run on higher blends of ethanol now represent 85 percent of all automobile sales in Brazil. Should the United States and other countries implement such aggressive policies, their capacity for biofuel consumption could significantly increase.

Industry Players: Who Invests and Who Benefits From Biofuels?

There are a few small, local operations that produce biofuel, because the technology for producing fuel from first-generation feedstocks (crops containing sugar or oil) is relatively simple. But given the amount of land required to grow these crops and the cost of a facility capable of producing millions of liters per year at commercial scale and prices, the biofuel industry is dominated by large agribusiness and energy corporations. (An average ethanol plant in the United States, for example, produces 416 million liters per year and costs roughly $.40 per liter—more than $160 million total—to build, and around $15 to $16 billion was invested in such biofuel refineries worldwide in 2008.) Countries with favorable climates and available land, such as Brazil (sugarcane) or Indonesia and Malaysia (palm oil), are the most common sites for production facilities, as are countries with established industrial agricultural systems like the United States (corn). Brazil and the United States currently produce 90 percent of the world's 67 billion liter supply of ethanol, while Germany, the United States, France, and Italy produce 90 percent of the world's 15 billion liters of biodiesel.

Investing in financially risky second-generation or advanced biofuel technologies, which rely on processes not yet proven for commercial scale, is also the provenance of well-funded companies or government programs, though many private companies are wary of investing in these technologies. In one of the largest research and development programs worldwide, the U.S. Department of Energy has devoted $400 million into improving the production of cellulosic ethanol (fuel from woody waste materials such as corn stalks and forestry by-products) and other second-generation biofuels. One cellulosic ethanol plant funded by this effort may be operational as soon as 2012; similar ventures in Europe will not be operational before 2018 or later.

Second-Generation and Advanced Technology Biofuels: The Energy of the Future?

Studies have shown that first-generation biofuels, in many cases, are not pristine renewable energy sources. First, they may in fact be detrimental to the environment: they may increase greenhouse gas (GHG) emissions by bringing additional land into energy-intense cultivation, strain water resources, and damage ecosystems with harmful agricultural runoff and habitat destruction. For many first-generation feedstocks, the period at which the GHG savings surpass the detrimental land use effects can be 50 to hundreds of years long. It may also be a more costly strategy to achieve sustainability than other environmental policies that reduce deforestation or promote more positive land uses.

Second, first-generation biofuels divert large amounts of farmland away from food production. Based on the policy targets of various countries, economic studies have predicted that biofuels could cause food prices to increase from nearly 10 to 30 percent or more by 2020, depending on the commodity. Though some economic models find these effects to be smaller, these negative effects on the environment and global food security give pause.

Second-generation and advanced biofuels, which are produced using more complicated processes from nonfood crops or from algae, may reduce such negative outcomes. Nonfood crops may still compete with food crops for land, however, and significant technological breakthroughs will be necessary to make certain processes feasible on a commercial scale, particularly for the production of algal oil.

As part of an effort to sustain automobile culture and bolster the agricultural sector, biofuel production may achieve some success, but it is not viable as a solution to climate change, small business development, and oil price spikes. Driven by the demand created by government targets, the biofuel sector will certainly grow in the next decades. But to reap large-scale environmental and social benefits, a focus on other sources of renewable energy may be a better investment.

See Also: Agribusiness; Alternative Energy; Biofuels (Cities); Biomass; Sustainable Development, Business.

Further Readings

Carey, John. "The Biofuel Bubble." *Bloomberg Businessweek* (April 16, 2009).

Carriquiry M. A., X. Du, and G. Timilsina. "Second-Generation Biofuels: Economics and Policies." The World Bank Development Research Group, Environment and Energy Team. Policy Research Working Paper 5406. Washington, DC: World Bank, 2010.

Naylor, Rosamond L., et al. "The Ripple Effect: Biofuels, Food Security, and the Environment." *Environment*, 49:9 (2007). https://wiki.brown.edu/confluence/download/attachments/9278/2007+Naylor+Biofuels-Food-Environment+with+citation.pdf (Accessed January 2011).

Timilsina, G. and A. Shrestha. "Biofuels: Markets, Targets and Impacts." The World Bank Development Research Group, Environment and Energy Team. Policy Research Working Paper 5364. Geneva: World Bank, 2010.

U.S. Department of Energy (DOE). "World Biofuels Production Potential: Understanding the Challenges to Meeting the U.S. Renewable Fuel Standard." Washington, DC: DOE, 2008. http://www.pi.energy.gov/documents/20080915_WorldBiofuelsProductionPotential.pdf (Accessed January 2011).

Sophie Turrell
Independent Scholar

Biofuels (Cities)

Biofuel is an umbrella term used to describe several kinds of liquid fuel produced from biological (usually plant) matter. Developed as an alternative to fossil fuels, biofuels are considered to be renewable sources of energy that can have significant environmental benefits due to their low toxicity and reduced overall emission of the greenhouse gases that contribute to climate change. For countries that import petroleum and other conventional energy sources, biofuels can also offer localized production that contributes to a secure domestic energy supply. However, some disadvantages of biofuels include the large amount of land needed to produce them, competing demands between energy and food production, and issues with scalability and end use. The expanding biofuel industry faces some challenges, but with careful management, it holds promise as part of a portfolio of renewable energy sources that will meet the energy needs of future generations.

Biofuels Overview

Liquid biofuels provide roughly 2 percent of road transport fuels worldwide, but production is growing at nearly 15 percent per year, a rate that is 10 times faster than that of oil. The most common kinds of biofuels are bioethanol and biodiesel. These are sometimes referred to as first-generation biofuels because of the relatively simple processes used to make them. Bioethanol, which accounts for 95 percent of biofuels, is an alcohol made by fermenting plants containing sugar and starch, most commonly sugarcane, sugar beet, corn, wheat, and barley. Biodiesel is a more viscous fuel that is made from plant oil, including palm, jatropha, rapeseed (canola), sunflower seed, castor bean, and soybean oil as well as from animal fats. A great deal of current research is focused on developing second-generation biofuels, which rely on more advanced processes to make ethanol by breaking down cellulose, a more complex carbohydrate that is a building block for plants like trees, wood processing by-products, and grasses. Known as cellulosic biomass, this category of nonfood plants may become an important energy source in the future, though the technology is currently 10 or more years away from being commercially viable. Finally, algae is being investigated as a source of biofuel. It has the advantage of needing less land than other biofuel technologies, but more research is needed to bring it to scale.

As of 2005, the leading producers of bioethanol are Brazil (sugarcane), the United States (corn), China (corn and wheat), the European Union (sugar beet, wheat, sorghum), and India (sugarcane); the leading producers of biofuel are Germany (rapeseed), France (soybean), the United States (rapeseed), Italy (rapeseed), and Austria (rapeseed). The United States and Brazil together produce 60–70 percent of the world's ethanol, whereas Germany and France produce and use almost 60 percent of the world's biodiesel. Price supports and encouraging regulatory environments have expanded biofuel production in several of these countries in recent years, and greenhouse gas mitigation strategies are expected to further its growth in years to come.

Greenhouse Gas Emissions, Sustainability, and Land Use

Bioethanol, biodiesel, and cellulosic biomass are all derived from plant sources that are originally carbon sinks. In a narrow assessment of the life-cycle emissions of these biofuels, they come out even on the balance sheet: the carbon that is released by burning the fuel is initially sequestered by the feedstock plants as they grow. But a substantial production of biofuels requires large-scale industrial agriculture methods that can release additional carbon dioxide. Current agricultural methods to produce corn ethanol in the United States, for example, contribute to additional greenhouse gas (GHG) emissions because they rely on inputs such as petroleum-based fertilizers and fossil fuels that power tractors and other machinery. Lowering GHG emissions from cultivating biofuel crops requires reducing these inputs. Even so, biofuels yield more energy than the fossil fuel energy used to produce them, and the use of biofuels as a substitute for fossil fuels promises to reduce GHG emissions significantly. A 2010 estimate by the U.S. Environmental Protection Agency (EPA) predicted that an increased use of renewable fuels by 2022 could prevent the release of 138 million metric tons of GHGs. Biofuels have the potential to save anywhere from 13 percent (corn ethanol) to 90 percent (sugarcane ethanol) of the fossil fuel GHG emissions that they replace. Despite the fact that they do release more nitrogen oxides than petroleum when combusted, biofuels can achieve significant emissions reductions overall.

With current first-generation technologies, however, biofuel production is extremely land-intensive. At 450 gallons of bioethanol per acre of corn, it will take a harvest of 20 million acres of land to produce just 6 percent of the total energy used in the United States in a year. If a portion of this land is converted from untouched or gently cultivated land to heavy industrial energy farms, it could have additional environmental consequences. The first major criticism of biofuels comes from this land use issue.

The demand for space is most likely to be met by less developed countries that have room to spare. This may encourage deforestation or the utilization of carbon-rich swamps in places like Brazil, Malaysia, and Indonesia. Monoculture cultivation in these and other areas can reduce the land's potential for carbon storage, shrink animal habitat, and leave crops more vulnerable to pests and blights that encourage the use of harmful pesticides, herbicides, and fungicides. In addition, the methods used to develop these lands, including the overapplication of fertilizers and heavy tilling practices, can release chemical pollutants into rivers and lakes and erode the land's topsoil. Fertilizer runoff in the U.S. Mississippi River watershed has well-documented negative effects in the Gulf of Mexico, where the excessive nitrogen and phosphorous loading has fed algal blooms that cut off the supply of oxygen to other sea life, creating a hypoxic dead zone. Should industrial agriculture flourish because of a boom in biofuels, it could replicate these negative impacts in new regions of the world.

Finally, an expansion of agriculture could introduce competition for freshwater in areas where water for drinking, cooking, and washing is in short supply. It is important to make sure that new croplands do not divert too much water away from other human and natural uses.

Food Versus Fuel: Biofuels and Food Security

With an estimated 850 million people experiencing hunger worldwide and a growing global population, the use of large amounts of cropland for energy rather than food is controversial. In the past decade, biofuel production has increased against a backdrop of rising global food prices. For those who live on less than $1 a day, these price increases can put much-needed calories out of reach. To ensure that the hunger for energy does not further disadvantage the hungry, biofuel's contribution to food prices, availability, and access must be assessed.

The interaction between biofuel and global food prices is complex. Many factors cause food prices to rise and fall. These include the price of oil, which is a significant agricultural input that also affects the cost of food distribution; weather conditions that affect harvests; market speculation; and international trade policies. It is difficult to predict how these factors will interact in the future to set prices. If there is a demand for energy from biofuels, however, first-generation feedstocks are likely to fetch higher prices as energy crops than as food crops. The United Nations (UN) Food and Agriculture Organization (FAO) projects that this demand will increase over the next few decades and will contribute to an overall rise in food prices.

But does this rise benefit the poor or harm them? If grain-exporting countries like the United States shift from producing corn for food to corn for energy, prices may rise and supply may decrease without any benefit to the global poor. But if implemented as an engine of rural development, biofuel production can stimulate local economies and raise incomes for the poor, allowing them to buy adequate amounts of food for their families. The problem of hunger has typically been seen not as a problem of supply—there is enough food in the world to supply every person with more calories than needed—but as a problem of economic access. If biofuels can increase the income of the local population enough to overcome a rise in food prices, then there will be gains for food security. Conversely, if the production of these fuels is controlled by large operations that do not employ a significant number of these people, then biofuel production may have negative impacts in this area. Steps must be taken to involve the local population so that the biofuel industry will produce social and economic advantages in addition to energy.

Biofuel End Use

Liquid biofuels are used primarily to power nonindustrial vehicles and are often used as additives to lower the sulfur content of petroleum-based fuels and facilitate engine lubrication. Cars, buses, and even some trains run on biofuel. However, a few limitations prevent biofuels from being used more extensively. First, they are more expensive than traditional fossil fuels. When the price of oil goes down, they are less attractive to consumers. They also have a slightly lower fuel economy and power (for example, pure biodiesel has 10 percent lower fuel economy as compared to oil), though most consumers do not notice the difference. Second, they can damage rubber and some types of metal inside engines (typically older ones). Switching back and forth between biodiesel and regular diesel can also cause problems for an engine, but the low availability of biodiesel makes this almost unavoidable for vehicles traveling longer distances. These issues can be overcome and in most cases are not too serious, but because of the possible legal risks that they pose, car companies will not offer warranties for the use of purer (more than 10 percent biofuel-to-petroleum) blends. This may affect a consumer's use of renewable fuels as well as the type of car he or she purchases. Cars with engines designed to run on E85 (85 percent ethanol, 15 percent petroleum), sometimes called flex-fuel cars, are available with warranties. The real reason that biofuels are not used extensively is structural: engines that can run on both biofuel and petroleum are not widely available, and the market for fuel as well as the cars that it powers is dominated by standards that are designed for fossil fuels.

But even if these structural problems were solved to bring biofuel blends into the fuel mix, the underlying transportation model that is contributing to GHG emissions, pollution, and climate change would not be radically altered. This is the third major criticism of biofuels: that they do not change the uses of energy or the demand for it and may even encourage the car culture that is contributing to many environmental problems, including climate change.

Because biofuels are versatile, made in many different climates from many different plants using relatively simple processes, they promise a stable, locally distributed source of energy for many countries. Even so, the major consumers of this technology are likely to be the developed, industrialized nations that have the extensive vehicle fleets and the infrastructure to use and distribute biofuels. These countries are beginning to stimulate demand and create opportunities for an international trade in biofuels. Such trade may allow some countries to benefit from decreased fuel emissions while others absorb the environmental costs of biofuel production, or it may prove to be a boost for rural incomes. Whichever the case, biofuels constitute an emerging market that is likely to become more dynamic in the coming years. However, protectionist trade policies may prevent the most efficient kinds of biofuels—those from sugarcane and sugar beets, which emit the least GHGs—from being produced at scale. The U.S. corn industry's opposition to the import of Brazilian cane ethanol is one example of these restrictions.

The Future of Biofuels

While the use of first-generation feedstocks remains the standard for commercial biofuel production, scientists continue to search for the magic bullet in cellulosic biofuels and algae feedstocks. If these two technologies can be perfected, they promise liquid fuel that is free from the most significant problems of first-generation biofuels: they do not use valuable arable land, and they do not compete with food crops. Cellulosic biofuels, made from woody plant material, can be grown in marginal lands and is very abundant; algae can be grown in water. Traditional forest and water resource issues will govern these technologies, but they are still several years away from implementation.

For now, biofuels remain an important but small part of the global renewable energy portfolio. Though renewable fuel targets will be essential for the transition to a more sustainable future for many countries, they must complement more significant efforts to change the industrialized world's patterns of energy use.

See Also: Agribusiness; Alternative Energy; Biomass; Ethanol; Sustainable Development (Business).

Further Readings

Blumberg, Carol Joyce. "A Comparison of EIA-782 Petroleum Product Price and Volume Data With Other Sources, 1999 to 2008." (2010). http://www.eia.doe.gov/pub/oil_gas/ petroleum/data_publications/petroleum_marketing_monthly/historical/2010/2010_04/pdf/ comparison7822010.pdf (Accessed September 2010).

Global Bioenergy Partnership. "Bioenergy: Facts and Figures." (2007). http://www .globalbioenergy.org/fileadmin/user_upload/gbep/docs/2007_events/press_G8/Bioenergy_ Facts_and_Figures_01.pdf (Accessed September 2010).

Gustafson, Cole. "Biofuels Economics: How Many Acres Will Be Needed for Biofuels? Part I." (2008). http://www.ag.ndsu.edu/news/columns/biofuelseconomics/biofuels-economics-how-many-acres-will-be-needed-for-biofuels-part-i (Accessed September 2010).

Naylor, Rosamond L., Adam Liska, Marshall B. Burke, Walter P. Falcon, et al. "The Ripple Effect: Biofuels, Food Security, and the Environment." *Environment*, 49:9 (2007).

U.S. Department of Energy Biomass Program. http://www1.eere.energy.gov/biomass/index .html (Accessed September 2010).

U.S. Energy Information Administration "Biofuels Overview." (2010). http://www.eia .doe.gov/cneaf/solar.renewables/page/rea_data/table1_6.html (Accessed September 2010).

U.S. Environmental Protection Agency. "Renewable Fuel Standard Program (RFS2) Regulatory Impact Analysis." EPA-420-R-10-006 (2010). http://www.epa.gov/otaq/ renewablefuels/420r10006.pdf (Accessed September 2010).

World Resources Institute, EarthTrends. "March 2007 Monthly Update: Global Biofuel Trends." http://earthtrends.wri.org/updates/node/180 (Accessed September 2010).

Sophie Turrell
Independent Scholar

BIOLOGICAL CONTROL OF PESTS

As pests became tolerant to synthetic pesticides, biological control agents, such as bacteria that grow naturally in soil or others that can be broken up into the soil, have been used to control pests.

Source: iStockphoto

Agricultural losses caused by plant pathogens and injurious pests can negatively affect growers' abilities to produce adequate crop yields necessary to meet increasing world food demands and achieve economic profits necessary for continued production. Until recently, growers turned solely to chemical pesticides to offset the negative impact caused by plant pathogens and pests. However, growing public concern over the past two decades about the potential health risks caused by synthetic chemical pesticides has helped energize interest in pest management alternatives, such as the use of biological control (biocontrol) agents. Biocontrol agents are an important component of efforts to reduce agricultural reliance on chemical pesticides (insecticides, miticides, bactericides, fungicides, nematicides, and herbicides) and increase agricultural sustainability.

Biological Control Versus Pesticides

Synthetic pesticide use increased following World War II. Their initial use dramatically reduced insect and pathogen populations, and consequently, food production increased. Unfortunately, over time and with continual use, pests and pathogens became tolerant and even resistant to individual or combinations of synthetic pesticides. If resistance levels become high enough, multiple treatments with numerous pesticides, or increasingly higher amounts, are needed to produce the same crop. In turn, this can lead to an increase in production costs that are passed on to the grower in profit loss and the consumer in increased food costs. Many growers may not be fully aware of the price they are currently paying to reduce pests and disease. For example, the cost of applied seed treatments, such as fungicides, is included in the price of seed.

Public concern about synthetic chemical pesticides has been growing for decades in the United States and around the world. Concerns raised by the public range from their initial production and manufacture safety (Bhopal, India), to their dispersal and poisoning of untrained retail and agricultural workers to potential negative impacts on human health and ecosystems. The number of commercial pesticides is large, and little is known about their combined effects over time. Currently, more than 16,000 pesticide products are marketed in the United States alone.

Biocontrol agents are living organisms or their metabolic by-products that are used to control pests or disease-causing pathogens. Part of their appeal is that they offer a natural means of control and can lessen dependency on chemical pesticides. Biocontrol agents can be employed individually or in combination with other biocontrol agents, pesticides, or management techniques as part of an integrated system of control known as Integrated Pest Management (IPM).

In the future, pesticide use may become limited due to strict use restrictions or public opposition. Due to the increase of chemical pesticide concerns, biocontrol agents have been steadily gaining attention for their potential use in the agriculture market. Several companies are conducting research toward marketing commercially practical products that include biocontrol agents.

Mechanisms of Biological Controls

Usage of biological control agents takes advantage of natural tendencies in communities composed of interacting populations, wherein certain organisms have developed strategies to limit competing or antagonistic population numbers. Potential biocontrol agents are prevalent in many communities, being common in soil, water and air and on plant surfaces. Gaining knowledge about their activities is advantageous because the interactions between organisms will continue whether we are aware of them or not.

There are several ways in which biological organisms act to alter the ecological functioning of pests and pathogens. Mechanisms through which biocontrol agents can antagonize plant pathogens or injurious pests can broadly be divided into four categories: (1) antibiosis, which involves the use of metabolized compounds such as antibiotics that are capable of killing or inhibiting the growth or reproduction of organisms; (2) competition for space and nutrients (niche exclusion); (3) parasitism/predation; and (4) induction of a host's defense response such as a hypersensitive response or induced systemic resistance, both of which can stimulate production of numerous defense-related compounds such as terpenoids and peroxidase.

Determining Commercial Viability of Biocontrol Products

Even though there are a great number of organisms that utilize some of the mechanisms mentioned above, most are not commercially viable to be useful as biocontrol agents. Reasons for this are many, such as the inability to be cultured outside living tissue, unusual culture demands, low propagule production, biocontrol activity against too narrow a host range, production of toxic compounds, or restriction to specific environmental area.

In order for a biological agent to have wide market appeal, it generally needs to have a large host range. Another consideration is timing of application to produce optimal results. Sometimes a biological agent may have specific environmental requirements, such as a necessary temperature range to germinate or maintain viability, or the agent may only be effective against a specific developmental stage (instar) of an insect's life. Typical additional factors affecting a biocontrol agent's short-term efficacy and long-term sustainability in the field with natural soils are soil pH and micronutrients and macronutrients, which may aid or inhibit a biocontrol agent. Additional factors that affect a biocontrol agent's ability to work optimally are moisture levels, level of sunlight, the crop itself, irrigation practices, agricultural practices such as till versus no till, and crop rotation. Finally, one of the main concerns in the development of a commercially viable biocontrol is formulation. It is important to find substances that do not hinder the viability or growth of the organism. Formulations are typically composed of inert compounds that are added along with the biocontrol agent to enhance its efficacy or extend its shelf life. Though these biocontrol considerations may seem burdensome, for chemical pesticides to work effectively, similar research must be done to seek optimal crop and pest targets, application times and amounts, and environments wherein their product can work most effectively.

Types of Biocontrol Agents

The types of biological controls are numerous and include bacteria, fungi, nematodes, and insects as well as other animals. Currently, many organisms are being examined for their ability to serve as biocontrol agents, either commercially or as naturally occurring epizootics. However, some hold more promise than others.

Bacterial Agents

Bacteria that grow naturally in soil are ideal for use as biocontrols because of their abundance. In the soil, they can protect plants from pathogens using a number of different mechanisms. Examples are fluorescent pseudomonads, a group of bacteria that produce a number of disease-preventing compounds, specifically phenazines, a derivative of a potent antibiotic. In addition, members of the genera streptomyces have been of particular interest because of their isolation ease and antibiotic production. Some strains, depending on environmental conditions, can make use of multiple mechanisms of biocontrol such as antibiosis, hyperparasitism, and the production of cell-wall-eroding enzymes. One of the most well-known biocontrol agents is *Bacillus thuringiensis* (Bt), which produces a crystalline toxin that is harmful to many insect pests of agricultural crops but is harmless to humans and several nontarget insects.

Fungal Mycoparasites

Trichoderma spp. are a well-known group of mycoparasites that have been used successfully as biocontrol agents on many crop plants against soilborne pathogens. As they are typically applied to plant parts where disease symptoms are present, direct antagonism with the pathogen is involved. Multiple mechanisms have been documented such as cell lysis, antibiosis, competition, and mycoparasitism with concomitant production of enzymes that degrade fungal cell walls. In some cases, the application of a biocontrol agent as a pretreatment can disrupt later infection development. For instance, prior treatment with *T. harzianum* and *T. viride* on plant parts spatially separated from the site of *Sclerotinia sclerotiorum* inoculation demonstrated consistently reduced infection development.

Entomopathogens

There are estimated to be more than 700 different types of entomopathogenic fungi. Fungal entomopathogens are unique among insect pathogens in that instead of depending on being eaten by the insect or by opportunity entering through an opening (wound or natural), they are capable of entering the insect directly through the cuticle by production of enzymes. Death is not instantaneous. Depending on the insect, virulence of the entomo-pathogenic strain, internal environmental factors, and host immune response, it may take two to eight days to actually kill the insect. During this time, the insect typically will continue to eat and move. With time, however, activity will slow, followed by death. After the fungus has utilized the insect, it will sporulate on the cadaver, and the infection cycle will begin again. If the insect is in molt, it may discard the exoskeleton before the entomopathogen can penetrate, thereby escaping infection. Two of the most well-known entomopathogens are *Beauveria bassiana* and *Metarhizium anisopliae*.

Beauveria bassiana's efficacy as an entomopathogen is well known, and it has an extensive insect host range. The fungus is ubiquitous in most soils, and it has been responsible for many natural epizootics. The insect disease it causes is called white muscardine disease. As a biocontrol agent, it can be used in all of its asexual reproductive forms: conidia, blastospores (although prone to desiccation), and mycelia, which can be dried, broken up, and applied to soil. It is currently marketed in many formulations with numerous application methods, including foliar conidial suspensions, granules, and soil amendments.

Metarhizium anisopliae is responsible for causing the insect disease commonly known as green muscardine disease, because of the green aerial conidia it produces on its mummified insect victims. This biocontrol agent has been produced commercially and has been used as a biocontrol in the United States against cockroaches, but it also has a wide host range with activity in several insect orders, including Lepidopteran, Coleopteran, Hemipteran, and Homoptera. *Metarhizium anisopliae* var. *acridum* is marketed under a commercial product name and widely distributed in Africa. Under favorable climatic conditions, it can cause local epidemics in grasshopper or locust populations.

There are a multitude of potential biocontrol agents, and only a small number are noted here. Additional biocontrol agents include viruses, nematodes, protozoans, and insects. Biocontrol agents are also viewed as potential treatment alternatives in the continual battle against invasive species.

Safety Issues

It is important to remember that just because something is considered natural does not mean it is harmless. Many biocontrol agents produce compounds and secondary metabolites that are toxic to competitors or prey. Any potentially hazardous toxin should be tested to determine if it is long-lasting, if it can have any adverse effects on organisms other than the target organism, and if the toxin has any potential of entering the food chain. The use of biocontrol agents can be further complicated by closely related species or strains or varieties of the species behaving differently in the field. For instance, *Burkholderia cepacia* complex (Bcc) are a group of closely related species that not only act as biocontrol agents but also as plant pathogens, saprophytes, or even human pathogens. Some of the *B. cepacia* complex organisms have been responsible for some nosocomial infections, and some strains have even led to fatal lung infections of individuals with cystic fibrosis. As concerns increase and stricter rules are placed on chemical insecticides, it is likely these will in turn lead to stricter rules being applied to new biocontrol agents. The most prudent action would be to commit resources and time to examine these concerns before fully committing to biocontrol programs.

For the most part, with some exceptions, biocontrol agents are not expected to be the sole source of treatment for a pathogen or injurious pest. They should instead be viewed as an integral part of an IPM program, which allows for utilization of all available resources to optimize crop production and provide protection of natural resources.

See Also: Agriculture; Antibiotic/Antibiotic Resistance; *Bacillus Thuringiensis* (Bt); Genetically Engineered Crops; Roundup Ready Crops.

Further Readings

Chin-A-Wong, Thomas F. C., Guido V. Blomberger, and Ben J. J. Lugtenberg. "Phenazines and Their Role in Biocontrol by *Pseudomonas* Bacteria." *New Phytologist*, 157 (2003).

Denoth, Madlen, Leonardo Frid, and Judith H. Myers. "Multiple Agents in Biological Control: Improving the Odds?" *Biological Control*, 24 (2001).

Feng, Ming-Guang, Tadeusz J. Poprawski, and George G. Khachatourian. "Production, Formulation and Application of the Entomopathogenic Fungus *Beauveria bassiana* for Insect Control: Current Status." *Biocontrol Science and Technology*, 4 (1994).

Handelsman, Jo and Eric V. Stabb. "Biocontrol of Soilborne Plant Pathogens." *The Plant Cell*, 8 (1996).

Pearson, Dean E. and Ragan M. Callaway. "Indirect Effects of Host-Specific Biological Control Agents." *Trends in Ecology and Evolution*, 18 (2003).

Raaijmakers, Jos M., Maria Vlami, and Jorge T. deSouza. "Antibiotic Production by Bacterial Biocontrol Agents." *Antonie van Leeuwenhoek*, 81 (2002).

Vega, Fernando E. "Insect Pathology and Fungal Endophytes." *Journal of Invertebrate Pathology*, 98 (2008).

Weller, David M. "Biological Control of Soilborne Plant Pathogens in the Rhizosphere With Bacteria." *Annual Review of Phytopathology*, 27 (1988).

Mary Ruth Griffin
Independent Scholar

BIOMASS

Biomass is organic matter such as aquatic and terrestrial plants and plant-derived materials, including agricultural crops and trees, wood and wood residues, animal waste, grasses, municipal solid waste, and other residue materials. Uses include energy production, landscape and agricultural use, and building production. Using woody biomass as an energy source is the most common form and includes utilization of residuals from hazardous fuels reduction, forest restoration, commercial timber harvest, forest products manufacturing, and plantation management. Despite controversy, it has a continuing and growing role as an energy source for millions of people globally. While considered by many as a relatively clean energy source that contributes to local economic growth and healthy forest management, there are concerns over human health impacts, relocation of certain forest species, and emissions contribution from burning and transportation. Despite fairly extensive research, the potential environmental and societal impacts from biomass removal and usage are complex and interrelated.

Both terrestrial and aquatic plants use light energy from the sun to convert water and carbon dioxide in the air into carbohydrates, fats, proteins, and minerals that they need to survive. When the carbohydrates (which contain complex compounds of carbon, hydrogen, and oxygen) are burned, they turn back into water and carbon dioxide and release energy stored from the sun. In addition to plant sources, biomass can include animal waste from agricultural and manufacturing industries as well as garbage and sewage city waste.

Biomass can be converted into liquid fuels, "gasified" to produce combustible gases, or burned with a mixture of coal in "co-firing" to substitute for the amount of coal needed in a boiler. While historically considered inefficient as a net energy loser, studies over the past 10 years have shown locally produced biomass to have either a positive energy balance (more energy released than used to produce) or at least energy neutral due to recent advances in technology. Research also shows that biomass contains less energy per pound compared to fossil fuels and therefore is only cost-effective if produced locally (within 50 miles) in rural communities or on individual farms.

The United Nations Food and Agricultural Organization (FAO) estimates that 15 percent of the world's energy comes from biomass, mostly as a heat and cooking source in developing countries. In the United States, biomass provides 83 percent of renewable energy use in the residential sector and 87 percent in the commercial sector, according to the Department of Energy. About 5 percent of total electric sales or 45 billion kilowatt-hours of electricity is supplied by biomass. The greatest source is wood, wood waste, and black liquor from pulp mills. In addition, biomass production supports 66,000 jobs in the United States. The amount of energy generated by biomass plants varies widely, from supplying individual homes to a 140-megawatt facility in Florida that powers 60,000 homes.

All combustion processes release exhaust or emissions of some kind as they are releasing carbon that has been locked away either in the ground or in the material. However, releasing oil, gas, or coal transfers carbon from the ground to the atmosphere as new carbon dioxide. Biomass is argued to be a clean renewable energy as there is no new carbon dioxide added into the atmosphere (as long as the wood is obtained from a forest that is sustainably managed), given that the carbon released was captured from the natural carbon cycle.

Burning Clean: Benefits of Biomass

Biomass has many positive benefits for the environment, society, and economy. Biomass production reduces forest density through selective thinning and clearing of understory. In many forests, fire suppression has allowed fuels to build up, which can lead to catastrophic fires. Biomass production removes many of these fuels, therefore reducing forest fire risk. Reduced density can also increase the resiliency of the forest to pests and invasive species that cause massive die-offs in dense forests. Understory plant biomass and biodiversity can increase through openings of stands, and conifers or other native plants may colonize the area. Less competition for water and soil nutrients can increase the vigor of the remaining trees, which also increases resiliency to pests and disease. Species that are fire dependent, species that prefer open habitats or canopy, or those associated with successional vegetation have all been shown to be successful in areas with reduced fuels. As long as dense patches of forest are retained, ungulates experience increased forage quality and quantity. Understory shrub growth following thinning can also promote greater bird abundance and diversity. Water and soil quality improve with sustainable harvesting of biomass when native species are used.

Modern biomass systems, such as wood chip and pellet, release virtually no visible emissions or odors and less particulate matter (which causes human respiratory problems) compared to wood stoves and other fuel sources, improving air quality and reducing health risks in local communities. Public safety is also improved with fuels reduction, as communities are less at risk of fires following thinning, and many communities report increased social cohesion. Communities with local biomass programs have a secure, competitive, renewable, and sustainable energy source that is unaffected by global markets and politics. In addition, forest-based infrastructure that already exists in many rural communities can be adapted and put to use. Individuals also experience reduced financial burden when using biomass for home and water heating as biomass fuel is about half the cost of fuel oil.

One of the greatest benefits of small-scale rural biomass development is creation of local jobs. Jobs associated with biomass production include logging, agriculture, forestry support, food services, industrial truck and equipment manufacturing, and financial and legal services. Some studies have shown that five jobs are created for every megawatt generated. These communities can successfully compete in the global marketplace, which is important for building a future green economy.

Burning Dirty: Costs of Biomass

Research on biomass production and use also demonstrates negative impacts on the environment and society. Removing portions of standing tress can reduce resiliency to wind events, leading to uprooting of trees and changes in fire behavior. During biomass removal, residual trees as well as soil may be damaged from the equipment used. Species that prefer closed-canopy forests or dense understory (e.g., marten and fisher) as well as preference for large tree snags (e.g., woodpeckers) are negatively affected by biomass removal. In addition, those with denning habits or that have already experienced loss of prey may be threatened from biomass removal due to cumulative negative impacts. Soil chemistry, fertility, and the growth of residual plants depend on decomposing woody biomass: by removing biomass, negative effects could be experienced in the ecosystem. In addition, unregulated production of biomass in some developing countries (especially southeast Asia and African countries) has led to massive deforestation.

Biomass production plants emit carbon dioxide and other pollutants such as sulfur oxides, nitrogen oxides, carbon monoxide, volatile organic compounds, and particulate matter, all of which affect those with respiratory illness and contribute to smog and acid rain. While the emissions are lower compared to fossil fuel plants, there is still debate about whether woody biomass power is "carbon dioxide neutral," given that emissions are still generated.

Negative impacts on communities include strain on existing transportation infrastructure, loss of raw material availability for the forest industry (possibly increasing the price of wood), and loss of natural or arable lands for harvesting crops for energy. Large-scale biomass production development is typically inefficient and unsustainable and relies on government support to be economically feasible. There is also concern over safety of individuals working in the biomass industry, as they work with heavy machinery on poorly maintained roads.

Conclusion

The costs and benefits of biomass as an energy source are dependent on the region, species, and communities living in the area of production as well as the scale of production. For communities transitioning to biomass, these production systems must use the most updated technology and ensure protection and enhancement of the environment, society, and the economy.

See Also: Alternative Energy; Biofuels (Business); Ethanol.

Further Readings

Almquist, W. "Environmental Group Perspectives on the Utilization of Woody Biomass Derived From Hazardous Fuels Reduction Activities." Master's Thesis, Department of Planning, Public Policy and Management, University of Oregon, 2006.

Biomass Energy Resource Center. "Air Emissions." http://www.biomasscenter.org (Accessed August 2010).

Domac, J., K. Richards, and S. Risovic. "Socioeconomic Drivers in Implementing Bioenergy Projects." *Biomass and Bioenergy*, 28 (2005).

Gan, J. and C. T. Smith. "Co-Benefits of Utilizing Logging Residues for Bioenergy Production: The Case for East Texas, U.S.A." *Biomass and Bioenergy*, 31 (2007).

Kullander, S. "Energy From Biomass." *European Physical Journal Special Topics*, 176 (2009).

Oregon Department of Forestry, Office of the State Forester. "Environmental Effects of Forest Biomass Removal" (December 2008). http://www.oregon.gov/ODF/PUBS/docs/ODF_Biomass_Removal_Effects_Report.pdf?ga=t (Accessed August 2010).

Union of Concerned Scientists. "How Biomass Energy Works." http://www.ucsusa.org/clean_energy/technology_and_impacts/energy_technologies/how-biomass-energy-works.html (Accessed August 2010).

U.S. Department of Agriculture, Forest Service. "Why Biomass Is Important—The Role of the USDA Forest Service in Managing and Using Biomass for Energy and Other Uses." http://www.fs.fed.us/research/pdf/biomass_importance.pdf (Accessed August 2010).

U.S. Department of Energy. "Biomass Energy Databook." http://cta.ornl.gov/bedb (Accessed August 2010).

Stacy Johna Vynne
Independent Scholar

BIOTECHNOLOGY

There are two distinct ways of looking at biotechnology—one as a continuous process of human intervention in nature and one as a radical discontinuity in the manner of that intervention. The continuous vision is based on looking at 10,000 years of human techno-cultural evolutions. Such advances involved selection and hybridization during the process of crop production, fermentation in the context of food processing or preservation, and extraction of medicinal qualities from plants and minerals. Over time, as

Biotech fields find solutions to health and environmental problems, but lay perspectives are often discounted because traditional technology has evolved through a laboratory-centered approach.

Source: Randy Wong/Sandia National Laboratory

these evolutions generated a common human pattern of living around the globe based on agricultural advances and gave rise to large cities, governance became an integral part of agriculturally based civilizations as much as distinctive cultures and identities associated with food and ways of living. Such governance gave rise to urban technocratic elites and philosophers who contemplated the metaphysics of ethics that was not related in any concrete fashion to the reality of those who toiled on farms to produce. This gave rise to a social hierarchy where leisure of consumption was tied to elitism and consumers had more of a say than producers.

The transition into the industrial age did not change this trend for need of governance and cultural identity, but it created a disjuncture between producers of technology and consumers of those technological products in a whole new manner. Now the "technical" elite has became the producers with backing from the technocrats and business elites, while the majority of urban consumers and farm producers who continue to live on older versions of experimenting with nature have been left out of the "loop" of governance based on private-interest science. Hence, there is a constant struggle over the normative ideals of costs and benefits or the "economics" of biotechnology, the very morality of undertaking such technology, and the impact on overall sustainability of human life and its modes of living. Given this situation, issues of impact of so-called scientific nonknowledge and claims of ethics of the "precautionary approach" for contending with emerging biological and environmental risks in the context of emerging technologies become important. This is of particular significance as rapid advances in genetics and genetically modified organisms (GMOs) affecting health/pharmaceutical and food/agribusiness sectors are becoming increasingly difficult for the lay population to keep up with or even to comprehend.

The precautionary principle is based on the assumption that scientific uncertainty is not a bar to preventive regulatory action, and decision makers are better off erring on the side of caution when dealing with issues of harm. However, such principles, while elucidating the inherent limitations of risk analysis within the rational framework of "science," fail to elucidate valid ways of estimating what is ethically and politically acceptable, leading to a moral paralysis of whether one should or should not intervene through regulative actions. There is a deficit of conceptual machinery that can distinguish the efforts of this principle from traditional risk management techniques. Hence, extremes of action or inaction take place in the empirical arena.

Critics of the precautionary approach point out that since evolution is defined by multifactorial, geometrically nonlinear, non-law-governed outcomes, it does not tend to produce "optimal, delicately balanced equilibria that generally coincide with what we value"

(Powell, 2010). Mother Nature is not a master engineer who creates optimized products through natural selection. Rather, she is defined by irrational and imperfect biological designs that take an enormously long time to come to "some" benevolent adaptations. Natural selection and value overlaps are thus sheer coincidences. Hence, there is a serious dearth in well-thought-out efforts at validating biotechnology based on a precautionary approach. One is better off taking a historically contingent approach to reality and its outcomes where sweeping regulations do not block possibilities on the basis of "precautionary approach." In short, the critics go beyond the risk factors of laboratory, enhanced by "precautionists," to point out that the "natural" laboratory is equally risky and non-predictive.

The critique of the precautionary approach seems to give room to the discontinuous vision of biotechnology where human engineering is given historical consideration. And in consequence, legal "intellectual property rights" are increasingly being granted to "discoveries" in nature (such as genes) rather than simply to industrial innovations that proceed from inventions. The discontinuous vision points out that there is a distinct pattern of change to how technology has evolved and how it has been understood and accepted in a modern era. The dominant rational paradigm of modern scientific technology involves deliberately constructed experiments within defined institutional cultures of academics, government, and business-sponsored academics. Within such "limited" institutional cultures it has become critical to document the ideas and processes or methodology of manipulating identified variables involved; demonstrate proof of cause and effect to an elite panel of scientific community; have one's study subjected to peer review within that defined community; implement the study on a slightly broader scale of clinical trials, and so on. This process is considered "superior" to laymen's understandings and, hence, all concerns of the latter are dismissed as not only emotional and irrational but also the product of people being Luddites or afraid of technology and its complexities.

This laboratory-centered, small-scale process takes a long time to provide fruits, to be approved by regulatory bodies of states, before being developed on an industrial scale. Therefore, it involves governance. However, the current biotechnical sciences' evolution is far faster (10–20 years in comparison to thousands) than the past slow pace of growth of technology where cultural evolution outpaced technological advances and people had time to absorb the long-term impacts of change. Hence, biotech's long-term impact is consequently hard to grasp for the common people as historical trajectory seems to have been short-circuited. Moreover, regulations typically tend to be based on the limited visions of "productive science" and thereby tend to exclude the concerns of the vast majority of consumers about those scientific products.

Common people's concerns include issues of health safety, environmental safety, cultural integrity in terms of food consumption, and so on. More critically, such issues are about anticipating and recognizing harm in good times in order to take preventive actions. In essence, concern about "ethics of biotechnology" has become a huge issue. There are strong ongoing struggles around the world as evident in push-backs for better governance from the societal sector (including nongovernmental organizations [NGOs]). There is also a huge transatlantic divide in "legal regimes" with the United States preferring a science-based risk assessment and management and the European Union preferring a precautionary approach. However, for all such efforts and differences, the "business-as-usual" scenario continues for the most part with laws being passed that overall are favorable to swift industrial outputs of "limited-tested" biotechnology. This is also evident in the sharp proliferation of gene patents and the increasing concerns over patents acting as barriers to further innovations and social and natural sustainable development. With this background in mind, let us take a comprehensive look at biotechnology and its outcomes.

Technologies have become embedded in a new paradigm of science as biotechnology and have become limited to issues of genomics—genomic decoding, understanding genetic mutations, genetic modifications (GM) or engineering, and so on. It involves a vast array of specialized sciences such as biochemistry, nutrigenomics, epigenomics, and pharmacogenomics. The main focus is recombination of DNA in a more precise fashion. Crossing the species barrier includes gene cutting and splicing with bacteria and viruses to enable these vectors to carry foreign DNA into living cells in order to recombine DNA. All biotech fields try to come up with solutions to health problems and environmental problems as well as increasing food production through creation of high-yield pesticide- and herbicide-resistant GM seeds.

These types of foci inevitably draw in various types of entities that are interested in biotech. Venture capitalists are drawn in by the expectation of making huge profits through selling of high-market-value niche innovations. For example, at a time of overharvesting of the oceanic fish crop, it is critical to find alternatives that provide the critical omega-3 fatty acid docosahexaenoic acid (DHA). DHA is an important component of brain activity that can help counter major diseases created by ingesting omega-6 acids (found in cooking oils such as soya, maize, etc.). Such diseases include depression, schizophrenia, obesity, and heart disease. DHA is also critical for intelligence. Hence, biotechnology can help by creating GM crops like soya beans with higher levels of DHA. The recent hype surrounding the U.S. Food and Drug Administration (FDA) certification of GM salmon injected with growth hormone to boost supply of seafood is also an effort in this direction.

In developing countries such as those in Africa, one alternative to other forms of growing sustainable food could be GM. Agribusiness or agro-food processing industries that want large-scale standardized production of foods are hence also drawn in by the lure of reaping the benefits of vertical integration of the supply chain. The justification provided by agribusiness is that biotechnology will lead to decreasing poverty. This will be achieved through cultivation of high-yielding GM crops that reduce costs by having drought- and pest-resistant characteristics. Agribusiness has also developed significant interest in development of biofuels (see below) as an alternative source of energy.

The pharmaceutical industry is drawn to biotech in order to ensure the prospect of making at least $20 billion from the patent life cycle of a blockbuster drug. For example, given the constant mutations in cancerous cells and no cure for cancer yet in sight, blockbuster cancer drugs like Gleevec can only provide a small lease on life. Here, biotech of decoding genetic mutations and harmful tumor-promoting proteins (the building blocks of life) has helped in discovering new classes of compounds/antibodies/drugs that can act as critical supplements to blockbuster drugs. Moreover, the big U.S. Pharmaceutical Research and Manufacturers of America (PhRMA) companies such as Pfizer are facing massive class-action lawsuits against some of their extant blockbuster drugs, such as Lipitor, because of their unrevealed deadly side effects. Most of these drugs are also expected to go "off-patent" soon and will face fierce competition from generic drug industries. Hence, there is a huge drive by big pharmaceutical companies to invest in customized biogenetics. To circumvent the patent system, they are ironically also invested in acquiring small biotech firms in order to acquire their patents and develop bio-similars—the alternative to generic drugs.

Academic institutions like Washington University, Professor Cooper, and his team have tried to tap into the "lay intelligence" sector's creativity for developing biotechnological

knowledge by creating a video game called FOLDIT. In this game, players follow unortho-
dox methods to try to squeeze proteins into the most chemically stable 3-D configurations.
This helps to figure out the underlying biochemical processes in order to enable creation
of drugs. Their fruits are publications in a top journal like *Nature*. Some academic studies
are turning conventional understandings of disease "on their head." A recent study claims
that metabolic syndrome (the term *syndrome* refers to a collection of symptoms whose
common cause is not properly understood) is caused by the toxic effect of over-creation of
lipids (a group of molecules that includes fats). Lipids are needed in small quantities to
make cell membranes. The body generates these syndromes as adaptive resistance when
there is over-generation of lipids. Hence, a metabolic syndrome like diabetes may be the
body's way of adapting to the consequence of lipids. Hence, lifesaving drugs like GM
insulin may not be of help if insulin-resistance may ironically be a protective mechanism
of the body. In such contexts, the recent discovery of a synthetic DNA sequence (of about
1,000 genes) created from preexisting laboratory chemicals by the J. Craig Venter Institute
may open new horizons.

Parasites (especially gut parasites) and pathogens have harmful impacts on the growth
of human capital. In order to ensure early childhood development of intelligence growth
and the later cycle of a sustainable workforce, it is necessary to prevent harmful diseases.
Hence, charitable foundations like the Gates Foundation have resorted to incentivizing
innovation through prize competition in the biotech sector for orphan diseases like
malaria. The irony is that rising intelligence is associated with increasing allergies and
asthma in developed countries. This is due to the fact that children's immune systems,
unchallenged by infection, are turning against the very body cells they are supposed to
protect. Hence, developed economies, like the industry-sponsored X Prize Foundation, are
also offering prizes as incentives.

Besides such industries, other mainstream industries like the oil and car industries, looking for alternative energy resources, are also drawn in by the possibilities of breakthrough
biotechnologies. Such breakthroughs include biofuel based on GM food crops like soya,
maize, corn, and so on. There are also second-generation biofuels like those based on processing of algae, nonfood feedstock, and biomass waste. There are also possibilities in the
realm of carbon-sequestration methods through breakthrough technologies. One such
breakthrough lies in recreating natural processes of photosynthesis through use of viruses
that are nonharmful to humans (a recent discovery at Massachusetts Institute of
Technology). Another lies in figuring out how to reproduce the organic by-products of
virally infected oceanic bacteria/microbes that have the capability of not turning back into
carbon dioxide (a recent discovery at Xiamen University in China). Industrial biotechnology's umbrella organization Bio, which is behind green chemistry that produces chemicals,
plastics, and fuels from biorefineries that use agricultural feedstock rather than petroleum,
foresees a new market of $300 billion by 2020. This industry has been unaffected by the
financial crisis of the end of the first decade of the 21st century.

However, despite the above-outlined benefits, consumers' fears are not allayed.
Scientific tests led by a team of biologists from the Norwegian University of Tromsø demonstrate how most of the antibiotic-resistant bacteria genes are found not in animals in
pristine environments but in domesticated species. This is not due to natural gene circula-
tion but to humans with antibiotic-resistant genes passing them on to animals. When
resistant bacteria hop between species, the rapid blending of evolutionary traits may lead
to dangerously resistant strains such as a possible combination of avian flu with swine flu.
Such superbugs are resistant to broad anti-spectrum antibiotics.

A similar story lies in the contamination of fields cultivated with farmer-selected seeds by spread of GM seeds. The latest news on the green revolution (the first step in biotechnology) shows that the high degree of use of pesticide led to severe environmental degradation. Producers and consumers who do not believe that herbicide- and pesticide-resistant GM food leads to better yields, tastes better, or is good for health because of its unpredictable nature are highly agitated by such developments. Humans have significant quantities of Paleolithic genome (that dominates 96 percent of human history) in contrast to their Neolithic gene (Neolithic time encompasses sedentary cultures and civilizations). We have a history of meddling with nature already, and our genome and brains can be plastic. However, there are enough studies to document how the brain can influence the body's reaction if it is not convinced about a cure. Moreover, the fact that we continue to have bodily genetic aberrations is certainly not ensuring evidence for consumers worried about genetic aberrations in food. Such aberrations are also feared from the possibility of development of resistant strains of superweed that can destroy agriculture. Skeptical consumers are also shocked by massive food safety scandals and the mismanagement of such problems. They are worried about loss of biodiversity due to changes in underlying genetic structure, loss of cultural taste and identity, and lack of sustainability.

The concerned consumers' viewpoint of man being what man eats/ingests tends to be waved off by technocrats as uneducated, emotional, lay knowledge that has no "scientific" corroboration. Leading green NGOs like Food First and Greenpeace have brought lawsuits against states and institutions in order to force them to reveal classified documents. Such organizations expect such documents to demonstrate negative results found in small-scale clinical trials done to assess risks. But the responses so far by the technocrats to such push-back efforts are far from positive and have failed to garner consumer trust. Mild sops, like labeling of GM food, have not been enough to soothe consumer concerns. Commodification of scientific knowledge has thus created a huge societal chasm. This chasm is the result of the extent and character of the decontextualization of the investigated object from real-world situations in terms of time, scale, epistemic object, and its context as well as the extent and character of the recontextualization of gained knowledge to real-world conditions. This situation has not been helped by institutional interests dominating media coverage and talking only about economic costs and benefits and routinely neglecting "unknown unknowns" in order to push through governmental laws by lobbying.

See Also: Agribusiness; *Bacillus Thuringiensis* (Bt); Genetically Modified Organisms (GMOs); Precautionary Principle (Ethics and Philosophy); Precautionary Principle (Uncertainty).

Further Readings

Bled, J. Amandine. "Technological Choices in Environmental Negotiations: An Actor-Network Analysis." *Business and Society*, 49:4 (2010).

Boschen, Stefan, Karen Kastenhofer, Ina Rust, Jens Soentgen, et al. "Scientific Nonknowledge and Its Political Dynamics: The Cases of Agribiotechnology and Mobile Phoning." *Science Technology Human Values*, 35:6 (2010).

Devos, Yann, Pieter Maeseele, Dirk Reheul, Linda Van Speybroeck, et al. "Ethics in the Societal Debate on Genetically Modified Organisms: A (Re) Quest for *SENSE AND SENSIBILITY*." *Journal of Agricultural and Environmental Ethics*, 21 (2008).

Pollack, A. Mark and C. Gregory Shaffer. *When Cooperation Fails: The International Law and Politics of Genetically Modified Foods*. Oxford, UK: Oxford University Press, 2009.

Powell, Rusell. "What's the Harm? An Evolutionary Theoretical Critique of the Precautionary Principle." *Kennedy Institute of Ethics Journal*, 20:2 (2010).

Priest, Susanna H. "Structuring Public Debate on Biotechnology: Media Frames and Public Response." *Science Communication*, 16:2 (1994).

Sheridan, Cormac. "Report Claims No Yield Advantage for Bt Crops." *Nature Biotechnology*, 27 (2009).

"Technology Quarterly." *The Economist* (June 12, 2010).

Zwart, Hub. "Biotechnology and Naturalness in the Genomics Era: Plotting a Timetable for the Biotechnology Debate." *Journal of Agricultural and Environmental Ethics*, 22 (2009).

Pallab Paul
Kausiki Mukhopadhyay
University of Denver

BROWNFIELD REDEVELOPMENT

Brownfields are former industrial or commercial sites located in city industrial zones and along railroad lines in suburban and rural communities. Abandoned and often environmentally contaminated, brownfields present both a problem and a promise in a time when environmentally conscious thinking is woven into many of our conversations about the United States. As we struggle to create a sustainable future, we are dogged by mistakes of the past, and the environmentally ravaged brownfields, often located in areas marked by poverty and crime, serve both as a reminder of our heady, reckless history and, perhaps, as a space to inscribe our hopes for the future. There are as many as 500,000 brownfields across the United States—sites that once accommodated various appendages of the United States' once-vast manufacturing economy, now worn out, polluted, and neglected—and many advocate redeveloping these lands for other uses. While it is difficult to find anyone entirely opposed to cleaning up abandoned, environmentally compromised lands, there are many who question the effectiveness of the cleanup measures and the veracity of the promises made regarding the benefits of such redevelopment.

Why Redevelop?

Champions of brownfield redevelopment cite the desire to solve the environmental and human health problems associated with brownfields as well as the potential for tax revenue that could be gained. Further, they point out that refusing to redevelop brownfields causes undeveloped "green fields" to be overused instead of preserved. Perhaps most compellingly, they argue that brownfields nurture inner-city blight, that they are potent markers of poverty and desperation, and that undeveloped, they symbolize urban decay, but redeveloped, they become sites of community growth. For these advocates, redevelopment is seen as a way to heal and revitalize the typically poor, minority neighborhoods where brownfields are concentrated.

Brownfield redevelopment is not a new idea, first gaining ground in the mid-1990s, when the U.S. Environmental Protection Agency (EPA) began a brownfield redevelopment program focused on 300 cities and other areas. Examples of recent brownfield redevelopment projects include Georgia's Atlantic Station. Nearly 140 acres of mixed-use land development, Atlantic Station includes office, retail, hotel, entertainment, and residential space, providing homes for 10,000 people and jobs for 30,000. This former site of a mammoth steel mill complex is the nation's largest brownfield remediation project. Pittsburgh, Pennsylvania, home to numerous abandoned steel mills and other defunct industrial ventures, also boasts several brownfield redevelopment projects. For example, in Pittsburgh's South Side neighborhood, where the Jones and Laughlin Steel Company's Southside Works used to sit, brownfield redevelopment has created "Southside Works," a mixed-use development that includes high-end entertainment, retail, offices, and housing.

A More Critical Look at Brownfield Redevelopment

Despite the apparent successes of many brownfield redevelopment projects, critics raise the following questions:

- Who do the jobs, services, and amenities created at brownfield sites actually benefit?
- Is redevelopment instigating gentrification?
- Are the environmental and institutional standards put in place to protect ecological and human health adequate? Do cleanup criteria lock sites into one type of use that may not serve the needs of the community in the future?

Chicago has been the site of numerous brownfield redevelopment projects, and a brief survey of the successes and limitations of those projects suggests that the criticisms listed above are warranted. For example, while many of the jobs created through redevelopment are accessible to the workforce residing near brownfields, many others are out of reach—such as jobs requiring technical and engineering skills not typically held by those living near brownfields. Further, of those jobs that do benefit area residents, there are questions about the long-term viability of such industrial-oriented positions in an economy that is rapidly de-industrializing.

As far as whether services and amenities benefit area residents, when local community groups are a part of the redevelopment process, there has been more success in tailoring services to area residents than when such community participation is absent. The high-end shopping and entertainment facilities that are found in many redeveloped sites are not beneficial to the original residents of such communities. This is closely tied to the gentrification that has too often accompanied brownfield redevelopment. For example, Chicago's Columbia Pointe redevelopment project included no rental units as part of its "affordable housing," and listed as "affordable" housing units that sold for $200,000 to $400,000. For area residents, 40 percent of whom live below the poverty line and only 28 percent of whom make more than $35,000 annually, this "affordable housing" was not viable. This kind of resident displacement is seen more often in larger redevelopment projects, while smaller projects have not been as detrimental to the original residents.

Beyond the significant debate about whether cleanup standards are sufficient to remediate brownfields and to protect area residents from future harm is the problem of limited vision. That is, most redevelopment projects need only remediate the land in order to comply with the next user's needs. This not only leads to relaxed environmental standards that many feel are inadequate, but it also restricts the land and its future uses.

Brownfield redevelopment, then, thus far has a mixed record. While there are examples of projects working to the benefit of both the land and the area residents, there are still many examples of redevelopment projects whose jobs and services are not open to original residents, just as there are many examples of projects causing resident displacement. Further, environmental standards used in brownfield redevelopments need to be closely monitored, and a broader vision should be employed when setting such standards. With community involvement and careful adherence to robust environmental standards, brownfield redevelopments have the potential to be beneficial to both the local residents and the environment, as well as to investors and other stakeholders.

See Also: Adaptation; Environmental Justice (Business); Green Jobs (Cities); Sustainable Development (Cities).

Further Readings

Carnegie Mellon University. "Brownfield Development: The Implications for Urban Infrastructure." http://www.ce.cmu.edu/Brownfields/NSF/index.htm (Accessed August 2010).
Sarni, William. *Greening Brownfields: Remediation Through Sustainable Development.* New York: McGraw-Hill, 2009.
U.S. Environmental Protection Agency. "Brownfields and Land Revitalization." http://epa.gov/brownfields (Accessed August 2010).

Tani E. Bellestri
Independent Scholar

CAPITALISM

Capitalism is an economic system where natural resources such as land and mineral deposits as well as capital goods such as factories, machinery, and equipment are primarily owned by private individuals rather than the government or society as a collective. These natural resources and capital goods are converted (inputs) to produce consumer goods and services (outputs), which are sold in the pursuit of profit.

Elements of Economic Theory

In capitalism, natural resources and capital goods as well as the consumer goods and services produced from them are traded in a free market. This is a system based on decentralized decisions by individuals and organizations where prices determine how and to whom resources as well as goods and services are allocated. A capitalist system emphasizes the benefits of self-interested pursuit of profit, competition among individuals to gain profit, and the incentives to pursue entrepreneurial endeavors to acquire profit. Some environmentalists see the capitalist system as antithetical to the protection of natural resources and promotion of sustainability, while others see the possibility of using a capitalist system to promote those same goals. Similarly, some capitalists, referring broadly to those organizational leaders favoring a free-market capitalist system, see the environmental movement as anticapitalist while others are more sympathetic to it, seeking to work with or even adopt environmental aims as their own. The positions among these groups can be viewed as part of a larger debate about the role of profit-oriented organizations in society and includes issues of corporate social responsibility and the role of ethics and morality in a system predicated on self-interested behavior.

Primary benefits of a capitalist system include its ability to generate economic growth, the promotion of wealth accumulation, and the protection of liberty through its emphasis on the protection of private property rights and individual choice. Perhaps the strongest argument for the benefits of capitalism is one of its oldest, Adam Smith's concept of the "invisible hand." Self-interested decisions by buyers and sellers generate mutually beneficial trade, creating value for individuals, organizations, and society. Despite the benefits, there are concerns about the inherent responsibility of capitalism, especially with regard to the environment. For example, Karl Marx believed that capitalism was as abusive to the soil as much as it exploited workers. It should be no surprise then that ecological conflicts may be inherent in a capitalist society. During the Great Depression in the United States, President Franklin D. Roosevelt introduced policies designed to reignite, reform, and regulate capitalism, including an environmental-based initiative called the Civilian Conservation Corps. His New Deal policies relied on the ideas of John Maynard Keynes that government economic stimulation can be used to increase private-sector growth and that government needs to regulate and curb the excesses of capitalism.

As 20th-century economist Joseph Schumpeter describes capitalism, it is a process of "creative destruction" where industries and their products become outdated and are replaced by new ones through innovation. Much like the biological concept, known in marketing as the product life cycle, creative destruction is a natural process of introduction, growth, maturity, and decline. Applying this to the capitalist framework, if natural resources used in industry cannot be fully recycled and are not fully renewable, then the total amount of those resources is reduced. Even the total amount of renewable resources can be reduced. As an example, forests are slowly renewable (i.e., trees take time to plant and grow). If trees are cut down at a rate faster than new trees grow, the total number of mature trees will decline, and there will be a shortage of wood. So even with renewable resources, if our pace of "creative destruction" (out with the old, in with the new) is faster than those renewable resources grow, we still face problems. Capitalists would argue, though, that the technological advances that come from the process of creative destruction can stretch the usefulness of existing natural resources, allowing us to do more with less.

Many situations exist where private interests do not align with social interests. When a by-product of production is pollution, the societal cost is not the same as the private cost of producing them. If left unchecked, pollution harms environmental quality and society over time. To some, government regulation is an appropriate response to this problem. To others, the solution lies in defining the property rights of pollution and allowing market forces to create incentives for innovative solutions.

Empirical evidence of the links between capitalism, economic growth, and the environment are uncertain and controversial. Given the long sweep of time and the numerous possible contributing factors, identifying cause and effect is difficult. Over the past 200 years, capitalist systems were widely adopted throughout the world. In that same time, average incomes increased roughly eightfold. To the extent higher income leads to better living standards, the economic conditions of a large part of the world have improved. At the same time, the level of carbon dioxide, a primary contributor to global warming, has increased by roughly 30 percent.

Examining much shorter time frames, empirical studies have found an interesting dynamic. For countries with relatively low income levels, economic growth leads to higher levels of many types of pollution (such as sulfur dioxide). At relatively high income levels, economic growth leads to lower levels of pollution. Robustness of these results as well as the possible causal factors is questionable. One explanation is simply that as a country becomes richer, it eventually crosses a threshold where resources can be diverted from basic economic needs to focus on economic wants such as environmental quality. Another explanation is that richer countries tend to be more heavily regulated and rein in the market forces that tend to produce more pollution.

Capitalism's role as a catalyst of economic growth influences this debate. Over the past 30 years, "globalization" (the integration of the world economy) is sometimes viewed as a triumph of capitalism and capitalistic thought. Some consider globalization as a way to reduce environmental harm, as capitalism's rise will increase average incomes and eventually afford more the luxury of reducing pollution amounts. Others worry that glo-balization will simply lead richer countries to relocate their most polluting industries to relatively poorer countries, increasing the total amount of pollution. As globalization and the spread of capitalism continue to unfold, the debate about the extent that self-interest coincides with social interest on a global scale will continue.

Shifting the Lens of Self-Interest

Today, many scholars suggest that for green capitalism to be successful, the foundation of capitalism needs to remain intact. Yet how capitalists approach the creation of new prod-ucts and services needs to be based on a broader definition of "self-interest." To create this shift in lens, the identity of individuals and organizations must shift from an independent view of self (what's in it for me and my shareholders) to a more collective view of self (what's in it for me and my stakeholders). Because the elements of capitalism are dependent upon the natural world, requiring human and environmental resources, there is an inher-ent responsibility to protect and conserve them en route to profit.

Some individuals and organizations have not assumed this responsibility of imposing some level of self-regulation as it pertains to respect for the environment. As such, govern-ments continue to play an increasingly critical role toward the creation and implementa-tion of regulatory compliance policies. Since 1995, the corporate stance on compliance has shifted from opposition to awareness, with some addressing environmental concerns head-on in an effort to "go green." But most still rely on the course set by their national govern-ments, typically waiting for compliance mandates before taking action. We saw how this can be controversial when the Kyoto Protocol was presented as the means to reduce global warming. This global treaty fueled the debate about quantitative evidence of global warm-ing and was ultimately not ratified by the United States and Australia.

Green capitalism calls for firms to operate via green criteria as represented by use of renew-able energy sources, avoiding toxic chemicals, repairing or recycling products, and minimizing reliance on long-distance shipments for either supplies or sales. Corporations implement such practices through a combination of self- and government-imposed regulations. Participation at the international level works on a broader scale to determine how economic growth pro-vides the conditions in which protection of the environment can best be achieved (see the World Business Council for Sustainable Development; WBCSD). The WBCSD has declared that business-as-usual cannot get us to environmental sustainability, suggesting that business will still need to do what business does best: innovate, adapt, collaborate, and execute. Activities for change to ensure a greener form of capitalism require partnerships between businesses, governments, academia, and nongovernmental organizations.

The world has an interest in the measures that may or may not be taken—whether by government or by the private sector—within any given country. International meetings advance global agreements for environmental strategies like emissions trading. Known as "cap and trade," it is an incentive-based approach to reduce emissions while offenders to pollution controls can "offset" such environmental harms rather than directly curtailing them (e.g., by planting trees). Further, organizations that do not use their full emissions quota may sell the unused portion to other, higher-polluting organizations in a free-market system. In this way, the mechanisms of capitalism are used to provide incentives to reduce pollution levels. The government's ability or lack thereof to set an optimal amount of pol-lution quotas is critical for such a system to function. Also, for high-level polluters, the system is effectively a tax on their activity that may distort the free market and hinder economic growth. In a free-market capitalist system, what is the appropriate balance between external regulation and self-regulation that is in the best interests of individuals and society? Most people think that some regulation is necessary to protect the public, but others believe that regulation does more harm than good.

Creating a Market for Green Products and Services

Ardent capitalists believe that only markets can make the necessary value-based decisions in ways that maintain the freedom of choice of the individual. Many governments treat sustainability as a "problem to be solved" by management experts and bureaucratic intervention. But free-market purists reject this technocratic approach, as environmental values cannot be reduced to objectively measurable monetary terms. This is because all value-based decisions have an ethical dimension. Therefore, any interventions in the market intended to address sustainable development are unavoidably political. Attempts to interfere will ultimately create unintended consequences, which may be counter to original goals and lead to additional problems. Despite these concerns, many see some intervention as helpful, not only in requiring environmental regulation, but in adding incentives for green or sustainable enterprise (i.e., a carrot-and-stick approach).

Corporate interests typically reflect the goals implicit in market competition, that is, profit maximization, growth, and accumulation. While green technologies, organic food and clothing, and sustainable transportation markets have gained much attention in the past decade, the basic driver of economic decision making does not require environmental responsibility beyond a basic compliance level. There are efforts to create an infrastructure for sustainable enterprise, which employs pressure from within the capitalist system itself to become more socially responsible. Focus on integration of corporate responsibility within companies' business models and the recognition of the impact of companies' business strategies and practices on stakeholders, societies, and sustainability drive these efforts. Moreover, new ideas on how to become more transparent in industry have led to innovative processes such as elemental accounting. This process has been introduced as a means to track which elements actually go into every product, not only revealing to consumers the materials and energy in what they buy, but also enabling more accurate and systematic resource tracking. Such efforts may help companies look for new business opportunities for the retrieved material and help regulators enforce sustainability standards.

Sustainable Enterprise

As we see capitalism react to the concept of "ecologically" sustainable development, it is being transformed into "economically" sustainable development. Some organizational leaders and investors are coming to share in the belief that with better management of natural resources, a larger supply and wider range of goods and services can be produced. This focus on sustainable enterprise will continue to move to establish cooperation toward change and innovation. Such beliefs may be held within the core identity of the business, and actions are implemented to be consistent with that directive. Sustainable enterprise can also be driven by strategically designed measures to save costs and to secure profits on emerging markets. But if the firm is authentically considering the long-term existence and development of humans and organizations, it will make decisions about the optimal composition of the economy's capital stock, including human capital, man-made capital, and natural resources. Green capitalism is based on the notion of genuine value creation rather than value transfer; that is, rather than taking resources to create something, there is a generative process to actually create value, to add rather than to exchange or deplete the net resources used in the process. This means organizations find ways to profitably meet unmet societal needs as a result of their business presence in the market. Through innovation processes such as design thinking, firms can create new business with benefits for society without economic, consumer, and environmental trade-offs. Whether green capitalism and sustainable enterprise can deliver benefits similar to those attributed to traditional capitalism remains to be seen.

See Also: Antiglobalization; Corporate Social Responsibility; Ecocapitalism; Sustainable Development (Politics).

Further Readings

Ali, S. H. "The Defining Idea of the Next Decade." *The Chronicle Review: Tenth Anniversary Issue* (September 3, 2010).

Arnaud A. and L. E. Sekerka. "Positively Ethical: The Establishment of Innovation in Support of Sustainability." *International Journal of Sustainable Strategic Management*, 2:2 (2010).

Berger, W. H. "Global Change and Global Warming: An Elementary Introduction to the Problems Surrounding Man Made Climate Change." http://earthguide.ucsd.edu/globalchange/index.html (Accessed August 2010).

Brown, T. *Change by Design*. New York: HarperCollins, 2009.

Copeland, B. R. and M. S. Taylor. "Trade, Growth, and the Environment." *Journal of Economic Literature*, 42:1 (2004).

Green, J. "The Illusion of Green Capitalism." *GreenLeft*, 364 (1999).

Greenwood, D. "The Halfway House: Democracy, Complexity, and the Limits to Markets in Green Political Economy." *Environmental Politics*, 16:1 (2007).

Kolk, A. and Jonatan Pinkse. "Integrating Perspectives on Sustainability Challenges Such as Corporate Responses to Climate Change." *Corporate Governance*, 7:4 (2007).

Lazslo, C. *Sustainable Value: How the World's Leading Companies Are Doing Well by Doing Good*. Stanford, CA: Stanford University Press, 2008.

Organisation for Economic Co-operation and Development (OECD) Development Centre. "The World Economy." http://www.theworldeconomy.org (Accessed August 2010).

Pfeffer, J. "Building Sustainable Organizations: The Human Factor." *Academy of Management Perspectives* (February 2010).

Potocan, P. and Matjaz Mulej. "Ethics of a Sustainable Enterprise—and the Need for It." *Systemic Practice and Action Research*, 20 (2007).

Waddock, S. "Building a New Institutional Infrastructure for Corporate Responsibility." *Academy of Management Perspectives* (August 2008).

Wallis, V. "Beyond Green Capitalism." *Monthly Review*, 61:9 (2009).

Weiss, J. W. *Business Ethics: A Stakeholder and Issues Management Approach With Cases*. Mason, OH: South-Western Cengage Learning, 2009.

World Business Council for Sustainable Development (WBCSD): Vision 2050. "The New Agenda for Business: Full Report." Geneva: WBCSD, 2009.

Derek Stimel
Leslie E. Sekerka
Menlo College

CARBON CAPTURE TECHNOLOGY

This gasification reactor will study the behavior of coal gases under pressure to aid in the development of new carbon capture technology.

Source: Bud Pelletier/Sandia National Laboratory

The technique of separating carbon dioxide from large point sources—especially fossil fuel–based power plants and large energy-consuming industries like cement, iron and steel, chemicals production, and oil refining—is known as carbon capture. Carbon dioxide (CO_2) is a greenhouse gas (GHG) that is generated out of human activity due largely to burning of fossil fuels. The increase in concentration of GHGs in the atmosphere is a primary reason for global warming. Thus, in order to mitigate global warming, the concentration of GHGs in the atmosphere needs to be stabilized. However, owing to the long lifetime of CO_2 among the GHGs in the atmosphere, stabilizing concentrations of GHGs at any level requires large reductions from current levels of global CO_2 emissions. Given that the "lion's share" of the world's commercial energy needs (nearly 85 percent) is currently met from fossil fuels and that this dependence is unlikely to come down drastically, this is indeed a challenging task. In this light, the great advantage of carbon capture technology over other possible options (like encouraging conservation, augmenting fuel efficiency, or shifting to renewables) lies in the fact that it would enable the world to continue to use fossil fuels but with greatly reduced emissions of CO_2. The unintended consequence, however, of pushing the technology forward lies in reducing the incentive for encouraging innovation and delaying investment in shifting the energy base from fossil fuels to renewable energy. Moreover, because the technology is still in its nascent stage and yet to be demonstrated on a large scale, there are a host of barriers, uncertainties, and knowledge gaps involved in its adoption and deployment. These relate primarily to cost and performance, efficiency loss, life-cycle emissions, and other environmental concerns.

Understanding the Technology

There are three main options for carbon capture from large point sources. These are (1) post-combustion capture, (2) pre-combustion capture, and (3) oxy-fuel combustion. Post-combustion capture involves separating and capturing CO_2 from the flue gas (i.e., exhaust gas) produced by fuel combustion at large point sources. Carbon dioxide is only a small part of the flue gas stream emitted to atmosphere by a power station. Other gases include nitrogen, oxygen (O_2), and water vapor. Due to this low concentration, a large volume of flue gas has to be handled to separate out CO_2, resulting in large and expensive equipment. Pre-combustion capture involves partial oxidization by allowing the fuel to react with oxygen and/or steam to give mainly carbon monoxide and hydrogen. The carbon monoxide is further reacted with steam in a catalytic reactor, called a shift converter, to give CO_2 and more hydrogen. The CO_2 is then separated, and the hydrogen is used as fuel in a gas turbine combined cycle plant. The process is, in principle, the same for coal, oil, or natural gas. In oxy-fuel combustion, the concentration of CO_2 in flue gas can be increased greatly by using concentrated oxygen instead of air for combustion, either in a boiler or gas turbine. If fuel is burned in pure oxygen, the flame temperature is excessively high, so some CO_2-rich flue gas would be recycled to the combustor to make the flame temperature similar to that in a normal air-blown combustor.

The advantage of oxygen-blown combustion is that the flue gas has a CO_2 concentration of more than 80 percent, so only simple CO_2 purification is required. Oxy-fuel combustion relies mainly on physical separation processes for O_2 production and CO_2 capture, thereby avoiding the use of any reagents and/or solvents that contribute to operating costs and the environmental disposal of any related solid or liquid wastes. The main disadvantage of oxy-fuel combustion is that a large quantity of oxygen is required, which is expensive, both in terms of capital cost and energy consumption. Another technology of carbon capture that is emerging is chemical looping combustion, in which direct contact between the fuel and the combustion air is avoided by using a metal oxide (like iron, nickel, copper, and manganese) to transfer oxygen to the fuel in a two-stage process. In the first stage, the fuel is oxidized in a reduction reactor by reacting with a metal oxide that is converted to a lower oxidation state. It is then transported to a second reactor, the oxidation reactor, where it is re-oxidized by reacting with O_2 in air.

The techniques that are used for separation of CO_2 from flue gases include chemical solvent scrubbing, physical solvent scrubbing, adsorption, membrane, cryogenics, and solid sorbents. Chemical solvent scrubbing is the most favored method for post-combustion removal of CO_2 from flue gases and involves use of chemical solvents (e.g., monoethanolamine), which absorb much of the CO_2 by chemical reactions. Physical scrubbing is preferred in pre-combustion capture and makes use of physical solvents (cold methanol, dimethyl ether of polyethylene glycol, propylene carbonate, and sulpholane), as they combine less strongly with CO_2. The advantage of physical solvents is that CO_2 can be separated from them in the stripper primarily by reducing the pressure, resulting in much lower energy consumption. In adsorption, flue gas is fed to a bed of solids with high surface areas like zeolites and activated carbon that adsorbs CO_2 and allows the other gases to pass through. When a bed becomes fully loaded with CO_2, the feed gas is switched to another clean adsorption bed and the fully loaded bed is regenerated to remove the CO_2. The membrane separation of CO_2 from flue gas relies on differences in physical or chemical interactions between gases and a membrane material (like porous inorganic membranes, palladium membranes, polymeric membranes, and zeolites), causing one component to pass through the membrane faster than another. In cryogenics, CO_2 can be separated from other gases by cooling and condensation. It is widely used commercially for purification of CO_2 from streams that already have high CO_2 concentrations (typically more than 90 percent). Sometimes in post-combustion capture systems, regenerable solid sorbents (like calcium and lithium) are used to remove CO_2 at relatively high temperatures.

The sorbents-based oxides can react with CO_2 to form carbonates, and the carbonates can be regenerated to oxides by heating to a higher temperature.

"Flip Side" and Barriers to Adoption

Although the technology of carbon capture reduces CO_2 emissions considerably, it results in a significant (thermal) efficiency loss in power plants and increase in costs due to increases in fuel consumption, capital cost, and overall costs of electricity generation. The extent of loss and increase in cost depends largely on the efficiency of the basic power-plant technology. The main technologies that are used currently to generate power from fossil fuels are pulverized coal-fired (PC) steam cycles and natural gas combined cycles (NGCC). Integrated gasification combined cycles (IGCC) technology is also being developed, although it is not yet considered to be economically competitive. In PC plants, pulverized coal is burned in a boiler, which raises high-pressure steam, which is then passed through a steam turbine, generating electricity. In NGCC plants, natural gas is burned in a gas turbine, which generates electricity. In IGCC plants, fuel is reacted with oxygen and steam in a gasifier to produce a fuel gas consisting mainly of carbon monoxide and hydrogen. This is then cleaned and burned to generate power in a gas turbine combined cycle. Carbon dioxide capture could be incorporated in all these types of plants. The overall efficiency losses due to CO_2 capture are, however, lower for an IGCC plant, which is most efficient as compared to a PC power plant. Furthermore, the energy losses in the CO_2 separation stage are higher for post-combustion amine scrubbing than for pre-combustion capture because of the high heat requirement for amine regeneration, and the power consumption for CO_2 compression is also higher because the CO_2 is recovered at a lower average pressure. According to estimates by the Intergovernmental Panel on Climate Change (IPCC), for new supercritical PC plants using current technology, the additional energy requirements range from 24 to 40 percent, while for NGCC plants the range is 11–22 percent, and for coal-based IGCC power plants it is around 14–25 percent. In other words, unless the carbon capture technology is deployed in modern, efficient power plants, the efficiency loss could be substantial. Similarly, retrofitting CO_2 capture technology to old, inefficient plants would not be appealing unless the power plant itself was substantially upgraded in terms of efficiency.

Recent estimates by the International Energy Agency's IEA Greenhouse Gas R&D Programme (IEAGHG) show that CO_2 capture increases the fuel consumption, and hence the cost of fuel, by about 25 percent. The increase is slightly lower (21 percent) for the IGCC process and is lowest (17 percent) for the natural gas–fired combined cycle plant with post-combustion capture. The increase in specific capital cost ($/kW) due to capture is 25–50 percent for coal-fired plants. The lowest increase is for IGCC. The percentage increase is higher, 70–85 percent, for natural gas–fired plants. The increase in specific capital cost is due partly to the reduction in efficiency and partly to the cost of the extra process units required for CO_2 capture and compression and increase in cost of solvent, especially in amine scrubbing cases. The percentage increase in the overall cost of electricity due to capture therefore lies between the percentage increases in the capital and fuel costs for each plant and depends largely on coal and natural gas prices.

A study by Naser Odeh and Timothy Cockerill (2008) shows that while carbon capture has the potential of decreasing life-cycle GHG emissions and consequently the global warming potential, there can be a significant increase in the concentrations of other pollutants like NOx and NH3. Furthermore, the solvents used to capture CO_2 gradually degrade in use and so there need to be suitable procedures for destruction/disposal of sludge produced from the decomposed solvent. There may also be some solvent carryover in the flue gas stream. These factors not only have potential environmental impacts but also increase costs considerably.

Ways Ahead

Most of the developing countries, especially in Asia, have ambitious programs to meet the daunting challenge of universal electrification primarily because the greater share of their rural population has little or no access to electricity for lighting. This massive unmet demand for electricity cannot be fulfilled by the power generated from existing plants, and new power plants need to be installed. It is also quite obvious that the new plants are largely going to be fossil fuel based, especially coal-based thermal power plants. Thus, it would be worth exploring if the new power plants could be made carbon capture ready by retrofitting with the capture technology. Otherwise, a large amount of CO_2 emission to the atmosphere will be "locked in," eventually exacerbating the process of global warming.

Currently, however, there are no commercial carbon capture systems available for power plants and most industrial applications. Retrofit of carbon capture technologies is also not proved. Furthermore, capital cost involved in making a new plant capture ready is prohibitive. Efficiency loss and increased energy consumption (referred to as energy penalty) further add to the cost. Thus, the near-term priority should be to reduce the energy penalty of using CO_2 capture in power plants. There is also enough scope for the development of improved solvents in absorption technology, starting at the laboratory scale and leading to use in commercial-scale plants. Investigation of improved separation processes (e.g., membranes, cryogenic separation, improved heat recovery to compensate for losses introduced by CO_2 capture, and novel concepts such as different methods of separating oxygen, enriched oxygen combustion, or a combined reactor/membrane separator for the decarbonization of fuel gases) should also be encouraged.

See Also: Carbon Emissions (Personal Carbon Footprint); Carbon Sequestration; Coal, Clean Technology.

Further Readings

Davison, John and Kelly Thambimuthu. "Technologies for Capture of Carbon Dioxide." In *Greenhouse Gas Control Technologies,* Vol. I, E. S. Rubin, D. W. Keith, and C. F. Gilboy, eds. New York: Elsevier Science, 2005.

Intergovernmental Panel on Climate Change. *Special Report on Carbon Dioxide Capture and Storage.* Prepared by Working Group III of the IPCC, B. Metz, O. Davidson, H. C. de Coninck, M. Loos, et al., eds. Cambridge, UK: Cambridge University Press, 2005.

International Energy Agency (IEA). *Technology Roadmap: Carbon Capture and Storage,* Paris: IEA, 2010.

Odeh, Naser A. and Timothy T. Cockerill. "Life Cycle GHG Assessment of Fossil Fuel Power Plants With Carbon Capture and Storage." *Energy Policy,* 36 (2008).

Kaushik Ranjan Bandyopadhyay
Asian Institute of Transport Development

CARBON EMISSIONS (PERSONAL CARBON FOOTPRINT)

Carbon footprints measure the amount of carbon dioxide (CO_2) or total greenhouse gas (GHG) emissions associated with a product, business, or individual. Carbon footprints of products come from a life-cycle analysis of how much CO_2 is produced during a product's life. This includes the emissions from a product's inputs and manufacture, transportation to market, use, and disposal. *Personal carbon footprint* is a term that refers to how much GHGs an individual person adds to the atmosphere over the course of one year as a result of his or her actions. A carbon footprint may be measured in pounds, tons, or metric tons either of CO_2 or of all GHGs as units of CO_2 equivalent.

Personal carbon footprints are composed of primary footprints and secondary footprints. Primary footprints are GHG emissions that are directly produced by the burning of fossil fuels as a result of people's actions. Ground and air transportation, electricity consumption, and home heating, hot water, and cooking all contribute to primary carbon footprints. Secondary carbon footprints include indirect emissions that result from the manufacture, transport, use, and breakdown of products that do not directly require fossil fuel use. This includes emissions associated with purchasing items such as food, clothing, nonelectric household items, books, and so forth. There are a number of online carbon footprint calculators that take different amounts of data supplied by a person and calculate his or her individual carbon footprint. Most calculators focus on the primary carbon footprint associated with driving, flying, and electricity use. Some calculators also take into account home energy use by asking about the size and type of house and about heating bills. Occasionally, carbon calculators also take into consideration a person's secondary carbon footprint by asking questions about diet, recycling habits, and composting efforts. Once someone has calculated his or her personal carbon footprint, there are several ways to use that information.

What Are Personal Carbon Footprints Used For?

Carbon footprints can be used to compare one's own emissions to a national average or a per person reduction target, to help people figure out how to reduce their own emissions, or to help people determine how much to "offset" their emissions. Per capita emissions are the average amount of GHG emissions produced by each citizen of a country over the course of a year. Per capita emissions may be a useful tool for comparing the emissions between different countries and also when trying to distribute emissions reductions between different countries. Personal carbon footprints help to differentiate those emissions. While on average China's per capita emissions may be only about one-third of those in the United States, there are undoubtedly some people in China with unsustainably large personal carbon footprints (as of 2009). The United Kingdom has considered implementing a policy by which every citizen would have an equal limit set for their personal carbon footprint, and people would need to keep their emissions within that personal carbon footprint by purchasing gas, electricity, and airline tickets using a sort of carbon emissions allowance.

A second way carbon footprints may be used is as a tool to help people figure out how they could reduce their own personal emissions. The very act of analyzing the main sources of carbon emissions may help people to become more aware of how they contribute.

People can easily add different numbers to the carbon calculators to see how much they could reduce their emissions by taking certain steps (such as driving fewer miles, switching to a more fuel-efficient vehicle, or taking fewer flights). Many carbon calculators link to websites with tips for how to reduce GHG emissions, especially by reducing one's electricity use.

A third way in which personal carbon footprints may be used is to figure out how many "carbon offsets" a person would need to purchase to reduce their GHG emissions or to become "carbon neutral." Many websites with online carbon calculators also sell carbon offsets. There are two types of carbon offsets: mandatory carbon reduction and voluntary emission reduction. Mandatory reductions are those that happen as a result of legal regulations or policies that require such carbon reductions. These reductions may be bought and sold by governments or businesses. Voluntary emissions reductions (VERs) are those undertaken on a voluntary basis, certified (usually in units of one ton of CO_2), and then sold. A number of independent agencies will certify that emissions reduction happened voluntarily. People can then buy these certified offsets. Ratings agencies differ in their standards for what constitutes "emission reduction." The best standards ensure that the projects are voluntary (that there is no legislation requiring people to undertake those emissions reductions), and also that the projects are dependent upon the funding that they received for an offset. Carbon offsets could come from anything from a group that is planting trees to a group that is changing people's incandescent to compact fluorescent light bulbs (CFLs) or perhaps paying to help install renewable energy when the cost would otherwise be prohibitive. The most stringent rating agencies also look at secondary and tertiary emissions associated with the projects and subtract those from the total emissions offsets that they will certify for a project. For example, these organizations will subtract the emissions associated with personnel who implement the project driving or flying to and from the project.

How Accurate Are Personal Carbon Calculators?

According to the United Nations' Intergovernmental Panel on Climate Change's (IPCC) Fourth Assessment Report, CO_2 accounts for 76 percent of all anthropogenic GHG emissions globally—as measured by how much those gases trap heat (global warming potential), not by volume. Most carbon calculators focus on transportation, especially driving and flying, as well as carbon emissions produced by electricity generation. Transportation accounts for just over 13 percent of global GHG emissions, and residential and commercial buildings and electricity supply accounts for around 8 percent and 26 percent of global emissions, respectively. This means that personal carbon footprints that account for electricity, home heating, and transportation account for fewer than 50 percent of global anthropogenic GHG emissions. Carbon calculators that include people's diets account an additional 13.5 percent of emissions that come from agriculture. Few personal carbon footprints accurately reflect the roughly 20 percent of emissions that come from industry, the 17 percent that come from forestry, or the 2.8 percent that come from waste and wastewater. Calculations have been made that would make it possible to account for the emissions associated with food choices. However, personal carbon footprints cannot accurately reflect the emissions associated with a person's general purchases because there is not enough data about the carbon footprints of different products.

Pros of Personal Carbon Footprints

The concept of personal carbon footprints is useful on two different levels of mitigating climate change. First, the concept has advantages for individuals, both in raising people's awareness of their personal impact on climate change and as a way of teaching people how they can reduce GHG emissions. The second way in which the idea of personal carbon footprints can be useful is to provide a new way of looking at issues of mitigating climate change on a national and international level. The idea of personal carbon footprints is one possible solution for ethical questions that arise about who should be responsible for reducing their carbon emissions. The idea of personal carbon footprints may provide a solution to Thomas Malthus's prediction that a growing human population increasingly puts pressure on common ecological resources to the point where the use of those resources becomes unsustainable.

Carbon footprints provide benefits for individuals in several ways. First, the concept helps raise people's awareness of life-cycle analysis of products, and this could help people make better choices in reducing their emissions. For example, understanding that the majority of a car's carbon footprint comes not from its manufacture but rather from its burning of fuel could help people decide whether it is better from an emissions point of view to purchase a new, more fuel-efficient car or to get more use out of an older car. Second, calculating personal carbon footprints can help people identify where a majority of their emissions comes from and can help people reduce their emissions across those main categories. People can calculate the impact of changing to a more fuel-efficient car, driving fewer miles, taking fewer flights, or reducing their electricity and home heating use. Raising awareness of the impact these activities have on producing CO_2 may help people become more conscientious and make efforts to reduce these activities. It may also encourage people to purchase more energy-efficient appliances, raise their thermostats in the summer and lower them in the winter, switch from incandescent lights to compact fluorescent light bulbs (CFLs), insulate their houses, replace their windows with better-insulating glass, and so on. Finally, calculating personal carbon footprints may encourage people to purchase voluntary emissions reductions to help offset their CO_2 emissions. If people purchase offsets that have been rigorously certified, and if purchasing those offsets does not cause people to actually act in ways that increase their CO_2 emissions (because as long as they continue to purchase offsets, they can continue to increase those emissions and still have a carbon footprint of zero), then purchasing offsets may help reduce the amount of CO_2 emitted into the atmosphere.

The concept of personal carbon footprints also provides a beneficial lens for examining questions of reducing GHGs on national and international levels. One of the most contentious issues surrounding climate change is who should be responsible for reducing emissions. When a country takes on national emissions reduction targets, there is no incentive for individuals to reduce their personal carbon footprints without also adding taxes to CO_2 or without government policies that require certain fuel-efficiency standards for vehicles, energy-efficiency standards for appliances, or requirements for utilities to produce a certain percentage of electricity from lower carbon sources such as solar, wind, geothermal, and hydroelectric. The problem with these enforcement methods is that they only restrict the emissions associated with any one entity but do nothing to ensure that individuals reduce their carbon footprints. For example, televisions may be getting more efficient, but people may also be buying ever-larger TVs, which offsets the increased efficiency.

Policies that attempt to restrict CO_2 emissions on an international level often encounter issues related to the allocation of emissions. One concept that is used to discuss which countries should reduce emissions is the idea of per capita emissions: the total emissions of a country over the course of a year divided by the number of people who live in that country. Looking at per capita emissions shows that, although China may currently be the world's largest polluter, just slightly above the United States, the United States produces nearly three times as much GHG emissions per person as China. Since per capita emissions are a ratio of emissions to population, a country's absolute emissions may actually increase while its per capita emissions stay the same or even decrease, if its population increases at a faster rate. Personal carbon footprints offer perhaps the most equitable measure of distributing responsibility for reducing GHG emissions. As the population increases, the personal carbon footprint allotted to everyone could be reduced, thus linking the effect of population growth and increased emissions. This concept would also prevent individual people with large personal carbon footprints from hiding behind national emissions numbers or per capita emissions data.

Cons of Personal Carbon Footprints

There are a number of arguments made against using personal carbon footprints. These arguments fall into three categories. First, there are arguments against the ways in which carbon footprints are calculated. Second, there are arguments against how personal carbon footprints are often tied to carbon offsets. Finally, there are arguments about the limits of what does not fall under the umbrella of personal carbon footprints but yet from which individuals benefit.

Carbon footprint calculators are limited in how completely and accurately they can calculate carbon footprints. First, there are many uncertainties and assumptions that go into calculations. For example, calculators may use your country or state to determine the average electricity production available in that state to calculate your emissions from using electricity, but even variations within states could change the composition of the electricity generated that an individual uses. While some calculations may take into account secondary and tertiary effects (such as the amount of CO_2 released from running a hotel room associated with emissions from flying, or the amount of CO_2 released during the manufacture and disposal of a vehicle), others may only calculate the primary emissions associated with activities. It is true, however, that the secondary and tertiary emissions are minor compared to the primary emissions from flying and driving. Another argument against the completeness of carbon calculators is that many leave out a person's emissions associated with agriculture and deforestation, which make up 13 and 17 percent of total global GHG emissions, respectively. Calculators may add an average share of industrial emissions to a person's footprint (based on an average consumption pattern for that individual's city, state, or household income), which does little to accurately measure an individual's consumption choices if he or she is making an effort to consume less or to produce less waste. Otherwise, calculators may leave out an individual's share from industry altogether, leaving the personal carbon footprint with an incomplete picture as to the total amount of emissions for which a person is responsible.

It is unlikely that the problem of calculating accurate personal footprints associated with industry will be resolved anytime soon. Currently, only a handful of companies or independent organizations have attempted to calculate the carbon footprints of individual products. Until a standard methodology for calculating the carbon footprints of products is developed, comparing alleged footprints of different products will be like comparing apples to oranges. There is a not-for-profit organization in the United Kingdom called the Carbon Trust that is working on increasing the use of a standardized carbon footprint label for consumer products. As of September 2010, it listed 20 companies that had certified a wide range of products with its carbon footprint label. Companies not using an independent calculating agency do not have incentives to accurately calculate or to publicize products that have large carbon footprints, making it more difficult to calculate the carbon footprint of a person's purchases.

The second argument against personal carbon footprints relates to the idea of carbon offsets. While some carbon offset projects may help to reduce the amount of CO_2 entering the atmosphere, other, less rigorously certified VERs do not reduce CO_2 emissions as much as they claim. Also, some projects that may qualify as VERs cannot guarantee that carbon emissions that were initially avoided will remain out of the atmosphere. For example, projects that plant trees and count the CO_2 that the trees take out the atmosphere do not necessarily account for the fact that if the trees burn or die and decompose, they will release that CO_2 back into the atmosphere. Even if carbon offsets contributed to reducing CO_2 emissions, there are arguments against coupling VERs with personal footprints. The ability to become "carbon neutral" online with a few minutes and a few hundred dollars may assuage people's desire to reduce their own footprints. The availability of easy offsetting may actually encourage some people to emit more GHGs because they may feel less responsible for environmental harm as long as they continue to increase their carbon offsets.

The third main con of personal carbon footprints is that even those that do include individuals' emissions associated with deforestation, agriculture, and industry still leave numerous emissions completely unaccounted for, even if individuals benefit from those emissions. For example, personal carbon footprints do not account for an individual's contribution to emissions associated with building roads or infrastructure. These footprints leave out many of the emissions associated with government (buildings, transport, military) or industry. Personal footprints do not account for the emissions associated with search engine server farms, even if people benefit, for free, from those emissions. If everyone used an extremely accurate personal carbon footprint calculator, there would still be unaccounted emissions.

Reducing Personal Carbon Footprints

Reducing primary carbon footprints can be achieved through relatively straightforward actions: purchasing a more fuel-efficient car, driving fewer miles, and reducing the number of flights taken each year. When comparing the emissions associated with different cars, the U.S. Environmental Protection Agency (EPA) estimates that cars release 19.4 pounds of CO_2 for each gallon of gasoline burned, and 22.2 pounds for each gallon of diesel fuel. When choosing which flights to reduce, short flights are less efficient than longer flights because a large part of an airplane's GHG emissions is created by taxiing and idling on the ground. Reducing personal carbon footprints from electricity and home energy use can be achieved through conservation (using less energy by turning off unused lights, changing the temperature setting for thermostats, turning off computers at night, and unplugging unused appliances) or through measures to increase a home's energy efficiency. Increasing energy efficiency can be achieved by the following measures. National Geographic correspondent Patty Kim suggests the following actions, which can help save more than a ton of GHG emissions every year: switching one 75-watt incandescent light bulb for an equivalent 19-watt CFL cuts 50 pounds of CO_2 emissions per year.

Cutting back from an eight-minute shower to a six-minute shower reduces 340 pounds of CO_2 per year. Replacing a refrigerator made before 2001 with an Energy Star–certified refrigerator saves more than 500 pounds per year of carbon emissions. Recycling all of a household's plastic, paper, and glass can reduce more than 1,000 pounds of emissions a year. Generally, for products that use a lot of energy, such as cars, electronics, and home appliances, the energy use over the product's life cycle is much more substantial than the emissions associated with manufacture. Therefore, it is better overall for emissions to buy new, more energy-efficient products than to keep older, less efficient appliances. People could do at least as much to reduce their carbon footprints by focusing on reducing their secondary emissions from agriculture, deforestation, or industry than they could by focusing on their transportation or their electricity use. However, it is not always clear how to do so.

Determining ways to reduce secondary carbon footprints is not always straightforward. For example, a study found that by switching from disposable plastic bags to reusable bags in Australia, a person could save 13.2 pounds of carbon emissions per year, which is the same amount of emissions that would be saved by forgoing between one and two quarter-pound cheeseburgers. For energy-intensive products such as jackets and shoes, the emissions associated with manufacture are much more substantial than those from transportation. For products with a relatively low energy intensity of manufacture, such as bottled water and soft drinks, the transportation emissions are often the most substantial emissions, and emissions are less for transportation by ship than for transportation by truck, train, or railroad. It may produce fewer emissions for a person living on the East Coast to buy wine from Spain or France than to buy wine from California. For products that need to be refrigerated, the refrigeration during transport and while the product is in the store may produce the most emissions for that particular product. Sometimes, the amount of energy needed to grow a crop varies regionally. This may lead to situations where, for example, flowers grown outside in Africa and flown to the United Kingdom may produce less GHG emissions than the same type of flowers grown domestically if those flowers must be grown in an energy-intensive greenhouse in the UK. Calculations suggest that if one is going to read more than a dozen books on an electronic reader, this may produce fewer GHG emissions than purchasing paper copies of those books.

Regardless of how one feels about the accuracy or thoroughness of personal carbon footprint calculations, the environment cannot support a world where personal carbon footprints are their current size or larger, especially in the developed world. The 1.6 billion people who currently lack access to electricity cannot be expected to reduce their carbon footprints. As the environmental effects of climate change are felt throughout the world, people may realize that their carbon footprints are too large to be sustainable.

See Also: Carbon Trading/Emissions Trading; Individual Action Versus Collective Action; North–South Debate; Tragedy of the Commons.

Further Readings

Ball, Jeffrey. "Six Products, Six Carbon Footprints." *Wall Street Journal* (March 1, 2009).

Dilli, Rae. *Comparison of Existing Life Cycle Analysis of Shopping Bag Alternatives.* Sustainability Victoria, Melbourne, Australia (April 18, 2007).

Herbert, Ian. "Carbon Footprint of Products to Be Displayed on Package." *The Independent* (London; March 16, 2007).

Kim, Patty. *How to Reduce Your Carbon Footprint at Home*. Transcript of video. www .howdini.com/howdini-video-6678398.html#transcripthere (Accessed September 2010).

Olivier, J. G. J. and J. A. H. W. Peters. *No Growth in Total Global CO₂ Emissions in 2009*. Bilthoven: Netherlands Environmental Assessment Agency, June 2010.

Ross, Lilla. "Do You Know Your Burger's Carbon Footprint?" *Florida Times-Union* (April 22, 2007).

U.S. Environmental Protection Agency (EPA). "Greenhouse Gas Emissions From a Typical Passenger Vehicle." Washington, DC: EPA, 2005.

Jesse Cameron-Glickenhaus
Independent Scholar

CARBON MARKET

The carbon market system allows technology providers to receive emission permits as payment in return for providing technology, equipment, or knowledge to projects in developing economies, such as these Brazilian wind turbines.

Source: iStockphoto

Participants in climate policy debates include many of the world's eminent scientists and institutions concerned with setting up a system of rules, policy instruments, and procedures to regulate carbon emissions. While these people and institutions may have common beliefs about future projections of perilous climate change, their opinions on climate change mitigation policies can be quite diverse.

The International Monetary Fund (IMF) focuses almost entirely on market instruments like carbon taxes or trading schemes to achieve the expected effect of emission reduction. It works on the premise that climate policy consists of getting the carbon price right and states that "an effective mitigation policy must be based on setting a price path for the greenhouse gas emissions (GHG) that drive climate change." According to the World Bank, the emissions trading policy is already proving to be successful, and its opinion is based on the results from two major markets for carbon emissions, the European Union Emissions Trading Scheme (EU-ETS) and the Clean Development Mechanism (CDM) contained in the Kyoto Protocol. They account for about $50 billion and $13 billion, respectively, of the $64 billion in worldwide carbon market transactions in 2007. Some consider this carbon trading to be a cost-effective approach to reducing emissions while, on the other hand, others favor command-and-control regulation, mainly through carbon taxes. The "Human Development Report," among others, emphasizes policies that call for carbon prices and selling of emission rights by developing countries in exchange for technology transfer and fiscal support. It also encourages government incentives for the development and deployment of renewable energy and low-carbon research, claiming they have an important role in the collective success of mitigation strategies. Thus, the emphasis laid on policy thrust differs according to the nature and aims of the institutions or experts, but carbon markets are mostly considered an inseparable part of the portfolio of climate change mitigation policies.

Effects of Direct Control and Incentives

Generally, increasing fossil fuel prices through taxation tends to increase the division between the wealthy and the poor. It works as a signal that leads consumers to choose among available alternatives but only affects those who are forced to make the change. The available cheaper alternatives may themselves be carbon intensive, or some consumers may be unable to access any alternative resources. Therefore, direct carbon taxes have, to an extent, the effect of bringing inequality in accessing resources among various income groups and slowing down market activity and may not necessarily achieve the expected discernible effect in carbon emission reduction. Resources may be directly linked to carbon emissions, like fossil fuels, or manufactured products from carbon-intensive industries. Since the success of any new product is critically dependent on market activity, carbon taxes alone are not sufficient to create new technologies needed to solve the climate crisis, especially as these new technologies may be carbon intensive and costly in the start-up or learning curve stages.

Manufacturers prefer the replication of tried and trusted consumer goods, even if carbon intensive, for low economic cost of production and thus affordability to the consumer. This makes it virtually impossible to break the supply of existing carbon and energy-intensive products in the market without government intervention. Hence, along with taxation on carbon emissions, government subsidies and investment in research and developmental for green technological alternatives are critical to diversify the types of products available to consumers in a bid to change consumer habits. Subsidies to firms include research and development grants, route to market support and soft loans, reduced taxes or tax deductions on energy-efficient equipment, depreciation, and similar fiscal incentives. Applying these in conjunction with carbon pricing by direct taxes on carbon emissions is a carrot-and-stick approach. The effectiveness of this top-down policy has to be evaluated in every case.

Cap-and-Trade Schemes

There is a school of thought that prefers carbon prices to be set by the market through emissions trading, rather than by top-down regulation through taxation. Emissions trading, also known as cap-and-trade, is a system where a cap or limit to emit pollutants within a set compliance period is allocated to entities, usually firms, that operate in the markets regulated by a central governing authority. The cap reduces over subsequent compliance periods, with its aim being to successively reduce carbon emissions over time. Firms hold the emission permits allowed by the cap and are required to keep the carbon emissions from their manufacturing or services activities within these set limits. Those that need to increase their emissions beyond their limit are allowed to buy permits from firms that emit less than their permit allows. This transfer takes place in a carbon market where the prices are dependent on the supply and demand mechanism that operates in normal trade. Therefore, the incentive is for firms to reduce their emissions in the most cost-effective way and benefit from the profit made by selling their excess permits through the carbon market. In theory, those who can reduce emissions most cheaply will be achieving the pollution reduction at the lowest cost to society, without hampering market activity or creating an artificial differential in access to resources. Demand from the market for lower-priced goods is expected drive firms to cut overheads by cutting carbon emissions, illustrating that the choice to cut emissions is made by the firms voluntarily, through market incentives.

Debates About Carbon Markets

The main issues debated about the working of carbon markets are concerned with whether there is equal opportunity for everyone to participate. For example, the African continent is now and has always been a low emitter of carbon and hence has limited potential for reduction of carbon emissions. Significant progress that could be made is for technology transfer to bring low-carbon technologies and products into the African marketplace so that it altogether avoids the traditional high-polluting activities that the industrialized economies undertook for many years. However, the required levels of investment for low-carbon innovation and production in the private sector cannot come from the emissions trading mechanism and must instead come from institutional and private investors. Investment activity is much less vibrant due to the perception of risk in these countries, and therefore, low-carbon technologies and products remain mostly out of reach of the consumers. Until they achieve the threshold that enables them to participate fully in the carbon emissions markets, the African economies and others in similar positions may be compelled to take the high-carbon path to development.

In continuation with this issue, some commentators suggest that carbon markets do not address the problem that it is a trading platform for those who have in the past caused the majority of greenhouse gases, that is, mainly the industrialized nations, and it is they themselves who are now able to benefit further by trading carbon emissions. Some claim the carbon market is, in fact, a source of income for polluters rather than a means of providing incentives for innovation. Furthermore, there is no mechanism within the carbon emissions trading schemes to make the polluters pay for the space they have used and polluted in the past.

The success of a cap-and-trade scheme relies on setting the cap at the correct level and effectively enforcing it. This sets the supply in the market because it determines the number of permits that will be available. The market is set in motion by allocating the permits among the polluters, either by an auction of the permits or a free distribution, often based on historical emissions or output. In 2005, politicians feared that burdening European industries with extra costs would undercut their ability to compete in a global marketplace, and hence handed out virtually all permits free, while much of their cost was paid for by the public in the price of goods and services, such as higher electricity bills. This aggravated the consumers and concerned authorities when it came to light that some companies had received permits about 3 percent in excess of their current emissions, so that they did not need to cut emissions in the first compliance period. An electricity company had also admitted charging customers for the emission permits, saying that while it may have received them free from the government, they still had value in the marketplace. This illustrates the point made by some that the carbon market is a platform for those who wield the most power and are perhaps the biggest polluters in the marketplace.

Notwithstanding these voices of dissent, there are many benefits of the carbon market system. The biggest advantage of the Clean Development Mechanism (CDM) market that operates in developing countries is its high flexibility. Firms can choose whether to use their permits in the current compliance period or save them for subsequent periods. This reduces the firms' risk to carbon price fluctuations and can also help them to plan their business activities in a way to maximize benefits from the emissions permits. CDM also has a potential role in facilitating exchanges made in the form of barter deals where technology providers from developed countries receive emission permits as payment in return for providing technology, equipment, or experience to projects in developing economies. Most CDM projects registered in 2005 were in Brazil (79 projects), India (64 projects), and China (13 projects) in a variety of sectors like hydropower, landfill, wind, and others.

There are still significant concerns regarding different aspects of the CDM process, including bottlenecks in the approval of methodologies, the application of additionality criteria, and the registration of projects. One of the developments that would improve confidence in carbon markets is the comparability of carbon prices among different markets, for example, the EU-ETS and Clean Development Mechanism/Joint Implementation (CDM/JI). Both these markets differ in that CDM/JI emission reductions (ERs) are vulnerable to registration and delivery risks, while the EU Allowance units (EUAs) are government-issued, compliance-grade assets and carry lower risk. Interoperability of certified credits between CDM/JI and EU-ETS is uncertain, and therefore projects are confined and opportunities to operate optimally on a global scale do not exist very easily.

Conclusion

The carbon market concept is a new and explorative form of combating climate change. Lessons are being learned from the experiences of the trading schemes, and policies are being adjusted to correct shortcomings in the systems. Changes are being made to allow for registration of smaller and bundled projects in a bid to enable smaller economies to participate and enhance their local capacity in carbon-saving projects. New sectors are being approved for potential projects that are relevant to many of the developing countries. However, it is important to remember that a carbon market is only a catalyst for climate change mitigation and a piece in the whole jigsaw puzzle of climate mitigation strategies.

See Also: Certified Products; Environmental Justice (Politics); Ethical Sustainability and Development; Kyoto Protocol.

Further Readings

Ackerman, Frank. "Carbon Markets and Beyond: The Limited Role of Prices and Taxes in Climate and Development Policy." *G-24 Discussion Papers*, 53 (2008).

Kruger, Joseph, Wallace E. Oates, and William A. Pizer. "Decentralisation in the EU Emissions Trading Scheme and Lessons for Global Policy: Discussion Paper." *Resources for the Future* (February 2007). http://www.rff.org/Documents/RFF-DP-07-02.pdf (Accessed September 2010).

United Nations Development Programme. "Human Development Report 2010." http://hdr .undp.org/en (Accessed September 2010).

Swati Ogale
Independent Scholar

Carbon Sequestration

The failure of the world's major economies to arrest the growth of greenhouse gas emissions is putting Earth on course for potentially catastrophic climatic impacts by the end of this century. Indeed, while the parties to the United Nations Framework Convention on Climate Change have embraced the scientific consensus that temperature increases of 2 degrees Celsius or more above pre-industrial levels will have serious implications for ecosystems and human institutions, we are on course for much higher temperature increases. A recent study concluded that it is increasingly improbable that we can stabilize atmospheric concentrations of greenhouse gases below a point that will yield temperature increases of 3 to 4 degrees Celsius above pre-industrial levels by the end of the 21st century. Temperature increases of this magnitude may overwhelm society's ability to adapt to many serious impacts and could push us past a number of sudden and irreversible climatic tipping points, leading to amplified warming, massive sea-level rise, and the fundamental transformation of many ecosystems.

Some believe that initiatives to capture and store carbon dioxide, the primary greenhouse gas, could prove to be an important complement to mitigation policies. In 1979, Cesare Marchetti of the International Institute for Applied Systems Analysis, a think tank based in Austria, proposed a system to bury CO_2. This method, carbon capture and sequestration (CCS), could prove to be a vital weapon in combating climate change. However, CCS remains an unproven technology, at least at the scale required to meaningfully address burgeoning levels of CO_2 emissions, and could also produce serious negative impacts that should give pause to policymakers.

Capture Technologies

In the context of large-scale power production facilities, carbon capture processes fall into three basic categories: (1) flue gas separation, (2) oxy-fuel combustion, and (3) pre-combustion separation.

In the flue gas separation process, flue gas (combusted exhaust gas in a power plant, comprising primarily nitrogen and CO_2) is bubbled through a solvent, most commonly monoethanolamine, in a packed absorber column, where the solvent removes the CO_2 from the flue gas. The solvent subsequently passes through a regenerator unit where the absorbed CO_2 is separated from the solvent, with the solvent recycled back to the absorption tower. Water vapor is condensed, leaving a highly concentrated (more than 99 percent) CO_2 stream, which may be compressed for commercial utilization or transport and sequestration. Post-combustion technologies can capture 80 to 90 percent of CO_2 emissions. This approach can also be retrofitted on existing coal-based power plants without changing their configuration. However, it is also the least efficient and most expensive option from the perspective of investment and cost of electricity generation.

When fossil fuels are combusted in air, the fraction of CO_2 in the flue gas is low, ranging from 3 to 15 percent. As a consequence, separation of CO_2 from the rest of the flue gases is capital and energy intensive. An alternative approach, oxy-fuel combustion, burns fossil fuels in pure or highly enriched oxygen. The flue gas then contains mostly CO_2 and H_2O. The water vapor can be readily condensed, and the CO_2 can be compressed and piped directly to a storage site. The oxy-fuel combustion process requires only simple purification of CO_2 and does not require reagents and solvents that add to costs and produce disposal issues. On the other hand, it requires large quantities of oxygen, which is expensive in terms of capital costs and energy consumption.

In pre-combustion CCS processes, fuel is reacted with oxygen, air, or steam in a high-pressure, temperature-controlled environment, producing a gas consisting primarily of hydrogen and carbon monoxide. The addition of water in a catalytic reactor called a shift converter produces highly concentrated CO_2 and additional hydrogen. The CO_2 can then be separated and prepared for storage, and the hydrogen can be used for fuel in a gas turbine plant. While this technology is usually used for coal gasification, it could also be applied to liquid and gaseous fuels. Because of the high concentration of CO_2 produced in the flue gas, pre-combustion capture technologies can be extremely efficient, potentially reducing CO_2 emissions by 90 to 95 percent and requiring substantially less energy than other capture processes. On the other hand, the capital costs of this process are very high.

Storage Options

Once CO_2 is captured from a facility, it can be transported via truck, rail, ship, or pipeline to locations that are geologically appropriate for long-term storage. There are three main options for geological storage of CO_2: injection into deep saline terrestrial or ocean aquifers, storage in depleted hydrocarbon reservoirs, and storage in deep, unmineable coal seams. Of these, deep saline aquifers have the highest potential capacity globally for storage.

Potential of CCS and Its Challenges

Several studies in recent years have concluded that there is an abundance of onshore and offshore storage space for CO_2. For example, it has been estimated that oil and gas fields could hold up to a trillion tons of CO_2, or 50 years' worth of emissions, at projected rates in 2030. Moreover, it has been projected that saline aquifers on land and in the oceans could provide storage for 10 trillion tons of CO_2, or 500 years' worth of emissions. However, sequestration is only economically and technologically feasible for a limited number of facilities. As a consequence, the Intergovernmental Panel on Climate Change (IPCC) has estimated that CCS could sequester anywhere from 10 to 55 percent of total anthropogenic CO_2 emissions through 2100.

However, there are many serious concerns associated with CCS. First, commercial-scale CCS plants likely will not be online before 2030, meaning that some critical thresholds associated with climate change may be passed before this technology could make any difference. Many analysts also express concern that the coal industry may now be touting "capture-ready" plants to maintain political support for new coal-fired facilities, despite the fact that most plants will be retired before the technology is available.

Second, it is far from clear that CCS technologies will be economically viable. According to the U.S. Department of Energy, incorporation of CCS technology into power plants could almost double operation costs, resulting in a 21 to 91 percent increase in the cost of electricity for consumers. Also, given the tens of billions of dollars that a commitment to CCS would require over the next few decades, there is a serious question as to whether such investments would be more fruitfully channeled into initiatives such as renewable energy and energy-efficiency projects to directly reduce greenhouse gas emissions.

A third issue is the so-called energy penalty associated with the deployment of CCS technology. CCS would increase the fuel needs of coal-fired power plants by 25 to 40 percent. Even a penalty of 20 percent would require the construction of an additional plant for every four constructed with CCS technology. As a consequence, more coal would need to be burned, and major localized environmental problems associated with coal extraction and transportation would be exacerbated.

Perhaps the most serious concern with CCS is the possibility of leakage of CO_2 from storage sites by lateral migration or to the surface. The IPCC has concluded that 99 percent of CO_2 could be retained in geological reservoirs over a period of 1,000 years. While this sounds like an impressive statistic, even a leakage rate of 1 percent of about 100 years' worth of stored carbon associated with fossil fuel emissions would release as much as 6 gigatons of carbon annually back into the atmosphere, roughly equivalent to current total global CO_2 emissions from fossil fuels. Even a highly unlikely leakage rate of only 0.01 percent annually would result in release of almost two-thirds of stored CO_2 within 10,000 years. Moreover, in the context of ocean sequestration, leakage could result in increased acidity, with serious implications for marine species, and oxygen-deprived "dead zones." While it may be possible to recapture some of the leaked gases, the cost would likely be extremely high.

The Future of CCS

The first offshore CCS plant, operative since 1996, is part of the Sleipner natural gas field in the North Sea. CO_2 is stripped from the gas and injected into a saline water–bearing structure composed of sands well below the sea bed. While Sleipner is a milestone in CCS technology, the amount of CO_2 that is captured is less than that from a large coal-fired power plant, and more than 1,000 Sleipner projects would be required to sequester just 4 percent of annual anthropogenic CO_2 emissions. At least four other projects are active, injecting about 1 million metric tons of CO_2 annually. There are also a number of CCS projects under way in Norway, Algeria, and Canada, and more are planned in the United States, China, Australia, and other European countries. CCS could prove to be the salvation of the fossil fuel industry or simply another unhelpful diversion from efforts to decarbonize the world's major economies.

See Also: Alternative Energy; Carbon Capture Technology; Carbon Emissions (Personal Carbon Footprint); Carbon Market; Carbon Tax; Carbon Trading (Cities); Coal, Clean Technology; Oil; Oil Sands.

Further Readings

Greenpeace. *False Hope: Why Carbon Capture and Storage Won't Save the Climate* (2008). http://www.greenpeace.org/raw/content/international/press/reports/false-hope.pdf (Accessed November 2010).

Klass, Alexandra and Elizabeth J. Wilson. "Carbon Capture and Sequestration: Identifying and Managing Risks." *Issues in Legal Scholarship*, 8:3 (2009).

Pearce, Fred. "Can Coal Live Up to Its Clean Promise?" *New Scientist*, 344 (March 2008).

Shaffer, Gary. "Long-Term Effectiveness and Consequences of Carbon Dioxide Sequestration." *Nature Geoscience*, 3 (July 2010).

William C. G. Burns
Journal of International Wildlife Law & Policy

CARBON TAX

Adding costs to emissions may provide an incentive to develop new technologies to reduce emissions, like solar technology, which here is measuring carbon dioxide exchange.

Source: Argonne National Laboratory, U.S. Department of Energy

A carbon tax is a tax on the carbon content of fossil fuels such as coal, natural gas, and oil. The tax is imposed in order to encourage people and firms to use less of these fuels, which contribute to greenhouse gases in the atmosphere. Carbon in this context is shorthand for carbon dioxide.

While legislation is aimed at directly changing the behavior of polluters by outlawing or limiting certain practices, economic instruments such as a carbon tax aim either to make environmentally damaging behavior cost more or to make environmentally sound behavior more profitable. Polluters are not told what to do; rather, they find it expensive to continue in their old ways. They have a financial incentive to reduce emissions. If they instead pass on the extra cost to their customers, their customers will have a financial incentive to seek alternative suppliers. The imposition of a carbon tax on competing goods or services can ensure that those that emit the least carbon will have a price advantage over more carbon-intensive products.

Economic instruments are supposed to be more economically efficient than legislative measures in that pollution reductions can be made for less cost. Firms that find it cheaper to reduce their carbon emissions than to pay a carbon tax can do so. Those for whom it would cost more than the tax to reduce their carbon emissions can choose to pay the tax instead. In this way, emissions are reduced most by those who can do it cheaply, and therefore, it is a more cost-efficient way of achieving a limited amount of pollution reduction.

Price-based policy measures are also market-based measures because, in economists' terms, a price is set, and the market determines the quantity of emissions that are released. This compares with carbon trading, where a limit is set on total emissions, and the market determines the price of carbon.

Because so much carbon is used in affluent countries, even a fairly small tax could raise large amounts of money. Governments can use the money raised in this way for environmental protection such as developing noncarbon energy sources, to compensate disadvantaged sections of the community affected by the tax or by global warming, or to reduce other taxes so that carbon taxes are revenue-neutral for the government.

Carbon taxes are levied in Belgium, Denmark, Finland, France, Italy, Luxembourg, the Netherlands, Norway, and Sweden. However, these taxes are not based on carbon emitted and are not applied to all sources of carbon emissions. For example, the United Kingdom's Climate Change Levy is a tax on energy rather than a carbon tax and excludes household energy use. In New Zealand, a genuine carbon tax was proposed in 2005, but this was abandoned because of concerns that its cost (estimated at NZ$4 per week per household) would not be justified by sufficient reduction in emissions.

Perhaps the most comprehensive carbon tax was introduced in British Columbia in 2008. Money raised is returned to taxpayers through personal and business income tax cuts. It started at $10 per ton and is supposed to rise by $5 per ton each year until 2012, when it will be $30 per ton. This is expected to add 7.24 cents per liter to the price of gasoline. The Liberal government that had introduced the tax was reelected in 2009, showing that voters did not resent the tax.

Internalizing Environmental Costs

A carbon tax is a price-based policy measure in that it seeks to internalize environmental costs that are normally external to a company's account books by making it more expensive for polluters to emit these gases. Laws can also force the polluter to take notice of these external costs by prescribing limits to what can be discharged or emitted, but economists argue that the market is better able to find the optimal level of damage: the level that is most economically efficient. The idea that there should be a level of pollution that is above zero but is called "optimal" is strange, and even repugnant, to many people. But it is a central assumption in the economic theory on which economic instruments are based.

If a carbon tax is equivalent to the cost of environmental damage, then the theory says that the company will reduce its carbon emissions until any further incremental reduction in pollution will cost more than the remaining tax, that is, until it is cheaper to pay the tax than reduce the emissions. This is said to be economically efficient because if the polluter spends any more than this, the costs (to the firm) of extra pollution control will outweigh the benefits (to those suffering the adverse affects of the pollution).

This might seem to be a less than optimal solution to the community, but economists argue that the polluter is better off than if it had paid to eliminate the emissions altogether, and the community is no worse off because it is being compensated by the firm for the damage through the carbon tax paid to the government. In theory, the carbon tax paid by firms can be used to correct the environmental damage they cause or to compensate the

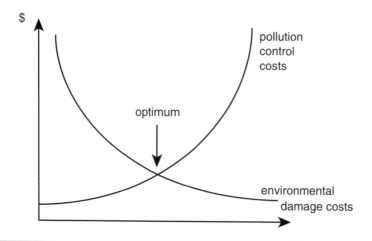

Figure 1   Economist's Graph on Incremental Costs and Benefits of Pollution Control

victims. All this supposes that the tax is in some way equivalent to the damage done, that environmental damage can be paid for, and that this is as good as, or even preferable to, avoiding the damage in the first place. This implies that the benefits that arise from the environment can be substituted for by other benefits that can be bought. However, environmental quality is not something that can be swapped for other goods without a loss of welfare. Also, there is considerable doubt about whether money payments can correct the consequences of global warming. More importantly, money collected from carbon taxes is unlikely to be used either to correct environmental damage or to compensate victims.

In practice, governments and regulatory agencies do not attempt to relate carbon taxes to the "external costs" of global warming, particularly when those costs are likely to be borne by people in other parts of the world. Additionally, environmental taxes are often promoted by economists and others as a way of replacing other charges and taxes that firms would normally have to pay.

The United Kingdom (UK) research organization Truscot estimated that if companies had to pay the actual cost of the economic damage caused by carbon emissions—estimated by the UK government to be some £20 per ton—then some companies would be paying around half their earnings. Overall, it would cost 12 percent of the earnings of the top 100 UK firms listed on the stock market, the FTSE 100. This is not something governments are likely to require because of the economic ramifications.

Innovation

Economists argue that the imposed costs, even if they do not internalize the real environmental costs of polluting activity, nevertheless provide an incentive for companies to reduce their pollution and thereby save money. The contention is that legal standards might ensure that firms meet particular targets, but that having met them, there is no incentive to go beyond them; whereas the financial incentives provided by a carbon tax give firms an ongoing motivation to save money by developing technologies that will further reduce their emissions.

Adding costs to a firm's operations may impose pressure on it to reduce its costs, but there is no guarantee that it will do so in the area where the cost is imposed. Such a firm may find it easier, cheaper, or even more profitable to apply new technology and methods in other parts of its operation or just to pass the increased cost on to the consumer—especially in sectors where there is little price competition between firms.

Advocates of economic instruments tend to assume that firms are managed by economically rational executives who are not subject to technical, financial, organizational, and perceptual constraints. This is not the situation in the real world. For example, a firm may not be able to afford the initial capital cost of changing production processes even if this would be cheaper in the long term than paying the carbon tax.

In practice, environmental taxes and charges tend to be too low to provide enough incentive to change production processes significantly. They are low because of political pressures from industries not wanting to pay and concerns that high taxes might encourage cheating and evasion of the charges.

Although economic instruments are supposed to encourage technological innovation, they often stifle it by allowing firms to pay for pollution rather than to reduce their emissions. It is often much easier to pay a carbon tax than to invest in research and development that may or may not result in emission reductions that are cheaper to put in place than the cost of paying the tax.

Substantial changes to technological paradigms require institutional changes that decision makers prefer to avoid. What is more, if money is invested in small reductions in emissions from existing coal-powered plants, for example, the owners are likely to want to keep them running for years longer to recoup that investment. And this provides a further obstacle to replacing them with renewable energy sources. In this way, the technological improvements are marginal rather than wholesale, and more radical innovations are avoided.

Equity

Economists argue that if the money collected from environmental taxes and charges is spent on something equally worthwhile, then the community is still no worse off. This is a view that those who suffer from the health and environmental impacts of pollution might find hard to accept, particularly since the money is seldom spent on them and is seldom used to clean up the pollution.

However, a carbon tax is a regressive tax. Taxing fuel- and energy-intensive products and services is particularly painful to lower-income households as these make up a higher proportion of a poor person's income. Increased energy costs aimed at encouraging people to use less energy by buying more energy-efficient appliances may most affect those who can least afford to replace such goods. Tenants are disadvantaged because landlords are less likely to spend money on energy-saving measures, such as roof insulation or solar water heating, than homeowners are, since landlords will not benefit from the energy savings themselves. In the UK, affluent households are almost twice as likely to have the more energy-efficient gas central heating.

Also, a tax will work as an incentive to change behavior only if there are alternatives available; otherwise, it just serves to penalize some sectors of the community. For example, a gasoline tax has most impact on people who have to travel long distances to get to work and do not have access to public transport. Since it is often the poor who are forced to live in the outer suburbs, because that is where the cheapest housing can be found, such a measure would impose its greatest burden on those least able to pay. People in rural areas and on the outskirts of cities will be also worse off because of the longer distances they have to travel. And rural industries will also be badly hit because of the longer distances and the heavy fuel requirements of agricultural machinery.

Although governments decide how much a carbon tax will be, the decision about whether to pay it or reduce emissions is made by individual firms. Such decisions are made on the basis of company economics, not on the basis of what is best for the community or the environment. In the end, it is the polluters who decide what trade-offs should be made between economics and environmental quality, rather than the community.

See Also: Carbon Market; Carbon Trading (Cities); Carbon Trading/Emissions Trading (Business).

Further Readings

Beder, Sharon. *Environmental Principles and Policies*. Sydney: UNSW Press: 2006.
Carbon Tax Center. http://www.carbontax.org (Accessed August 2010).
"The Reality of Carbon Taxes in the 21st Century." South Royalton, VT: Environmental Tax Policy Institute and the *Vermont Journal of Environmental Law*, 2008.

Robins, Nick. "The Coming Carbon Crunch." *Ethical Corporation*, 21 (July 2005).

U.S. Environmental Protection Agency (EPA). "International Experiences With Economic Incentives for Protecting the Environment." Washington, DC: National Center for Environmental Economics, EPA, 2004.

Sharon Beder
University of Wollongong

Carbon Trading (Cities)

Over the past 20 years, the Intergovernmental Panel on Climate Change (IPCC; established in 1992 by the United Nations Framework Convention on Climate Change [UNFCCC]) has produced a series of extensive reports that have left little uncertainty that human activities—primarily the combustion of fossil fuels—have forced and are forcing a rise in the average global atmospheric temperature resulting in the increasingly rapid destabilization of Earth's climate. International negotiations sponsored by the UNFCCC resulted in the adoption (in 1997) of the Kyoto Protocol (it became effective in 2005), which set binding targets for the reduction of greenhouse gas (GHG) emissions by its expiration in 2012.

International Carbon Markets

The Kyoto Protocol introduced several strategies for linking national GHG reduction targets to local action. As a free-market alternative to the centralized governmental regulation that had been prevalent in environmental policy circles of the 1960s and 1970s, Kyoto offered signatories several mechanisms to help meet their emission reduction targets. Prominent among these was "emissions trading" (or a "carbon market"). This type of mechanism had been successfully implemented in response to international concern over the environmental consequences of acid rain (e.g., deforestation and freshwater acidification).

Popularly known as "cap and trade," an emission trading scheme is established as a market for the distribution (or sale) and subsequent trading of permits authorizing an operator to emit a specified amount of the target pollutant; it would be prohibited from emitting more than its permits allowed. The carbon exchange would purchase allowances from members that have emitted less than they were permitted, and it sells allowances to those members that have been unable to meet their emission reduction targets. The operator could either invest in processes and technologies that would reduce its emissions to a level consistent with its permits or purchase additional emissions permits from the market.

The demand for the number of permits available on the market would determine their price at any given time. The total number of permits would be gradually reduced (i.e., withdrawn from trading) over time, placing an upward pressure on the price of an individual permit. Operators would continue to have strong financial incentive to invest in more carbon-efficient processes and technologies as the permits became more expensive relative to those pollution-control technologies. Within regulatory guidelines, exchange participants would be encouraged to determine for themselves the most cost-effective emissions reduction strategies.

Subnational Markets

The failure of the Copenhagen round of international climate negotiations to forge a successor agreement to Kyoto has severely compromised the capacity of the international community to meet emission reduction goals set out by the IPCC. Though unregulated and largely uncoordinated, GHG emissions reduction initiatives undertaken by subnational authorities (e.g., states, regional consortia, and municipalities) have assumed a much more important role in achieving global GHG reduction targets.

Several regional groups in the United States have formed allowance-trading schemes to meet regional emissions reduction targets. Ten northeastern states formed the Regional Greenhouse Gas Initiative (RGGI), the first mandatory, market-based program in the United States. Subject to a regional cap negotiated by RGGI, the member states commit to reducing emissions from their power-generation sectors by specified amounts through the trading of allowances. Proceeds from the periodic auction of these allowances are designated for investment in strategic energy programs. The Western Climate Initiative is a collaboration of seven states and four Canadian provinces. Six Midwestern states and one Canadian province form the Midwest Greenhouse Gas Reduction Accord. The members of these initiatives negotiate binding GHG reduction targets that are then formalized in state or provincial legislation and, ultimately, municipal GHG reduction commitments. California is in the process of developing its own market designed to reduce GHG emissions to 1990 levels by 2020.

To meet the demand for the creation and trading of commoditized carbon reduction instruments from international, national, and subnational cap-and-trade agreements, numerous private-sector carbon exchanges have been established worldwide. The Chicago Carbon Exchange (CCX) manages the only North American cap-and-trade system for GHGs. Its members make voluntary but legally binding commitments to meet annual GHG emission reduction targets. CCX also trades project-based offsets. The European Climate Exchange (ECX), launched by CCX in 2005, now accounts for more than 80 percent of the traded volume in the European Trading Scheme. The first environmental products market in Canada, the Montreal Climate Exchange, was formed as a joint venture between the Montreal Exchange (a financial market) and CCX. Likewise, the Tianjin (China) Climate Exchange is a joint venture between CCX and the municipality of Tianjin. Founded in 2007, the Australian Climate Exchange (ACX) is that country's first climate exchange.

Local Participation in the Carbon Market

A growing—but still relatively small—number of cities are electing to reduce their carbon emissions voluntarily through elective organizational agreements or by virtue of state mandate. There are numerous strategies a city can employ to achieve these reductions. First, it can elect to undertake projects that either remove GHG from the environment (e.g., by capturing or flaring the methane generated by municipal landfills) or prevent them from being generated (e.g., by implementing energy-efficiency and conservation initiatives that reduce the amount of electricity consumed in municipal buildings). Depending on its unique financial profile, the city might elect to finance the projects through the direct cost savings attributable to the completed projects. In this case, it might borrow the cost from the bond market and repay interest and principal with proceeds from the implemented project.

Alternatively, a municipality might elect to finance a project by selling *carbon offsets* on a brokered market. Offsets are tradable financial commodities that represent a specific quantity of carbon or carbon-equivalent emissions that have been captured or avoided as a result of a specific project. The City of Aspen joined the Chicago Climate Exchange in 2005, legally committing to reduce GHG emissions from its municipal operations by 1 percent annually. By the end of 2006, City of Aspen operations had reduced its GHG emissions by 11.5 percent below 2005 levels.

An Alternative to Cap and Trade

While there is significant international and domestic support for pricing carbon through a free-market cap-and-trade system (discussed above), many favor national "carbon taxes." The carbon content of various fossil fuels on which a carbon tax is levied is a quantity that can be precisely calculated. The carbon tax is most generally described as being levied "upstream." That is, producers of various types of fuel—oil, coal, or natural gas—would be taxed on the carbon content of the fuel produced. As with other production costs, that tax would likely be passed on to consumers of those fuels. Depending on its size and the ultimate characteristics of the legislation, a carbon tax may be so small as to go practically unnoticed by consumers while generating the necessary revenues for meaningful clean energy investments. Alternatively, the carbon tax could be designed to be "revenue neutral." That is, the vast majority of taxes generated in the program would not be used to fund government programs, but would be returned to taxpayers in the form of rebates or credits.

Carbon Pricing Debate

There has been considerable controversy over the relative strength of these two approaches to pricing carbon. In general, these debates center on the effectiveness of a particular approach in reducing carbon dioxide emissions, the ease with which the approach could be implemented and managed, the political feasibility of developing and implementing the pricing model, and the ease with which such a national program could be integrated into international pricing systems.

Effectiveness

Proponents of a carbon cap and market claim that one of the greatest benefits of it sets a clear and legal goal for carbon emissions reductions. A market participant may only emit a specified amount of carbon dioxide, and any further amounts must be either abated by investment in cleaner production technologies or offset through the purchase of additional permits. In theory, the total number of permits made available by the carbon market sets the maximum amount of emissions allowable.

A carbon tax, by contrast, does not specify the amount by which carbon dioxide emissions will be reduced. It achieves emissions reductions by taxing the production of fossil fuels, making them more expensive to consume. This encourages producers to invest in such a way as to minimize their carbon tax exposure, not necessarily their carbon dioxide emissions. Given the urgency with which the global economy must move to reduce its greenhouse emissions, a carbon tax may not provide sufficient incentive for the rapid reductions required.

Simplicity

One of the most appealing aspects of a carbon tax is the ease with which it could be implemented. The mechanisms to collect (and rebate) the tax already exist and could be leveraged to implement the tax much more quickly than could a cap-and-trade system. Furthermore, the tax would be levied upstream on relatively few operations (e.g., at the wellhead, the refinery, or the port) instead of across a broad spectrum of downstream operations that burn the fuel in their business activities.

A cap-and-trade approach, though, requires the creation of new markets and a new set of openly traded commodities. In spite of the fact that various regional carbon markets exist—or are in development—in the United States, the creation of a national market (and the integration of that market with other global carbon markets such as the European Trading Scheme) would require the creation of new regulatory frameworks and institutions.

The tax is also a more widely, if not popularly, understood mechanism than is a permit market. Whether or not one likes the tax, it is relatively easy to understand how it is applied and who pays. On the other hand, the carbon market is not only complex from regulatory and institutional perspectives, its workings are also poorly understood by the general public. How it prices carbon and to whom the revenues flow from the sale and purchase of carbon permits are not widely understood.

Political Feasibility

Some proponents of a carbon tax argue that a cap on carbon emissions would lead to significantly higher energy prices for the consumer that would be politically unsustainable. As a result, politicians would look for ways in which the regulated industries could circumvent the cap and thereby keep energy prices down to a politically acceptable level. And while profits from cap-and-trade activities remain in the private sector, a carbon tax produces a revenue stream that can be allocated to public programs.

Another political aspect to this debate is the certainty with which the costs of either approach would be known. Given its upstream implementation, the relatively small number of taxed entities, and the certainty with which the tax rate is known, a carbon tax would produce a reasonably reliable revenue stream that could be used as policy dictated. By contrast, a cap-and-trade system would result in costs to industry—and ultimately consumers—that would fluctuate with the market value of the emissions permits being traded.

International Harmonization

The global nature of climate change threats must ultimately be recognized in a coordinated international effort to achieve IPCC-specified greenhouse emissions reductions. Many cap-and-trade proponents argue that the approach encourages the evolution of a global market in carbon that would provide the mechanisms and flexibility to meet these emissions targets most efficiently and at the lowest cost. In spite of its shortcomings and growing pains, the European Trading Scheme has demonstrated the feasibility of such a system.

On the other hand, proponents say carbon taxes can be equalized across borders by levying duties on goods imported from countries with lower or nonexistent carbon taxes.

Cap or Tax?

While there are fundamental differences in the perspectives of advocates of the two approaches, there is also widespread agreement that pricing carbon will be necessary to achieve the reductions in carbon dioxide emissions necessary to blunt the worst effects of climate change.

The effectiveness of either approach, though, relies on how the respective systems are configured. The carbon cap must be sufficiently low (i.e., the permit prices must be sufficiently high) so that the price of the traded permits incentivizes investment in clean technologies rather than purchasing of additional permits. Should the permits become too expensive, regulators could introduce additional permits into the market in order to drive the permit price down to a "reasonable" level. Similarly, a carbon tax must be set at such a level that it encourages the transition to alternate energy sources.

See Also: Carbon Market; Carbon Tax; Carbon Trading/Emissions Trading.

Further Readings

Beinecke, F., J. D. Sachs, F. Krupp, R. A. Pielke Jr., et al. "Putting a Price on Carbon: An Emissions Cap or a Tax?" *Yale Environment 360* (May 7, 2009). http://e360.yale.edu/content/feature.msp?id=2148 (Accessed November 2010).
Carbon Tax Center. http://www.carbontax.org (Accessed November 2010).
City of Aspen. *Canary Initiative Action Plan* (2009). http://aspenpitkin.com (Accessed November 2010).
Ellerman, A. D. and P. L. Joskow. "The European Union's Emissions Trading System in Perspective" (2008). http://www.pewclimate.org/docUploads/EU-ETS-In-Perspective-Report.pdf (Accessed November 2010).
"Policy Options for Reducing CO₂ Emissions." Washington, DC: Congressional Budget Office, 2008.
Weisbach, D., G. Wagner, J. E. Milne, K. R. Richards, et al. "Carbon Tax vs. Cap-and-Trade." *Bulletin of the Atomic Scientists* (November 24, 2008). http://www.thebulletin.org/web-edition/roundtables/carbon-tax-vs-cap-and-trade (Accessed November 2010).

Kent L. Hurst
Independent Scholar

CARBON TRADING/EMISSIONS TRADING

The term *emissions trading* refers to a market-based approach to environmental pollution problems in which a government or some other private or public entity creates a market in permits for an environmentally harmful activity. Polluting entities are allotted, either through free allocation or auctioning a set amount of permits, commonly determined from a baseline of past emissions; entities can buy, sell, or trade permits depending on whether their emissions are less or more than their allotted allowance. If their emissions are greater than their allowances, entities must pay to pollute by buying additional credits through the market. Such a system is often referred to as "cap and trade": entities that are able to reduce their emissions below their allotted amount can sell their credits to other parties who cannot reduce their emissions or find it more cost-effective to purchase credits than to make emissions reductions. While not the first cap-and-trade or emissions trading program, carbon trading (including carbon dioxide and other greenhouse gases [GHGs]) represents the largest emissions trading market in the world and represents one way signatory nations to the Kyoto Protocol can meet their emissions reductions targets. Advocates of carbon trading view it as the most politically feasible way to cut GHG emissions and as a viable alternative to a carbon tax or government regulation of emissions ("command-and-control" approaches). Critics view carbon trading as a false solution that gives polluters a "license to pollute" and fails to address the fundamental structural problems driving climate change. This article briefly describes the history of emissions trading and the expansion of carbon trading, followed by a discussion of the viewpoints of supporters and critics of carbon trading and emissions trading in general.

Emissions Trading and Carbon Trading: Background

The U.S. Environmental Protection Agency (EPA) Acid Rain Program, established as an amendment to the Clean Air Act, created one of the first cap-and-trade programs. In this program, the EPA put an overall "cap" on total sulfur dioxide (SO_2) emissions from power plants, and starting in 1995, existing power plants were allocated allowances to emit SO_2 that could be bought, sold, or banked for future use. This allowed companies to choose whether to buy additional permits and proceed with current emission levels or to lower their emissions to stay within their cap. By 2005, SO_2 emissions declined 35 percent from 1990 levels, resulting in improved air quality and reduced sulfate deposition in the United States. This program succeeded in reducing emissions above and beyond legislative requirements and cost much less than anticipated, leading to the widespread view that the cap-and-trade approach was more effective and economical in reducing pollutant levels than the command-and-control approach would have been.

The cap-and-trade model that successfully helped to reduce sulfur dioxide emissions in the United States has been applied to reducing emissions of carbon dioxide and other GHGs and is broadly referred to as "carbon trading." Carbon trading often includes a range of GHGs, but normally trades in units of metric tons of CO_2 equivalent (tCO_2e). Many carbon trading programs allow parties that cannot or do not wish to reduce their emissions to within their permit allocations to "offset" emissions by paying another entity to reduce carbon emissions in order to make up for the excess emissions. Existing carbon markets can be broadly divided into mandatory (or regulatory) and voluntary markets. Regulatory markets, currently valued at $143,897 million, are those that have primarily emerged from efforts of ratifying nations to reach emission reductions targets set by the legally binding Kyoto Protocol. Voluntary markets include all carbon trading not legally mandated and currently comprise 1 percent of the overall carbon market, valued at $387 million, with expectations for substantial growth over the next decade. Primarily due to the economic downturn and to political uncertainty surrounding climate legislation in the United States and beyond, the price of carbon declined substantially between 2008 and 2009. Despite this, most regulatory markets grew in value by 7 percent; however, the voluntary market declined by 47 percent overall, with an 87 percent decline in the Chicago Climate Exchange.

Regulatory Carbon Markets

The 1997 Kyoto Protocol, a legally binding international agreement linked to the United Nations Framework Convention on Climate Change (UNFCCC), sets binding emissions reduction targets for 37 industrialized countries and the European Union. These ratifying industrialized countries agreed to GHG emissions reduction targets of an average of 5.2 percent below 1990 levels over the period of 2008–12. These targets led to the establishment in 2005 of the European Union Emissions Trading Scheme (EU-ETS) that covers more than 11,000 installations, with a market valued at $108 billion, trading 5,499 $MtCO_2e$ in 2009. A number of other national and regional cap-and-trade initiatives have developed around the world, including New Zealand's Emissions Trading Scheme, the United Kingdom's Carbon Reduction Commitment Energy Efficiency Scheme, Australia's New South Wales Greenhouse Gas Reduction Scheme, and Tokyo's regional cap-and-trade scheme. While the United States has not ratified the Kyoto Protocol and there is currently no government-sanctioned carbon market, a large number of regional initiatives have been established to reduce greenhouse emissions and meet Kyoto targets despite the national government's refusal to ratify the protocol. These include the Regional Greenhouse Gas Initiative, the Western Climate Initiative, and the Midwestern Greenhouse Gas Reduction Accord.

Voluntary Carbon Markets

The voluntary market includes all carbon trading not legally mandated and can be divided into the Chicago Climate Exchange (CCX) and the voluntary over-the-counter (OTC) offset market. Launched in 2003, the Chicago Climate Exchange is a small, private, and voluntary, but legally binding, climate market that seeks to reduce emissions of all six major GHGs. CCX allows up to 50 percent of emission reductions to be achieved through the purchase of offsets, but as of yet only 15 percent have been achieved through offsets, with the vast majority through actual emissions reductions. The other half of the voluntary market consists of a wide range of voluntary transactions not controlled by a cap and rarely conducted as a formal exchange. The majority of the OTC market is composed of carbon offsets or verified (or voluntary) emissions reductions (VERs). Pure voluntary buyers are individuals or businesses that offset their emissions for ethical reasons or for aspirations of corporate social responsibility. Pre-compliance buyers, on the other hand, are those who purchase credits in the voluntary market either to be able to use in a future compliance market (primarily polluting facilities) or those that aim to sell them at a higher price in a future regulatory market (primarily financial firms).

Pros and Cons of Carbon Trading

Those who support carbon trading suggest that it is the most politically feasible way of regulating carbon and other GHG emissions and believe that businesses respond to the economic incentive to reduce emissions. Some prefer carbon trading over a carbon tax as companies or industries may be able to avoid taxes through lobbying, negating the incentive to reduce emissions. Advocates in the environmental field also suggest that carbon trading is a way to redirect greed toward saving the planet through including carbon and other GHGs in market transactions.

Critics of carbon trading can be divided into those who oppose it based on concern that any limits on carbon emissions could weaken national and global economies and those who want to reduce GHG emissions but do not support carbon trading as the most effective or ethical way to do so. The first group generally opposes cap and trade, a carbon tax, or any type of control over carbon and other GHG emissions and is against any type of regulation that could threaten economic growth. Others, however, oppose cap and trade but support a carbon tax as a more equitable, transparent, and predictable way to reduce carbon emissions. Many economists support a carbon tax over carbon trading because it would provide a more predictable long-term price on carbon and not be subject to manipulation by special interests. Others claim that freely distributing permits to polluters is essentially giving a license to pollute, can lead to corrupt behavior, and that allocating permits based on a past baseline can lead to perverse incentives, where an entity would have no reason to cut emissions for fear that it will have fewer emissions allocations in the future. Those in support of emissions trading counter this by claiming that the regulating policies are more likely to be effective if those most affected by the policy (i.e., polluting facilities) receive some benefits, given the likely liabilities they will be required to assume, and that it would be politically impossible to pass a carbon tax high enough to substantially reduce GHG emissions.

The inclusion of carbon offsets in carbon trading is another point of disagreement between advocates and critics. While some argue that by offsetting one's emissions, a person or entity becomes more aware of its own emissions and thus may attempt to reduce them, critics argue that offsets are a false solution designed to allow people and entities to claim carbon neutrality and alleviate a guilty conscience without making any meaningful changes in their carbon emissions. Fundamental to this debate is how rigorous and effective offset programs are in actually reducing carbon emissions or sequestering carbon. The UNFCCC's Clean Development Mechanism (CDM) is an example of an offset program that has been both heavily criticized and highly lauded. The CDM was designed to help ratifying industrialized countries meet their emissions reduction targets under the Kyoto Protocol by offsetting some of their GHG emissions through funding projects such as reforestation or hydropower development that theoretically reduce emissions in developing countries. While there are many examples of successful CDM projects, many have come under scrutiny both in terms of their actual efficacy in offsetting carbon emissions and in terms of "trade-offs," where carbon is sequestered by activities that simultaneously degrade other ecosystem services such as provision of water or soil quality.

Some oppose carbon trading on ethical grounds. For example, groups such as Climate Action Now and Carbon Trade Watch suggest that carbon trading is a form of colonialism under which rich, developed nations are allowed to maintain unsustainable levels of consumption and carbon emissions by purchasing inefficient offsets. These critics also point out that the caps in most cap-and-trade programs are far from sufficient in terms of achieving real, scientifically based reductions to avoid dangerous climate change. They argue that this would require the United States and other developed nations to cut emissions by at least 25–40 percent below 1990 levels by 2020 (as opposed to, for example, the U.S. cap and trade proposed in the Waxman-Markey Bill, which aims for a 4 percent cut below 1990 levels). It has been further suggested that carbon trading puts too much emphasis on individual lifestyles and carbon footprints and not enough attention on the broader, systemic, political changes that need to take place to create a lower or zero carbon economy. It has also been suggested that while markets may find the easiest way to reduce emissions in the short term, they will not necessarily facilitate the development of long-term solutions that will lead to real reductions.

In summary, carbon trading and emissions trading in general are viewed by many as a politically and economically effective way to reduce emissions of GHGs and other pollutants. This approach, however, is not without its critics, particularly in regard to whether this approach is the most effective and ethical way to substantially reduce GHG emissions.

See Also: Carbon Market; Carbon Sequestration; Carbon Trading (Cities); Carbon Trading/ Emissions Trading (Business); Kyoto Protocol.

Further Readings

Adams, J., V. Arroyo, M. Barger, J. Boyce, et al. *Allocating Emission Allowances Under a California Cap-and-Trade Program* (March 2010). http://www.climatechange.ca.gov/eaac/ documents/eaac_reports/2010-03-22_EAAC_Allocation_Report_Final.pdf (Accessed November 2010).

Ellerman, D. A. and P. L. Joskow. *The European Union's Emission Trading System in Perspective.* Washington, DC: Pew Center on Global Climate Change, 2008.

Hamilton, K., M. Sjardin, M. Peters-Stanley, and T. Marcello. *Building Bridges: State of the Voluntary Carbon Markets 2010.* Report by Ecosystem Marketplace and Bloomberg New Energy Finance, June 2010.

Hanemann, M. "Cap-and-Trade: A Sufficient or Necessary Condition for Emission Reduction?" *Oxford Review of Economic Policy*, 26 (2010).

Reyes, O. *Carbon Trading: A Brief Introduction. Carbon Trade Watch*, 7 (September 2009). http://www.carbontradewatch.org/articles/carbon-trading-a-brief-introduction.html (Accessed September 2010).

Wittneben, B. B. F. "Exxon Is Right: Let Us Re-Examine Our Choice for a Cap-and-Trade System Over a Carbon Tax." *Energy Policy*, 37 (2009).

Leah L. Bremer
San Diego State University/
University of California, Santa Barbara

Kathleen A. Farley
San Diego State University

Carbon Trading/Emissions Trading (Business)

Emissions trading is a policy measure aimed at reducing pollution. Originally referred to as "tradeable pollution rights," it enables firms to trade the right to emit or discharge specified amounts of particular pollutants. It has been used to control sulfur dioxide that contributes to acid rain, nitrous oxides and volatile organic compounds that contribute to smog, and nutrient discharges into waterways. Carbon trading is a form of emissions trading used to reduce greenhouse gases.

Open-market emissions trading allows companies to earn emission reduction credits for voluntary reductions in a particular time period of specified pollutants discharged from their plants. These can be either reductions from the usual emission rates for a particular facility or reductions below the regulated standards that the facility is required to meet, whichever is less. Firms that reduce their rate of emissions from a particular facility can then sell the credits they earn to other firms for whom buying credits is cheaper than reducing their emissions to comply with those regulations. Trading is usually open to all firms.

Some open-market emission trading schemes allow firms to gain credits from reducing pollution from a variety of small mobile sources such as old cars, leaf blowers, and lawn mowers. Credits can be exchanged between different types of sources and industries. Sometimes, different types of pollutants are covered in one scheme so that reductions in emissions of one pollutant can be used as credits for increased emissions of a different pollutant.

With "cap-and-trade" emissions trading, a limit is set for total emissions of a specific pollutant or set of pollutants that are allowed to be emitted over a particular period— usually a year—by specific industries in a particular region. This cap is then divided into allowances that are allocated to firms, normally limited to large firms in a particular industry sector with significant emissions, for example, electricity-generating plants. A firm can sell any allowances that are surplus to its requirements to another firm that needs extra allowances, or it can save them up for the future when they might be needed. In other words, the allowances become tradeable pollution rights.

The two main ways of initially allocating allowances are usually referred to as "grandfathering" and "auctioning." Grandfathering involves allocating allowances to firms on the basis of their past emissions. Firms that polluted more in the past would be allocated a larger share of allowances. Alternatively, a pre-specified number of allowances can be auctioned off to polluters. In either case, the total allocation is supposed to be within the estimated capacity of the environment to assimilate the specified type of pollution, or at least a step toward achieving that goal.

The first cap-and-trade emissions trading program was established in March 1993, when the U.S. Environmental Protection Agency (EPA) auctioned rights to emit sulfur dioxide (SO_2), which is a primary cause of acid rain. The program set a cap that required the total amount of SO_2 discharged by power stations to be reduced by 2010 to half the levels discharged in 1980. Each allowance gave the owner the right to emit one ton of SO_2. The SO_2 allowances were auctioned every year by the Chicago Board of Trade.

According to the EPA, an emissions trading program is a suitable policy when the following is true:

- The environmental and/or public health concern occurs over a relatively large area.
- A significant number of sources are responsible for the problem.
- The cost of controls varies from source to source.
- Emissions can be consistently and accurately measured.

Carbon Trading

In December 1997, the United Nations Framework Convention on Climate Change (UNFCCC) was held in Kyoto, Japan, to discuss a treaty to reduce the emissions of greenhouse gases. The Kyoto Protocol came into force in February 2005, although the United States and Australia did not ratify it. Nations can pay to exceed their agreed targets using a range of mechanisms. These are the following guidelines:

- Emissions trading, which allows countries to buy the rights to discharge emissions above their agreed target from countries that reduce their emissions below their agreed targets

- Joint implementation (JI), which allows countries to offset their excess emissions by paying for emissions reductions or carbon sinks in other countries that have agreed to the protocol
- Clean development mechanism (CDM), which allows countries to offset their excess emissions by paying for emissions reductions or carbon sinks in countries that are not signatories to the protocol, that is, developing nations

Offsets can include the creation of carbon sinks, such as tree planting, that absorb carbon dioxide. CDM allows for afforestation, reforestation, and avoided deforestation to offset greenhouse gas emissions. Offsets could also be generated by providing renewable energy generation projects and energy-efficient technologies to developing countries or the closing down of old, dirty plants in eastern Europe.

The emissions trading system under the Kyoto Protocol is a cap-and-trade system. It began in 2008 and covers the 38 nations that are signatories to the protocol. The cap for each nation is the emissions target it agreed to in the Kyoto agreement.

There are now several active emissions trading markets, including the European Union Emissions Trading System, which began in 2005 and covers some 13,000 companies, including electricity and heat generators and producers of cement, ceramics, ferrous metal, glass, and paper. Credits are bought and sold through six exchanges and three electricity-trading companies. Individual European countries have also established trading programs. For example, Denmark has set up a cap-and-trade program to cover its electricity sector. The Netherlands has also set up a domestic greenhouse gas emissions trading system.

Pros

Emissions trading is an economic instrument. While legislation is aimed at directly changing the behavior of polluters by outlawing or limiting certain practices, economic instruments aim either to make environmentally damaging behavior cost more or to make environmentally sound behavior more profitable.

Environmental economists argue that environmental pollution and damage are externalities that should be "internalized" by adjusting prices so that the person buying the goods or services causing the external cost is obliged to pay for it. This can be done by means of a tax or charge—for example, the firm discharging the waste into the river might be charged a fee to cover the cost of lost recreational amenity and fish life, thus making external costs part of the polluter's decision. Internalization of environmental costs can also be done through emissions trading if allowances have to be bought to cover emissions discharged.

By giving carbon emissions or other pollution a price, markets can find the optimal level of environmental damage, that is, the one that is most economically efficient. The optimal level of pollution is the level at which the costs to the company of cleaning up the pollution equal the cost of environmental damage caused by that pollution. If the allowance price is equivalent to the cost of environmental damage, then the theory says that the company will reduce its emissions until any further incremental reduction in pollution would cost more than the remaining charge, that is, until it is cheaper to pay the charge than reduce the pollution. This is said to be economically efficient because if the polluter spends any more than this, the costs (to the firm) of extra pollution control will outweigh the benefits (to those suffering the adverse affects of the pollution).

Even where the allowance price does not internalize the full cost of environmental damage, emissions trading is believed to stimulate technological change and provide an incentive for polluters to reduce their emissions because they can save money and sell their excess allowances to others. The money that can be earned from selling credits encourages firms to come up with innovative ways to reduce their emission rates and earn credits.

Environmental economists also argue that because the environmental commons is not privately owned and access is open, there is a tendency to overuse it—this is referred to as the "tragedy of the commons." Legal sanctions in the form of environmental laws and regulations are the modern way of preventing tragedy-of-the-commons situations. Economists prefer to incorporate the commons into the market system through the use of economic instruments that create artificial property rights. They believe this ensures that public goods are allocated in a more "efficient" manner.

Rights-based economic instruments such as emissions trading create ownership or property rights for individual firms to discharge a certain amount of pollution into the environment and allow these rights to be bought and sold. Firms that can reduce their pollution more cheaply than others can sell their excess rights to firms for whom it would be expensive to reduce their pollution. In this way, economists argue, a given level of air quality can be achieved more efficiently with a lower aggregate cost to the firms involved.

Emissions trading is based on the idea that it is cheaper for some firms to reduce their emissions than others and therefore more cost-effective to allow the market to decide where emission reductions will be made than for governments to require uniform reductions across an industry. It is often argued by economists that markets are more efficient than centralized government decision making because they automatically gather information and ensure that supply and demand are balanced and resources allocated efficiently.

In this way, economic instruments are more economically efficient than legislative measures in that pollution reductions can be made for less cost. Polluters are not told what to do; rather, they find it expensive to continue in their old ways. Individuals or firms can then use their superior knowledge of their own activities to choose the best way to reduce emissions.

The U.S. Acid Rain Cap and Trade scheme is consistently cited as a success because it achieved emissions reductions at minimal cost to the firms involved. Before 2005, SO_2 allowances cost around $150 to $200/ton, much cheaper than paying for flue gas scrubbers to remove SO_2 from plant emissions. It was claimed that this program saved industry hundreds of millions of dollars each year compared with having to comply with legislation.

Cons

Opponents argue that it makes only little sense to set up markets to ensure that the cheapest reductions are made first if reductions can be limited to what can be done cheaply. However, if more expensive reductions have to be made, then there is little point in setting up markets that enable some firms to avoid making those expensive reductions so as to minimize overall costs.

This became evident in Germany when it considered implementing an acid rain emissions program. The aim of the German program was a 90 percent reduction in SO_2 between 1983 and 1998. In comparison, the aim of the U.S. emissions trading program for SO_2 permits was only a 50 percent reduction between 1990 and 2010. This meant that in the United States, there was much more scope for power stations to find cheaper ways to reduce their emissions, whereas in Germany, every power station had little choice but to retrofit its plant with flue gas desulfurization and selective catalytic reduction for nitrogen oxides. Consequently, there was no scope for trading in Germany.

The reductions achieved by the U.S. Acid Rain Cap and Trade scheme, although often cited as an example of the success of emissions trading, do not compare well with what was achieved elsewhere with traditional regulation. The United Kingdom Environmental Agency noted in 2003 that sulfur emissions in the United States exceeded those from the EU member states by 150 percent.

Emissions trading can cause some neighborhoods to get a lot more pollution than others because the companies in their area are buying up allowances rather than reducing their pollution. In the case of mercury trading in the United States, power plants that buy up mercury emission credits put their neighbors at risk of brain damage. Even the trade of nonhazardous gases like carbon dioxide can cause "hotspots" of pollution because such gases have toxic co-pollutants that increase with the increase in carbon dioxide emissions.

Even proponents of trading admit that there will inevitably be a conflict and an implicit trade-off between the goals of reducing costs to firms for reducing emissions and improving environmental quality. This conflict can be seen in the setting of baseline standards or caps for tradeable emissions programs. In practice, baselines and caps tend to be based on economics and politics rather than on what is technically feasible to protect the environment and human health. For example, carbon trading under the Kyoto Protocol aimed to reduce the emissions from industrialized nations by an average of 5 percent rather than the 50–70 percent that is thought to be necessary to prevent global warming.

What is good for encouraging trade in a market is not necessarily good for the environment. If the cap is too low and too few allowances are issued, or the baseline standard is too low, there will be few allowances or reduction credits for sale—because few firms will be able to reduce their pollution levels below the allowances they are allocated or the emissions standards set. Yet such a low cap may be necessary to protect the environment. If the cap is too high and too many allowances are issued, the price of allowances will be low and there will be little incentive to reduce emissions.

When the EU emissions trading scheme for greenhouse gases was introduced, many governments were overly generous in allocating permits to local firms because they feared their local industries would be at a competitive disadvantage if they had to buy extra permits. A study by Ilex Energy Consulting for World Wildlife Fund examining six EU countries found that none of them had set caps that challenged business as usual. Because allowances were not in great demand, the market for allowances opened at €8 per tonne and settled around €23 a few months later, far less than necessary to provide an incentive to reduce emissions. When the recession hit in late 2008, the price went down again to around €10. This is the one reason why the EU emissions trading scheme has failed to reduce emissions significantly or to promote investment in new low-carbon technologies.

Are Reductions Real?

A major problem with awarding carbon credits for emissions reductions is ensuring that those reductions would not have occurred without carbon trading. For example, by 2000, some countries in Eastern Europe were already emitting 30 to 45 percent less carbon dioxide than in 1990 because of lowered production, yet they could sell their rights to emissions they were not going to make, to the United States or Japan in return for hard currency, with no net benefit to the planet. The reductions that would have occurred without emissions trading are now available to affluent countries to avoid their own emissions reductions. They are referred to as "hot air" or "phantom" emissions reductions.

In New South Wales (NSW), Australia, the Greenhouse Abatement Scheme issues certificates to those who reduce greenhouse gas emissions that can then be sold to electricity retailers who have to meet mandatory emissions reductions. However, a study by researchers at the University of NSW found that 95 percent of the certificates issued in the 18 months leading up to June 2004 were for projects established before the introduction of the scheme, and more than 70 percent were awarded for emissions reductions that would have occurred anyway.

The problem with carbon offsets is that it is up to those claiming carbon credits to explain how they are reducing greenhouse gas emissions and why these reductions would not have occurred without their investment. This means the carbon offsets can be rather debatable and often would have occurred anyway. The company getting the credits is able to freely estimate emissions that they might have otherwise made without the carbon offsets. If it is a gas-fired power plant, it can argue that the alternative would have been a coal-fired power plant, and there is no onus on project proponents to prove that the coal plant would have been built or that the gas-fired plant would not have been built without the CDM credits. Nor does it matter that a wind farm would have reduced CO_2 emissions far more. Using the credits gained with "imagined" reductions, a company or country can increase its emissions back home. However, the benefit to the environment is doubtful, particularly since a gas-fired power plant may have been built in the absence of a carbon market.

In some cases, projects that are already under way belatedly claim CDM credits, even though it is obvious they would have gone ahead without them. An example is the Esti Dam in Panama, which was more than half complete when the Dutch government applied for 3.5 Mt of CDM credits for it. In these ways, CDM credits are claimed where there has not been any genuine emission reduction. This nonreduction then allows more emissions in another country than would otherwise have occurred, and in fact diverts funds from genuine reductions to subsidize business as usual.

Offset projects favor cheap methods of reducing carbon emissions rather than renewable energy projects in developing countries. The use of tree plantings for carbon offsets is particularly problematic. First, there is no accepted method for calculating how much carbon is temporarily taken up by growing trees. Such trees may release their carbon early as a result of fires, disease, or illegal logging, but the necessary long-term monitoring is often not carried out. In many situations, plantations are not sustainable because they use so much water and agrochemicals. Such plantations reduce soil fertility, increase erosion and compaction of the soil, and increase the risk of fire.

Opponents argue that markets are not more efficient at gathering information, and in artificial markets such as those created for pollution rights, the need for monitoring and enforcement remains and is, in fact, arguably greater. For emissions trading to work properly, the regulator needs to know if a firm deserves emission reduction credits and whether firms are keeping within their allowances. Too often, inspection and verification does not happen, even though emissions trading increases the incentive to cheat because claimed reductions are worth money.

Often, in emissions trading schemes, companies do not report actual measured emissions but estimated emissions based on models that are far from accurate. Monitoring many small and medium-sized firms is also difficult in an emissions trading scheme, which is why many such schemes include only large companies. Some analysts go so far as to dismiss emissions trading as inappropriate for small and medium-sized enterprises.

The difficulties of monitoring and enforcing statewide emissions trading programs is multiplied many times when it comes to monitoring emissions and claimed reductions worldwide, especially given the involvement of consultants, brokers, and others in far-flung transactions around the world. This is particularly the case in developing countries where the regulatory infrastructure and skilled personnel to measure and monitor emissions reductions may not be well developed.

Where emissions reductions are verified, it is often done by transnational corporations such as PricewaterhouseCoopers that are also consultants and accountants to the companies whose emissions they are auditing. This opens the way to conflicts of interest, and even fraud, meaning there is no real assurance that the emissions reductions being claimed are actually real.

Emissions trading tends to protect very polluting or dirty industries by allowing them to buy emission allowances or offsets rather than meet higher environmental standards. In this way, trading can reduce the pressure on companies to change production processes and introduce other measures to reduce their emissions.

Various technologies of contested environmental benefit are being promoted as eligible for credits in the JI and CDM schemes. Even aluminium producers are claiming that the use of aluminium in cars should earn credits because it reduces carbon dioxide by making cars lighter. This is despite the fact that aluminium is very energy intensive to manufacture. The CDM also acts as an effective subsidy for the nuclear industry by rewarding it with carbon credits despite the known hazards associated with operating nuclear plants and storing nuclear waste. The CDM is providing an incentive for the construction of nuclear power plants in developing nations, particularly China.

See Also: Carbon Market; Carbon Tax; Carbon Trading (Cities); Carbon Trading/Emissions Trading; Kyoto Protocol.

Further Readings

Beder, Sharon. *Environmental Principles and Policies.* Sydney: UNSW Press, 2006.

Brohé, Arnaud, et al. *Carbon Markets. An International Business Guide.* London: Earthscan, 2009.

Carbon Trade Watch. "Carbon Trading." http://www.carbontradewatch.org/issues/carbon-trading.html (Accessed August 2010).

"Carbon Trading—How It Works and Why It Fails." *Critical Currents,* 7 (November 2009).

Clifton, Sarah-Jayne. *A Dangerous Obsession: The Evidence Against Carbon Trading and for Real Solutions to Avoid a Climate Crunch.* London: Friends of the Earth, November 2009.

Lohman, Larry. "Carbon Trading: A Critical Conversation on Climate Change, Privatisation and Power." *Development Dialogue,* 48 (September 2006).

Sorrell, S. and J. Skea, eds. *Pollution for Sale: Emissions Trading and Joint Implementation.* Cheltenham, UK: Edward Elgar, 1999.

Tietenberg. T. H. *Emissions Trading: Principles and Practice,* 2nd ed. Washington, DC: RFF Press, 2006.

Sharon Beder
University of Wollongong

CERTIFIED PRODUCTS

The U.S. Department of Agriculture's National Organic Program certifies that produce and other foods comply with organic farming standards, like this organic wheat farm in Minnesota.

Source: Bruce Fritz/Agricultural Research Service/U.S. Department of Agriculture

Programs were developed to certify products as a means to identify that a product meets an acceptable standard for environmental quality or an ethical or social benefit, and to differentiate it from other similar products. The certification symbol or label is used for communicating to consumers the unique value of the product and to acknowledge that the maker is continually working to meet certain standards. To receive certification, products must voluntarily set and meet internationally or domestically recognized standards and prove continual adherence to those standards through auditing. The objective is to differentiate producers and products, and by providing a financial incentive as a price premium for products (often 10–20 percent more than standard products), social and/or environmental performance is improved. However, as the government has remained fairly "hands off," a number of certification programs have emerged that have brought both positive and negative impacts for consumers, industries, and the environment.

According to Stephen Hamilton and David Zilberman, there are thousands of certified products in more than 20 countries, and environmental or social certification accounts for 9 percent of all new products introduced in the United States. Certification programs come in a variety of forms: single attributes for single products, multiple attributes, and product scorecards, to name just a few. They have been developed by a number of organizations as well as manufacturers, and in the United States alone, approximately 400 environmental certification programs exist (as of 2010).

Product certification did not take off until the late 1990s, when educational and governmental institutions began mandating the purchasing of "green" products. Public consumers eventually gained interest as the number of certified products increased, large retailers such as Walmart and Home Depot adopted programs to build awareness, and the organic food movement raised interest in other products and certification programs.

While the market for certification programs remains relatively unregulated—certification programs can be developed and monitored by anyone—the U.S. Federal Trade Commission (FTC) and the Environmental Protection Agency (EPA) have begun developing guidelines to reduce false environmental marketing claims and to help inform consumers. Few claims have been brought against companies for false green claims ("greenwashing"), but claims are increasing. The term *greenwashing* was coined by Jay Westerveld in 1986 to refer to organizations or industries that spend more time and money on advertising claiming they are green than they do in ensuring their products are green. It is intended to provide consumers with an impression that the company is environmentally responsible, when in fact it may be doing very little that is environmentally beneficial.

The almost 400-and-counting certification programs cover a wide variety of products and issues: organic food, sustainably harvested wood, green renewable energy, bird-friendly coffee, and life-cycle analysis of supply chains. Below is a summary of some of the more common labels and their credibility:

- *Cradle-to-cradle*: This certification promotes products that do not create waste as part of either their production or their life cycle. This measures products' all-around sustainability, from manufacturing to recycling and reuse. A four-tier program is used to indicate whether the company is intending to abide by standards or whether it has achieved standards.
- *100% Organic, Certified Organic, and Made with Organic Ingredients*: Organic food and wine certification is determined by the U.S. Department of Agriculture (USDA). Products with the "100% Certified Organic" label can contain no nonorganic ingredients. "Certified Organic" products must contain 95 percent or more organic products, while "Made with Organic Ingredients" requires that the product contains at least 70 percent organic ingredients. Foods must be grown with no conventional pesticides and limited fertilizers, and animal products may not receive antibiotics or growth hormones and must be fed organically grown feed.
- *Natural*: The USDA only labels "natural" meat, implying that it does not contain artificial flavors, preservatives, or synthetic ingredients, but does allow for antibiotics and hormones.
- *Leaping Bunny*: This certification is for cosmetics and cleaners that do not contain ingredients tested on animals and is sponsored by the Humane Society. The "Cruelty Free" label, on the other hand, has no set standards.
- *Biodegradable*: Products with this label must "return to nature" (an FTC rule), but this is not enforced. However, there is a third-party certification program for biodegradable soaps and cleaners that will not hurt fish and have been proven to break down quickly.
- *Forest Stewardship Council*: Certification by the Forest Stewardship Council (FSC) was developed through a collaboration between environmentalists, loggers, and consumers and requires timber and paper companies to monitor their supply chain.
- *VeriFlora*: This label is certified by a third-party investigator and is for flowers grown with good labor practices, without heavy-duty chemicals, and on farms that are—or are going—organic.
- *Energy Star*: Certified by the EPA, these products are identified as energy-efficient without sacrificing performance, and financial savings can be significant.
- *Carbon-free or certified carbon offsets*: This certification does not imply that no emissions were generated in a product's production, but that the manufacturer bought offsets.
- *Green Seal*: This certification is by a nonprofit that provides science-based environmental certification for a range of products, including construction products, household products, and food packaging.
- *Fair Trade*: This certification ensures that fair conditions, a fair price, and environmentally sustainable practices are used for coffee, tea, and other products, including chocolate and sugar.

Benefits of Certification for Consumers, Producers, and the Environment

Consumers receive a benefit from product certification as they have a better understanding of what they are buying, as long as the label can be trusted. The certification is a means for informing the consumer about the environmental or social impacts of various products and allowing him or her to make a choice. The consumer has the option to discriminate between products that are harmful (or at least have not undergone the requirements for being certified) and those that meet certain environmental or social standards. The clear labeling of certified products through logos helps to educate the consumer and can also increase competition among manufacturers, incentivizing them to shift production practices if certified labels are favored by consumers. Certification removes the guesswork or the need for individual research by consumers, and even the mere presence of a certification logo can provide comfort and encourage purchase over a noncertified product. With labeling of certification programs with clear logos, it is easier to monitor claims, and it will become easier to bring charges against producers should a claim emerge as false.

The U.S. government's taking a fairly "hands-off" approach to certification has been beneficial, as receiving and auditing certification is likely cheaper compared to government intervention and regulatory control. Regulation is minimized as customers as well as producers are given the power to drive the market by making environmentally supportive decisions. This leads to increased pressure on companies to create more environmentally friendly products. Companies have reported increased revenue, an improved image among consumers and shareholders, benefits to corporate culture, better communication and connection with the public as well as with retailers, and stronger positioning of the company and its products in the marketplace. In addition, certification can introduce products to new market niches or retailers. Innovation can also be inspired by providing a filter through which new and innovative ideas or programs must be passed prior to implementation.

When certification increases sales, companies may redesign products so they have less of an impact on the environment and meet even stronger standards. The market begins to shift toward greater environmental awareness and reduced environmental impact. Companies continue to strive toward more stringent environmental claims, which may lead to a decline in the environmental and social impacts of products over time. Certification, in general, promotes energy efficiency, waste minimization, and product stewardship.

Negatives of Certification for Consumers, Producers, and the Environment

While there are many obvious benefits of certification, there are also a number of drawbacks for consumers, producers, and the environment. A study by TerraChoice found that 98 percent of labeled natural and environmentally friendly products on the shelves of U.S. supermarkets are likely to be making false or at least misleading claims. In addition, it found that 22 percent of products making "green" claims are using claims that have no inherent meaning. Of nearly 4,000 products studied, greenwashing was found in almost every product category. Even if the majority of these labels are not officially certified products, this can lead to confusion among consumers.

Additional confusion among consumers is likely to emerge from the high number of certification programs as well as from the lack of regulation. Certification programs can claim almost anything and in many cases can be misleading and fraudulent. Consumers have to rely on trust or detailed research on the certification's credibility. Consumers may become more discouraged and skeptical, even if claims are legitimate and third-party certification occurs, simply due to the flooding of markets with advertising campaigns.

Companies must bear additional costs for certification labeling, reporting, and compliance as well as conducting outreach to prospective consumers to promote their certified products. Smaller businesses may be at a disadvantage because they cannot pay for certification without charging extra premiums, even if their products meet the standards.

Companies can also face increased risk and less flexibility in their manufacturing or marketing and may need to undergo significant changes to production and training. This comes at a cost and may come without revenue gain. Companies may face unfair competition from overseas ecolabeling programs that may not meet the same standards as domestic programs, or from companies that misrepresent their products with uncertified "environmentally friendly" statements. Companies that are making internal changes and taking on the cost of certification may be at a disadvantage. In addition, certification programs are always evolving and becoming more thorough, which means that companies must be continually making (often costly) additional improvements to their sourcing, production, packaging, distribution, and marketing.

As long as people are purchasing products that are certified "green," there is concern that consumerism will rise or at least remain steady. Many environmentalists believe that consumerism is one of the major environmental concerns and that by applying green certification to products, there will be little economic or social change that is necessary to reduce the impact on the planet. Purchasing green products, many have argued, is a way for people to feel less guilty about being unwilling to give up a high standard of living. In addition, people become "turned off" to genuine environmental messages when almost any type of product can now be purchased in "green" form.

Conclusion

Some argue that in order to gain more credibility, certification programs need to either have increased regulation (which some claim is more effective than voluntary standards) or have the USDA or the FTC increase their involvement in the programs. There may also need to be standards set for certification methods: for instance, whether the entire life cycle is considered, techniques used to measure environmental impacts, criteria for rating, and which impacts are the most critical to assess. This action may need to be taken by the government, or by collaboration among industries, or by individual companies (such as Whole Foods, which will be assessing certified beauty products to ensure they meet claims).

See Also: Ecolabeling; Ecolabeling (Consumer Perspective); Organic Consumerism; Organic Foods; Organic Trend.

Further Readings

Clarren, Rebecca. "Is Your Eco-Label Lying?" *Mother Jones*, 34:6 (2009).
Federal Trade Commission. "Sorting Out 'Green' Advertising Claims" (April 24, 2009). http://www.ftc.gov/bcp/edu/pubs/consumer/general/gen02.shtm (Accessed August 2010).
Frankel, Carl. "Green Labeling Facing Crisis of Faith." Matter Network Green Business (January 12, 2009). http://featured.matternetwork.com/2009/1/new-era-green-labeling.cfm (Accessed August 2010).
Hamilton, Stephen and David Zilberman. "Green Markets, Eco-Certification, and Equilibrium Fraud." *Journal of Environmental Economics and Management*, 52 (2006).
Orange, Erica. "From Eco-Friendly to Eco-Intelligent." *The Futurist*, 44:5 (2010).

Stacy Johna Vynne
Independent Scholar

Citizen Juries

As opposed to a forum or public hearing, a citizen jury limits the number of participants and allows them to deliberate and produce recommendations after hearing from experts.

Source: iStockphoto

A citizen jury is a method for participatory action in public policy issues, based on the assumption that deliberative democracy has failed to take into account all relevant perspectives that exist within a community. Among others, two main arguments have been used to justify the use of this and other similar participatory methods in public policy decision making. First, a policy, plan, or project will have greater success if it includes all relevant stakeholders in the decision-making process, and second, all stakeholders have a right to be informed and to participate in the decisions that affect them. It was within this rationale that the concept later named "citizens juries" was first proposed and applied by Peter Daniel in Germany in 1969, under the name of *planungszelle* or "planning cell," and by Ned Crosby in the United States in 1974.

In these initial versions, planning cells and citizens juries were used to collect data on citizens' opinions and perceptions about complex public policy issues. In the four decades since it was first proposed, the method has been increasingly used in different parts of the world, namely, in Australia and in the United Kingdom, and in different social and cultural contexts, with slightly different names and procedures, to address different types of problems and to allow the participation of citizens in the decisions that to some extent will affect them. Among other policy areas, the citizen jury, as a participatory method, has been applied in environmental decision making and environmental management, namely, in the governance of national parks, in decision making related to wilderness conservation, and associated with the use and exploitation of natural resources. There is also evidence that it has been used in woodland restoration projects, in the allocation of land resources for competing uses, in the construction of waste incineration plants, and in debates on the advantages and disadvantages of genetically modified organisms, to mention just a few examples of potential application of this participatory method in the field of environmental policy.

A citizen jury is a participative tool based on the idea of a legal trial by a jury and can be applied to controversial issues where decisions have not yet been made or where alternative policies have been designed or are under discussion. It can be described as a small group of nonexpert citizens who meet during a short period of time to be informed and to discuss a certain problem, concluding with recommendations. The jury, a panel composed of a sample of citizens demographically representative of that community, is asked to answer a number of charges, taking into consideration the arguments presented by the relevant public authority (e.g., municipality), after listening to opinions from a series of experts (witnesses) who come, for example, from universities or nongovernmental organizations or who are stakeholders in that particular field or issue. At the end of the process, a report with recommendations is provided by the jury to the public and to the public authority that commissioned the work. These informed opinions and recommendations are considered relevant and as such are supposed and expected to be incorporated into the policy process. In short, a citizen jury allows citizens to make informed recommendations within processes of public participation in various fields of public policy decision making.

The organization of a citizen jury varies along a number of dimensions. There are differences in the method of selection of the members of the jury, which includes the identification of the universe from which the sample will be taken; the size of the sample or number of participants, ranging from 12 to 24, as is frequently referred to in the literature; as well as the sampling method, random or stratified random sampling. It has been argued that members of the jury should be selected not only on the basis of demographic characteristics of that particular territory or population but also considering the different points of view on the issue under examination. Although the most frequent practice seems to be to run just one citizen jury, in some cases the citizen jury or the planning cell can be replicated in different areas or in different sectors of the same area, improving, for that reason, the representativeness of the sample.

The structure and organization of the discussions, the agenda, and the preparatory activities, among other aspects, differ from case to case. The issue to be addressed is usually a topic of general interest and not a problem that concerns only a segment of the population. Members of the jury are selected randomly or through a stratified random sampling process among the residents in the community. The jury receives a specific and clear charge (accusation) to examine, and this charge is usually presented as a single statement that is subsequently divided into several other questions, which the jury must answer at the end. The duration of the experiment is usually enough for the jury to consider the main facets of the problem that is being examined as well as to prepare its verdict and recommendations, typically between two and four days, although the use of longer and shorter periods has also been reported in the literature, depending on the complexity of the issue and charge being discussed. In some cases, it has been reported that the jury added and answered other charges not initially given to them. The information necessary to answer the charges is presented to the jury by the witnesses, usually experts in the field or stakeholders with different or opposing perspectives. The role of witnesses is to inform the jury about their arguments on that particular issue. The deliberative process includes a facilitator. The jury's final report can include an evaluation of the process itself. A key issue highlighted by most researchers is the need to guarantee the accuracy of the information given by the witnesses to the jury, a problem that can be worked out by a neutral expert who can be present during the process.

A citizen jury has numerous advantages and some weaknesses that need to be considered. Among the advantages, as has been reported in the literature, is the fact that in this method, participants examine the problem or issue from a citizen's point of view, having access to all relevant information prior to their decisions and recommendations. In addition to this, participants deliberate as a group, replicating the usual context in which a community learns and decides. The small size of the jury facilitates the deliberative process, and participants have the opportunity to develop a deep understanding of what are usually complex policy issues. On the contrary, the issue of representativeness is perhaps the main shortcoming that has been pointed out about this method, since such a small sample cannot adequately represent all social and cultural dimensions that exist in that particular population or territory. For that reason, these findings and recommendations cannot be generalized to the entire population, a problem that the "planning cells" developed by Peter Daniels in Germany tend to answer by carrying out several juries in different locations at the same time. Besides the representativeness issue, it is usually not possible to use inferential statistics due to the small size of the sample. It has also been argued in the literature that the citizen jury does not reflect the opinion of the common citizen, since what it offers is the perception or the view of informed citizens.

Notwithstanding some apparent similarities, a citizen jury differs from other participatory research methods, namely, from opinion surveys, public hearings, focus groups, and group interviews, in the sense that it is intended to allow participants to conduct a deliberation and to produce policy recommendations, while in a focus group and in other similar methods, the objective is purely to get information and the views of participants on a given issue.

See Also: Environmental Justice (Business); Environmental Justice (Ethics and Philosophy); Environmental Justice (Politics).

Further Readings

French, D. and M. Laver. "Participation Bias, Durable Opinion Shifts and Sabotage Through Withdrawal in Citizens Juries." *Political Studies*, 57 (2009).
The Jefferson Center. *Citizens Jury Handbook* (2004). http://www.jefferson-center.org/index.asp?Type=B_BASIC&SEC=%7B50312803-2164-47E3-BC01-BC67AAFEA22E%7D (Accessed November 2010).
Price, David. "Choices Without Reasons: Citizens' Juries and Policy Evaluation." *Journal of Medical Ethics*, 26 (2000).
Wakeford, Tom. "Citizens Juries: A Radical Alternative for Social Research." *Social Research Update*, 37 (2005).

Carlos Nunes Silva
University of Lisbon

COAL, CLEAN TECHNOLOGY

The term *clean coal* has developed in tandem with technologies to reduce atmospheric emissions from burning coal. Although this term has traditionally been linked to technologies for reducing pollution from coal-fired power plants, primarily sulfur oxides (SOx) and nitrogen oxides (NOx), it has come to include the reduction of carbon dioxide (CO_2) emissions. Matched with carbon capture and storage (CCS) technologies, clean coal is one of many mitigation strategies proposed for reducing anthropogenic climate change, and like all possible solutions, there are both positive and negative views on the technology and its implementation. Some of the pros and cons include the appropriate role of coal in developing nations versus developed nations, energy independence and security, the availability of coal and its place in the energy infrastructure, and phases of coal use not addressed by clean coal technologies, starting with its extraction and ending with waste storage.

Coal was formed as the remains of prehistoric plant matter turned into rich underground deposits of hydrocarbon over the course of millions of years. It comes in many forms that vary in their chemical makeup of carbon, oxygen, hydrogen, and other trace elements with purer forms being the most efficient for energy generation. Coal is the most abundant nonrenewable energy resource in the world. Since the 1800s, coal has been used to fuel the ever-expanding fires of industry and to power businesses and residential areas. With dwindling oil and natural gas reserves, coal is the only fossil fuel exhibiting long-term staying power for future energy production (approximately 130 years' worth of coal compared to approximately 42 to 60 years for oil and natural gas, according to 2009 World Coal Institute calculations) with 847 billion metric tons still available in underground deposits around the world. Coal is also being depleted at a faster rate than oil and natural gas, which could drastically affect reported reserves-to-production ratios.

Even with such massive reserves available, the future of coal is uncertain due to its contribution to anthropogenic climate change and other environmental issues related to its extraction and processing. In the process of converting coal to electrical power, for instance, coal-fired power plants emit large quantities of greenhouse gases (GHGs) such as CO_2 into the atmosphere, contributing to an overall increase in planetary temperatures. According to the U.S. Environmental Protection Agency (EPA), the United States emitted 7,100 million metric tons of GHGs in 2006, with 3,800 million metric tons from stationary sources. Of those stationary source emissions, electricity generation contributed 83 percent. The other 17 percent was composed of agricultural processing, cement plants, ethanol plants, fertilizer, petroleum and natural gas processing, chemical and other refineries, and other random industrial and nonindustrial sources. The coal industry is attempting to use clean coal technologies to secure its future in an increasingly carbon-constrained energy system intended to mitigate anthropogenic climate change.

Clean coal describes technologies that enable a reduction in pollution emissions such as particulate matter, sulfur trioxide (SO_3), sulfur dioxide (SO_2), NOx, mercury (Hg), and CO_2 to near-zero levels. Though some of these pollutants such as SOx, NOx, and mercury are already regulated through the Clean Air Act, clean coal technologies would be used to capture additional pollutants that would otherwise be released out the smokestacks, mainly CO_2. Clean coal technologies include atmospheric and pressurized fluidized bed combustion systems, integrated gasification combined cycle (IGCC) systems, dual-fuel hybrid combined cycle systems, fuel cell combined cycle systems, magnetohydrodynamics, and retrofit applications for existing pulverized coal (PC) systems that make up the majority of coal-fired power plants in operation today. Though clean coal technologies are capable of "cleaning up" the energy generation process (i.e., coal to electricity), it should be noted that these "clean" technologies do not extend to the process of extraction and are also contingent upon public acceptability of storage of captured pollutants.

With an international push toward clean power and an ever-increasing demand for energy, the continued use of coal in our energy future is open to debate. These arguments are political, economic, and environmental in nature. Outlined below are some of the major pros and cons of the clean coal debate.

The Rise of Developing Nations and the Concept of Energy Independence

The majority of the world's coal deposits reside in the United States, China, India, Russia, South Africa, and Australia. Though many have other energy options such as hydro, geothermal, solar, wind, biofuels, and nuclear, not all of these countries or their respective regions have the resources or capabilities to develop these alternative fuels at a level that meets their energy needs. For example, of the six countries listed above, two are developing nations whose economies and industries are in a state of perpetual growth and are highly dependent on coal to meet their newly emerging energy needs. India, for instance, reported generating 72 percent of its utility-supplied electricity in 2004 from coal. Coal-utilization technologies continue to comprise a large portion of its energy infrastructure due to its availability as a domestic resource, thus affording India energy independence during development. In the development of an energy technology policy, policymakers also acknowledge environmental impacts, financial restraints, and other long-term challenges and constraints in their policy design for coal. This favors the short-term use of high-efficiency, low-pollution PC technologies while following developments of such technologies as IGCC, oxy-fuel combustion, and CCS for possible future use.

China, as another developing nation heavily dependent on coal, is going through similar energy-induced growing pains. It is experiencing rapid urbanization and demands for a higher standard of living, leading to increasing demands on its energy supply and a plethora of environmental and public health issues resulting from current energy practices such as air pollution, acidification, and climate change. Its energy strategy includes the use of coal as the main source of energy, importation of oil and gas for liquid fuels, and expansion of alternative energy technologies such as wind and biomass, though in a restricted capacity. This strategy demonstrates a preference for energy independence by focusing efforts on local resources (i.e., coal and a mixture of renewable energy sources) over imports, which also require a modernization of energy systems and more in-depth participation in the global energy market. This continued focus on coal would therefore make clean coal technologies a practical solution for addressing both environmental and economic issues, assuming that funding is made available for investing in such technology.

Though developing nations such as India and China could potentially benefit from the use of clean coal technologies because of their rate of growth and energy needs, they are not the only countries heavily dependent on coal. The United States, Russia, and Germany rank in the top five along with China and India for highest coal consumption by nation. Europe, having once contained large coal deposits that have now been mined for more than a century, is experiencing a rapid decrease in the quality of mined coal (i.e., hard coal), which is resulting in the extraction of poorer quality coals (i.e., brown coals) and increasing production costs. Russia, Kazakhstan, and Ukraine, on the other hand, hold 20 percent of the world's hard coal reserves, but much is not economically feasible to retrieve. The United States still has large reserves of both hard and brown coal but is moving from a net exporter of coal to a net importer, thus negating some of the claims of coal being a plentiful and domestically available resource. These trends call into question the long-term viability of clean coal technologies as a means of achieving energy independence in the developed world.

The Availability of Coal and Its Place in the Energy Infrastructure

As mentioned above, projections for coal reserves indicate it will remain available more than 70 years beyond other fossil fuels, but these projections also have yet to adjust for the growing rate of consumption. China's increasing coal consumption over the course of five years alone (2000 to 2005) caused the world's reserves-to-production ratio to drop by a third (277 years to 155 years). As energy demands continue to increase, further pressure will be put on key exporters such as Australia, South Africa, Indonesia, China, and the United States. Unfortunately, energy demands in China and the United States are already starting to exceed production, leaving the market in search of other exporters for coal procurement, many of which suffer from their own set of logistical issues.

Even with constant recalculation of reserves-to-production ratios and ever-decreasing predictions, a clear advantage of coal-based energy is that coal is already part of the energy infrastructure. Unlike wind, solar, and other renewable energies, transmission systems already exist for coal, not to mention that it is currently a more stable source of energy (i.e., without the fluctuations in output exhibited by renewables such as wind). Clean coal technologies, assuming the passage of GHG-reduction policies, would allow both developed and developing nations to drastically reduce harmful emissions, while allowing continued utilization of coal until other energy sources are successfully incorporated into the energy grid. An unintended outcome to this continued reliance on coal, however, may weaken the political resolve and delay economic investment needed to encourage sufficient innovation to support the transition from fossil fuels to renewable energy.

Further complications with the implementation of clean coal technology are its energy requirements for emissions capture (10–40 percent), which decreases net output, and its dependence on the successful development and implementation of carbon storage technologies. Currently, CCS is still in the research and development stage. Predictions indicate that commercial-scale deployment of the technology will not occur for another 10 years, making it too late for dates set by the Kyoto Protocol for major emissions reductions by participating parties. Other considerations are the construction of new power plants with clean coal technology, costs of retrofitting older coal-fired power plants, cost of transportation and storage of pollutants, increased water consumption, determination of liability, and public acceptance of these projects. Decision makers in the United States, Denmark, and the Netherlands have already met with some resistance to CCS technologies because of the associated risks and uncertainties, whereas Norway and Germany seem to be experiencing broader acceptance.

The Dirty Side of Clean Coal

When determining the future use of coal as a source of energy, the entire industrial process—of which clean coal technologies are just a part—should be taken into consideration. The majority of coal in use today is for the generation of electricity. Focusing specifically on power generation, the process includes the procurement of mineral rights, extraction either by surface or underground (deep) mining, and transportation to the power plant via rail, barge, truck, or imported by ship. After processing, electricity is transmitted through power lines, marketable by-products such as ash/slag and sulfur are trucked away, and waste/pollutants such as CO_2 are transported predominantly by pipeline but also by ship, rail, and truck to storage facilities for proper disposal. In the case of CO_2, storage entails injection into underground formations such as oil and natural gas reservoirs, deep unmineable coal seams, deep saline formations, and oil- and gas-rich organic shale and basalt formations, though it should be noted that these formations are also used for CO_2 storage from other sources besides coal (e.g., ethanol plants and cement and steel refineries).

The "clean" aspects of clean coal technologies focus on power generation with linkages to storage. These technologies have little to no impact, however, on extraction, which carries its own suite of health and environmental issues. Health issues associated with coal extraction include but are not limited to asthma, cancers, and ingestion of heavy metals.

Environmental issues, often linked with health issues, include impacts to air and water quality, health and biodiversity of terrestrial and aquatic systems, and the climate. Mining practices such as mountaintop removal for coal have come under heavy fire in the United States because of the dumping of mining debris into streams and watersheds, contaminating drinking water and disrupting ecological systems. In 2009, the EPA stated intentions to investigate permits for this practice because of violations of the Clean Water Act. This is just one step in the EPA's attempt to strengthen previously relaxed standards for the extraction of fossil fuels. According to power companies such as Duke Energy, restrictions placed on practices such as mountaintop removal, strengthening of environmental regulations, and resistance to CCS technologies present significant limitations to continued use of coal for future power generation.

It is therefore important to clarify that the term *clean coal* refers only to emission-reducing technologies for the generation of electrical power from coal. Within this limited context, proponents cite its value for developing economies, such as China and India, which use and will continue to use coal as their primary energy source, and late-industrial economies, such as the United States, to mitigate anthropogenic climate change during the shift to a system based in renewable energy sources. Opponents of the technology argue that continued use of coal, however clean the power generation process, merely delays the necessary change in the systems used to produce energy and encourages continued use of fossil fuels.

See Also: Alternative Energy; Carbon Capture Technology; Carbon Sequestration.

Further Readings

Chikkatur, Ananth P. and Ambuj D. Sagar. "Towards a Better Technology Policy for Coal-based Power in India." *Advances in Energy Research* (2006).

Greenpeace. *False Hope: Why Carbon Capture and Storage Won't Save the Climate.* Amsterdam: Greenpeace International, 2008.

Kavalov, B. and S. D. Peteves. *The Future of Coal.* Petten, Netherlands: Directorate-General Joint Research Centre, Institute for Energy, European Commission, 2007.

McMullan, J. T., et al. "Clean Coal Technologies." *Proceedings Institute of Mechanical Engineers*, 211:part A (1997).

U.S. Department of Energy. *Carbon Sequestration Atlas of the United States and Canada, Second Edition.* Washington, DC: U.S. Department of Energy, Office of Fossil Energy, National Energy Technology Laboratory, 2008.

Weidou, Ni and Thomas B. Johansson. "Energy for Sustainable Development in China." *Energy Policy*, 32 (2004).

Andrea M. Feldpausch-Parker
Texas A&M University

COMMODIFICATION

Traditional governmental responses to environmental degradation and protection have involved a legislative rather than economic approach. In the late 20th century, however, economists began to promote the commodification of natural processes and resources that

Commodification assigns economic values to resources that traditionally had no market cost, such as oxygen generated by plants, the natural filtration of water, and bee pollination (shown here).

Source: Elizabeth A. Sellers/U.S. Geological Survey Library of Images From the Environment

have traditionally been viewed as free goods, defined as those goods that do not have an associated direct market cost. Proponents argue that the commodification of nature will aid in the development of a sustainable society, the overall goal of green politics. Critics argue that commodification is not feasible and does not consider nature's intrinsic value and will not result in better government decision making with regard to environmental protection.

Economists, environmentalists, and politicians debate the usefulness and practicality of the commodification of nature, defined as assigning a natural resource or process an economic cost. Examples of free goods include the oxygen generated by trees and plants, the natural filtration of water, natural barriers to flooding, honeybee pollination, and open natural spaces where waste is dumped. There has traditionally been no method for calculating the costs in terms of economic value when human activities such as construction or industry alter or damage ecosystems or their components. Thus, when a 1997 *Nature* article provided an early attempt at such an estimate, placing the total economic value of the world's global ecosystems at $33 trillion or greater, controversy arose.

Since the rise of the environmental movement in the 1960s and 1970s, businesses have faced increased public criticism and demands for accountability for their role in environmental degradation. These calls were renewed in the 1980s, when concerns about potential climate change such as global warming arose. Businesses have utilized a variety of strategies to counter such criticism and loss of consumer confidence, including the commodification of green culture through the sale of products labeled green or environmentally friendly and the advertisement of sustainable business practices such as the use of recycled materials or the elimination of product testing on animals.

Businesses have also cooperated with government in the development of sustainable policies. The development of a sustainable development movement in the late 20th century was based on the use of technological change and the market system to incorporate sustainable costs, goals, and incentives into the economy and showing that economic profit and sustainability are not mutually exclusive. Thus, cooperation between governments, businesses, and environmental groups began to emerge. Many environmental economists supported sustainable development as a way to incorporate the external costs of free goods into the internal marketplace.

Some opponents argue that the economic value of nature cannot be accurately assessed while others argue that it should not even be attempted. The former note the inherent difficulty in determining, for example, the value of a particular species, clean air or water, or other non–privately owned natural resources. Even more importantly, all natural resources and ecosystems are ultimately essential to the survival of the human species, which is invaluable. Economists refer to these types of goods as intangibles precisely because of the impossibility of assigning them a monetary value. They also note that assigning individual values to specific natural resources or processes fails to account for the fact that the collective value of a species or ecosystem can be more important than the component values of its parts, such as individual animals.

Those people who argue that economic values should not be assigned to nature feel that reducing nature to its economic value, even if feasible, denies its intrinsic or other noneconomic values and proclaims the supremacy of economic value over other types of value. Thus, the main goal of sustainability would be the environment's use value to humans, an anthropocentric view that fails to adequately consider its intrinsic value or a moral human duty to protect nature in its own right.

Economists and green politicians who support assigning economic values to nature (commodifying nature) state that the benefits will include holding businesses and others accountable for their use of formerly free goods and added incentives for environmental protection that will operate more efficiently than reliance on government regulations. This idea extends commodification to include negative effects on the environment, such as the use of toxic fertilizers or the dumping of waste into rivers or oceans. They also argue that commodification will allow for the extra costs involved, such as contributions to pollution or environmental degradation, to be factored into a natural resource's market price. These are known as external costs as businesses have not historically had to include them in the cost of doing business.

In the late 20th century, some environmental economists began to propose alternatives to government regulation as methods of holding businesses accountable for external costs. Some environmental economists such as John Shanahan of the Heritage Foundation have argued that privatization and free markets with little government regulation of business will best provide for the efficient and equitable distribution of environmental goods, thus promoting conservation. Restrictive government environmental regulations, however, will result in economic downturns and increased unemployment.

Proponents note that traditional government environmental legislation has its own drawbacks and limits, such as the corporate ability to lobby politicians for favorable legislation or use stalling tactics to delay policy implementation. Free-market environmentalism states that political control of resource conservation will not result in efficient optimization of environmental costs and benefits, a process proponents feel is best left to market forces. For example, political solutions to problems such as pollution will result in the production of excess clean air or water because there are no associated costs to consumers of those resources, just as corporations that do not have to pay an extra cost for creating pollutants results in too much air or water pollution.

Pricing natural resources will allow for a true accounting of their cost to businesses that use them or contribute to their degradation, lessening the likelihood that businesses will simply abuse natural resources because they are free goods whose cost does not have to be accounted for in the bottom line. For example, if a resource becomes scarce, its market price will rise, and businesses will search for a less expensive replacement material. There is a financial incentive to the conservation of non–privately owned resources. Some economists are also seeking ways to assign property rights to natural resources such as air or water.

Commodification of negative environmental consequences such as pollution would serve a similar function. Businesses would then have an economic incentive to develop green technologies and products on their own rather than rely on government regulations to force them into compliance. If they fail to do so, they will have to bear the higher costs involved. Rather than changing the market system through government intervention, such an approach would work within the system and harness it for environmental protection and sustainability. Another idea in this area is that of tradable pollution rights in the same way that businesses can currently purchase renewable energy certificates as compensation for their use of nonrenewable energy.

Environmental economists who support the commodification of nature disagree as to how that commodification should be calculated, adopting a variety of different methods. Environmental scholars Marino Gatto and Giulio De Leo enumerated a number of these methods. The conventional market approach relies on the specific assignment of market prices to those natural processes or resources affected by industrial or government activity. For example, a business would be required to pay a replacement cost equal to the determined amount assigned to the loss of clean air caused by their pollutants. The household production function approach relies on basing environmental costs on related expenditures, such as the cost of traveling to a national park in terms of gas, tolls, and related expenses.

The hedonic pricing method determines the price of natural processes or resources based on the change in market value associated with environmental factors. For example, using the amount that pollution or proximity to toxic waste would lower a house's market property values as the cost of that pollution or dumping. Finally, the contingent valuation method relies on conducting surveys of affected populations based on hypothetical choices to determine the price of a natural commodity. For example, surveying residents near a factory to determine how much they would be willing to pay to prevent that factory from polluting the water or, conversely, how much they would accept as compensation for allowing such pollution.

Critics have pointed out the problems associated with these methods, such as the fact that some species or resources have less aesthetic appeal and may not be as highly valued. The public may also have little understanding of a species' ecosystem functions and subsequently may undervalue them. The data required to undertake the various calculations involved in determining value are expensive, difficult, and time consuming to gather. Finally, people who indirectly benefit from a particular resource or ecosystem may not be considered in the valuation process.

Proponents argue that governments will save money through the commodification of natural resources. For example, New York City spent more than $1 billion to build sewage treatment plants and otherwise protect and restore the natural watershed through the Catskill Mountains because environmental changes such as deforestation were threatening the quality of the city's drinking water supply. The cost of restoring the natural filtration was much lower than the potential cost of building and maintaining the water filtration plant they would have been required to build under U.S. Environmental Protection Agency (EPA) regulations. Other economic incentives to governments include the development of ecotourism, as natural areas are popular among travelers.

Governments could also use the economic values assigned to nature when conducting cost-benefit analyses and environmental impact assessments of proposed projects with environmental consequences such as dam construction or flood control measures. Other recommendations for the use of economic values include government incentives to private owners of such natural commodities (capital) for their preservation or sustainability efforts that also benefit governments and the public and a better determination of punishments or costs for violators. Many green politicians also propose increased consumption and sales taxes on items that are resource intensive, meaning they require large amounts of natural resources in their manufacture.

Opponents have argued that economic methods of sustainable development rely too much on businesses to incorporate environmentally friendly practices by offering market incentives for such practices but failing to actually require them. An economic approach to environmental degradation may result in companies' simply paying for the ability to continue polluting the environment, whereas government regulations and legislation force companies to stop polluting the environment. Some opponents also fear that smaller companies or low-income populations may bear a disproportionate burden of the new costs associated with the commodification of nature and that economic difficulties could result.

See Also: Anthropocentrism Versus Biocentrism; Common Property Theory; Corporate Social Responsibility; Intrinsic Value Versus Use Value; Sustainable Development (Politics); Tragedy of the Commons.

Further Readings

Brannstrom, Christian. *Territories, Commodities and Knowledges: Latin American Environmental History in the Nineteenth and Twentieth Centuries*. London: Institute for the Study of the Americas, 2004.

Easton, Thomas A., ed. *Taking Sides: Clashing Views on Environmental Issues*, 12th ed. Contemporary Learning series. New York: McGraw-Hill, 2007.

Gatto, Marino and Giulio A. De Leo. "Pricing Biodiversity and Ecosystem Services: The Never-Ending Story." *BioScience* (April 2000).

Harvey, David. *Spaces of Capital: Towards a Critical Geography*. London: Routledge, 2001.

Hoffman, Andrew J. and Marc J. Ventresca. *Organizations, Policy and the Natural Environment: Institutional and Strategic Perspectives*. Palo Alto, CA: Stanford University Press, 2002.

Lefebvre, Henri. *The Urban Revolution*. Minneapolis: University of Minnesota Press, 2003.

Milbourne, Paul, Lawrence Kitchen, and Kieron Stanley. "Commodification: Re-Resourcing Rural Areas." In *Handbook of Rural Studies*, Paul J. Cloke and Terry Marsden, eds. Thousand Oaks, CA: Sage, 2006.

Morrison, Jim. "How Much Is Clean Water Worth?" *National Wildlife* (February/March 2005).

O'Neill, John. *Ecology, Policy, and Politics: Human Well-Being and the Natural World*. London: Routledge, 1993.

Ratneshwar, S. and David Glen Mick. *Inside Consumption: Consumer Motives, Goals, and Desires*. London: Routledge, 2005.

Schmidt, Charles W. "The Market for Pollution." *Environmental Health Perspectives* (August 2001).

Stevis, Dimitris and Valerie J. Assetto. *The International Political Economy of the Environment: Critical Perspectives*. Boulder, CO: Lynne Rienner, 2001.

Strasser, Susan and Charles McGovern. *Getting and Spending: European and American Consumer Societies in the Twentieth Century*. New York: Cambridge University Press, 1998.

Tokar, Brian. "Trading Away the Earth: Pollution Credits and the Perils of 'Free Market' Environmentalism." *Dollars & Sense* (March/April 1996).

Urry, John. *Consuming Places*. London: Routledge, 1995.

White, R. D. *Crimes Against Nature: Environmental Criminology and Ecological Justice*. Portland, OR: Willan, 2008.

Marcella Bush Trevino
Barry University

Common Property Theory

Common pool resources refer to resources like wildlife, forests, oceans, and watersheds that do not belong to individuals or states, but to large numbers of users, often belonging to contiguous communities. These resources are characterized by, first, subtractability—that is, each user reduces the amount of finite goods available to the community as a whole—and, second, excludability—that is, it is difficult and costly to enforce exclusion of outsiders through physical or institutional means. The common pool resources can be managed publicly, privately, or communally. The typology of property, based on the ownership, consists of four different kinds: private, commons, state property, and open access.

In 1968, Garrett Hardin, a biologist, published a seminal paper, "The Tragedy of the Commons." The paper was to prove extremely influential in spurring research into the role of property rights in driving environmental and resource degradation. According to the argument presented by Hardin, common pool resources are prone to overexploitation and eventual collapse unless they are privatized or enclosed to limit the number of users and hence the damage. The assumption underlying Hardin's claim is that, due to the benefits of a good exceeding the cost of its extraction, users will exploit the resource until its eventual extinction. Employing a rational-actor paradigm, Hardin claimed that the individual user will find it in his self-interest to overexploit a common resource, since the resulting benefit only accrues to him while the costs are shared by the group as a whole. Elinor Ostrom, a political scientist, challenged Hardin's thesis by empirically demonstrating the existence of self-governing collective institutions for different common pool resources in diverse social and political contexts around the world. Ostrom's work and that of others has proved that there are numerous communities that have historically managed their resources collectively and sustainably without intervention from the state or resorting to privatization.

The common property theory provides a description of the conditions under which a common pool resource comes to be managed through successful self-organization of the group. The commons differ from properties owned by the states and individuals in the sense that a group of users has customary rights to the goods associated with the property. They include resources over which the group or community as a whole exercises control and the norms and rules of exploitation, which are endogenous in origin. The common property theory proposes that the sustainable management of a common pool resource will hinge on the rules governing its preservation, maintenance, and exploitation. In other words, the scenario of overexploitation and collapse predicted by Hardin for nonprivate, non-state-owned resources is by no means inevitable.

Variables Affecting Successful Self-Governance
of Common Pool Resources

Sustainable management of common pool resources depends on the characteristics of the resource in question, group attributes, institutional arrangements, and the external environment. Resource characteristics that increase the probability of successful management include small size, clear boundaries, steady stream of benefits from the resource, and easy monitoring and being fixed. Socioeconomic features of the group favorable to its sustainable management include small group size, high level of mutual trust, homogeneity of interests, low levels of poverty, history of successful leadership, and high degree of interdependence. Institutional arrangements that rely on locally devised and enforced rules are more successful. The rationale underlying the superiority of local management, especially in terms of community rights, is based on two interrelated assumptions: one, the local people have comparatively greater knowledge and understanding of the local resources, other users, and the environmental conditions affecting the resources; and two, the in-depth knowledge of and dependence on resources will create incentives for their sustainable management. External environment can facilitate sustainable management if it grants autonomy to the institutional arrangements and the presence of markets and roads do not create pressures for destruction of resources. Another beneficial aspect is access to low-cost judicial dispute resolution mechanisms.

Threats to Commons

The self-governing community institutions that once were globally dominant are either increasingly being appropriated by the state or privatized. One of the biggest threats to community-based property rights arises from the attempts of the nation-states to claim ownership over indigenous lands and commons areas. Often, the indigenous areas have significant natural resources, and the legal classification helps the state stake claim to them, while treating the inhabitants of these lands as squatters. Generally, the perception that collective or communal management of natural resources is inefficient and likely to lead to overexploitation and degradation is prevalent in policy circles. In an echo of Hardin's argument, an individualized system of property rights is seen as most efficient and even natural, and certainly most appropriate for promoting economic development.

The penetration of the markets and the decline of traditional cultural norms, especially those that contributed to high levels of trust and interdependence, have also taken a toll on the abilities of communities to manage their resources. The ever-deepening state control has also displaced the traditional leaders and their ability to forge consensus and enforce rules. Therefore, the decline in the traditional common property arrangements is due more to the loss of autonomy and interference from powerful socioeconomic and political forces, rather than intrinsic shortcomings in the system itself.

Mixed Property Regimes

The research into property rights shows that multiple property regimes can coexist. Often, these property regimes allow for different groups of users to stake their claim to the goods being produced. Legal pluralism, or existence of contending legal claims, can lead to fuzzy property rights. Many societies have traditions of permitting public access to private lands (e.g., forests) for recreation and other specific uses. Conservation easements, for instance, can create a situation of multiple owners of the same property, in which each has circumscribed rights. There is thus no requirement that common pool resources can only be managed by common property regimes.

Revival of Common Property Regimes

The past two decades have seen some recognition of the potential for the community management of common property resources. The impetus for the change has come from a realization that neither the highly centralized, top-down management associated with the state-owned resources nor the markets-based approach associated with private property had been successful in stemming the widespread degradation. Moreover, conservation programs have run into significant opposition from communities resenting their loss of control over resources as well as the erosion of their livelihoods. Another factor in favor of the move toward communal management has been the cost of centralized management that a number of states, due to their fiscal problems, find difficult to sustain. Internal donor agencies and multilateral institutions have also pushed the states toward recognizing community property rights.

See Also: Commodification; Ethical Sustainability and Development; Intrinsic Value Versus Use Value; Tragedy of the Commons.

Further Readings

Acheson, James and Julianna Acheson. "Maine Land: Private Property and Hunting Commons." *International Journal of the Commons*, 4:1 (2010).
Agrawal, Arun. "Forests, Governance, and Sustainability: Common Property Theory and Its Contributions." *International Journal of the Commons*, 1:1 (2007).
Hardin, Garrett. "The Tragedy of the Commons." *Science*, 162 (1968). http://dieoff.org/page95.htm (Accessed January 2011).

Neeraj Vedwan
Montclair State University

CORPORATE SOCIAL RESPONSIBILITY

Corporate social responsibility is the self-regulation of a company's social impacts as built into its business model. It moves beyond the traditional business model of being accountable only to shareholders into the realm of accountability to all stakeholders. Corporate social responsibility requires that businesses integrate the public interest, including environmental interests, into their decision making.

Ethics are a fundamental building block of corporate social responsibility. When employees act in an unethical manner, negative social impacts can result. Because of the importance

Corporate social responsibility is not a definitive indicator of environmental performance, as the Deepwater Horizon oil spill indicates.

Source: U.S. Coast Guard via National Oceanic and Atmospheric Administration IncidentNews website

of employee ethics, many businesses are integrating ethics training into their social responsibility efforts. This training prepares employees to do the "right" thing, even when it is not required.

Corporate social responsibility requires some knowledge of the public's needs, interests, and desires. To better understand how to positively contribute to the local communities in which they operate, some companies practice community-based development as a part of their corporate social responsibility programs. Community-based development gathers stakeholder feedback to guide the development of social programs to address local issues.

Sustainability focuses on accounting for the economic impacts of a business as well as the environmental and social impacts. Today, investors are increasingly demanding that businesses not only act in an ethical and sustainable manner but disclose their actions to the public in voluntary reports. These reports are called by many names, including corporate social responsibility reports, sustainability reports, citizenship reports, ecological footprint reports, environmental social governance reports, social reports, and triple bottom line reports. Many large companies today are producing these reports in an effort to be unceasingly transparent and accountable to the public.

It is difficult to tell the differences between the reports, and the terms *sustainability* and *corporate social responsibility* seem to be used almost interchangeably in some circles. Corporate social responsibility implies acting in society's best interest and integrating the concept of sustainability into standard business decision making. Sustainability focuses more on managing the economic, environmental, and social impacts of a business. Sustainability efforts, including measuring and reporting on sustainability, are a part of a company's overall corporate social responsibility strategy. Existing reporting frameworks, such as those developed by the Global Reporting Initiative, assist companies with sustainability reporting and stakeholder involvement.

Beyond Shareholders

In the traditional business world, only economic impacts of business were measured. Because of the focus on profits, shareholders were viewed as the primary group affected by business decisions and actions. In the 1984 book *Strategic Management: A Stakeholder Approach*, R. Edward Freeman used the term *stakeholder* to include anyone affected by business activities. Stakeholders may include shareholders, but it is not necessary for an individual to have financial interest in a company to be affected by its actions. Stakeholders also include neighbors, consumers, employees, and society at large.

Impacts to stakeholders may be environmental. Neighbors near an industrial facility may suffer health effects from increased air or water pollution. Any impacts to the environment can be considered impacts to society at large, as the whole society benefits from a healthy environment. Corporate social responsibility encourages companies to examine more than just environmental impacts beyond stakeholders, but to examine the social and economic impacts of society at large as well.

Critics of corporate social responsibility argue that a company's obligation and purpose is to make a profit. Proponents of the division of labor that occurs in most capitalistic societies argue that businesses and other organizations contribute the most to society when they focus on what they do best. An automobile manufacturer, for example, can create the most profits when it focuses on just that, rather than worrying about social and environmental impacts beyond those regulated by government. Nonprofit organizations meant to benefit society or the environment, such as religious organizations, Goodwill, and the Sierra Club, can achieve their best results when focusing on their activities, rather than trying to make a profit for owners or shareholders. It is argued that those activities meant to improve the environment for the good of the whole society are best left to environmental organizations.

It has even been argued that corporate social responsibility is a threat to capitalism and free trade. Some critics view the inclusion of environmental and social concerns in company goals and metrics as a threat to an economically competitive free-market society.

Ethics

Many companies begin their internal social responsibility programs with internal ethics training. The intent of this training is to guide employee choices, helping them to make the ethical choice when a situation may be morally ambiguous. As social responsibility begins with making the right choices, on a daily basis, ethics training can help prevent adverse or undesired social impacts by encouraging employees to do the "right" thing, going beyond the minimum of what is required by law. Many companies offer anonymous ethics "hotlines" where employees can internally report nonethical or illegal behavior within the company.

This ethical approach to environmental performance is particularly important for companies doing work in the developing world, where environmental laws may be lax or not enforced. In these situations, employees must often make decisions to work as if they were governed by the commonly accepted and regulated practices of developed nations. In short, they do the "right" thing for the environment, even though it may not specifically be required in that country.

For example, laws governing proper disposal of waste streams are not the same in every country. In most of the developed world, there are laws governing the segregation of electronic wastes (such as computers and monitors) and universal wastes (such as fluorescent lamps and batteries). These laws are meant to appropriately segregate wastes that may be high in metals and other contaminants from the general refuse stream for recycling or proper disposal.

An ethical company operating in a country that does not specifically require the segregation of these waste streams would do so because it is the "right" thing to do for the environment and public health.

Critics of ethics programs argue that they are nothing more than smoke and mirrors, allowing companies to make external claims that their employees receive ethics training, without effectively changing employee actions or decisions. Training itself may not change an individual's actions or decisions if the overall company's cultures, goals, and metrics for success remain the same. Individuals may fear internal repercussions for imposing voluntary measures on a company that are not specifically required and affect their financial bottom line. Similarly, even though most internal hotline programs guarantee that reporters of nonethical behavior remain anonymous, many employees do not report unethical behavior, for fear that the report will be tracked back to them. Some environmentalists are skeptical of ethics training and ethics programs on the basis that they can make a business appear to be acting in an ethical and responsible manner but may not actually affect the overall environmental or social performance of a company.

Community-Based Development

Increasingly, companies are forming focus groups or panels of local representatives to guide their efforts to manage their impacts on local communities. Businesses seek input from community members regarding ways in which the community could be improved and then voluntarily provide services to improve education and health for local communities. For example, the gold-mining company Barrick has focused its community-based development efforts around improving access to health services and preventing the spread of disease. But these efforts are adapted for regional concerns. For example, in Tanzania, Barrick has health facilities and programs to combat human immunodeficiency virus and acquired immune deficiency syndrome (HIV/AIDS) and malaria. HIV/AIDS is an issue that affects Africa more than any other continent today. Other continents do not have the same urgency with regard to HIV/AIDS work. In Peru, Barrick is focusing on child malnutrition and intergenerational poverty. Community-based development requires companies to adapt their social programs to address significant local issues.

Not only is community-based development the ethical thing to do, it often makes good business sense. A concept that is gaining acceptance in the business world is the shared value model, which asserts that the success of businesses and social welfare are interdependent. In this model, educated and healthy workers make better, more reliable workers in the long run, creating shared value. Obviously, this model has its limits, as some efforts to improve employee welfare may not realize returns to the business itself but be performed purely out of a sense of corporate social responsibility.

Critics argue that community-based development programs look good, but are no actual predictors of the environmental performance of a company. One example is BP's program for social investment, which seeks to "find the join" between a community's interests and the company's, by joining with local partners to build business skills, share technical experience with local governments, and support education and other community needs. BP also publishes a sustainability "review." But none of these efforts prevented the 2010 oil well leak into the Gulf of Mexico.

The Triple Bottom Line

Corporate social responsibility requires the intentional inclusion of the environment's and society's interest in decision making. To do this, businesses need a new bottom line, expanded to consider other factors beyond economics when making decisions. In his 1998 book *Cannibals With Forks: The Triple Bottom Line of 21st Century Business,* John Elkington coined the phrase *triple bottom line* (TBL) as a new means of accounting, expanding the traditional reporting framework to take into account environmental and social performance in addition to the traditional business metric of financial performance. TBL represents the three elements (or pillars) of sustainability as environment, economy, and society, which must all be balanced for truly sustainable development to occur. A product or design that balances social and environmental impacts, but doesn't account for economic impacts, may not be profitable in the long run. Products that balance environmental and economic impacts may fail due to lack of social acceptance or undesired social impacts. And products that balance economic and social impacts, but do not account for the environmental impacts, may cause unacceptable environmental damage. Sustainable solutions and products exist only at the intersection of economic, social, and environmental concerns. When all three of these impacts are accounted for and the triple bottom line is considered, the most sustainable solution can be selected. Commitment to this triple bottom line is considered a step beyond environmental awareness of corporations into the realm of corporate social responsibility.

In the context of sustainable business, the environment is considered natural capital. This capital is uniquely renewable, and as such, it must be adequately protected for the use of future generations. In this case, the business is managing its environmental impacts because it is the sustainable thing to do, not because of economic drivers in the short term. Businesses with a commitment to the triple bottom line make a commitment to conduct their business in the most environmentally sustainable manner as a standard part of doing business.

There are many critiques regarding TBL as a business concept. Environmental concerns are often viewed as a luxury of the rich, or at least the economically comfortable. In countries where many are starving, dwindling forests may be cut down to make room for agriculture. The need for food may take immediate importance over long-term environmental needs, such as biodiversity and carbon. Similarly, businesses that are profitable can better afford to examine their triple bottom line, while struggling and developing businesses may be more likely to examine only the economic bottom line.

Performance Metrics

Corporate social responsibility includes taking the public interest into account when making decisions and integrating a basic concept of sustainability, the triple bottom line, into standard business practices. Corporate social responsibility includes acting in a sustainable manner. For a business to be able to manage and balance its TBL, there must be some metric by which the impacts on each of the three pillars can be measured. Economic impacts can be quantified using currency, such as U.S. dollars. But social and environmental impacts do not have a similar common unit by which to measure all impacts. A business may affect both the environment and society in a myriad of ways.

Indicators are measurable effects that can be tracked over time. Common indicators have been developed for each of the three pillar areas, which are usually selected and tailored to serve as meaningful metrics for a unique business. There are many standard environmental indicators, covering impacts on all environmental media, and selection of the appropriate metric to capture the unique environmental impacts of a specific business is important. For example, let us compare meaningful metrics for an urban dry cleaner and a rural trout farm. For the urban dry cleaning operation, the annual emissions of hazardous air pollutants is one of the most meaningful environmental metrics to monitor. But for a rural trout farm, there are likely no emissions of hazardous air pollutants, and one of the most meaningful metrics may be the annual freshwater withdrawn from surface water.

This example demonstrates the importance of selecting meaningful environmental indicators for a specific business.

Critics of environmental metrics, and their reporting, argue that the selection of which metrics to monitor and report can allow opportunities for greenwashing. For example, if the trout farm were to report on its hazardous air emissions and compare itself to other industries, it would appear to have comparatively no environmental impacts.

Global Reporting Initiative

The Global Reporting Initiative (GRI) is an international organization with the goal of standardizing sustainability indicators and reporting internationally. The GRI has developed a sustainability reporting framework, including general sustainability reporting guidelines and special guidance for different industry sectors, called sector supplements. Both the sector supplements and the general sustainability reporting guidelines have been created through consensus-seeking processes involving representatives from business, academia, governments, labor, professional societies, and others worldwide. However, all of these efforts are voluntary, and the GRI does not verify that companies that use their protocols for sustainability reporting actually create goals and objectives based on that information.

Critics of voluntary sustainability reporting argue that because the content of each sustainability report is determined by the reporting organization, situations where negative social and environmental impacts occur are often not highlighted within these reports. In essence, a reporting organization is only as transparent as it chooses to be.

See Also: Capitalism; Ecocapitalism; Ethical Sustainability and Development; Precautionary Principle (Ethics and Philosophy); Socially Responsible Investing; Sustainable Development (Politics).

Further Readings

AccountAbility. http://accountability.org (Accessed August 2010).
British Petroleum (BP). "Gulf of Mexico Response." http://www.bp.com/bodycopyarticle .do?categoryId=1&contentId=7052055 (Accessed August 2010).
Elkington, J. *Cannibals With Forks: The Triple Bottom Line of 21st Century Business.* Gabriola Island, British Columbia, Canada: New Society Publishers. 1998.
Elkington, J. "Towards the Sustainable Corporation: Win-Win-Win Business Strategies for Sustainable Development." *California Management Review,* 36:2 (1994).
Freeman, R. E. *Strategic Management: A Stakeholder Approach.* London: Pitman, 1984.
"Sustainability Reporting Guidelines Version 3.0." Amsterdam, Netherlands: Global Reporting Initiative, 2006.
United Nations. "Universal Declaration of Human Rights." December 10, 1948.
Wood, D. "Corporate Social Performance Revisited." *Academy of Management Review,* 14:4 (1991).
World Commission on Environment and Development. *Our Common Future.* Oxford, UK: Oxford University Press, 1987.

Michelle E. Jarvie
Independent Scholar

E

Earth Liberation Front (ELF)/ Animal Liberation Front (ALF)

The Animal Liberation Front (ALF) and Earth Liberation Front (ELF) take direct action, often in the form of illegal activity, to prevent or expose harm to animals and to the environment. Both organizations have grown from less radical movements in the United States and the United Kingdom and have gained increased attention since the late 1990s with progressively high-profile damage inflicted by anonymous members. These radical movements have been both beneficial and damaging for their cause: while they help to raise awareness of the treatment of animals or environmental destruction, they do so using illegal tactics that invoke fear and economic costs.

The Animal Liberation Front (ALF) is committed to ending the abuse and exploitation of animals. Its short-term objective is saving as many animals as possible from harm and disrupting the practice of animal abuse, while the ultimate goal is to end all animal suffering by forcing those using animals for testing or food production (e.g., research laboratories, restaurants, retailers) out of business. ALF is working to expand the definition of the moral community and to reduce "speciesism," defined in Peter Singer's *Animal Liberation* as the belief that nonhuman species exist to serve the needs of human species, that animals are inferior to human beings, and therefore human interests are favored over nonhuman interests due to species status. ALF grew from anti-hunting and animal liberation movements in the United Kingdom that existed in the 19th and 20th centuries. The British ALF was founded in 1976 by animal rights activist Ronnie Lee; the American branch began operations a few years later. ALF is now active in many countries, including Australia and Canada. It is a loosely organized movement: membership is claimed not through attending meetings or paying dues, but by taking direct action to destroy property or cause economic loss to companies or individuals using animals for research or financial gain. Direct action is typically in the form of criminal or illegal activity, such as vandalism, arson, or theft. Common targets include fur companies, mink farms, restaurants that engage in factory farming, and animal research laboratories. ALF has published the following objectives: to inflict economic damage on those who profit from the misery and exploitation of animals; to liberate animals from places of abuse, for example, laboratories, factory farms, or fur farms, and place them in good homes where they may live out their natural lives, free from suffering; to reveal the horror and atrocities committed against animals behind locked doors by performing nonviolent direct actions and liberations; and to take all necessary precautions against harming any animal, human and nonhuman. In addition, those who are vegetarians or vegans and wish to carry out activities that fit within ALF guidelines can identify as part of ALF.

The Earth Liberation Front (ELF) was founded by members of more mainstream environmental organizations who were frustrated with the lack of progress in the movement. The original ELF (which previously stood for Environmental Life Force) was started in 1977 by John Hanna. After a number of guerrilla actions in California and Oregon, the group disbanded a year later following Hanna's arrest. A few years later, in 1980, the radical group Earth First! formed and engaged in protests and civil disobedience events. The movement drew from Edward Abbey's novel *The Monkey Wrench Gang,* in which members practiced limited vandalism and destructions such as "tree spiking" (insertion of metal or ceramic spikes in trees in an effort to damage saws) as a tactic to thwart logging. The Earth Liberation Front emerged in 1992 in the United Kingdom (UK) with members of Earth First! who refused to abandon criminal activities. Elves or pixies (as the acts by ELF members were attributed) continued to carry out criminal activity, referred to as "eco-tage" (ecologically motivated acts of sabotage, often in the form of a covert operation) or "monkey-wrenching" (acts of sabotage and property destruction against individuals or companies that are believed to be damaging the natural environment). The acts are often not reported and are aimed at stopping actions that are harmful to the environment and promoting a more environmentally sustainable society through a strong message. In addition to tree spiking, other common actions include arson, sabotage, and vandalism to equipment or property. Like ALF, there is no leadership, centralized organization, or official membership: activities are carried out by anonymous individuals or small groups of people. ELF gained attention in 1998 with its first major attack against a Vail, Colorado, ski resort that caused $12 million in fire damage.

ELF and ALF have informally collaborated since 1993, although greater unity and partnership has been promoted since a formal recommendation for joining forces at the 1998 National Animal Rights Conference. There is believed to be joint leadership and underground training for activists from both movements. They work on gaining intelligence on animal or environmental businesses and share information at rallies, protests, and through the web. Those acting on their behalf often work to gain employment with a potential target to conduct pre-activity surveillance (e.g., video, documents), gather inside information, and plan the operation. One of the more common and destructive practices of both movements is arson, and their websites post instruction manuals on building simple incendiary devices for those carrying out direct action. Other major actions include releasing animals (such as 5,000 mink at a Michigan fur farm) and spray-painting (such as anti-meat slogans at McDonald's).

ALF and ELF have been identified as a serious terrorist threat within the United States, given their more than 600 criminal acts between 1996 and 2002 and damage in excess of $43 million. They are also classified as special interest extremists, which are those who conduct acts of politically motivated violence to force segments of society, including the general public, to change attitudes about issues considered important to their causes. The passage of the USA PATRIOT Act on October 26, 2001, provided the government with unprecedented powers of surveillance, search, and seizure. The Patriot Act has led to further classification of ELF and ALF activities as acts of terrorism, providing the government with almost unlimited power to conduct surveillance and imprisonment of those committing direct acts. Members of ALF and ELF who are sentenced to jail are supported by the Earth Liberation Prisoners Support Network (ELP) and North American ALF Supporters Group. Support is granted on a case-by-case basis and is not provided to those who provide information to law enforcement or information about codefendants. The support networks provide defense for imprisoned activists, educate the public on direct action, publish newsletters, and raise funds for those imprisoned and for educational activities. They claim they are not part of ELF or ALF but instead offer an opportunity for those who do not wish to or are unable to take part in direct action.

How ELF and ALF Support the Movement

The actions carried out by ALF and ELF have helped to advance the animal liberation and environmental movements in many ways. For example, in a number of developed countries, including Britain, Germany, Switzerland, the Netherlands, and Australia, awareness of intensive animal rearing (especially around confinement of hens, pigs, and veal calves) has been raised and laws have been passed for improved conditions.

Actions are intended to educate the public about institutionalized animal abuse and environmental destruction "masquerading" as science, and in many cases, the movements have been successful. They have sparked public debate about animal testing, attracted media attention that may evoke some sympathy from the public, brought about welfare reforms, released animals from research laboratories, stalled or stopped developments that are environmentally destructive and harmful to wildlife, and in some cases shut down operations. For instance, a film made by ALF on primate testing and teasing by scientists led to the U.S. Secretary for Health and Human Services announcing that the scientists had failed to comply with guidelines for the use of animals, and their funding was suspended.

ALF and ELF activists have aimed to cause economic hardship to those who use animals or create major environmental impacts in order to reduce and ultimately end their practices. In this sense, they have been successful, especially among the research community. Laboratories have increased security such as fences and guards, reducing the amount of funding available for animal testing. Ultimately, researchers may need to turn to computer models or other means of testing, which would be a great success for the animal liberation movement.

ELF and ALF may also help to make less radical organizations such as People for the Ethical Treatment of Animals (PETA), the United States Humane Society, or Greenpeace seem less extreme and more reasonable to work with on policy or regulation reform.

How ELF and ALF Hurt the Movement

The direct actions taken by ALF and ELF also hurt the movement, as illegal action can undermine the rule of law and threaten political order. These movements may increase polarization and fuel the antienvironmental rhetoric, leading to further harm to animal liberation and environmental movements. The environmental and animal rights movements may be undermined as well as the real nature of terrorism. While no direct human injuries have resulted from actions taken by ELF and ALF, harm to an employee or firefighter may occur in the future, which could polarize the public even further.

Actions taken by ALF and ELF have also drawn some negative media and political attention to more mainstream environmental and animal rights organizations because of ties between staff, financial contributions, and media provided on behalf of the activists.

The actions invoke fear in communities whose property is destroyed or vandalized and whose workers are threatened. The scientific community experiences setbacks in research, as historical records and laboratory infrastructure are destroyed. Medical studies that could benefit humans may be disrupted. In addition, some researchers who have been targeted are conducting research for environmental or wildlife benefit.

Given that anyone can claim that a direct action is on behalf of ALF or ELF, there could be individuals who act for the wrong reason: taking action for animals or the environment may be simply a way for them to cause destruction and violence or gain media attention. That the action taken is illegal is enough to draw opposition, even among some animal rights or environmental activists who feel that using violence in the fight plays into the same mind-set that ELF and ALF are challenging and reproduces destructive social dynamics.

The illegal actions of the movement mean that people, often young adults, end up in jail. In at least one circumstance, an individual arrested for taking direct action on behalf of ELF committed suicide while in prison. With the passage of the 2001 Patriot Act, sentences have become more severe as activists are charged with domestic terrorism.

Conclusion

The illegal activities carried out by ELF and ALF on behalf of the environmental and animal liberation movements have both advanced and retreated their cause. The awareness built by the movements has led to reduced harm to animals and to reconsidering environmental impacts; however, the vandalism and the fear these tactics incite also undermine the movement.

See Also: Animal Welfare; Environmental Justice (Ethics and Philosophy); Social Action; Veganism/Vegetarianism as Social Action.

Further Readings

Abbey, Edward. *The Monkey Wrench Gang*. New York: HarperCollins, 1975.

"Eco-Terrorism Specifically Examining the Earth Liberation Front and the Animal Liberation Front." *Hearing Before the Committee on Environment and Public Works United States Senate* (May 18, 2005). http://www.access.gpo.gov/congress.senate (Accessed August 2010).

Linton, Cathy. "The History of Animal Liberation Activism." http://www.helium.com/items/343779-the-history-of-animal-liberation-activism (Accessed July 2010).

Newkirk, Ingrid. *Free the Animals: The Untold Story of the U.S Animal Liberation Front and Its Founder, "Valerie."* Chicago, IL: Noble Press, 1992.

"North America and United Kingdom Earth Liberation Front Prisoners Support Network." http://www.ecoprisoners.org/ (Accessed August 2010).

Petty, Terrence, "Earth Liberation Front, Stepping Up Arson Campaign." *Associated Press* (June 2, 2001). http://www.mindfully.org/Heritage/ELF-Stepping-Up.htm (Accessed August 2010).

Singer, Peter. *Animal Liberation*. New York: Random House, 1990.

"The Threat of Eco-Terrorism." Testimony of James F. Jarboe, Domestic Terrorism Section Chief, Counterterrorism Division, FBI Before the House Resources Committee, Subcommittee on Forests and Forest Health (February 12, 2002). http://www.fbi.gov/congress/congress02/jarboe021202.htm (Accessed August 2010).

Stacy Johna Vynne
Independent Scholar

ECOCAPITALISM

Ecocapitalism is the term used to describe market-based solutions to environmental issues. Aside from a small yet vocal minority, most people accept that Earth's resources are finite, that our environment is in jeopardy, and that human activity causes changes in Earth's climate. Efforts to reverse these changes and protect the environment have emerged from diverse camps, including environmentalists, ecoanarchists, ecosocialists, and ecocapitalists. Ecocapitalists believe that ecofriendly regulatory schemes, business models, and economic policies can be used to mitigate environmental degradation while providing a positive economic return. In other words, ecocapitalists recognize both that Earth and its natural resources are in peril and that there are profits to be made from embracing a sustainable, ecofriendly future. They see not a disconnect between capitalism and environmentalism, but instead argue that capitalistic mechanisms are the best methods for restoring and protecting the environment.

Ecocapitalism: People, Planet, Profits

Ecocapitalism encompasses a variety of methods and practices from altering mining, manufacturing, and packaging practices to creating and selling ecofriendly products to regulatory schemes such as the "cap-and-trade" system of swapping pollution credits to achieve emission reductions. Most point to Paul Hawken's 1999 book *Natural Capitalism: Creating the Next Industrial Revolution* as the defining text of the ecocapitalist movement. The central argument of *Natural Capitalism* is that the capitalistic system evolved dysfunctionally by failing to recognize the value of the ecosystems and natural resources as well as of the social and cultural systems that form the basis of all human wealth. In attempting to reshape this flawed capitalistic framework, ecocapitalists focus on four broad goals:

- *Industrial efficiency*: Fundamental changes in production design and technology can help stretch natural resources up to 100 times further than in the past, slowing resource exhaustion at one end and lowering pollution at the other.
- *Biomimicry*: By modeling industry after nature, "closed-loop," zero-waste production systems can be created that reduce dependence on nonrenewable resources, eliminate waste, and allow for more efficient production.
- *Service and flow*: Ecocapitalist manufacturers see themselves not as "sellers of products" but as "deliverers of a continuous flow of service," leasing instead of selling consumer products so that consumers get the service without buying the hardware; because these products such as cell phones and air conditioners will be returned for repairs and to salvage parts for new products, manufacturers are compelled to produce the most long-lasting and serviceable merchandise possible.
- *Investing in natural capital*: Ecocapitalism dictates restoring nature where degraded, sustaining it where healthy, and expanding it where sparse; a good example of this is brownfield redevelopment.

For ecocapitalists, these four broad goals exist within a larger framework built around the "triple bottom line" or a focus on "people, planet, and profits." One of the most potent examples of just how far ecocapitalist ideas have crept into the business world is that of the purchase of power utility TXU by two of the United States' largest private equity firms, Kohlberg Kravis Roberts (KKR) and Texas Pacific Group (TPG), for

$45 billion—the largest private equity transaction in history. This enormous deal was heralded as an "environmental watershed," as KKR and TPG plan to scale back TXU's construction of coal-powered plants, increase investment in wind and solar power, drastically reduce greenhouse gas emissions, and support a $400 million energy-efficiency program. For many, this deal also represents a significant shift in how business approaches environmental progress; instead of thinking of the environment only in terms of government-mandated regulations and fines to be avoided, the private sector is becoming environmentally proactive by focusing on innovation and technology development guided by the knowledge that embracing ecocapitalistic ideas is good for people, for the planet, and for profits.

Ecocapitalism: A Critical Look

While ecocapitalism has animated a significant portion of the business community as well as mainstream environmentalists and many citizens, others view it with a critical eye. Ecosocialists and green anarchists see capitalism as an inherently destructive force whose consequences cannot be mitigated through the kinds of regulations and industry changes that ecocapitalists advocate. They frame ecocapitalism as nothing more than a way for capitalists to increase revenue, viewing ecocapitalists as profiteers whose interest in protecting natural resources is not only self-serving but also insufficient for truly challenging the ecological devastation that capitalism creates. Beyond this broad rejection of ecocapitalism, more nuanced criticisms have been raised about its efficacy.

Ecocapitalism is based on the notion that nature can be taken into the capitalistic accounting framework so that natural resources have "prices." This commodification of nature is deeply troubling to ecocapitalist critics, who see it as not only distasteful, but as just a different pathway to exploitation of finite resources. Critics argue further that the growth/profit/consumption-oriented framework of capitalism remains the framework for ecocapitalism and that ecocapitalists refuse to recognize that the measures they propose do not sufficiently diminish the anti-ecological nature of capitalism.

One of the main ecocapitalist arguments is that externalizing costs leads to a malfunctioning market and that a more transparent market will lead consumers to make more rational choices; this in turn will encourage ecologically sustainable production. To achieve this, ecocapitalists advocate cost/price integration; this involves factoring into the prices of items the full price of creating that item—like clear-cutting forests, fossil fuel refining, and all the bigger processes that are part of making almost everything with which we interact. Ecocapitalists believe that cost/price integration will lead to a more transparent market and that because the fewer resources needed to create something, the cheaper it will be, consumers will be motivated to buy the most ecofriendly products. Because of the laws of competition, other manufacturers will follow suit and also create products that require fewer resources. In this way, consumers will be compelled to make choices that will protect the environment and human health, all via market mechanisms. The problem, say critics, is that implementing cost/price integration is incredibly difficult and would likely make consumer goods prohibitively expensive. Where does the process begin for food, for instance? With the chemical corporations that sell seeds and fertilizers? In farmers' fields? With the factories that produce farm equipment? In the factories that process crops into the foods we eat? And who has the authority and know-how to establish and monitor such measurements?

Another criticism of ecocapitalism is that it does not challenge our commitments to hyperproduction and hyperconsumption, which ecocapitalist foes see as unsustainable. In fact, not only does ecocapitalism not challenge these notions, it contends that we can multiply our current levels of consumption by as much as eight times and still restore and protect the environment and human health and rake in huge profits, all without need for regulations—because the ecocapitalist, transparent market will naturally foster ecofriendly progress. On one hand, to anyone concerned about the environment, many of ecocapitalism's ideas are welcome; but on the other hand, its celebration and elevation of hyperconsumption is unacceptable to those who believe our hunger for "things" is problematic and should be challenged.

Ecocapitalists are motivated not just by altruism or concern for the environment—and indeed, those factors may not motivate some ecocapitalists at all—but also by recognition that higher energy costs, vanishing resources, and pressure from environmentally conscious stakeholders dictates a more rigorous focus on conservation and efficiency. Success in today's marketplace is tied intimately to a proper environmental strategy, and ecocapitalist corporate leaders see that environmental issues represent not just a problem to be remedied but an opportunity to increase profits while satisfying employee, customer, and community demands. On the other hand, those who challenge ecocapitalism believe that in order to achieve sustainability, we need to embrace measures such as living more simply, structuring our communities to be as economically self-sufficient as possible, and embracing a cooperative, participatory way of life, all of which is the antitheses of what ecocapitalism dictates.

What is enticing, and even captivating, about ecocapitalism is that it seems achievable, and it does not require changes to our culture or our economy that feel complicated or threatening. It appeals to not just political and corporate leaders but to many citizens and to the mainstream environmental movement. Alternatives to ecocapitalism, for those who recognize that the environment is imperiled, may seem difficult and unsettling, but their advocates insist that these alternatives offer enormous benefits in terms of quality of life. Ecocapitalists, however, insist that we need not take drastic measures to restructure our economy or our culture and that the fixes they suggest will allow the market to function in such as a way as to protect the environment while still returning robust profits.

See Also: Carbon Trading/Emissions Trading; Commodification; Corporate Social Responsibility; Ecosocialism; Green Anarchism.

Further Readings

Esty, Daniel and Andrew S. Winston. *Green to Gold: How Smart Companies Use Environmental Strategy to Innovate, Create Value, and Build Competitive Advantage.* Hoboken, NJ: John Wiley & Sons, 2009.

The New Atlantis. "The Greening of Capitalism." http://www.thenewatlantis.com/publications/the-greening-of-capitalism (Accessed September 2010).

Sarkar, Saral. *Eco-Socialism or Eco-Capitalism? A Critical Analysis of Humanity's Fundamental Choices.* London: Zed Books, 1999.

Tani E. Bellestri
Independent Scholar

ECOFASCISM

Britches, a baby stump-tailed macaque, is held by an Animal Liberation Front (ALF) activist after being removed from a university. The Earth Liberation Front (ELF) and ALF have committed more than 600 criminal acts at a cost of more than $40 million in damages.

Source: Wikimedia

Ecofascism is vaguely defined as a combination of ecological concepts and goals with a political approach and organization based on ill-defined fascist philosophy and methods. The common person most often identifies it with terrorist or totalitarian behaviors that are designed to induce fear to support a cause or goal. The term *ecofascism* is used as a pejorative accusation by ideological extremes on the political right and left to attack other ecological movements.

Ecology, the science that studies the complex relations between living organism and environment, has documented the impact of the growth of human population and the Industrial Revolution on various ecosystems. This knowledge has led to philosophical and ethical discussions about man's role and responsibility to other life on Earth. Political and economic systems as well as their underlying values come to the forefront when impacts of capitalism, communism, democracy, socialism, and fascism are examined using moral reasoning.

The term *ecofascism* has been used by the social ecology movement to criticize the radical antihumanist extremism of the deep ecology movement. As a political epithet, ecofascism is used as an insult to discredit deep ecology, and even some mainstream environmentalism, as fascist ideologies. Fascism is commonly viewed as a system of government characterized by rigid one-party dictatorship (totalitarianism) and suppression of the opposition. In actual practice, it may have some or all of several core tenets: extreme materialism, authoritarianism, social Darwinism, indoctrination, propaganda, anti-intellectualism, and economic interventionism, including state population control policies. Fascist systems have supported racism and eugenics as well as romantic admiration of nature, as oneness with nature. Nazis in Germany adopted ecological positions such as sustainable forestry and were at the forefront of conservationism before World War II. However, ecological concerns were not central to its totalitarianism approach but only one of a number of issues used to mobilize political support. Other historic examples of governments considered to be fascist include Mussolini's Italy, Franco's Spain, and Pinochet's Chile. Since fascism was not a monetary movement, ecofascism, in general, is a meaningless term. When *ecofascism* or *ecofascist* is used to describe today's environmental or green movements, it is considered attack terminology with only negative connotations of a repressive state or terrorism.

Today, the word *fascism* often becomes entangled with the term *terrorist*. Terrorism is a current topic of debate in many areas of philosophy, including environmental ethical analysis and inquiry. Traditionally, two approaches are taken for analysis: a human-centered principle-based (beneficence, autonomy, justice) or a case-based (comparative) approach. Historically, environmental ethics evolved from the Enlightenment focus of man's role and provides support for the argument that environmental stewardship is essential to both current human populations and future generations. Principles of justice or equity have been key additions over the years in discussions of sustainable uses and in identifying the available options as well as methods of proper dissent. Environmental activists have used many strategies to stimulate change, including letter writing, media campaigns, lobbying for legislation, and economic boycotts. Nonviolent civil disobedience is also an acceptable action that has been used to call attention to political problems and force policy changes by governments. The diverse groups in the global environmental movement have used these nonviolent direct actions to fulfill their self-described ethical obligations. Nonviolence is an effort to achieve a goal of stopping injustice without hurting the opponent.

Radical ecoterrorist groups are increasingly using violence to halt activity that they believe destroys or exploits the natural environment. In the name of the concepts of conservation, preservation, or sustainability, ecoterrorist groups have burned, bombed, and destroyed both private and public and governmental properties. In Europe, Earth First! members felt that bombing, arson, and tree spiking should be used to protest environmental issues. In the United States, the Earth Liberation Front (ELF) defined a mission to ensure that all forms of life on the planet can flourish. This organization holds that anyone participating in earning wealth by threatening the ability of all life on the planet to exist should be stopped—a radical ecocentric position. The ELF and the Animal Liberation Front (ALF) have committed more than 600 criminal acts at a cost of more than $40 million in damages. Ecoterrorist groups act in strict secrecy and use violent fear-generating techniques similar to those used by fascist political parties in Germany and Spain. Their violent and authoritarian actions are associated with the development of the term *ecofascism*. However, others use it as a smear to either discredit ideological rivals within the green movement or proponents of human priority access to natural resources. They hope to call attention to their issue and win public relations battles. A November 18, 2010, headline in the *Anchorage Daily News* regarding the Pebble gold-copper mine development project illustrates this: "Pebble backer blasts legal terrorism." The article describes a meeting where a pro-industry representative decried the outside influences on Alaska's "salmon versus mining" debate.

The term *ecofascism* has been used in regard to issues such as land use and natural resource development, animal rights, organic farming, or vegetarianism, when actions to reform current economic and political systems are discussed. Earlier, it had been used to characterize some political movements that used either the term *green* or *green ideas*. In the United States, Greenpeace and other social justice groups like the Environmental Defense Fund have been criticized for using these terms. While the arguments are usually environmental justice topics related to economic and social justice, the underlying issue in the debate is related to ethical values as applied to sustainability issues at both local and global scales as well as generational justice. Outlined here are both the purported evolution of ecofascist terminology from the deep ecology and social ecology debate as well as some current examples related to sustainability where the pejorative terms *ecofascism* or *ecoterrorist* might arise.

Deep Ecology

Believers in deep ecology feel that crisis demands a basic change of values from human centered to ecosystem centered, or in more general terms, a respect for a holistic material world through an ecocentered value system. Some scholars have traced the introduction of the term *ecofascism* to critiques of the deep ecology movement by social ecologists. This movement is often referred to in environmental justice literature as more extreme than biocentrism, which focused only on living organisms in the ecosystem. Some deep ecology advocates promote going beyond individual consciousness raising to creating an alternative cultural vision to that of the current industrial capitalistic society. They are motivated by a perceived ongoing violent onslaught against nature and its nonhuman life forms, as seen in the 2009 film *Avatar*. Deep ecology believers justify their actions through their vehement opposition to the capitalistic position that economic progress is a valid reason to disturb natural systems. Some ideas associated with deep ecology support the love of nature, spiritual transformation, and population reduction, which reminds critics of fascist beliefs. Their advocacy for government-led activities, such as population and immigration control, implies to some the creation of a repressive state that would be required to implement their policies. Some advocates in the deep ecology movement and the German Green Party re-examined the national socialist movement in Germany (Nazism), thus raising to others the image of fascist methods to create a spiritual, ecopsychological transformation in society similar to the Nazi movement of the 1930s and 1940s.

The deep ecology movement believes that all living things have an equal right to live and have intrinsic value. Three of the movement's virtues include (1) biological egalitarianism, humans have no more value than any other organism, (2) holism, connectivity of processes (ideas) that are emergent in a complex system such as a niche, and (3) self-awareness of individual personal identity that includes the other (or nature). Their value of organic unity is reminiscent of Native American belief systems and even the Gaia hypothesis. All organisms and their relationships are equal in intrinsic worth and part of an interrelated whole. Some deep ecology advocates feel they have an obligation to either directly or indirectly try to implement policy and even social changes. Their philosophy is considered "deep" because of its redefinition of humanity's role or place in existence was a departure from the classical human-centered ethic system. Deep ecology's ecocentrism reduces the value of humans, so it is considered to be an anti-human ideology that is based on emotion rather than reason.

Deep ecology is believed to have influenced the creation of direct-action environmental groups such as the ALF and Greenpeace. It is often criticized by conservationists, ecosocialists, and environmental justice advocates for falling prey to the fallacy that "since something appears to exist, it ought to exist." Environmental justice scholars worry about the environmental injustice to people created when protecting academic ecological concepts about nature. In the United States, eugenics and social Darwinism led to social injustice.

Social Ecology

Social ecology is a more bio-aware social justice movement that competes with deep ecology. It is a vision for the future that rejects notions of hierarchy, power, and place in order to advocate a restructuring that will resolve basic social issues of economic, gender, and environmental imbalance. It is based on an understanding that our present ecological issues are related to sustainability and the result of social, economic, and cultural history. Social ecology's tenet is that the main source of destruction of ecosystem services, needed by both humans and all living organisms, is the capitalist system and its products, such as overconsumption, consumerism, and concomitant economic growth. This is similar on the surface to deep ecology. Since other systems such as socialism and communism have also led to ecological destruction, advocates see that a social ecology society would need to link ecological regeneration with social regeneration. Such an ideal society would form eco-communities that adopt ecotechnologies that create an intersection between technology and human spirituality. Their world would be holistic, and humans would act as moral agents who foster aesthetic appreciation of living organisms and a healthy community. Humans would play a key role in maintaining the integrity of the planet's ecosystem by establishing community-based ethical systems. This is also similar to the holistic concept seen in the Gaia hypothesis and cultures of many indigenous groups. However, social ecology still accepts the basic philosophical position that rights are solely human unless conferred by humans on animals or on environmental entities.

Social ecology subscribers often attack capitalism, believing that it is amoral and corporate self-interest is inherently ecologically destructive. In general, they see that private and personal interests lead to a human hierarchy that causes forms of domination that are incompatible with ecological restoration and sustainability. Also, capitalistic personal belief systems are generally not favorable to the environment so collective interest needs to be imposed. The anticapitalism and holistic, romantic view of nature can also evoke a sense of fascism by some critics. In social ecology there is a sense of a special role for humans as master interventionists. Since Earth First! promoted restoration ecology, some social ecologists could identify with some concepts and values that imply governmental authoritarian imposition. However, social ecologists have used the term *ecofascism* to attack the deep ecology movement.

Sustainability Debates

The biggest debate today that involves charges of ecofascism is sustainability. Commonly discussed as "sustainable development," this issue involves renewable resources and the impact of developing nonrenewable resources. The concept of sustainability is capacious and also carries weight in socioeconomic discussions because the developed world increasingly perceives that many current practices are unsustainable. Current issues of climate change, air pollution, water rights, fisheries, forests, food supply, and military conflicts between nations have an effect on the sustainability of Earth's life-support systems. Sustainability means changing the way societies do business so that dire consequences can be avoided and healthy communities created.

In the north, sustainability underlies controversies related to fisheries and marine mammals. The endangered status listing of both bearded and ringed seals by the National Oceanic and Atmospheric Administration (NOAA) is tied to the issue of climate change, and the NOAA action is supported by the Center for Biological Diversity but opposed by industry and the state of Alaska. The Canadian ground fishery industry has in the past supported hunting harp and hooded seals, and their rhetoric often included hateful diatribes. Greenpeace has been actively interfering with whaling and research vessels in the Antarctic region. Likewise, disputes on the use of forests in Alaska and Canada have led to road blockades. There have even been reports of physical attacks. Blockade strategies have also been used to thwart the development of mining operations, while governmental agencies have used armed guards to monitor access to parks and public lands. The most recent controversy involves the Pebble Mine in Alaska, where village corporations want to stop the mine's development to protect natural watersheds in the region. Currently, the battles occur in the nonviolent forum of the court system. Another type of direct action is the refusal to participate in government programs. The tribal government in the Yukon River village of Fort Yukon will not participate with a state agency's salmon study, because the U.S. Fish and Wildlife service has been arresting or fining some Alaska Native subsistence fishermen. The low salmon returns are related to by-catch by commercial interests and unfair government distributions, an environmental justice issue.

In conclusion, while many industries have moved from a "profit only" to a more sustainable "profit, people, and planet" model with an industrial ecology approach, other industries still act in a way to encourage environmental opponents to take a direct approach, sometimes leading to inappropriate violence. The damage generated by the so-called ecofascist approach usually hurts innocent people and animals; the purveyors of this violence typically rationalize the necessity of their actions by referring to it as "collateral" damage.

See Also: Animal Ethics; Animal Welfare; Environmental Justice (Business); Ethical Sustainability and Development.

Further Readings

Agyeman, J., P. Cole, R. Haluza-DeLay, and R. O'Riley, eds. *Speaking for Ourselves: Environmental Justice in Canada*. Seattle: University of Washington Press, 2010.

Coughlin, S. S., C. M. Bonds, J. Malilay, and J. Araujo. "Ethic." In *Encyclopedia of Global Warming and Climate Change*, S. George Philander, ed. Thousand Oaks, CA: Sage, 2008.

Enger, E. D. and B. F. Smith. *Environmental Science*. New York: McGraw-Hill, 2009.

Orton, D. "Ecofascism: What Is It?" *Green Web Bulletin*, 68 (2000). http://home.ca.inter .net/~greenweb/Ecofascism.html (Accessed January 2011).

Vaugh, L. *Doing Ethics: Moral Reasoning and Contemporary Issues*, 2nd ed. New York: W. W. Norton & Company, 2009.

Wenz, P.S. *Environmental Justice*. Albany: State University of New York Press, 1988.

Lawrence K. Duffy
University of Alaska, Fairbanks

Ecolabeling

Ecolabeling is the practice of using voluntary or mandatory labeling systems for consumer products as a way of indicating a certain standard of sustainability or environmental responsibility—varying by the labeling system—in order to guide the choices of environmentally minded consumers. Systems vary not only in the type of product they cover and the green issue they are concerned with, but the nature of the label itself. Some labels may be "pass/fail," so to speak: either a label is present, indicating that a certain standard has been met, or it is not, indicating that it has not or the manufacturer has not applied for a label. Other labels give more specific qualitative information, analogous to a report card. Some labeling systems are operated by governments, others by industry and consumer groups. Unlike labels indicating contents for dietary reasons—disclosing the presence of potential allergens or announcing vegetarian, vegan, or kosher status—ecolabels when used for food or other products are meant to aid "moral purchasing." Common motivations behind moral purchasing include environmental concerns, human rights and workers' rights issues, a desire to support or not support a particular political entity or persuasion, a desire to buy local (or domestic), and animal rights.

Moral purchasing—also termed *green consumerism* when the morality at work is that of environmentalism and sustainability—is not the same as charitable donation, though it can include the purchase of products that support charities. It is essentially the broad umbrella under which boycotting—avoiding a purchase for moral reasons or to send a moral message—is one option. Buying responsible goods, or goods that in some way or another meet moral demands, is the other. Moral purchasing is a behavior many people engage in without delving too deeply into their reasons or philosophy behind it—they may buy candy bars to support a local school charity or drink fair-trade coffee without being deeply engaged in the issue, which of course is one of the benefits to the availability of such products. They may become more involved than that, spending disproportionately on products and services from companies with ethical practices that reflect their own, or avoid spending money with companies that support undesirable political causes, engage in unethical practices, or fail to engage in ethical ones. The line between moral and practical purchasing sometimes blurs, as with the perception that organic food is both better for the environment (a moral value) and better for one's health (a practical value). Technically, moral purchasing would also include food choices made according to religious laws such as the need to keep kosher or halal.

Labeling laws exist in part to make moral purchasing possible without constraining the marketplace by banning the goods moral consumers are expected to avoid. Goods are required to be labeled with their country of origin, for instance, "Made in China," "Made in the U.S.A.," and so forth, which enables positive "Buy American" purchasing or boycotts of Chinese-produced goods in the aftermath of safety scares without restricting international trade through any means other than market forces. Consumers generally assume that there is a vetting or verification process for any labels on the items they have purchased. The truth of this varies considerably. In some cases, an industry group is responsible for overseeing labeling; in others, a government or (in the case of optional labels) third-party group oversees it; in still others, companies are left to their own devices, on threat of reprisal if mislabeling should be discovered. Ecolabels exist in all of these categories.

Related to moral purchasing is conscious consuming, a social movement that began in 2003 in Boston, Massachusetts, at an alternative gift fair called Gift It Up. While moral purchasing is often done with little thought, conscious consuming encourages consumers to consider the impact of their purchasing decisions beyond the appeal and utility of the product or service on offer. While conscious consuming is primarily interested in environmental concerns and the environmental impacts of purchasing, the overall message is not so narrow but is rather to encourage consumers to become engaged and aware, to use the information resources at their disposal in the 21st century, and to demand more information when what is provided is not sufficient.

The Dolphin Safe Label

The first ecolabel most Americans became familiar with was the "dolphin safe" label established by the U.S. Department of Commerce in 1990, following the previous decade's swell of concerns over the fate of dolphins netted in the course of tuna fishing. Several species of tuna socialize with dolphins, the dolphins swimming alongside them and protecting them (with their presence) from sharks. The capture of dolphins in nets, especially purse seines (large nets that are drawn shut by a rope passed through rings, like a purse's drawstring), was not incidental: fishing vessels in search of tuna would seek out dolphin pods, which were easier to find, and then encircle the pods to capture both the dolphins and tuna at once. This practice has largely stopped, with fish-aggregating devices such as buoys and floats now put out at sea to attract fish instead. The Commerce Department's dolphin safe label is available for fish caught using purse seine fishing methods in the Eastern Tropical Pacific Ocean when an observer from the National Marine Fisheries Service, present on the ship at the time the fish were caught, can verify that no dolphins were harmed. The observer is not necessary for fishermen using other methods or fishing in other parts of the sea. There is a great deal of information thus not conveyed by the dolphin safe label. For one thing, because skipjack and albacore tuna do not socialize with dolphins, they may be considered dolphin safe regardless of methodology or provenance. For another, in many cases dolphin-friendly fishing methods simply harm other by-catch species (species caught along with the targeted species) instead, typically sharks, albatrosses, and turtles. Fish caught and canned in other countries and imported may bear either different ecolabels with different standards—there are two labeling systems in Australia, for instance, one in New Zealand, and one in the Netherlands—or leave the issue entirely unaddressed, as with most imported Italian tuna.

The dolphin safe label demonstrates the strengths and weaknesses of ecolabeling. On the one hand, a government standards program—however narrow—resulted in response to consumer outrage and demand, and industry practices shifted accordingly, and the program is transparent enough that it is very easy to tell which canneries take part in it and which do not. On the other hand, consumers are prone to misunderstanding, and the tendency of labels to boil down information to a quick phrase—dolphin safe—certainly does not disambiguate. Eighteen years after the program's inception, a Greenpeace study (of dolphin safe labels in general, not just the Commerce Department's) found that most consumers believed that dolphin safe labels meant the canned tuna was more environmentally responsible. In fact this is not true, and the environmental impact of tuna fishing is significant regardless of the effect on dolphins. Of the 23 species of fish sold as tuna, nine are classified as fully fished; four as overexploited or depleted; three as critically endangered; three as endangered; and three as vulnerable to extinction. The term *peak tuna* is used, although the usage, paralleling peak oil, is tongue in cheek and not a true analogue: tuna fish are a renewable resource and can be expected to replenish their numbers as long as fishing ends when they become rare enough to no longer be cost-effective. But it seems certain that peak tuna catches are behind us and that a long, gradual decline will be ahead before numbers recover.

Tuna labels typically leave off the most key, the most fundamental, piece of information about their contents: the species. Not all tuna are even in the same genus: though the best known are part of the *Thunnus* genus, most come from elsewhere in the Scombridae family, including skipjack (*Katsuwonus pelamis*), dogtooth (*Gymnosarda unicolor*), and frigate (*Auxis thazard*). In addition to the 23 commercially sold varieties, there are 25 more that may be commercially exploited in the future, as better-known varieties dwindle in numbers. The commercial seafood industry has long traded on the ability to sell multiple species by the same "common name": the "redfish" made famous by Louisiana's "blackened redfish" may be any of a dozen species, with several others likely to show up in restaurants as substitutes. So long as consumers are conditioned to think of the commodity they are buying as "tuna," the industry has much deeper stocks to rely upon without the market effects of introducing a "new" product. Furthermore, consumer awareness of the endangered status of several tuna species has little effect on purchasing habits when consumers are unable to tell which species they are buying. But the dolphin safe label assuages some of their concerns: it presents, even if this is not its intent, the allure of environmental responsibility and makes the consumer feel as though his/her choice is an environmentally sound one.

The alternatives used, instead of netting dolphins, however, have proved just as detrimental to the environment. For every 10 pounds of tuna caught by purse seines around fish-aggregating devices, there is one pound of by-catch, including sharks, turtles, rays, and other species. Much of the shark fin on the market—a valuable delicacy in Asia—enters it when sharks are caught in tuna nets, only to have their fins sliced off before they are thrown back into the sea. Tuna by-catch has exacerbated the dangers faced by many threatened ray and turtle species as well. Juvenile fish are killed in large enough numbers to exacerbate population decline. There is even speculation that the fish that are not caught are still put at risk, as the devices attract fish to depart from their migratory routes and thus face malnourishment. But since the shift away from fishing by netting dolphin pods, purse seine fishing has actually increased, considerably overtaking other forms of tuna fishing and accounting for 70 percent of commercial catches.

Are Ecolabels Enough?

In short, the problem with the dolphin safe label is that it presents the appearance of information without conveying everything and without making available sufficient information to guide the choices of consumers who do understand the full picture, and it presents the appearance of environmental responsibility despite having no correlation to sustainable fishing practices or fishing practices with responsible environmental impact. An association is created in the consumer's mind between the broad cause of environmentalism and the can of tuna, having some of the same intended effect of "cause marketing" (in which corporate entities associate themselves with a nonprofit cause for mutual exposure, such as Cheerios breakfast cereal's use of the American Heart Association's "Heart Check" icon, which raises awareness of heart disease while bolstering positive perceptions of the cereal) without the benefits of actually providing support for the cause. Because the label exists, there is no longer a demand for a label or for standards to be imposed on the industry, and it is actually harder to mobilize consumers to demand change than it was when there was no label at all.

There are 347 ecolabels tracked by the Ecolabel Index, the most complete global database thereof, in 212 countries, representing 40 sectors of industry. Many are as specific and narrowly focused as the dolphin safe label. Some are quite broad, like the Green America Seal of Approval, maintained by nonprofit organization Green America, which screens applicant businesses to determine if their practices are socially and environmentally sound. Green America also publishes *The National Green Pages*, *The Guide to Socially Responsible Investing*, and a bimonthly magazine called *The Green American*.

One of the big potential problems with ecolabels is their implication in *greenwashing*, a term coined in the 1980s combining *green* and *whitewashing* to describe a company's deceptive association with green ideals to promulgate the perception that the company's goods or services are more environmentally friendly than they are. Shell, for instance, was criticized in 2008 for an advertising campaign in the British market that referred to an oil sands project as a "sustainable energy source" despite relying on a nonrenewable resource. Often, greenwashing is less clear. The Rainforest Alliance, a nongovernmental organization (NGO) in the United States, makes its Rainforest Alliance Certified ecolabel available to products that meet certain standards of sustainable agriculture. The problem is twofold: first, that "Rainforest Alliance Certified" is a vague phrase that hints at but does not actually indicate the nature of the standards in effect. Consumers might associate the rainforest with environmentalism and assume a broader set of standards than are applied; they might assume it indicates some portion of the ingredients came from a rainforest; and they might assume the applied standards deal with deforestation and reforestation. Beyond this, the NGO's standards of sustainability are not strict enough, according to critics. Coffee is eligible for the seal if at least 30 percent of the beans come from sustainable plantations—which means most of them do not. Other products may bear the seal even when only one of its ingredients meets the ecolabel's standards: a chocolate bar may use the seal when the cocoa beans are certified, which a consumer will take to be an endorsement and certification of the entire candy bar, even if the other ingredients include GMOs, hormone-laden milk, trans fats, and palm oil from nonsustainable plantations, all ingredients that green consumers can typically be expected to avoid. Consumers are also very likely to assume that the label indicates certain values that it does not: Rainforest Alliance Certified coffee is not fair-trade coffee, for instance, and because the certification process does not guarantee a minimum price the way the Fair Trade certification process does, Rainforest Alliance Certified coffee growers receive, on average, less money than Fair Trade coffee farmers do. This has actually been a point of pride for some, though—in the view of *The Economist*, refusing to guarantee a minimum price means that the farmers are incentivized to compete through quality, while Fair Trade coffee varies widely in its ratio of quality to price.

Organic certification is problematic in this regard as well. Because of the expense of certification, small or family-run farms cannot afford to refer to their products as organic, regardless of the means of production. Indeed, the cost is so prohibitively high in an industry where subsidies already disproportionately fall to the largest businesses that many argue that organic certification is of the greatest benefit to large agribusinesses. Exemptions are made only for producers selling less than $5,000 a year, who need not go through the formal application process (but must comply with an audit if asked). Placing the exemption threshold so low, however, means that it really only applies to those for whom organic growing is a minor sideline or not-for-profit activity: families who sell pumpkins from their backyard gardens at town square farmers markets or children on summer break picking and selling raspberries. Small but full-time farming operations, even with only one or two employees, go through the same certification process as multinational corporations like Dole and General Foods. Green customers, on the other hand, typically want to support smaller businesses and are derisive of factory farms. Furthermore, though the results are contested by industry representatives, at least some studies have found that organic farming has a greater environmental impact than conventional farming, at least in some respects: notably, organic agriculture burns more fossil fuels. Given the contribution of the agricultural sector to greenhouse gas emissions as it is, this may arguably offset—or at least undercut—the environmental benefit of avoiding pesticides and other chemical inputs.

Furthermore, in the United States, the U.S. Department of Agriculture's organic seal suffers from the same problem as the Rainforest Alliance's, in that products need not be made purely of ingredients meeting certification standards to use the label: those that do not qualify to call themselves 100 percent organic may still use the word *organic* and the seal if 95 percent of the ingredients are organic, while products that are only 70 percent organic ingredients may label themselves "made with organic ingredients."

Ecolabeling may well be an effort like utility companies' green pricing or Earth Hour—a way to keep the conversation going, to keep green issues in the public dialogue, but of little actual effect in and of themselves. If nothing else, it can certainly be said that some labels are better than others.

See Also: Agribusiness; Certified Products; Ecolabeling (Consumer Perspective); Green Pricing.

Further Readings

Paul, John. *The Value of Eco-Labelling*. New York: VDM Verlag, 2009.
Ward, Trevor. *Seafood Ecolabelling*. New York: Wiley-Blackwell, 2008.

Bill Kte'pi
Independent Scholar

ECOLABELING (CONSUMER PERSPECTIVE)

Labels certify that a product or service is environmentally preferable to others of the same type, but they may not give the whole picture.

Source: iStockphoto

Ecolabels inform consumers which products or services comply with particular environmental criteria or are environmentally preferable to competing products. They are a way of authorizing a manufacturer's claims that its products have environmental virtues. Consumers who want to be environmentally responsible are able to choose products or services with an ecolabel, and therefore, the label confers a competitive advantage to those products. This provides an economic incentive to product manufacturers and service providers to change their production processes and redesign their products to improve their energy efficiency, reduce their wastes, or do what is required to earn an ecolabel.

In this way, ecolabels are a market-based instrument that seeks to utilize market forces to achieve environmental goals through the use of consumer power. However, the ability of ecolabels to do this depends on the desire of consumers to be environmentally responsible, the willingness and ability of individual firms to achieve significant environmental improvements at a reasonable cost, and the existence of competitive markets as well as the credibility, accuracy, and environmental meaningfulness of the labels and consumer awareness of them.

Ecolabeling schemes are sometimes run by government authorities, sometimes by environmental groups, and sometimes by industry associations. Some schemes include independent verification by third parties, but many do not. They are generally voluntary; that is, manufacturers and service providers choose whether to participate, although some, particularly energy labels, are mandatory in some countries. Some labels give a rating, some provide comparative information in a standard format, and others are just a symbol of approval.

The criteria of the U.S. Consumers Union for "good" ecolabels include the need for labels to make clear, meaningful, and consistent claims that are independently verified and the organization behind the label to be open about itself and its certification standards and to provide opportunities for public comment on those standards, particularly from consumers, environmentalists, and others. The people setting the standards and certifying products should be independent and not have any conflict of interest, which means they cannot be affiliated in any way with the companies seeking certification.

Similarly, the Global Ecolabelling Network (GEN) claims that good ecolabels should not be awarded to products and services that do not comply with the relevant local and national environmental legislation or are not good-quality products in other nonenvironmental respects. They should be based on life-cycle assessments, but criteria should be reasonably attainable and measurable by products seeking certification. They should be run by an independent organization, and a "credible program must be based on an open and accountable process that can be observed, monitored and questioned at any time." Finally, the programs need to be reviewed and changed as environmental criteria and technologies change.

The International Organization for Standardization (ISO) has formulated ISO14024, a voluntary standard to cover ecolabels that are supposed to be based on life-cycle assessments of products and services (from raw material extraction through production, distribution, and disposal) and are authorized by a third party. Such labels indicate that the product or service is environmentally preferable to other products or services of the same type.

History

During the 1980s, a series of media reports and books such as *The Green Consumer Guide* gave many people the impression that the environment could be protected if individuals were responsible in their shopping habits and bought only environmentally sound products. At the time, surveys showed that a significant proportion of consumers, particularly young mothers, made an effort to buy green products such as pump packs, unbleached papers, and items made of recycled paper. They were willing to pay more for safe aerosols and biodegradable plastic products as well as for natural foods that were not produced using pesticides.

This prompted a surge of advertisements and labels claiming environmental benefits. Green imagery was used to sell products. Caring for the environment became a marketing strategy. Often, the claims had very little substance. For example, Greenpeace claimed that Chevron, a multinational oil company, spent about five times as much publicizing its environmental actions as it did on the actions themselves.

The problem for consumers was being able to tell which green claims were genuine and which were merely marketing gloss. Green consumers tend to prefer natural fibers rather than synthetics, but the cotton industry and large-scale sheep grazing also result in significant environmental problems. The debates over whether plastic packaging is better or worse than paper packaging for the environment, or whether milk bottles are better than cartons, were no less confusing.

Any judgment about whether a product is ecologically sustainable is extremely complex, requiring long-term assessment from manufacture to disposal and taking consideration of how long the product will last, whether it can be reused or recycled, whether it is biodegradable, how much energy it consumes, and how efficiently it uses resources. Other matters that need to be considered include the way the product will be used, transported, distributed, marketed, and packaged.

Few consumers were in a position to do the research necessary to investigate each product thoroughly. What is more, the obvious "hype" surrounding green claims and their discrediting by environmental groups led many consumers to become skeptical of green marketing. Official systems of ecolabeling were introduced to combat the use of false claims by marketers and to give consumers the confidence to go on shopping responsibly.

Examples of Ecolabeling Schemes

Ecolabeling schemes may cover specific products or services (such as automobiles), product categories (such as white goods), or a range of product categories. Ecolabels can also cover a range of environmental issues, including energy and resource efficiency of electrical appliances, reuse and recyclability of packaging or products, chemical content of products, water usage of products, energy or water used during manufacture, pollutants produced during manufacture, or impact of the product on biodiversity, habitats, or natural resources. Sometimes ecolabels are specific to one of these; less often they attempt to cover several of them.

Perhaps the most well-known comprehensive ecolabel is the European Ecolabel, which was established in 1992 to "encourage businesses to market products and services that are kinder to the environment." It is a voluntary scheme and awards products meeting its environmental criteria with a flower logo. It covers cleaning products, appliances, paper products, textile and home and garden products, lubricants, and services such as tourist accommodation. For example, "Ecolabelled detergents exclude substances that may cause cancer or heritable genetic damage, are very toxic to aquatic environments, or may impair fertility."

To be able to use the European Ecolabel, products and services must meet environmental criteria based on life-cycle analysis and set by the European Union and be independently verified. In 2010, more than 1,000 EU Ecolabel licenses had been issued: 37 percent are for tourist accommodations, 11 percent cleaners, and 9 percent textile products. The scheme is managed by the European Commission and regulated in each country by the relevant national authority, for example, by the Department for Environment, Food and Rural Affairs (Defra) in the United Kingdom.

Many individual nations also have their own ecolabel schemes. These include the following:

- Nordic Swan, covering Denmark, Finland, Iceland, Norway, and Sweden
- Blue Angel in Germany

- NF Environnement in France
- Green Seal in the United States
- Good Environmental Choice in Australia
- EcoLogo Program in Canada

Many nations also have specific energy labels for electrical products and automobiles. These include the following:

- *Canada Household Appliance EnerGuide Label*: household appliances, heating and cooling equipment, even new houses, have to show annual energy consumption and energy efficiency relative to similar models.
- *Canada Automobile EnerGuide Label*: for fuel consumption ratings and annual fuel usage.
- *Energy Star*: United States, Canada, Australia, New Zealand, Japan, Taiwan, and the European Union; based on international standards for energy efficiency in major appliances, office equipment, lighting, home electronics, and sometimes new homes and commercial and industrial buildings.
- *European Energy Label*: white goods in the EU are required to show a label indicating their relative energy efficiency on a scale of A to G, with A being the most energy-efficient product in its class. Car labels also indicate carbon dioxide emissions.

Sector-specific labels apply to one kind of product. For example, there are various labels for "sustainable" timber, including those of the Forest Stewardship Council, the Rainforest Alliance, and the Programme for the Endorsement of Forest Certification (PEFC) Council. Similarly, there are labels for tourist accommodation and computer and information technology.

There are also food ecolabels, particularly for seafood that is obtained from a sustainable source. These include labels from the Marine Stewardship Council and Friends of the Sea. Dolphin safe labels are used for tuna (in the United States, Europe, Australia, and New Zealand), but these are usually not independently verified and do not mean tuna have been fished sustainably or without environmental damage.

Some labels indicate waste disposal options. For example, the Mobius Loop is a voluntary international recycling symbol to indicate that a product is recyclable or includes recycled content (with percentage shown within the loop). Use of this symbol is covered by International Organization for Standardization rule ISO7000. Green Dot labels indicate compliance with and financial contribution to an authorized packaging recovery scheme in some European nations. However, it does not mean that the packaging will be reused or recycled.

There are also a growing number of labels related to climate change, including the Carbon Reduction Label, developed by the Carbon Trust in the United Kingdom (UK), which shows how much greenhouse gas emissions are associated with the product from manufacture through to disposal; the UK's Quality Mark labels, intended to certify carbon offsets; and the international voluntary carbon standard (VCS), an international certification for voluntary carbon offsets.

Problems

Ecolabeling schemes are often run by or unduly influenced by an industry association that is not altogether unbiased. There are hundreds of schemes around the world of varying degrees of authenticity, independence, and comprehensiveness. Any organization, including a private company, can set up its own ecolabeling scheme. Examples include UK retailer Tesco and U.S. footwear manufacturer Timberland. Sometimes ecolabeling schemes achieve recognition by having glossy websites, endorsements by paid experts, and association with celebrities.

Often, logos are not what they seem. The Panda symbol of the World Widelife Fund (WWF), for example, merely indicates that the company has given money to WWF, not that the product is necessarily environmentally beneficial.

In 2008, the Inspector General in the United States found that the Energy Star program was inaccurate with respect to greenhouse gases and claims about energy savings: "Deficiencies included the lack of a quality review of the data collected; reliance on estimates, forecasting, and unverified third-party reporting; and the potential inclusion of exported items."

Even genuine certifications may not give the whole picture. Take recyclable symbols. Just because there is a recyclable symbol on the package does not mean that the local waste disposal authority can or will recycle it. Also, the symbol will probably relate to the packaging rather than to the product inside.

Even if a product is recyclable or has a good energy rating, that does not mean it is good for the environment. A washing machine may use less energy, but it might also use more water, and its manufacture may create toxic pollution. A detergent with an ecolabel for being free of phosphates might be overpackaged and use high amounts of energy in its manufacture. This is why environmentalists call for a whole life-cycle analysis of products.

Even ecolabels that purport to include life-cycle analysis are not beyond criticism. They usually only consider some phases of a particular product in terms of their environmental performance in comparison with similar products and omit other phases such as the impacts of raw material extraction, particularly where similar products require the same raw materials. This process is not transparent to the consumer.

Rather than identifying products that are environmentally benign, ecolabels tend to recognize products that are merely better than competing products. For example, an SUV may be labeled as being fuel efficient but still be less fuel efficient than other types of cars. Consumers tend to assume an ecolabel means that the product does not harm the environment, but this is not necessarily the case.

Labels that endorse carbon offsets do not take account of the fact that tree plantations are not always good for the environment. They suck up all the water in an area, reduce soil fertility, increase erosion and compaction of the soil, and increase the risk of fire. They require heavy use of agrochemicals, including fertilizers, chemical weeding, and herbicides, and can lead to a loss of biodiversity. Timber ecolabels such as the Forest Stewardship Council also certify plantation timber despite these problems.

The main rationale for ecolabeling is that it will encourage production and sale of greener products. However, should environmental protection depend on each shopper choosing the environment-preferred option rather than basing his or her decisions on other considerations such as price? The popularity of green choices waxes and wanes. It reached a peak around 1990 and then declined for a number of years. Even though it has been on the rise again recently, the EU ecolabel still does not have much influence over European consumers.

If manufacturers are able to reduce the environmental impact of their products, should it be something they choose to do when and if it confers a competitive advantage, or should it be something they are required to do by legislation? Often the ecolabeled product is just one more option in a range of models and marketing strategies.

Even more importantly, ecolabels do not address the question of wasteful and needless consumption. Most environmentalists argue that the consumer's first priority should be to buy less, to ask, "Do I really need this product?" Yet the efficacy of ecolabeling depends on marketing that encourages us to buy more by suggesting we should buy a certain product because it is good for the environment.

There is a danger that those who follow ecolabeling advice in the supermarket or at the shopping mall will believe that their actions are all that is required to protect the environment, while the need to change attitudes toward consumption, values, and institutional structures and to regulate manufacturers and suppliers will be ignored. The danger is that the power of citizens to demand significant environmental reform will be reduced to consumer choices provided by profit-driven businesses.

See Also: Certified Products; Ecolabeling; Green Pricing; Individual Action Versus Collective Action.

Further Readings

Consumer Reports. "Eco-Labels Center." http://www.greenerchoices.org/eco-labels (Accessed August 2010).

Department for Environment, Food and Rural Affairs, UK. "Pitching Green—Green Labels and Credentials: A Guide to the Options." http://www.defra.gov.uk/environment/consumerprod/ecolabel (Accessed August 2010).

Global Ecolabelling Network (GEN). "What Is Ecolabelling?" http://globalecolabelling.net/whatis.html (Accessed August 2010).

Lavallée, Sophie and Sylvain Plouffe. "The Ecolabel and Sustainable Development." *International Journal of Life Cycle Assessment*, 9:6 (2004).

Zaman, Atiq Uz, Sofiia Miliutenko, and Veranika Nagapetan. "Green Marketing or Green Wash? A Comparative Study of Consumers' Behavior on Selected Eco and Fair Trade Labeling in Sweden." *Journal of Ecology and Natural Environment*, 2:6 (2010).

Sharon Beder
University of Wollongong

ECOLOGICAL IMPERIALISM

Ecological imperialism, a thesis first proposed by Alfred Crosby in 1972, advances the theory that European colonization, beginning in the 16th century, was facilitated by the ecological damage inflicted due to the transfer of germs, plants, and animals to the New World. Further, ecological imperialism proposes that the most drastic effect of European colonialism in the New World was not in the realm of social and political change but in the natural world. More specifically, the transfer of people, plants, animals, and germs from Europe, and vice versa, had a transformative and hugely disruptive effect on the local cultures and their economic viability in the Americas. This unequal exchange has been termed by Crosby as the *Columbian Exchange*, and had ramifications that continue to reverberate to the present day. For instance, far more indigenous inhabitants of the

Americas died of the germs brought over by the Europeans than by direct warfare. Similarly, the land, rivers, and forests on which the local economies depended were transformed in accordance with the European cultural preferences and economic interests. For instance, the bison, which was the mainstay of the Great Plains Indians, was driven nearly to extinction, dealing a devastating blow to the cultures dependent upon it. In more contemporary terms, ecological imperialism describes the dominance of the European ideals of nature and wilderness, and the resulting adverse consequences for environmental conservation and indigenous peoples, especially in the non-Western countries.

Portmanteau biota, a term coined by Crosby, refers to the humans, plants, animals, and diseases transported by the Europeans to the New World. Apart from their economic function, these elements of the European environment, when transferred to the New World, helped create "Neo-Europes." For instance, the transplanting of the wheat and livestock agriculture to Australia, North America, and South America almost completely transformed the local environment. However, a very high cost was often inflicted on the local inhabitants and their environment in the colonial endeavors to make the environment conform to the European notions of ideal landscapes.

Ecological Imperialism and Global Environmental Problems

John Bellamy Foster has argued that global environmental problems, due to their unprecedented scale and scope, have brought the planet to a turning point. No longer can the lifestyles and unsustainable consumption levels in the core countries be maintained by the ecological degradation in the countries on the economic periphery. Global warming and loss of biodiversity are but two examples of how ecological imperialism, which helped sustain the extravagant standard of living in the so-called first world, is no longer a viable strategy. Foster argues that the crisis of capitalism, brought on by running against ecological limits to economic growth, cannot be solved by technology alone. What is required is a fundamental reordering of social relations and a lessened focus on growth and a greater emphasis on equity to allow for greater redistribution.

Implications for Historiography

The idea of ecological imperialism represents a watershed moment in the writing of environmental history. Crosby broke new ground not only in terms of the topic but also in the kinds of data and methods that were utilized for this seminal work. Environmental history despite its undeniable importance has traditionally attracted relatively little attention within the discipline of history. When the environmental conditions were described, it was often as a backdrop to social and political events unfolding. There is little sense of the interplay of the social and environmental forces that work as the engine of history. The role of weather and climate or soil conditions, despite being critical to the maintenance of the productive base of society, has often been neglected. At the very least, inclusion of environmental factors in the analysis of social change can only correct the inherent bias toward anthropocentric explanations in history writing.

Colonial States and Transfer of Plants and Animals

The role of the colonial states in the 19th century toward environmental conservation has been described by scholars as mixed and contradictory. While the colonial states certainly engaged in large-scale extractive activities, they also implemented strategies and policies for conservation. Often, the colonial state in this period was particularly receptive to scientific arguments in favor of conservation. Physicians and other scientists were hired by the colonial states to conduct geological surveys, ethnographic studies, and inventories of the flora and fauna. Due to the state's interest in long-term political stability, it was often willing to override the short-term profit motive of private operators who were engaged in unsustainable exploitation of natural resources. The highly interventionist nature of these policies carried out in the colonies would not have been possible in Europe.

The scientific pioneers advocating conservation in the colonial period were remarkably prescient in foretelling the emerging global environmental problems. Their concerns for environmental degradation were partly rooted in their belief that such degradation would eventually render the land uninhabitable by Europeans. Although their influence was constrained by the policies of the colonial state, and the scientists were often limited in their knowledge of the environment, they left behind an important legacy. The botanical gardens established during the colonial period spurred the cataloging of flora and fauna. The national parks and reserves established—although often entailing exclusion and displacement of indigenous peoples—set models for conservation efforts. The dominance of Enlightenment values, with their belief in the beneficial social impact of scientific knowledge, helped to augment the authority of the conservation researchers.

Colonialism, especially in the islands such as Mauritius and regions such as south Asia, provided an unprecedented opportunity for European thinkers and scientists to observe nature and learn about its workings. Many of these ideas and the knowledge learned from local traditions and management of the environment found their way back to Europe. For instance, the Dutch gardens were modeled after Mughal gardens, which in turn were based on the Edenic notions of nature going back to Zoroastrian culture.

Criticism

Crosby's thesis of ecological imperialism has been criticized for its central claim that the destructive impact of colonialism was partly a result of the ecological dislocation and rupture, due to the transfer of plants, animals, and peoples. According to the critics, this view downplays the centrality of the political and economic interests, motivations, and deliberate actions to colonialism's massive human and social toll.

From a methodological perspective, the thesis presents a useful although overly generalized and one-sided characterization of the colonial encounter. Environmental historians have pointed out the colonial use in some cases of the local knowledge of the environment in conservation policy and practice. Sometimes the recognition of traditional ways of natural resources management was spurred by a political need for stability, but in other cases, it grew out of a realization of the merits of long-standing ways of managing the environment.

One of the central assumptions of the ecological imperialism thesis is that the plant species of the Old World were in a unique position to thrive in the New World environment. This in turn was based on the assumption that the plants had originally evolved along with the flora native to North America, before the process of continental drift separated Europe from the Americas. Ecologists, however, have discovered no difference in the rate at which invasive species of plants and animals are able to establish themselves in Europe or America.

The ecological imperialism thesis also overestimates the role of scientists, colonialists, explorers, and other elites in the transmission of plants and animals. Historical accounts, however, indicate that ordinary people were active agents of these transfers. The transfers were also multidirectional, with the success of the transfer depending upon multiple contextual factors. For instance, the mere transfer of seeds was not a guarantee that the plant species would spread successfully. The outcome would also depend on the transfer of the associated knowledge, practices, and technology for the propagation of the plant species. Another factor that would determine the success of the transfer would be the usefulness to the receiving population. Often the fit of the plant or animals within the existing economic and social system would decide the fate of species.

Evolution of Colonialism and the New Face of Ecological Imperialism

The colonial empires became over time loosely organized political–economic systems geared toward exploiting a wide range of natural resources from all over the world. The modes of regulation employed toward use of natural resources were specific to the mode of exploitation. The contemporary phase of global environmental politics has been described as an empire, but without imperialism. The critical difference is in the noncentralized and diffuse working of the contemporary economic system. The improvements in transportation and communication and the penetration of market forces have all made the world's natural resources, no matter how seemingly remote and inaccessible, highly available for mobilization. These technological and economic advances have also made obsolete the need for direct political subjugation and control of colonies. The traditional control and uses of natural resources like forests and rivers have been weakened and delegitimized. In their place has emerged a regime of management and control that draws its power from the twin forces of the state and markets. The ecological imperialism arising out of the colonialism of the 18th and 19th centuries has thus been replaced by a subtler but perhaps more effective form of resource exploitation, with none-too-dissimilar results in terms of the displacement and dispossession of the indigenous peoples and their progressive marginalization.

Globalization, in the view of many scholars, has led to a softening of the national boundaries, with some even describing the current world system as being post-national and post-sovereign. There is no denying the emergence of global institutions, treaties, and agreements, especially in the economic and the environmental realms. However, simply the existence of transnational institutions does not automatically imply a weakening of the system of nation-states or the domination of certain countries. Another perspective on the current state of political affairs is provided by the global flow of material resources. The flow of natural resources from the periphery (consisting of third-world countries) to the core (comprising countries in Europe and North America) still continues unabated, although emerging economies like China and India render the picture somewhat less clear. In many ways, the situation represents a continuation of the ecological relations, predicated on the exploitation of the resources of Africa, Asia, and Latin America, established by the colonial powers in the heyday of the colonial encounter.

In the contemporary debates surrounding conservation and development, ecological imperialism is seen as a form of neo-colonialism and refers to the imposition of a conservation agenda, typically in non-Western or third-world contexts, that is implicitly or explicitly influenced by Western cultural values, preferences, and biases. Thus, ecological imperialism is no longer exercised through direct political control, but through the programs, ideologies, and administrative techniques of global environmental institutions, alliances, and treaties. In other words, a new global regime of environmental control and management has attained a dominant position. This regime emphasizes consensus, market forces, and individual property rights as the preferred tools and processes for achieving the conservation goals. The ongoing international attempts to ban whaling, led by some Western nations, are a case in point. Countries like Japan that have protested the ban on commercial whaling, on the grounds that it is a deeply rooted traditional activity, have accused the advocates of a global ban as engaged in ethnocentrism and racism. Another more subtle, but perhaps insidious, instance of ecological imperialism is the global conservation politics, as manifested in the conservation policies and practices of international environmental organizations that are guided by the Western cultural notion of wilderness. In this conception, nature and culture are seen as mutually exclusive and incompatible, making therefore the expulsion of local and indigenous people a prerequisite for conservation and restoring forests and other ecosystems to their pristine nature. The displacement of local people, erosion of their property rights and livelihoods, and outright criminalization of their culture are some of the consequences of the imposition of Western conservation practices. These exclusionary policies have often been resisted by the affected people, leading to actions ranging from a preemptive destruction of forests and wildlife—to forestall establishment of protected areas—to a general disillusionment with and rejection of conservation.

Some Consequences of Ecological Imperialism

Consequences of ecological imperialism are discussed below.

Disappearance of the Commons

Among the prominent casualties of ecological imperialism has been the decline of traditional systems of management of resources that belonged to communities and were managed collectively. The forests, wildlife, grasslands, marine resources, and rivers have often been appropriated by states directly or privatized for commercial profit. The underlying reasons are the perennial quest for profits (in capitalistic economies), equation of collective forms of property with inefficiency and waste, and the marginalized condition of the indigenous peoples who inhabit and control the commons. The conversion of the commons has had a destructive effect on the livelihoods of the indigenous groups, rural populations, and women—all groups that disproportionately depend on the commons for subsistence.

Biopiracy

Vandana Shiva, a feminist scholar and activist, has described the appropriation of plant and animal genetic resources in the third world by first-world governments and corporations as biopiracy. The transfer of germplasm of such commercially important crops as the tomato, which has been domesticated and bred by indigenous peoples, has been unilateral, with no compensation provided to the communities or nations of origin. Partly this situation has arisen from the dominance of the Western notion of intellectual property rights, which only recognizes individual inventions, with virtually no mechanism for the recognition and rewarding of collective intellectual property. Another reason has been the lack of awareness and resources in the third-world countries and more specifically the communities, which often do not have the money or the expertise to file the required documentation.

The total monetary value of the uncompensated transfer of plant and material resources has been estimated to be in the billions of dollars.

Climate Change and Ecological Imperialism

Climate change mitigation policies aimed at slowing global warming typically call for a reduction in national greenhouse gas emissions. The emphasis on equal sharing of the burden of emission reduction has meant that countries at very different stages of economic growth are expected to undertake similar reductions. This one-size-fits-all approach does not take into account the requirement for economic growth in developing countries that are struggling to raise the living standard of their desperately poor populations. The notion of ecological imperialism, in this context, has been used to describe the proposed policies of the developed countries, which if implemented, would result in slowing down industrial development and hence economic growth in developing countries. The net result would be a considerable setback for poverty alleviation efforts. This approach, aimed at uniform reductions in greenhouse gas emissions, suffers from two significant flaws. First, it does not take into account the historic contributions of the developed countries, beginning with the Industrial Revolution, to the buildup of greenhouse gases in the atmosphere. Second, as the argument goes, it fails to account for the differing underlying causes for increasing emissions in widely divergent socioeconomic contexts. In developed countries, the emissions of greenhouse gases can be considered as arising from the manufacture of goods that are not strictly needed ("luxury emissions"). In the developing countries, on the other hand, the emissions take place in the course of manufacturing basic necessities ("survival emissions"). A solution proposed to help achieve a more equitable solution to climate change involves using per capita emissions, and not total emissions, as the basis for setting the quotas for greenhouse gas emissions. This arrangement would make allowance for the large populations of countries like China and India, while determining their targets for emission reduction.

See Also: Common Property Theory; Ecocapitalism; Ecofascism; Environmental Justice (Ethics and Philosophy); North–South Debate; Precautionary Principle (Ethics and Philosophy); Tragedy of the Commons; Utilitarianism.

Further Readings

Agarwal, A. and S. Narain. *Global Warming in an Unequal World: A Case of Environmental Colonialism.* New Delhi: Center for Science and Environment, 1991.

Beinart, W. and K. Middleton. "Plant Transfers in Historical Perspective." *Environment and History*, 10 (2004).

Brockington, D., J. Igoe, and K. Schmidt-Solatau. "Conservation, Human Rights, and Poverty Reduction." *Conservation Biology*, 20:1 (2006).

Brockway, L. *Science and Colonial Expansion.* New York: Academic Press, 1979.

Conca, Ken. "Ecology in an Age of Empire (A Reply to and an Extension of Dalby's Thesis)." *Global Environmental Politics*, 4:2 (2004).

Crosby, Alfred. *Ecological Imperialism: The Biological Expansion of Europe, 900–1900.* Cambridge, UK: Cambridge University Press, 2004.

Foster, John B. *The Ecological Revolution: Making Peace With the Planet.* New York: Monthly Review Press, 2009.

Grove, Richard H. *Green Imperialism: Colonial Expansion, Tropical Island Edens, and the Origins of Environmentalism, 1600–1860.* Cambridge, UK: Cambridge University Press, 1995.

Jeschke, J. and D. Strayer. "Invasion Success of Vertebrates in Europe and North America." *Proceedings of National Academy of Sciences*, 102:20 (2005).

West, P., J. Igoe, and D. Brockington. "Parks and Peoples: The Social Impact of Protected Areas." *Annual Review of Anthropology*, 35 (2006).

Neeraj Vedwan
Montclair State University

ECOSOCIALISM

Ecosocialists promote common worker ownership of the means of production (such as this railway and factory) and reject capitalistic solutions to ecological concerns.

Source: Wikimedia

The most general definition of "socialism" is that it is a movement that advocates replacing capitalism with a system that promotes common ownership of the means of production and its supporting infrastructure such as factories, mines, and railroads. In a socialist society, workers control and manage industry and social services through a democratic government. Socialism does not mean government or state ownership of industry but is, instead, a system that places economic control into the hands of workers and of democratically elected representatives. Socialists view the capitalistic model as profoundly flawed, and they cite the yawning gap between the richest and poorest among us as evidence of capitalism's exploitative nature that benefits only a handful at the expense of the majority. Ecosocialism shares this anti-capitalist view and adds to socialism's overarching goals the preservation and restoration of ecosystems; in fact, ecosocialists argue that this preservation and restoration should be central to all human activity. One of the main contentions of ecosocialism is that ecological devastation is not an accidental feature of capitalism but is, instead, an inherent feature of capitalistic growth.

From its earliest modern roots, which reach back to the late 18th century, and up to the present day, debate about both the nature and viability of socialism has taken place among both its proponents and its critics. Because it encompasses such a wide array of political and economic beliefs, many who call themselves "socialist" differ in their views on everything from how to create a socialist society to how that society should function. Typically, less attention is paid to this insular debate than is given to the wider clash between those who call themselves socialist and those who are typically capitalist, who view socialism as an illegitimate political framework, though both levels of the debate inspire fierce, passionate deliberation. Ecosocialism follows this pattern, being critiqued by both other kinds of socialists and by those who identify as capitalist. Beyond these multilevel debates, there are also disputes among various groups and individuals who identify as ecosocialist, yet who differ in their beliefs as to what this means.

Rise of Ecosocialist Thought

Traces of what we today call "ecosocialism" can be found in the works of late 19th-century British socialist William Morris, who espoused preservation of natural resources and protection of the natural environment from pollution and industrialism. By the 1970s and the rise of the modern environmentalist movement, more and more socialists began incorporating into their critiques of capitalism the ecological devastation wrought by an economic system predicated on notions of endless growth. The 1980s saw the creation of the journal *Capitalism, Nature, Socialism,* and by the 1990s, ecosocialism had taken root as a distinct movement, as evidenced by David Pepper's 1993 book *EcoSocialism: From Deep Ecology to Social Justice.* Finally, in 2001, Joel Kovel and Michael Lowy published their "ecosocialist manifesto," and in 2007, the Ecosocialist International Network was formed by ecosocialists from Australia, Belgium, Canada, Greece, Italy, and the United States, among many others. Ecosocialism evolved from the rather placid desires of William Morris to a movement that seeks to replace capitalism with a system that is built upon the following:

- Ecological prudence
- Social equality
- Democratic control and planning
- The dominance of use-value over exchange-value

Ecosocialists see capitalism as not just the enemy of the poor and working classes but also of nature itself and of all human life. They roundly reject "market ecology" or capitalistic solutions to environmental protection such as the "cap-and-trade" system of trading pollution credits to achieve emission reductions and the advent of clean development mechanisms (CDMs), which allow developing countries to continue emitting high levels of greenhouse gases if they pay for reductions made in developing countries. Ecosocialists also reject "productivist socialism," which they say ignores natural limits and associates with Marxism in general and, more specifically, with socialist regimes such as in the Soviet Union and China as well as with Democratic Socialist nations in western Europe. Further, socialists in general argue that regimes in the Soviet Union, China, and Cuba, for instance, were and are not socialist but represent, instead, forms of "state capitalism."

In relation to its critiques of productivist socialism, ecosocialism criticizes the wider socialist movement for treating environmentalism as an afterthought and for not recognizing the links between the domination of labor and the domination of nature; ecosocialists are particularly scathing in their treatment of contemporary socialists who believe there is a positive side to capitalism's self-expansion that can be used to meet the needs of emancipated labor and who argue that socialism cannot come at the expense of human satisfaction.

Beyond their criticisms of other socialist-based movements, ecosocialists also find fault with various "green politics" movements, for a variety of reasons, including the following:

- Not being sufficiently anticapitalist
- Working from within "the system"
- Relying on technological advances to solve ecological problems
- Failing to demand the emancipation of labor
- Failing to advocate the separation of producers from the means of production
- Failing to advocate societal transformation

Achieving the Ecosocialist Vision

Ecosocialists envision a nonviolent revolution that results in the dismantling of not just capitalism but of the state as we know it. They seek to achieve a world guided by an economic policy founded on nonmonetary criteria that fulfills social needs while protecting ecological equilibrium. Factories and other means of production would be collectively owned by the workers themselves, and natural resources, instead of being considered private commodities, would be considered as part of "the commons." Ecosocialists are also deeply committed to the emancipation of women, seeing the domination and degradation of women and of nature as similar processes. While ecosocialists see a role for the working class in achieving this societal transformation, they generally advocate a revolutionary process that embraces agency from both autonomous individuals and grassroots groups, no matter their class or status. Ecosocialists believe that this process of transformation will begin with widespread, intense critiques of capitalism coupled with challenges to the notion that there is no alternative to the capitalistic system. Some of the steps that ecosocialists believe can be taken right now in order to begin ushering in an ecosocialist reality include the following:

- Organizing labor
- Forming cooperatives
- Creating localized currencies
- Inducing radical changes in the energy system, the transportation system, and consumption patterns

Ecosocialists see democratic planning as central to achieving their vision. They contend that control of investment and technological change should not be relegated to banks and other capitalist enterprises but should instead be controlled by democratic public decisions. While some see in this notion the potential for the kind of tyrannical control that has marked other supposedly socialist regimes, ecosocialists see this kind of democratic planning as the very epitome of freedom. In seeking to create an egalitarian, democratic civilization that meets the needs of all citizens while protecting the environment, ecosocialists argue that a transformation of society, culture, and attitudes is necessary. Without this transformation, and without the active support of most of the population, the ecosocialist vision will remain mere conjecture.

Points of Contention

Ecosocialism has been met with criticisms from within socialist circles as well as from capitalists. The criticisms of socialists regarding ecosocialism include the following:

- As Marxism is intrinsically ecological, there is no need for an ecosocialist movement.
- The ecosocialist movement ignores Marx's class-based principles and ignores the working class.
- Ecosocialism has the potential to nurture fascism.
- Ecosocialism focuses on the far-off future instead of on fixes that can be embraced now.
- Ecosocialism neglects contemporary and future population problems.

Capitalists also challenge the viability of the ecosocialist vision. Some of their criticisms include the following:

- The free market is the best way to meet the needs of citizens.
- Ecosocialism will breed totalitarianism.
- A centrally planned economy, without market prices or profits, lacks effective incentive mechanisms to direct economic activity.
- Controlled or fixed prices convey ambiguous information about scarcity of resources and commodities.
- Disregard of the role of private property in creating incentives that foster economic growth and development is irrational.
- Privatization of "the commons" is the best way to protect it.
- Market-based solutions to ecological problems are the best method for remedying an ailing environment.

Ecosocialism, then, faces from capitalists a deep and entrenched hostility to its notions of cooperation and democratic planning as well as to its solutions for protecting the environment and fostering sustainability. From the wider socialist community, ecosocialism is seen as unnecessary at best and, at worst, at odds with Marxism's class-based revolutionary model. For their part, ecosocialists contend that without the type of revolution they envision, the gap between the richest and poorest among us will continue to widen while the Earth suffers irreversible damage.

See Also: Capitalism; Carbon Trading/Emissions Trading; Ecocapitalism; Intrinsic Value Versus Use Value; Tragedy of the Commons.

Further Readings

Kovel, Joel. *The Enemy of Nature: The End of Capitalism or the End of the World?* London: Zed Books, 2007.

Sarkar, Saral. *Eco-Socialism or Eco-Capitalism? A Critical Analysis of Humanity's Fundamental Choices.* London: Zed Books, 1999.

Workers' Liberty. "Review of Joel Kovel—The Limits of Eco-Socialism." http://www .workersliberty.org/node/9574 (Accessed September 2010).

Tani E. Bellestri
Independent Scholar

ECOTAX

An ecotax has been defined as a tax intended to promote ecologically sustainable activities via economic incentives. In practice, ecotaxes have also been referred to as environmental taxes and sustainability taxes. Two parts of this definition bear examination in order to fully understand the concept of ecotaxation. First, they are intended to promote ecologically sustainable activities. Ecotax policies are intended to affect sustainability behavior with both positive and negative incentives. Ecotax policy may be designed to encourage environmentally friendly activities, for example, tax credits for the purchase of hybrid electric vehicles. On the negative side, ecotaxes are designed to discourage environmentally harmful activities. An example of this would be a tax on the use of plastic bags for purchases in stores.

The second part of this definition focuses on "economic incentives." Again, we see that this can involve incentives such as credits or disincentives such as an environmental charge or fine. But the use of ecotaxes frequently goes beyond a strict interpretation of the definition of tax and moves into the areas of fees, fines, or other such mechanisms.

Ecotaxes are examples of Pigovian taxes, a concept developed in the early part of the 20th century by the economist A. C. Pigou. He drew a distinction between the private and social value of economic activities, stating that the social costs must be included in determining the ultimate worth of an economic activity. Social costs are the "externalities" that are not included in the private costs. If the social costs exceed the private value, authorities should introduce "restraints" such as fees or taxes so the private costs are equal to the private values. For example, the construction of an airport will benefit those who utilize air travel. At the same time, homes and businesses may be displaced and inconvenienced by the construction. The reverse is also true—if an activity has social value exceeding the private value, "encouragements" in the form of government subsidies should be included to ensure an adequate supply of these public goods. Recreational parks, street lamps, and other public goods are difficult projects to charge for, so the encouragements are needed to ensure an adequate supply of these public goods.

Taxation in general can be revenue motivated or regulation motivated. Ecotaxes should not be viewed as a continuing source of revenue. A good revenue tax is one that has a robust tax base—the tax does not change the behavior of the people paying the tax. This is not the goal of ecotaxes. In addition, a successful ecotax would reduce the incidence of the activity being taxed and, hopefully, exist only at a minimal level. For example, Ireland introduced a tax of 33 cents on each plastic bag. In a matter of weeks, usage of plastic bags declined by 94 percent.

Ecotaxes fall into the realm of regulation-motivated taxes. As seen with the example of the Irish plastic bag tax, the revenue stream was short lived, but the intended purpose of reducing use of the bags was accomplished. Since ecotaxes should not be relied on as a continuing source of revenue, one area of debate has been what to do with the revenue from these taxes.

It is obvious that there are both pros and cons to any system of ecotaxation. The issues may be insurmountable, due to differing viewpoints. Anytime there is change, particularly on a large scale, there will be those who are harmed by the changes and those who benefit from them. It can be hoped that all interested parties can reach agreement on what should be done in this area.

Proponents of sustainability and, by association, ecotaxation make the argument that strong action in this area will lead to a cleaner environment. While some may deny the need for us to be better stewards of this planet, most today agree that our collective environmental indifference is not a position that can continue forever.

Pros and Cons of the Double Dividend

Advocates of ecotaxation have declared a double dividend in relation to these taxes. In addition to the above-cited environmental improvements, they claim that the tax collections obtained through ecotaxation can be utilized to reduce payroll taxes on employees and employees.

A related argument for ecotaxation is that it will result in better overall national performance. The argument is made here that by refocusing on production of economic "goods" rather than "bads," the economy will prosper as less effort is devoted toward less efficient production activities. As the economy improves and production focuses on human rather than natural resources, the result will be increases in employment and an economic boost.

There are some serious doubts about the validity of the double dividend. Empirical tests have not proved it does not hold. There is little argument that a regime of ecotaxation will result in a cleaner environment. It should not be overlooked, however, that not everyone agrees to the need for a cleaner environment. The main concern with the double dividend arises with the second component. Reducing payroll taxes is a laudable goal. It can lead to higher employment and give a boost to the economy. However, as has been observed, ecotaxes are regulatory in nature, not an intended revenue source. If the ecotax is successful, the revenue stream will dry up. What then happens to payroll taxes? Either they are raised to their previous levels, or other sources of revenues are sought.

Pros of Ecotaxes

Generally speaking, five advantages are cited for an ecotax system.

First, ecotaxes provide incentives for behavior that protects or improves the environment. History has shown that businesses, consumers, and governments will take action based on their own self-interest. If they are not compelled to pay for the full cost of their activities, they have incentive to pursue those activities, regardless of the effect on others. As an example, most municipalities have laws regarding excessive noise. If someone wants to play their "boom box" at top volume, they do it without regard to the fact they may be disturbing others. However, if they are fined for that activity, it is hoped that the incidence of excessive noise will decrease.

Agreement on the desired behaviors is somewhat of an issue at this point. What behaviors protect the environment? What harms it? While many feel they know the answers to these questions, there is by no means a consensus of opinion.

Second, ecotaxes enable environmental goals to be reached in the most efficient manner. By specifying those goals, a first step is taken. This alerts everyone regarding the specific environmental goals that are sought. The desired behavior then becomes known. In taking the next step, taxes can provide the incentive or disincentive needed to achieve the goals. The bag tax is an excellent example in that the use of plastic bags is virtually eliminated when a tax is placed on their use.

Some would argue here that efficiency decreases when government gets involved and that the most efficient approach would be a free-market solution. This does not seem to be a valid argument, as it is clear to most that the free market has failed in this instance.

Third, ecotaxes can help signal structural environmental changes that are needed for a more sustainable economy. Taxation can give a clear signal about what changes must occur in order to achieve the desired goals. This is not always overt, but taxation can and does change behavior. One only needs to look at the income tax system in the United States to realize that taxes can cause people to change their behavior.

Two issues arise regarding this advantage. First, while it is hoped that the ecotax regime would provide a clear signal, that is not always the case. The Ontario government recently enacted an "eco fee" on certain consumer goods. It was designed as a pass-on fee to consumers who pay Ontario for recycling their products. The fee was fraught with inconsistencies in collection, overcharges, and other issues and was ultimately scrapped in less than two years. No clear signal was given. Second, there must be agreement about which structural changes are needed.

Fourth, the existence of taxes can encourage innovation and the development of new technologies. If, for example, a heavy tax were levied on the use of fossil fuels, it is highly likely that research into alternative technologies would increase. This becomes a stronger argument if tax credits were given for this type of research.

Finally, revenue from ecotaxes can be used to reduce other taxes. This goes beyond the reduction of payroll taxes cited as a part of the double dividend. Since ecotaxes tend to be regressive, an argument is made to return a portion of these taxes to those in the lower-income levels to offset the distortions from the ecotax.

The main problem with this point is the same as with the double dividend of lowering payroll taxes. If ecotaxation accomplishes its purpose, there will be dwindling amounts of revenues from the tax and a smaller pool of funds to offset the distortions. However, the distortions may have been reduced or eliminated by this time.

As has been shown, those who oppose ecotaxation do not accept all of these advantages. An overriding issue for all of these advantages is the increase of government regulation in this area. Libertarians and others may cite the desire for smaller government, but ecotaxes can be intrusive in their level of regulation.

Cons of Ecotaxes

There are three primary disadvantages that are frequently cited for an ecotax system. The first of these are the displacements that occur through job losses and company closings due to an emphasis on sustainability. Although these are likely to be short-term displacements, when someone loses a business or a job, it becomes a personal crisis. From a macro view, ecotaxes should provide increased employment and new business opportunities, so it is expected that such a system will produce significant economic growth.

This objection to ecotaxes could be minimized by taking a macro view that, in the long run, it will be better for the economy. One solution to the short-term displacements is government intervention. This could take the form of extended unemployment benefits, retraining for new job skills that would be in demand in a sustainable economy, or assistance in forming a new business.

The second drawback to an ecotax system is that a successful ecotax regime cannot exist in isolation. To succeed, cooperation and coordination among all nations is necessary.

Otherwise, the benefits realized from those enacting ecotax systems would be diminished. While not minimizing the difficulty, this can be overcome by an international agreement not unlike the existing Kyoto Protocol. Such a treaty could be crafted to coordinate a global approach to a sustainable tax policy. However, it is not likely that even this would result in universal acceptance. It should be noted that the Kyoto Protocol lacks the participation of two of the largest producers of greenhouse gases—China and the United States.

The third disadvantage relates to efforts to circumvent the tax. As with any tax legislation, efforts will be taken by some to attempt to avoid complying with the statutes. This is also true in the environmental area. It is not inconceivable that an entire underground economy will spring up, devoted to avoiding the ecotaxes. Unfortunately, the best answer to this issue is to attempt to make the legislation comprehensive enough that it becomes difficult to circumvent the laws and to couple this with aggressive enforcement.

See Also: Capitalism; Ecocapitalism; Environmental Justice (Business); Kyoto Protocol; Sustainable Development (Business).

Further Readings

Beauregard-Tellier, Frederic. "Ecological Fiscal Reform" (March 17, 2006). http://www2.parl .gc.ca/Content/LOP/ResearchPublications/prb0595-e.html (Accessed August 2010).

Pillet, Gonzague. "Environmental Fiscal Reform." *Environmental Economics* (2006). http:// www.meso-platform.org (Accessed August 2010).

Robertson, James. "Eco-Taxes." *New Internationalist* (2005). http://www.newint.org/ issue278/taxes.htm (Accessed August 2010).

Stancil, John. "Building Sustainability Into the Tax Code." 2010 Southeastern Chapter of InfORMS Annual Proceedings. http://www.roanoke.edu/business/SEINFORMS%20 2010%20-%20Proceedings/Start.HTM (Accessed August 2010).

Stoner, James A. F. and Charles Wankel. *Global Sustainability as a Business Imperative.* New York: Palgrave Macmillan, 2010.

Sutton, Philip. "Eco-Taxation" (2000). http://www.green-innovations.asn.au/eco-tax.htm (Accessed August 2010).

John L. Stancil
Florida Southern College

ECOTOURISM

Ecotourism—with its emphasis on minimal environmental impact, empowering local populations, financing environmental protection initiatives, and educating tourists and hosts alike on their environmental as well as cultural surroundings—offers remarkable potential to link conservation interests to economic development, particularly in the developing world. While by no means a panacea, much of the promise of ecotourism centers on its ability to bridge northern and southern nation-state interests by providing an environmentally benign funding mechanism for economic development in the southern hemisphere.

Yet the term remains problematic due the lack of independently verified global certification programs; this has allowed elements within the tourist industry to claim they are full-fledged ecotourist operations even when they merely adopt practices that save them money. Better labeled as *ecotourism lite* or even outright *greenwashing*, such practices mimic some of the very difficulties faced within the larger umbrella term of *sustainable development*, especially when the reputation of legitimate ecotourist operations is compromised by those simply attempting to capitalize on the public popularity of green practices and design.

Problems start first and foremost with the very definition of *ecotourism*. Origins of the term range from work by Mexican ecologist N. D. Hetzer in 1965 to Kenton Miller and his 1978 explanation of national park planning in Latin America, but renowned Mexican architect and international ecotourism consultant Hector Ceballos-Lascurain is most often credited with coining the term and developing its initial working definition in 1983 while heading up Pronatura, a highly influential conservation nongovernmental organization (NGO) in Mexico. Used and misused since its emergence in the 1980s, the term *ecotourism* includes activities ranging from hiking and mountain climbing to bird watching and nature photography to river rafting and canoeing and fishing to safari and camping and botanical study. Worldwatch Institute notes at least seven related categories of tourism, which complicate matters further, including adventure tourism, geo-tourism, mass tourism, nature-based tourism, pro-poor tourism, responsible tourism, and sustainable tourism. Despite some notable overlap in terms of common destinations and objectives, ecotourism is distinct with its emphasis on nature and assisting local populations at the same time.

Founded in 1990 and serving as the oldest and largest association in the ecotourism industry today, The International Ecotourism Society (TIES) offers perhaps the most succinct definition as well as one of the most cited, at least within industry circles. According to TIES, ecotourism is responsible travel to natural areas that conserves the environment and improves the well-being of local people. Noted authors such as Martha Honey take this description a step further and spell out well-being for locals in both economic and ecological terms as well as adding the need for environmental and cultural awareness on the part of both visitors and hosts. Posadas Amazonas, Tambopata Research Center, and Refugia Amazonas in the Peruvian Amazon are cases in point. Operated by Rainforest Expeditions, the lodges are actually owned by the indigenous community of Infierno. Along the Rio Tambopata, one of the headwaters of the Amazon, this operation emphasizes environmental education and directly exploits the economic merits of the majestic macaws, whose value scholars such as Honey estimate as $4,700 per free-flying bird.

Over the years, Honey's definition of ecotourism combined with the TIES wording have guided the growth of the concept into a significant force within the larger tourism industry. Indeed, TIES estimates ecotourism is growing at a rate three times that of tourism. This is all the more notable when one considers tourism in general as one of the largest industries in the world, employing some 235 million people and accounting for approximately 10 percent of world GDP. Over the past three decades, tourism has become the principal export in 83 percent of the developing world and the leading foreign exchange earner in the poorest one-third of the world.

But within these numbers lie some notable difficulties. For one, there is the obvious danger of overreliance on one form of foreign export, one that is especially susceptible to economic recession. Second, profits from conventional tourism operations typically bypass much of the local population, with monies from all-inclusive resorts such as those popular in Puerto Plata, Dominican Republic, finding their way back to international headquarters, not locals. Within this neo-colonial type of relationship, dissatisfaction with the negative aspects of traditional tourism began to coalesce in the 1980s, and looking to Africa and Latin America, two parallel tracts developed. In Africa, early proponents saw ecotourism as an alternative to preservationist park management policies that had painfully marginalized local communities. In Latin America, advocates saw the concept as an opportunity to increase the value of natural capital and improve understanding of conservation priorities.

Conventional sun-and-sand-type operations that cater either to the masses or to the wealthy are still a major force within the tourist industry, of course, but their growth rate has not matched that of ecotourism over the past decade. TIES and the United Nations World Tourism Organization provide support mechanisms and global attention here. Initiatives such as the 2002 United Nations International Year of Ecotourism are also an important start, but only a start. Tourism in "hotspots" grew at a staggering rate the past two decades, and there is a real danger ecotourism might become a victim of its own success. Too many visitors both ecologically and aesthetically threaten carrying capacity, not to mention the possibility of cultural commodification and continued neocolonial relationships that promote dependency with an overreliance on one export.

Yet what really threatens to undermine ecotourism to date, before these key latter-stage issues of carrying capacity and dependency, is the lack of a genuine global certification regime. In fact, moving ecotourism from merely a good concept to actually good practice depends fundamentally on certification. Certification is a well-established concept in the tourist industry, dating back to at least 1900 and the Michelin guidebooks that the French tire company first published as a rating of hotels and restaurants. Soon thereafter, the American Automobile Association followed suit with its rankings based on a system of one to five stars.

Industry insiders insist the ecotourism industry needs a green equivalent. Admittedly, since the Earth Summit in Rio in 1992 there has been a burst of efforts, with 60 to 80 such initiatives now circulating. Most of these are based in Europe, though, and cover one nation only, or at best, one region. The challenge then remains to incorporate large stretches of Africa and Asia, which to date have been left out, and to standardize existing ratings around the globe. Beginning with the three-day summit in May 2002 in Quebec City, this need was publicly recognized by the industry, and the Quebec Declaration on Ecotourism calls for approved guidelines on ecolabels backed by measurable criteria.

Costa Rica provides perhaps the best model to transfer to the global level with its national program of Certification for Sustainable Tourism (CST). Beginning in 1997 with its first edition for lodges and hotels, Costa Rica recognized that its reputation as a world leader in ecotourism was becoming muddied by new coastal tourist developments that claimed to be green but were merely greenwashing as well as by heightened competition from nearby states such as Ecuador, Panama, and Nicaragua. The solution, a set of 154 yes–no questions in four general categories, had initial success but then struggled when government funding for the initiative was slashed. In the past several years, it has enjoyed a resurgence and gained recognition for the degree to which its system of one to five green stars strikes a familiar chord to past tourism practices such as the five-star system. The CST system has also been applauded for using the lowest of the four question categories as the one that determines the rating (instead of merely averaging the four together), and thus encouraging continued improvement in all four areas of physical–biological environment, hotel facilities, guest services, and socioeconomic environment (instead of allowing good practices in one area to cancel out poor practices in another).

At its core, ecotourism is about power, positioning local concerns as the driving force, or at least equal partners to the brick-and-mortar of the tourist operation itself. This is not easily accomplished and often takes years to develop sustainable initiatives. As with the larger concept of sustainable development, successful ecotourism requires capacity building within local communities. It also needs partnerships among the private sector, nongovernmental organizations (NGOs), and local as well as national and state authorities.

Still, the proliferation of legitimate ecotourist operations over the past three decades provides increasing evidence that ecotourism is much more than a passing fad. Almost every United States–based conservation organization incorporates ecotourism at some level, including Conservation International, Environmental Defense Fund, the National Geographic Society, The Nature Conservancy, World Resources Institute, and World Wildlife Fund (WWF), to name but a few. NGOs such as the Audubon Society, Earthwatch, Sierra Club, and WWF run ecotourist field trips. Similarly, the Smithsonian Institution and American Museum of Natural History run travel programs in ecotourism, as do a number of universities and colleges across the United States, for both their current students and alumni.

See Also: Carbon Emissions (Personal Carbon Footprint); Certified Products; Ecolabeling; Sustainable Development (Business).

Further Readings

Conservation International. "Developing Ecotourism." http://www.conservation.org/learn/culture/ecotourism/Pages/ecotourism.aspx (Accessed September 2010).

D'Orso, Michael. *Plundering Paradise: The Hand of Man on the Galapagos Islands.* New York: HarperCollins, 2002.

Duffy, Rosaleen. *A Trip Too Far: Ecotourism, Politics and Exploitation.* Sterling, VA: Earthscan, 2002.

Honey, Martha. *Ecotourism and Sustainable Development: Who Owns Paradise?*, 2nd ed. Washington, DC: Island Press, 2008.

Honey, Martha. "Treading Lightly: Ecotourism's Impact on the Environment." *Environment*, 41:5 (1999).

The International Ecotourism Society. http://www.ecotourism.org (Accessed September 2010).

Leakey, Richard and Virginia Morell. *Wildlife Wars: My Fight to Save Africa's National Treasures.* New York: St. Martin's Griffin, 2001.

McLaren, Deborah. *Rethinking Tourism & Ecotravel*, 2nd ed. Bloomfield, CT: Kumarian Press, 2003.

National Geographic. "Center for Sustainable Destinations." http://travel.nationalgeographic.com/travel/sustainable/index.html (Accessed September 2010).

Pattullo, Polly. *Last Resorts: The Cost of Tourism in the Caribbean*, 2nd ed. New York: Monthly Review Press, 2005.

Michael M. Gunter Jr.
Independent Scholar

EDUCATION, FEDERAL GREEN INITIATIVES

Here, a "death row" of vehicles being traded in during the government's "Cash for Clunkers" program will be replaced with more fuel-efficient vehicles.

Source: Wikimedia

Federal green initiatives are broadly defined as programs or policies forwarded by the federal government of the United States to promote environmental conservation, management, and sustainability efforts in both the public and the private sectors. In this way, the federal government educates citizens through the dissemination of information and the implementation of programs that intend to help all people gain access to notions of sustainability and an appreciation for how their consumer choices affect the environment. In addition, the consumer-oriented initiatives often help citizens understand how they can save money by making green choices. The initiatives emerge as attempts to ameliorate the rising levels of greenhouse gases (GHGs) in Earth's atmosphere, the proliferation of slowly deteriorating plastic and other petroleum-based products in the environment, and the growing problem posed by human activity as it affects the sustainable state of Earth. Presumed in these initiatives are the contentious claims that (1) there is a growing problem regarding the sustainability of natural resources, and (2) human beings are, at least in part, responsible for the environmental changes. Additionally, debate continues over whether there is work that can be done to alter the kinds and amounts of environmental impact humans have on the world. If so, is it the responsibility of the federal government to use resources to work toward these goals? That the federal government does forward green initiatives opens debate regarding the quality and types of initiatives. What could be done differently? Finally, debate surrounds the effectiveness of past, current, and future initiatives. Sustainability as a guidepost for future action and green initiatives as means to make change contain political elements, making the debates around such topics layered and complicated. This article presents a synopsis of initiatives and discusses the differing perspectives held regarding federal green initiatives.

Federal Government Green Initiatives

According to the Environmental Protection Agency (EPA), whose mission is to protect human health and to safeguard the natural environment—air, water, and land—upon which life depends, the federal government is moving forward with green initiatives in these areas: taking action on climate change, improving air quality, ensuring the safety of chemicals, cleaning up our communities, protecting America's waters, expanding the conversation on environmentalism and working for environmental justice, and building strong state and tribal partnerships. Titles from a representative sample of federal laws and Presidential Executive Orders (EOs) relating to green concerns include the Atomic Energy Act (1946), the National Environmental Policy Act (1969), the Clean Air Act (1970), the Clean Water Act (1972), the Endangered Species Act (1973), the Safe Drinking Water Act (1974), the Resource Conservation and Recovery Act (1976), the Toxic Substances Control Act (1976), the Ocean Dumping Act (1988), the Oil Pollution Act (1990), the Pollution Prevention Act (1990), the Federal Food, Drug, and Cosmetic Act (2002), the Energy Policy Act (2005), and the Energy Independence and Security Act (2007). On October 5, 2009, President Barack Obama signed Executive Order 13514, Federal Leadership in Environmental, Energy, and Economic Performance, which requires government offices and agencies to meet sustainability goals in their operations and purchases. In addition to requiring agencies to purchase greener products, the EO also spells out requirements for green services and vehicles.

The federal Environmental Education Act (1970) established the Office of Environmental Education to develop curricula for incorporation into elementary- and secondary-level environmental education programs. This marked the first time the federal government created curricula to be used in schools. The EPA provides resources through its website for educators, including education grants, student fellowships, and classroom curricula. In addition to disseminating information to students and teachers, the EPA also has separate initiatives for businesses and nonprofits, concerned citizens, media, scientists and researchers, state and local governments, and tribes. Beyond the headquarters operation of the EPA, there are also 10 regional offices, and most states have EPA branches that work closely with the public and private sectors of each state to disseminate information, award grant money, and host online blogs for open discussion. In addition to the EPA, these federal departments also undertake green initiatives: Department of Energy, Department of Defense, Department of Agriculture, Department of Commerce, and Department of the Interior.

The Energy Star initiative, begun in 1992 as a joint project between the EPA and the Department of Energy, is a voluntary labeling of products that meet green standards spelled out by federal guidelines. Programs like Energy Star help to disseminate information regarding power usage and monthly bill reduction as means to educate the American public about how to save money and use fewer resources. Another federal program intended to help citizens save resources and money was the Car Allowance Rebate System (CARS) or "Cash for Clunkers" program. Lasting through the months of July and August 2009, the multibillion-dollar program gave rebates to citizens who traded in low-efficiency vehicles for more fuel-efficient new vehicles. This initiative exposed many Americans to the stark reality of cost savings to be gained from efficiently operating vehicles and was a means to help them live less resource-dependent lives.

As of 2010, three federal tax credits for consumer energy efficiency are available for qualifying homeowners when making specific energy-saving product purchases. Included in the tax credit program are biomass stoves; heating, ventilating, and air conditioning (HVAC) units; insulation; metal and asphalt roofing; windows and doors; geothermal heat pumps; residential wind turbines; solar energy systems; and residential fuel cells. Federal tax credits are also available for purchases of qualifying fuel-efficient and hybrid vehicles.

The following sections portray the contention surrounding green initiatives forwarded by the federal government. Much debate occurs around whether the federal government should be involved in environmental activism. Much of this debate pertains to the discussion of how much power the federal government should have regarding regulation of and control over the private sector (i.e., corporations) and individual states (states' rights).

Green Initiatives as Fixes?

Federal green initiatives are the government's attempts to foster awareness of and work against climate change. Within the perspective that acknowledges dangerous elements of climate change, there is debate over whether federal (or for that matter, any other social or governmental group's) initiatives can have a positive impact. What if it is too late? To believe that climate change is real is not necessarily to believe that spending federal tax dollars on initiatives to help curb it is necessary or even desirable. One position within the debate is occupied by people who believe that the federal government, or any group or individual, can in no way affect the impending doom awaiting Earth's inhabitants; we could never stop what has already begun. Another position is occupied by people who believe government-led social movements to protect and conserve the environment can have lasting impact and even healing effects upon Earth. A third position, and this is a position common to debates concerning federally related programs, is occupied by people who believe that the federal government should not be spending hard-earned taxpayer money on such expensive activities as these.

Do Green Initiatives Work, and If So, Which Ones?

Within the movement that believes federally funded green initiatives can have a useful and measurable impact on environmental conditions and health, there is no consensus on which direction to head or if green initiatives have the desired effect. Long-standing debates between conservationists (those who believe in the need for management of natural resources) and environmentalists (those who believe that nature should be left in or returned to its natural state) are often invoked during discussions of how to be green. Is heavy regulation the answer? Are incentives the means by which to engage a capitalist economy? Should businesses simply use less energy? Or should they purchase Renewable Energy Certificates? Do federal initiatives that are regulatory in nature have the power to demand different practices from corporations? Can the federal government make demands of states? Or is the federal government too weak to exact its initiatives with any real effect? Can green initiatives save individuals and corporations money? Or does the green movement require Americans to rethink how much they should have to pay to purchase and use nonrenewable resources? The only matter to be sure of is that the debate over the federal government's involvement in environmental matters will continue as long as there is a federal government and an environment in which it acts.

See Also: Carbon Emissions (Personal Carbon Footprint); Certified Products; Corporate Social Responsibility; Education, State Green Initiatives; Environmental Justice (Politics).

Further Readings

Environment, Health and Safety Online. "All U.S. Government Web Sites." http://www.ehso
.com/govtenvirsites.php (Accessed September 2010).

Environmental Defense Fund. "Science of Global Warming." http://www.edf.org/page
   .cfm?tagid=54192 (Accessed September 2010).
National Center for Policy Analysis. "Myths of Global Warming." http://www.ncpa.org/pub/
   ba230 (Accessed September 2010).
U.S. Environmental Protection Agency. http://www.epa.gov (Accessed September 2010).

Chad A. Becker
Indiana State University

EDUCATION, STATE GREEN INITIATIVES

State green initiatives can be defined as programs or policies created by the state governments of the United States to promote efforts of management, conservation, and sustainability for public organizations, private industry, and individual citizens. Through these initiatives, state governments educate citizens via information dissemination and the implementation of programs and policies that intend to educate about sustainability and give an appreciation for how individual choices can have immediate and lasting effects upon the environment. In addition, consumer-oriented initiatives assist purchasers in making environmentally friendly choices that will help save them money. The initiatives are born from organized government attempts to affect the rising levels of greenhouse gases (GHGs) in Earth's atmosphere, the collection of toxic substances in our environment and our bodies, and the increasing concern that human activity is creating an imbalance in Earth's sustainable existence.

Topics of debate regarding these initiatives include contention about whether changes in the environment are happening at all and discussion about whether, if there is climate change, human activity is the cause. Additionally, debate wages over the possibility of state green initiatives having an influence on the size and scope of environmental impact by human activity. If it is possible, are state governments the venue through which the environment should be protected and conserved? The use of taxpayer money for what is often considered a political agenda includes within the debate over green initiatives an element that scrutinizes the function of state government and government in general. As state governments do forward green initiatives, debate regarding the quality and types of initiatives ensue, and criticism is often leveled at the way that data are reported and how the "environment" becomes a political tool. This entry contains a summary of state government green initiatives and an overview of the debates and positions held within discussions of state-level green initiatives.

State Government Green Initiatives

The U.S. federal government, through the Environmental Protection Agency (EPA), promotes green initiatives that improve air quality, ensure the safety of chemicals, protect America's waters, clean communities, work to facilitate discussions of the environment and environmental justice, take action on climate change, and build strong partnerships with states. The EPA, whose mission is to protect human health and to safeguard the natural environment—air, water. and land—upon which life depends, is a federal branch with some purview and influence over individual states. Federal laws under which states must operate include one of the first federal acts attending to the environment from the 1940s—the Atomic Energy Act. The late 1960s and early 1970s were a time of productive work regarding federal law and the environment. During this period, the National Environmental Policy Act, the Clean Air Act, the Clean Water Act, the Endangered Species Act, the Safe Drinking Water Act, the Resource Conservation and Recovery Act, and the Toxic Substances Control Act were all enacted. The late 1980s and early 1990s saw another round of federal acts adopted, including the Ocean Dumping Act, the Oil Pollution Act, and the Pollution Prevention Act. The mid-2000s saw the passage of the Energy Policy Act and the Energy Independence and Security Act.

In 1970, the Federal Environmental Education Act created the Office of Environmental Education. The purpose of this office was to devise teaching materials for use in K–12 environmental education programs. The EPA provides resources through its website for educators, including information about education grants, student fellowships, and classroom curricula. In addition to making information available to teachers and students, the EPA also has separate initiatives for businesses and nonprofits, concerned citizens, media, scientists and researchers, state and local governments, and tribes. The EPA operates 10 regional offices with locations in Boston, Massachusetts; New York City; Philadelphia, Pennsylvania; Atlanta, Georgia; Chicago, Illinois; Dallas, Texas; Kansas City, Kansas; Denver, Colorado; San Francisco, California; and Seattle, Washington. These regional offices coordinate and facilitate the activities in surrounding states to offer cohesion to initiatives and organizational support to the efforts of state governments. Areas in which state representatives can gain support from regional offices include air and radiation, climate change, complying with and enforcing environmental laws, emergency management, energy efficiency and resource conservation, financing environmental systems, grants, performance and accountability, pesticides, reducing pollution and toxins, regulations and reporting, transportation, waste and cleanup, and water.

In addition, most states have government environmental offices that work closely with both the public and private sectors of each state to transmit information, award grant money, and provide online venues for further discussion of issues. A number of states have large and directed environmental programs and initiatives. California's Green Building Initiative and Green Building Action Plan require state buildings to be 20 percent more energy efficient by 2015 and encourage the private sector to do the same. In New York, the Green Initiative has the state comptroller's office inquiring into whether state agencies, public authorities, and local governments are complying with environmental and energy requirements. The Michigan Green Jobs Initiative helps to ensure green businesses and industries have the workers needed to prosper through skills training and readiness certification. Pennsylvania's Growing Greener II Initiative works to preserve natural areas and open spaces, keep rivers and streams clean, attend to contaminated sites, work on state park infrastructure, and repair habitat. The Vermont legislature created the Vermont Sustainable Jobs Fund to develop markets where sustainable practices and products proliferate. Their current efforts are related to biofuels and renewable energy. The Washington, D.C., Green Collar Jobs Initiative helped small businesses and citizens with policies to increase green building, reduce carbon emissions, improve the water and sewer infrastructure, and upgrade public schools. In Colorado, Executive Order # D005 05 adopted LEED certifications for all state buildings, old and new. The order also created the Colorado Greening Government Coordinating Council, which is charged with creation of environmental policy.

As of 2004, Massachusetts requires LEED certification for all new construction or major renovation to state buildings. New Hampshire requires its Department of Administrative Services to implement a tracking system to measure how efficient its green initiatives have been. Indiana's Alternative Power and Energy grant program assists nonprofit, for-profit, and public groups in their own green initiatives for reducing the use of carbon-based fuel. Iowa's state government requires all agencies to use a minimum of 10 percent of their energy from renewable sources. In Arizona, three out of every four state vehicles must be clean-fuel ready. California and Massachusetts require low-emissions vehicles to be employed in the state fleet. Connecticut and Maine require state vehicles to exceed miles-per-gallon requirements. Many states have committed to reducing government-run facilities energy use by 10 to 20 percent.

Thirty-six states have requirements for energy efficiency in their public facilities. Almost as many states have requirements for purchasing energy-efficient appliances for state buildings. California and Virginia are two states that require Energy Star purchases for state facilities. In addition, many states require their facilities to incorporate green initiatives that conserve resources: these include turning off lights when not needed, installing automatic controls to keep heating, ventilating, and air conditioning (HVAC) systems operating minimally, switching to energy-efficient lighting and bulbs, and utilizing outdoor light inside buildings. Through green initiatives, state governments can have a large impact upon how consumers conceive of their consumer choices and how markets are created and driven. Energy efficiency, clean energy, and alternative fuel production policies continue to be an important part of state-level green initiatives. States incentivize the creation and use of renewable resources and the practice of efficient use of energy through rebate programs, grants, and other publicly funded practices. Green incentives include green building, utility grant, loan and rebates, corporate exemptions and tax credits, industry recruitment support, property and sales tax incentives, and personal tax credits. Most states also have incentives for the purchase of hybrid, and other energy-efficient, vehicles.

Much debate occurs around whether state government should be involved in environmental activism, with much of it pertaining to the discussion of how much power state governments should have regarding regulation of and control over the private sector (i.e., corporations) and individuals.

In 2010, the report "Climate Change Indicators in the United States" was published by the EPA. Its intent was to show that climate change was real by presenting indicators that might be explained by theories of global warming and climate change. Details from the report include such indicators as increases in radiative forcing, heat waves, drought, growing amounts of precipitation, changing levels of ocean acidity, snow cover loss, an increase in deaths from heat, plant hardiness zone shifts, and the disruption of bird and animal migration patterns.

While these indicators present the case that there is a drastic cause to these changes, some still hold the position that climate change is a myth. Proponents of green initiatives point to the indicators of climate change and argue that the warming of the Earth is a real, shared problem that will have profound effect on all living creatures. Proponents of green initiatives also see the need to work to counter the effects of climate change, even if that means only to slow the decline.

Debate Over the Effect of Green Initiatives

Seeing a problem with climate change usually reflects a desire to reduce the effect of the crisis. Personal attempts are made (recycling, water and energy consumption practices monitored, and so forth) to limit the impact individuals have on the environment. Businesses often engage in events or campaigns to live more sustainably or help raise money for environment-related causes. Municipalities and county extension services offer periodic recycling events where less common items like batteries, paint, and appliances can be brought for proper disposal and recycling. State green initiatives are attempts by state governments to counter climate change while fostering understanding of the root causes of it. With the spending of taxpayer money on initiatives meant to delay or counter the effects of global climate change, not all citizens agree that state governments should be focusing spending on environmental concerns. The use of taxpayer money for politically charged activities can invoke the anger of some taxpayers. Additionally, there is always a concern when taxpayer money is spent on expensive programs and initiatives, and due to the scope and size of environmental advocacy efforts, many of the initiatives forwarded to improve the environmental crisis are expensive.

For those who believe that initiatives created by local, state, and federal governments can be productive measures to improve the environment, and the way humans act within it, there is much disagreement over which initiatives should be enacted. One position is that which supports the regulation of polluters. This is more of a coercive approach to social change. Another position is to give incentives to reward movement in a green direction. Some feel that incentives can be an effective way to encourage greener practices while making them less burdensome. Green initiatives can help all citizens reflect upon the true costs of nonrenewable resources by focusing attention on the connectedness of human beings, their choices, and the world they live in.

See Also: Carbon Emissions (Personal Carbon Footprint); Certified Products; Corporate Social Responsibility; Education, Federal Green Initiatives; Environmental Justice (Politics).

Further Readings

Environmental Defense Fund. "Science of Global Warming." http://www.edf.org/page .cfm?tagid=54192 (Accessed September 2010).

National Governor's Association. "Greening State Government: 'Lead by Example' Initiatives." http://www.nga.org/Files/pdf/0807GREENSTATEGOVT.PDF (Accessed September 2010).

U.S. Environmental Protection Agency. "State Environmental Agencies." http://www.epa.gov/ epahome/state.htm (Accessed September 2010).

Chad A. Becker
Indiana State University

Environmental Education Debate

As interest in sustainability and environmental issues has grown, many schools have begun to offer instruction that addresses these issues. Many teachers and parents have welcomed these green education initiatives, believing that they broaden the curriculum and offer students valuable experience with the knowledge and the tools to make the powerful changes that are necessary to reverse the damage that has been done to the Earth and to prevent more harm from occurring. An environmental education debate, however, has been initiated by others who believe that the focus on green issues is overblown, detracts from the core curriculum, or is not adequately grounded in science. Despite these concerns, a growing number of classrooms, schools, and school districts offer environmental education to their students. Environmental education programs are strengthened, and their place in the curriculum is ensured when such instruction is aligned with best practices in science education and adheres to appropriate content standards. While environmental education will continue to stimulate debate regarding appropriate content and methods of instruction, its popularity is increasing.

Criticism of Environmental Education

Many believe that science education in the United States is stagnant, a condition that imperils future economic growth and threatens historic American leadership regarding scientific research and development. In 2005, for example, a group of business leaders, scientists, and college and university faculty published a report, *Rising Above the Gathering Storm*, which detailed problems facing science education in the United States. Included in these criticisms were specific instances wherein this group believed science and mathematics instruction in U.S. schools was lacking. For example, when standardized test scores are compared, U.S. K–12 students rank 48th among all nations with regard to information related to science and mathematics. Many adults also are unable to correctly answer simple questions related to science; for example, nearly half of U.S. adults cannot correctly identify how long it takes the Earth to circle the sun. Poor understanding of scientific and mathematical concepts has led to dramatic changes that are linked to this problem, such as China's overtaking the United States to become the world's leading high-technology exporter. The decline in scientific knowledge is linked to a lack of rigor in many K–12 science programs, which in turn has been linked to an overemphasis on "softer" science in the classroom, especially that which has resulted from increased environmental education offerings.

Proponents of environmental education counter that many of these criticisms are overblown and emphasize that the United States produces three times as many college graduates holding engineering and science degrees as there are jobs available. Increasing interest in the environment simultaneously intensifies interest in science—at least, it does if environmental education programs are rigorous, grounded in the best practices related to the scientific process, and make explicit how topics studied relate to core science disciplines. These advocates of a more project-based approach argue that environmental education and work that promotes sustainability can provide both a greener and more scientifically aware population. In order for such benefits to come to fruition, K–12 teachers and administrators must ensure that environmental education adheres to the rigorous standards of science education. To that end, teachers and administrators must be familiar with and willing to use best practices related to science instruction when designing environmental education experiences.

Science Education

A quality science education has become a keystone of 21st-century schooling. Provisions in the No Child Left Behind Act mandated assessments in science, joining those in reading and mathematics, beginning in the 2007–08 academic year. Additionally, a changing world—one that is experiencing unprecedented globalization, technological advances, and environmental threats—has a greater need for scientific knowledge than ever before. Teaching science—and teaching science well—is of paramount importance to students for four primary reasons. First, science is an enterprise that can be harnessed to improve quality of life globally. Scientific advances, such as cleaner sources of energy or improved food production, have the ability to benefit billions, including those who live far from the location of the scientific discovery. Second, science may be used to furnish a basis for developing language, logic, and problem-solving skills in the classroom. Projects that center around science often include ancillary skills that will assist children's learning. Third, democracies demand that their citizens be able to make informed personal, local, and national decisions involving scientific information. As issues that demand a strong grasp of scientific understanding—such as the threat of global warming, the use of stem cells, and the benefits of recycling—are increasingly important, citizens must be able to analyze and process information related to these issues in order to make decisions. Finally, some students will decide to pursue science as their lifelong vocation or avocation as a result of science classes in their schools. Allowing these students to first get excited about science is one of the most valuable contributions schools make to society.

For any sort of science education to be valuable and productive, it must focus on what scientists really do. Rather than burden children with meaningless and time-consuming busywork, environmental education optimally should incorporate the language of science into lessons and projects. As much as possible, students—including very young children—should be clear about specific scientific usage to avoid confusion or misunderstandings. Common scientific language with which students should become familiar includes such terms as *scientific theory*, *datum* or *data*, *prediction*, *observation*, and *evidence*. All such terms may be used, and understood, by even the youngest students. Engaging students in activities that encourage them to think about the natural world around them builds upon natural interest children have in their environment and builds upon the knowledge and skills that students bring with them to the classroom.

Science Education Standards

Although frequently discussed in the media and common in other fields such as business and engineering, standards that relate to education are a fairly recent development. In 1989, President George H. W. Bush called a national education summit, held in the Rotunda at the University of Virginia. In addition to President Bush and officials from the U.S. Department of Education, the summit was attended by the governors from most states. Although many issues related to education were discussed, the summit focused on establishing academic achievement goals for American schools, with an emphasis on creating common curricula to be covered in classrooms across the nation and particular states, with a desire to standardize excellent content as much as possible. The summit resulted in several initiatives, including the National Educational Goals Panel (NEGP) and the National Council on Education Standards and Testing (NCEST). The NEGP and NCEST sought to establish the curriculum that schools should teach, the types of testing that should occur to assess that learning, and the student performance standards that should enable evaluation of this. These initiatives have led to the standards movement, which was further supported by the No Child Left Behind Act of 2001.

The repercussions of these efforts were influential and far reaching. As a result of the standards movement, at least 50 different state content standards have been created, each with a separate means of aligning curriculum, assessment, and evaluation. In addition to state content standards, a variety of organizations such as the National Council of Teachers of Mathematics (NCTM) and the National Research Council (NRC) have promulgated standards that cover a myriad of topics, including mathematics and science. Recently, support has grown for the implementation of common core standards, those that are consistent across state borders. Proponents of rigorous environmental education have embraced content and learning standards as a way of both ensuring rigorous scientific content and legitimizing the broader push for sustainable programming.

Advocates of standards-based instruction assert that standards ensure a certain level of quality and rigor in every classroom. Using a standards-based system of instruction, every child receiving a high school diploma is essentially warranted to know and understand certain key concepts and is able to perform specific tasks, especially those related to reading, writing, mathematics, science, and social studies, that will make him or her able to succeed in college or in the workforce. As much as possible, content standards attempt to emulate the practices and beliefs of members of the disciplines, therefore allowing students access to work that is engaging, meaningful, and appropriate. This is especially important for advocates of environmental education, as linking green education to standards legitimizes what occurs in the classroom.

Content standards must be developmentally appropriate, written in a manner that supports both equity and excellence, and sensitive to the needs of both advanced and struggling learners so that all students receive an appropriate degree of challenge in the classroom. As with schools in general, environmental education classes and programs have become increasingly interested in assessment instruments that measure student mastery of content as a means of demonstrating the content learned. As the proponents of standards-based instruction believe in accountability, students' mastery of particular standards is often measured by means of a criteria-referenced test. In addition to teacher-created tests and standardized exams, criteria-referenced assessments are of particular use in evaluating the efficacy of environmental education. Criteria-referenced tests are those that translate test scores into an assessment of student mastery of certain behaviors. The objective of criteria-referenced tests is to see if the student has mastered certain material, not to determine how the student is performing in relation to his or her peers.

Environmental Education Standards

Content standards are popular for several reasons. Students and teachers benefit when content standards lay out a scope and sequence for a discipline using input from college and university professors, master teachers, and curriculum specialists. Content standards also tend to legitimize a field of study. As a result, a push has developed to create content standards pertaining to environmental education. Despite this, many teachers across the United States are left to develop environmental education programs through their own devices. Although environmental education is increasingly popular in K–12 classrooms, only 18 states currently have standards that address environmental issues.

Of the states with environmental standards, only 12 have mandated some form of environmental education, and of these, most do not have the funding to enforce the instruction or to allow testing of these topics. Nearly all states that have mandated sustainability topics incorporate them into the existing science curriculum rather than outlining them as a separate subject. When a state has not designated sustainability as a topic to be included in the content standards, many teachers, schools, and school districts have created their own curricula for teaching sustainability. Although sustainability instruction can be incorporated into any subject area, including English/language arts, mathematics, and social studies, many teachers, schools, and school districts that have created their own sustainability curriculum have chosen to do so through their science curriculum. By applying accepted principles of science learning to environmental education, teachers and other advocates for such programming help to make the case that such programs are rigorous and legitimate.

Four Strands of Science Learning

In an effort to improve science learning in K–12 settings, the National Research Council, a group sponsored by the National Academies, identified four strands of science learning that assist teachers in linking content and process in the classroom. These four strands assist those interested in environmental education, as they assist teachers in thinking about what it means to be proficient in science and provide a framework for moving toward and achieving proficiency. The framework supported by the four strands views science both as a body of knowledge and as an evidence-based, model-building enterprise that asks participants to constantly extend, refine, and revise knowledge. Rather than drawing a distinction between content and process, the four-strands approach asks that students acquire knowledge and reasoning skills that allow them to be proficient in science. This approach easily encompasses environmental education goals and objectives.

The four strands of science learning have certain overlapping similarities. As a result, one might conceptualize each of the four strands as supporting and promoting the goals of the others. The four strands of science learning are the following:

- *Strand 1*: Understanding scientific explanations
- *Strand 2*: Generating scientific evidence
- *Strand 3*: Reflecting on scientific knowledge
- *Strand 4*: Participating productively in science

Quality environmental education incorporates each of these four strands and allows students to engage in respectful work to explore issues regarding sustainability, pollution, and other issues in a manner that integrates content and process.

The first strand, generating scientific evidence, asks that students know, use, and interpret scientific explanations regarding the natural world. This practice permits students to understand the interrelation among central scientific concepts and to use them to build and critique scientific arguments. The second strand, generating scientific evidence, recognizes that generating and evaluating evidence as part of building and refining models is at the heart of scientific practice. It is heavily involved with what is commonly thought of as "process," but expands that concept to include the theory- and model-building facets of science. The third strand, reflecting on scientific knowledge, emphasizes that our understanding is continually evolving and that ideas and insights related to the world around us change and evolve over time. Key to this appreciation is a keen awareness of the scientific process, including how evidence is gathered and how arguments based on this evidence are created. The fourth and final strand, highlighting the need for students to participate productively in science, envisions classroom practice as a social enterprise that is grounded in a core set of values and norms that govern behavior and participation. This strand emphasizes the centrality of skillful participation in a scientific community and the necessity of representing ideas, using scientific tools, and interacting with peers as an effective means of interacting with peers about science.

Within the context of environmental education, it is vital that adherence to the four strands of science learning are present and honored. Certainly in practice the strands overlap to an extent, and quality classroom instruction might not implement them in the same way. That being said, for teachers, administrators, and others interested in using environmental education as a means of fulfilling certain needs for scientific instruction within the curriculum, tying environmental education to science content standards helps to ensure that programs deliver quality instruction and provide quality information.

Standards-Based Environmental Education

The National Science Education Standards provide guidance regarding how to use what is understood about learning theory to best guide classroom learning experiences. Quality instruction related to environmental education thus should emulate practices that are aligned with these guidelines. Quality environmental education will address each of the following:

- Understanding the environment and related sustainability issues concerns more than knowing facts.
- Students build new knowledge and understanding on what they already know and believe.
- Students formulate new knowledge through modifying and refining current concepts and adding new concepts to those already known.
- Learning is mediated through a social environment through which learners interact with others.
- Effective learning allows students to take control of their own learning.
- Transfer of learning, that is, the ability to apply existing knowledge to novel situations, is affected by the degree to which students learn with understanding.

Teachers, schools, and school districts desiring to build environmental education into their curriculum would be well served to allow for investigations of environmental issues and sustainability concerns that are pertinent to students and their lives.

Expertise is built upon a deep foundation of factual knowledge, understanding of these facts in a conceptual framework, and organizing knowledge in ways that allow for retrieval and application. Thus, exposure to and use of the scientific inquiry process while exploring sustainability topics permits students to do more than know facts. Educational experiences where the teacher probes and discovers students' prior knowledge allow misconceptions to be addressed and new concepts to be acquired. Students who engage in hands-on work are more likely to make such a change, as students are willing to accept new ideas when they discover ideas that seem plausible and appear more useful than the ideas they previously held. Learners construct their own knowledge about the environment best when they have opportunities to articulate their ideas to others and challenge their peers' ideas and, in doing so, construct their own ideas. Allowing for project-based, group investigations allows students to grapple with environmental topics. A project-based approach also permits students to evaluate the type of evidence they need to support certain claims, build and test certain theories of phenomena, and utilize the metacognitive skills of monitoring and regulating their own thoughts and knowledge. Finally, students addressing environmental topics must achieve an initial threshold of knowledge and then have opportunities to apply that knowledge to new and unique experiences.

A deep focus on inquiry thus allows classroom investigations of environmental topics to address science standards. Building questioning and analysis into lessons strengthens students' ability to ask questions, formulate hypotheses, speculate, inquire, and develop answers. As part of the inquiry process, environmental topics encourage students to acquire knowledge of various processes and systems, allowing students to demonstrate an understanding of natural systems, habitats, and various changes that occur in the physical environment. Working with peers to investigate these issues provides students with both the learning environment best suited to acquire the inquiry and metacognitive skills that will allow optimal learning and the social skills necessary to build a warm and vibrant classroom community. Thus, aligning environmental education with the operable science standards will build student mastery of content and process.

State Mandates for Environmental Education

Some states have encouraged environmental education, recognizing that encouraging students to investigate and interact with their communities has many benefits for them as well as for society at large. Unfortunately, there is little mandated curriculum for environmental education. Only 18 states have formal environmental education learner objectives and outcomes or are in the process of creating them. Of these states, 12 have mandated some form of environmental education. Even if a state has not designated sustainability as a topic to be included in the learning standards, school districts have the option of creating a curriculum for teaching sustainability. Districts that have done this outlined their intentions for becoming more sustainable and have assembled resources for teachers to use when including these topics in the classroom. Student enthusiasm for environmental education can often justify the expense of these resources, as many schools that have introduced environmental education programs have witnessed increases in student attendance and graduation rates.

Wisconsin, Maryland, Vermont, and Washington are a few states that have established environmental education and sustainability learning standards. Furthermore, nongovernmental organizations (NGOs) like the North American Association for Environmental Education (NAAEE) and the U.S. Partnership for Education for Sustainable Development have published learning standards that provide educators with a framework for teaching sustainability topics. All of these sources vary in the number of principal learning standards that are set forth, ranging from two to six. They also differ in how the standards are organized; some dictate standards that should be reached by the completion of 4th, 8th, and 12th grade, and others offer standards for each grade level. Regardless of organization and slight variations in content, all of these sets of standards provide educators with the necessary framework to implement sustainability topics into the classroom.

Community-Centered Learning

Sustainability education helps students apply knowledge learned about the environment to their lives, and it is also important to consider the environmental impact of one's actions on the lives of others around the world. Focusing on the planet as a whole, or as a "global community," provides students a different view of sustainability issues by allowing them to see how the actions of a small group can affect the well-being of millions. While many American students may have difficulty grasping some concepts of sustainability and linking cause and effect if they have not experienced the results of climate change, widening the scope of sustainability education can allow students to see the interconnectedness of the world and its many inhabitants. For example, students in the Pacific Northwest can more easily comprehend the implications of a drought after studying cases in sub-Saharan Africa. Likewise, students from the Midwest can have a better understanding of famine and crop failure after examining how China has been affected by floods. All American students can conceptualize the importance of the rainforest on life all over the world by studying what products come from the rainforest, how the rainforest affects the climate and weather patterns, and the rich biodiversity that exists there. Environmental education also benefits from a focus on local issues that affect the local community.

Opponents of environmental education are often not opposed to the notion of sustainability concepts per se but rather its implementation. Encouraging involvement of multiple stakeholders in the process—including students, teachers, administrators, community members, and local businesses—may assist in alleviating some of these criticisms. Certainly not all constituencies can be expected to embrace environmental education, but allowing as diverse a group as possible to participate in the planning and implementation of projects will increase the chances for their success and reduce unfounded criticism.

Broadening Environmental Curriculum

Once environmental education offerings are initiated in a classroom or school, the focus can turn to ways to expand these opportunities. Broadening the scale on which students learn about the environment is important to consider when determining topics for environmental education. Practically all aspects of the curriculum can incorporate aspects of environmental issues. Coupling an environmental education curriculum with resources like the state standards or learning guidelines from the NAAEE can help teachers and curriculum designers to decide which topics to approach at which grade levels. Furthermore, the standards serve as a practical guide and a helpful tool for educators to create assessments of students' knowledge of the environment. Many of the standards coincide with science, social studies, math, and language arts instruction, so teachers can introduce environmental education topics across the curriculum. Typically, environmental education standards outline learning goals for students by the time they have completed 4th, 8th, and 12th grades, so there is flexibility in the topics that can be taught and the order in which they are introduced. Schools will benefit if teachers across grade levels develop a scope and sequence of environmental education topics to be addressed at specific grade levels. A scope and sequence identifies precisely when a topic will be addressed, thus eliminating duplication of coverage and permitting instruction to build on the foundation introduced in lower grades.

Because environmental education looks critically at economic, manufacturing, recycling, and other practices in the community at large, it will always have its critics. When offered within a rigorous and research-based course of study, however, environmental education can allow students to become more familiar with the scientific process, better able to engage in critical thinking, and ultimately better-informed citizens. As students grow older, their continued experiences with natural science both inside and outside the classroom lead to a better understanding of how habitats and various life forms interact. This essential knowledge enhances the students' continued appreciation of nature. The philosophy behind environmental education is that students will want to protect nature if they learn about its diversity, rarity, and importance to human life. As more teachers, parents, administrators, and students are exposed to quality environmental education, questions about its efficacy will diminish and such programs will grow in importance and popularity.

See Also: Antiglobalization; Capitalism; Education, State Green Initiatives; Green Community-Based Learning; Individual Action Versus Collective Action; Social Action.

Further Readings

Baarschers, W. A. *Eco-Facts and Eco-Fiction: Understanding the Environmental Debate.* London: Routledge, 1996.

Davis, J. M. *Young Children and the Environment: Early Education for Sustainability.* New York: Cambridge University Press, 2010.

Gray, D., L. Colucci-Gray, and E. Camino, eds. *Science, Society and Sustainability: Education and Empowerment for an Uncertain World.* London: Routledge, 2009.

Hewitt, T. W. *Understanding and Shaping Curriculum: What We Teach and Why.* Thousand Oaks, CA: Sage, 2006.

Michaels, S., A. W. Shouse, and H. A. Schweingruber. *Ready, Set, Science! Putting Research to Work in K–8 Science Classrooms.* Washington, DC: National Academies Press, 2008.

National Academy of Sciences, National Academy of Engineering, and Institute of Medicine. *Rising Above the Gathering Storm: Energizing and Employing America for a Brighter Economic Future.* Washington, DC: National Academies Press, 2007.

National Research Council. *Inquiry and the National Science Education Standards: A Guide for Teaching and Learning.* Washington, DC: National Academies Press, 2000.

Robottom, I. "Critical Environmental Education Research: Re-Engaging the Debate." *Canadian Journal of Environmental Education*, 10:1 (2005).

Stephen T. Schroth
Jason A. Helfer
Diana L. Beck
Barry L. Swanson
Knox College

Environmental Justice (Business)

As awareness of the environmental degradation occurring around the world as a direct result of global business enterprises began to spread, a related factor became acutely pertinent. Minorities (in particular, African Americans, Latin Americans, and Native Americans) and low-income families in the United States as well as other disadvantaged socioeconomic groups and minorities around the world were facing multiple adverse effects from this environmental damage, especially from pollution, pesticides, and resource depletion. This catalyzed the creation of the notion of environmental justice, which has been at the heart of a discernible confluence of the international human rights and environmental fields.

Companies that expel by-products into the environment or deposit toxic waste into the air, water, or land are among the primary degraders of the environment.

Source: U.S. Environmental Protection Agency

The term *environmental justice* has a number of different connotations. At the root of all of them is a positive requirement that the interests of all individuals of any color, race, national origin, income level, and/or gender should be adequately represented with respect to the development of environmental law and policy. It also carries with it the inherent converse requirement that these groups should not be the ones to shoulder all the detriments from environmental use, or misuse, in their respective communities—the negative of the concept, or environmental "injustice." Thus, environmental justice signifies the involvement of marginalized and often vulnerable groups such as women, children, and minorities in the lawmaking process in regard to the environment, ensuring their full participation in the process, including access to information and, as a result, their protection.

It has also become apparent that corporations and other businesses are among the primary degraders of the environment. This conclusion is based on the dual realization that their operations are often located in the areas where these negative environmental impacts and effects are being recorded and that these can be tied in many cases to their business enterprises. The argument is that since corporations and businesses have a fundamental responsibility for the conditions that give rise to the issues of environmental justice, they should have a role to play in promoting it. Considering the global influence of corporations, this article will examine the advantages and disadvantages of corporate self-regulation to integrate environmental justice into their business models and practices. It recognizes that government regulations, laws, and other policy formulations are important—and necessary—components of the solution, but solely considers the role (and whether it may be a self-enforceable role) of businesses in this endeavor.

For the purposes of this article, business and corporations are interchangeable and refer to those companies that are engaged in operational practices that would have a potentially negative impact—direct or indirect—on the environment. This includes companies that expel by-products into the environment in the form of pollutants or deposit toxic waste by any other means into the air, water, or land.

The Right to a Healthy Environment

Briefly, there is an argument to be made that all persons should have the fundamental right to a healthy environment, and thus environmental justice is merely an articulation of that right. A debate on this issue is outside the scope of this article, but there are several human rights treaties that codify this right. Although the Universal Declaration on Human Rights and the Covenants on (i) Civil and Political and (ii) Economic, Social, and Cultural Rights do not mention environmental protection being important for a human's full enjoyment of his or her rights, a number of later human rights and humanitarian treaties make direct reference to it. For example, the Additional Protocol to the American Convention on Human Rights in the Area of Economic, Social, and Cultural Rights states that all persons have a right to live in a healthy environment and that states have the duty to promote these rights by protecting, preserving, and improving the environment. Similarly, the African Charter declares that all people should have the right to a satisfactory environment that is favorable to their development. Environmental treaties have also begun to refer to human rights, including the Stockholm Declaration, which (although nonbinding in nature) states that healthy natural and man-made environments are necessary for the enjoyment of human rights—rights as fundamental as the right to life. Environmental justice helps to ensure that these rights are realized.

It is therefore considered axiomatic that businesses should act with awareness of the rights of those people who may be adversely affected by their actions and should take steps to ensure that their business activities do not violate those rights. The following discussion identifies some of the advantages and disadvantages of businesses' incorporating environmental justice policies and practices of their own accord.

The Pros of Businesses Taking the Lead in Environmental Justice

There are a number of positive if intangible benefits that adhere to the corporate entity that self-regulates, enhancing its reputation and quite possibly its success. Philosophically, and probably most notably, the international community has made a value judgment about the rights and status of individuals where the actions of business or corporate entities impinge upon those rights. There is a growing expectation that corporations should be cognizant of those rights. In this regard, the most influential factors are the notions of corporate social responsibility and sustainable business practices, which have recently gained traction. This progression to socially minded business practices would create a natural link to environmental justice. It could be one of many considerations that businesses have in mind when evaluating their processes and practices.

The precautionary principle, developed primarily to temper environmental harm, should also be a guiding principle for corporate activity that may affect people. The underlying objective of the precautionary principle is caution: the absence of evidence of harm does not mean that proactive steps should not be taken to stop or abate the potential harm. This principle shifts the burden of proof onto those wishing to conduct the questionable practices to show that there will be no detrimental effects—in this case, onto the corporations. Linked to this is the concept of intergenerational rights. Just because the actions of today may not have negative current effects on a population, it does not mean that future generations will not suffer.

In addition to incorporating a sense of business ethics into their practices, corporations realize a number of tangible benefits from operating under a socially responsible umbrella. In addition to the peer pressure from other socially minded businesses, this may also translate into a positive or negative reaction from their consumers, who may actually be a quite influential force. Corporations are more often than not singled out for condemnation if their practices are negatively affecting local communities. The contrary is also increasingly true—corporations that act responsibly are rewarded with increased patronage. The growth of corporate social responsibility has demonstrated that perhaps businesses should have a more comprehensive "bottom line" in mind, recognizing that their obligation to their shareholders may well be enhanced, not just ethically but also financially, by a corresponding obligation to the individuals in the communities in which they operate.

Apart from the moral and financial considerations, there are a number of pragmatic reasons for corporations to adopt self-regulating environmental justice principles in their operations. The first is that they are at the heart of the issue—it is their actions that may adversely affect a population. They are best equipped to identify, understand, and correct the problems that their operations may be causing and to deal with the repercussions. This holds true for all corporations, from the smallest to the largest, and the positive changes they could make would likely be proportional to the scope of their operations.

Another advantage to the self-regulation of businesses is that the states in which these corporations operate globally may be unable or unwilling to put the necessary safeguards in place to prevent harm to their respective citizens. Many of these countries do not have the ability to control environmental problems themselves. Furthermore, for those companies that are headquartered in the United States, there is no jurisdictional reach to force them to abide by U.S. laws when they operate in other countries. Therefore, the power of self-regulation to address issues of environmental justice can be immediately realized if corporations take this responsibility upon themselves.

Finally, linked to this is the further advantage that businesses also have the financial wherewithal to address these problems. They are likely to conduct a cost-benefit analysis of what it would mean to make changes to the ways in which they operate and the impacts of their activities on the environment. If there is a fiscal benefit in doing so, it is quite probable that the business will change its manner of operations anyway. If there is no advantage or an actual disadvantage, then the moral/ethical considerations will come into play.

The Cons of Businesses Taking the Lead in Environmental Justice

The last of the advantages of this approach is also effectively the first disadvantage. While corporations can best afford the cost of ensuring environmental justice through the adoption of certain principles and practices, implementing such processes will undoubtedly be costly. With the corporate eye always on the bottom line, this may be money that businesses are not willing to invest unless there is a clear economic benefit, and the financial impact may be too great for them to sustain. Many businesses would concede that there is a moral or ethical element in their business choices insofar as those choices might cause harm, and many would contend that if they could, they would take action, but then offer certain realities—primarily monetary ones—that preclude them from doing so.

The problem of cost is compounded if there is no or questionable scientific proof that a specific corporation's actions are causally linked to the negative effects the human population in the area is suffering. The primary objective of a business is to sell goods or services to make a profit (and to increase the profit of its shareholders). Therefore, if the cost is too great to implement proactive measures, these businesses may not be willing to make the necessary financial commitment to remedial measures on the mere possibility that they would make a difference. At best, they may be willing to correct harm already caused, either through voluntary action or compulsorily by order of a judicial or administrative body. In addition, requiring businesses to include environmental justice interests into their business plans may be considered overregulation, hampering the efficient conduct of their business enterprise at considerable financial cost.

A further, even more acute disadvantage is that a causal link to potential future harm may be impossible to prove, and therefore the onus would be on the business to take pre-emptive action. In such cases, should businesses have to prove that their practices would not affect the population for years to come, even if there is no immediate discernible harm? There may never be such scientific proof, and the precautionary principle is often not enough of an incentive for action. Furthermore, how much harm must be demonstrated? If there are hundreds of thousands of people living in the area but only a few are affected, is there an obligation to act? What does the ratio have to be before action clearly has to be taken? These are all fundamental questions that businesses may argue are outside their scope of responsibility. Many corporations may argue that they do not have the in-house expertise to conduct the necessary research into the potential impacts and effects of their actions in this regard. They may argue that they are not the best placed to conduct such research, and that being too cautious might well mean that they would have to cease operations altogether, which is unacceptable.

Finally, there is no binding international legal framework to ensure that corporate social responsibility is being incorporated into business practices. Therefore, regulation is left to the laws of the nations in which these businesses operate. Like the advantage/disadvantage of cost mentioned above, this may constitute both a disadvantage and an advantage. Countries have the ability to monitor and regulate the activities of corporations operating within their national boundaries, but if there are no or ineffective regulations in place, businesses would have to take the initiative. Some may be willing to do so, but others may not see the benefit. Some corporations may have actively chosen to operate in certain countries for this very reason (the absence of effective regulations) and they may not care to change the status quo.

Conclusion

Corporations do not operate in a vacuum—their actions, or lack thereof, have conse-quences for the communities in which they operate. Thus, the antagonism between eco-nomic benefits and ethical considerations, viewed in the past as a deep-seated dichotomy, is changing. There is no international enforcement mechanism that specifically has jurisdic-tion over or is in existence primarily to regulate corporations that operate worldwide. If there are also few or no effective local enforcement mechanisms, then perhaps self-regulation may well be the best option until change occurs. The best interim mechanism may well be for businesses to take socially responsible action themselves and incorporate environmen-tal justice into their business and strategic plans as an inherent component of corporate policy rather than an afterthought. It is apparent that there has been a shift in the dynamic of corporations over the past several decades toward more ethical operation, and it may well be in this climate that positive action is possible.

However, it would be naive to think that businesses can do this alone. The actions of corporations are only a piece of the puzzle. It is necessary that government regulations, laws, and other policy initiatives be in place to address the problems of environmental justice holistically. For instance, environmental justice policy making became a clear objective of the U.S. federal government in 1994 when President Bill Clinton signed Executive Order 12898 (government agencies must take steps to identify and remedy disproportionally high and adverse human health or environmental impacts of its programs, in particular on minorities or low-income populations). Ethical business practices, if applied in tandem with other policies, may be the most effective means of dealing with the issue of environmental justice.

See Also: Environmental Justice (Ethics and Philosophy); Environmental Justice (Politics); Precautionary Principle (Uncertainty); Sustainable Development (Business).

Further Readings

Ash, Michael and James K. Boyce. "Measuring Corporate Environmental Justice Performance." Working Paper Series. Political Economy Research Institute, University of Massachusetts, Amherst, No. 186, June 2009.

Emeseh, Engobo, Rhuks Temitope Ako, Patrick Okonmah, and Lawrence Ogechukwu Obokoh. "Corporations, CSR and Self Regulation: What Lessons From the Global Financial Crisis?" *German Law Journal*, 11:2 (2009).

Partridge, Ernest. "Environmental Justice and 'Shared Fate': A Contractarian Defense of Fair Compensation." *Human Ecology Review*, 2:2 (Winter/Spring 1996).

Shrader-Frechette, Kristin Sharon. *Environmental Justice: Creating Equality, Reclaiming Democracy*. New York: Oxford University Press, 2002.

Waits, Juliann and Kathryn Hicks. "Environmental Justice and the Effect of the Industrialization in the Delta Region of the Southern United States." *Business Perspectives* (July 2009).

Weiss, Joseph W. *Business Ethics: A Stakeholder and Issues Management Approach*. Mason, OH: Cengage Learning, 2009.

Briony MacPhee Rowe
Independent Scholar

Environmental Justice (Ethics and Philosophy)

Environmental justice is a social movement that began in the late 1970s and early 1980s in response to the concentration of environmental risks and hazards in poor and minority communities. It focuses on the distributional effects of current land use, pollution control, trade, and resource use policies. The central argument proposes that access to a healthy and clean environment is increasingly distributed by power, class, and race, and citizens of different races and classes experience disparate environmental quality.

This article first defines the concept of environmental justice and differentiates it from related concepts with which it is often used interchangeably. After giving a brief overview of the history and development of the movement, it examines the criticisms. The final section explores the future of the movement.

Concepts and Definitions

The term *environmental justice* has been used in a broad sense to encompass responsibility to other species and future generations and the rights of the environment. Environmental justice as discussed here is a movement distinct from mainstream environmentalism, which argues for eliminating the inequitable distribution of the environmental "goods" and "bads" and disproportionate impacts on certain groups in the society. Environmental injustice occurs when vulnerable sections of the society are exposed more to negative environmental effects and have disproportionately less access to environmental "goods." It is exclusively anthropocentric, not about ecology but procedural and distributive social justice. Its main goal is to empower communities marginalized by race, ethnicity, and poverty to effectively protect their health and defend and manage their territories and resources. The term incorporates environmental racism and environmental classism. When the environmentally disadvantaged groups are correlated with ethnic minorities, environmental injustice is parallel to environmental racism; when it is solely an issue of economics, it is termed *environmental classisim*.

Another related concept is that of environmental equity, which requires equal treatment and protection of environmental laws. The idea of equal distribution of risks raised in the early days of the movement received criticism because it may imply equalized pollution rather than risk reduction and avoidance. This misleading terminology that allowed interpretation of equity only in distribution terms eventually gave way to the more common use of the broader term *environmental justice*, which includes procedural justice as well. Environmental justice emphasizes the right to a safe and healthy environment for all people and is less provocative than environmental racism, which suggests discrimination in policy making, the enforcement of laws, and targeting communities of color for disposal sites and polluting industries. It is more difficult to argue against claims of justice compared to equity or racism.

History and Development

The environmental justice movement is commonly dated to the first court case challenging the siting of a waste facility on violation of civil rights grounds in Houston, Texas, in 1979 (*Bean v. Southwestern Waste Management Corp.*). Another key event was a protest against a proposal to site a polychlorinated biphenyl (PCB) landfill in a predominantly African American community in Warren County, North Carolina, in 1982. During the dispute, Reverend Benjamin F. Chavis Jr., director of the United Church of Christ's Commission for Racial Justice, coined the term *environmental racism* and linked the two popular social movements of the late 20th century: environmentalism and civil rights. Following these events, the interest aroused in environmental justice issues produced several studies that confirmed race as the single best variable able to predict the siting of hazardous waste facilities in a community. These studies were not taken on board by everyone, however. Some stressed the importance of economic class instead of or in addition to race, while others questioned the extent to which the placement of toxic facilities was clearly intentional. Following decades continued to produce contradictory evidence and arguments.

Taking some of the criticisms on board, the movement began to widen its scope. The 17 Principles of Environmental Justice adopted by the First National People of Color Environmental Leadership summit in Washington, D.C., in 1991 extended the movement's focus on race to include other concerns, such as class and nonhuman species. Still, issues central to environmental justice remained mostly the siting of locally unwanted land uses (LULUs) such as municipal landfills, hazardous waste facilities, nuclear waste dumps, and polluting industries. Recently, there are moves toward expanding the scope of the movement to include the distributional effects of the extraction, use, and preservation of natural resources. There is also a push for making its ties with more fundamental injustices brought on by economic and social disparities more explicit.

Criticisms of Environmental Justice

Criticisms of environmental justice revolve mainly around three broad areas. The first area deals with the existence and nature of injustice, the second with its philosophical basis and scope, and the final area critiques its approaches to solutions and its use in environmental decision making.

Existence and Nature of Injustice

The issue that received most attention from the scholars is gathering evidence on the existence or lack of injustice. Arguments in this area can be summarized in three topics: (1) whether there is injustice in the distribution of environmental goods and bads, (2) whether injustice is on the basis of race or class, and (3) whether injustice is intentional or unintentional and whether how it is produced matters.

The first point of dispute on the environmental justice movement is whether environmental injustice can be empirically demonstrated or not. For some, injustice is clearly demonstrated by the studies; for others, the evidence is less than convincing. Here, the disagreement originates from the research design, methods and measurement, and unit of analysis of the studies that confirm the existence of injustice as well as studies that found noncorroborative results. Opponents argue that variation displayed may be attributable to many other factors, and the scope of confirming research is too limited to indicate broader patterns. Another line of argument by opponents discredits what they label as advocacy research on the grounds that it has been conducted by policy advocates who have a specific agenda and is therefore unlikely to be of scientifically acceptable quality. These advocates are also accused of failing to recognize the distinction between proximity to a hazard and risk.

Another point of disagreement centers on whether injustice is on the basis of race versus class. Studies that find race to be a stronger factor in determining distributional inequalities are criticized on methodological grounds. Some environmental justice advocates claim that the "race versus class" debate has been unproductive, and focusing on the racism aspect has resulted in excluding inequalities by class within communities of color. They disagree with emphasizing one form of inequality to the exclusion of others with the claim that environmental injustices affect humans unequally along the lines of race, gender, class, and nation.

The third point of dissent is related to intent. Is a policy or a siting decision discriminatory only when discrimination is intended? Some people look for discriminatory intent in rules, regulations, policies, or decisions that produce unequal protection from environmental ills to suggest environmental racism while others find disparate impact sufficient. Especially in cases in which environmental racism is alleged, a common defense is the claim of lack of any discriminatory intent. This argument proposes that if there is no intention, there is no injustice. For the proponents of the movement, policies and practices that affect communities or groups of color disproportionately are racist even if they were not intended to be so.

Nevertheless, lengthy discussions take place on which came first at a facility location: the facility or the minority community? The answer to this question has serious implications. If the disparate impact is due to discriminatory siting, the issue is comparatively easier to address. However, if people moved into the area after the LULU was built, in other words, if the disparate impact is a product of discrimination in the housing market and poverty, achieving justice becomes more complicated and difficult. Opponents explain the siting decisions as a function of market dynamics. They claim that if the host communities were not disproportionately minority at the time of the site selection, there is no discrimination. If persons of low income and noxious facilities are both economically drawn to the same cheap land, it is possible that a minority neighborhood locates or expands near those facilities. They claim that it distorts the truth to describe rational business practices of firms as racist location decisions. For the proponents, the issue of justice can be explained with the haves and have-nots frame. Corporations and governments have a tendency to follow the path of least resistance in facility siting. The haves are those with the resources and political and economic power who can avoid these ills and afford to move out. Disadvantaged groups are left behind.

Lack of direct evidence of discriminatory intent prevented courts from challenging polluters on this ground in the United States, but for proponents of the movement, whether the hazardous waste sites were there before or after the people does not matter. What matters is that certain groups are disproportionately overburdened with toxins.

Criticisms to the Philosophical Basis and the Scope of the Movement

Philosophical criticisms of the movement come from opposing sides on (1) its anthropocentric nature, (2) its focus on social justice by a new name, (3) its resemblance to NIMBYism (not in my backyard), and (4) its scope as either too broad or too narrow.

First, environmental justice is not embraced by everyone sympathetic to the environmental movement as might be expected, but criticized by mainstream environmentalists for diluting and shifting the focus of the environmentalism movement from the ecocentric to the anthropocentric with demands of more integration with the social needs of human populations. This view regards the more anthropocentric concerns of racism, classism, and sexism as less important than environmental issues. The problem originates, of course, from the belief that social justice and environmental sustainability are not always compatible objectives. Opponents argue that environmental justice can be achieved without solving environmental problems, whereas ecological justice must address both. In return, proponents criticize mainstream environmentalism for separating nature from humanity and charge mainstream environmental organizations and environmental policy for showing greater concern for preserving wilderness and animal habitats than protecting the homes and workplaces of humans.

Second, the movement's attempts to redefine environmentalism as much more integrated with the social needs of human populations resulted in another criticism from the mainstream environmentalists for being a social justice movement rather than an environmental movement. This claim is supported by the fact that in communities of color, the environmental justice movement was characterized by the reformulation of the goals of existing civil rights and community organizations to include environmental concerns. As a result, the movement has been accused of being a tool of community advocacy seeking a more democratic and egalitarian society and enhanced role for the community and power politics rather than concern for the public health of minority and low-income communities.

A third claim by opponents of the environmental justice movement is that it is just a form of NIMBY syndrome. This is essentially labeling the local opponents of a LULU proposal as selfish and shortsighted people who are unconcerned about the regional and national needs. Proponents of environmental justice argue that oppositions on environmental justice grounds are based on regional and national comparisons. Although all opposition to a particular project or siting decision might look the same in terms of "we don't want this here," proponents of the movement argue that the difference lies in their genesis. They claim that the difference is between the "we don't like this" attitude of the NIMBY syndrome more likely to occur in economically and politically elite communities versus the "this is yet another burden this community must bear" attitude of the WIMBY (Why in my backyard?) syndrome generally found in less-affluent and minority communities. The second claims equal treatment rather than indicating a preference like the first. The opposition in the environmental justice case is explained within the context of a larger critique about environmental decision making and the substantive content of environmental protection: there are communities that share specific characteristics such as high poverty levels and/or large populations of people of color that are made to bear a disproportionately larger share of the environmental burdens.

Fourth, with its origins in multiple related movements, multinational scope, and multiethnic and multiracial composition, the environmental justice movement has been criticized for its lack of focus and coherency. On the other hand, it has also been criticized for not being broad enough to confront the root political economic causes of environmental racism or deal with the role of natural resource exploitation in the production of environmental inequalities.

Role of Environmental Justice in Environmental Decision Making

Another point of disagreement relates to the approaches to solving the inequity problem, suggesting the approach advocates of the movement take to bring about justice is (1) ineffective, (2) philosophically objectionable, and (3) does not deal with the necessity of setting priorities.

First, the movement's framing the issue in terms of injustice and its strategy of seeking justice through the existing legislative, judicial, and regulatory systems is criticized because these are the systems that produced that injustice in the first place, and they will only serve to maintain the status quo.

Second, the movement has also been criticized for commodification of justice through the use of the tort system to seek redress for past environmental injustices, with the outcome of people of color trying to purchase justice like whites have done for centuries.

Third, a group of environmental justice opponents questions the value of using environmental justice in environmental decision making. They argue that in environmental justice claims the risks to communities are exaggerated, and focusing on relatively low or unlikely risks just because they affect low-income citizens threatens to worsen the problem of environmental policy's missing priorities. They propose risk assessment as an alternative way to make environmental decisions. The argument here is that by directing community attention away from problems posing the greatest risks, environmental justice may undermine public health as well as cause economic inefficiency. The citizens are accused of making limitless demands without considering the cost and avoiding confrontation of inevitable trade-offs between economic opportunity and environmental risks, which are said to be mostly relatively low and manageable. The opponents claim that the argument that no community should be allowed to become a sacrifice zone prevents prioritizing. The proponents of the movement label the dichotomous jobs-versus-the-environment choice forced upon disadvantaged communities as "environmental blackmail." The danger here is that disadvantaged communities may become so desperate for local jobs that they are willing to accept any type of economic activity regardless of its implications for environmental quality. The counterargument claims that opposing jobs on these grounds produces its own victimization of minorities. This brings about the ethically controversial suggestions of compensation of a community for hosting hazardous facilities. The rhetoric in these discussions is charged. Environmental justice opponents call the risk assessment approach "rationalizing," implying the environmental justice alternative is irrational; and by labeling it democratizing, they reframe it as social justice rather than environmental. The proponents of environmental justice respond to this push of risk assessment method in environmental policy by suggesting it is using a quantitative method and language of analysis to avoid consideration of value judgments or qualitative decisions.

The environment-versus-jobs argument is used on a global scale while exporting polluting industries or hazardous wastes. The global south is seen as a "pollution haven" due to lack of strong regulation, politically powerless residents, and lower costs of production. Misguided environmental movements in developed countries can do more harm than good globally. Global justice advocates claim that the mainstream environmental movement and white communities are partly responsible for influencing the shift in waste dumping into communities of color in the United States and abroad through antitoxic mobilizations and the passage of more stringent and costly environmental regulations. An environmental solution for one community often becomes a problem for another community as polluting industries move from community to community. Ironically, if not carefully thought out, the environmental justice movement may contribute to the globalization of environmental inequality in the same manner. There are calls for the environmental justice movement to be global and hold northern residents responsible to the global south because sources and causes of inequality are global.

The Future of Environmental Justice

The debate on environmental justice is likely to continue in years to come. With the expanding scope, the points of disagreement will likely increase. However, as incidents like Hurricane Katrina continue to demonstrate the disparate impacts of environmental hazards on different socioeconomic and ethnic groups in society, issues of injustice become more and more visible.

Climate change may result in the greatest environmental injustice of all. Resource shortages increase prices for basic commodities such as housing, food, and energy. These increases can be absorbed more easily by the wealthy. This phenomenon can be observed at different scales: locally, nationally, and globally. Climate change will exacerbate many of the existing resource scarcities such as food and water and increase extreme environmental events. The poor and the disadvantaged will be disproportionally affected by this. Its impacts may result in mass exodus of people from flooded and dry areas in search of food, housing, and employment. The poor will pay a higher percentage of their income for the basics of life as well as healthcare as tropical diseases move farther north. Taking into consideration disparate impacts of environmental policy will be increasingly more important with these new challenges.

See Also: Environmental Justice (Business); Environmental Justice (Politics); Intergenerational Justice; North–South Debate.

Further Readings

Kuehn, Robert R. "A Taxonomy of Environmental Justice." *Environmental Law Reporter*, 30 (2000).

Pellow, David Naguib and Robert J. Brulle, eds. *Power, Justice, and the Environment: A Critical Appraisal of the Environmental Justice Movement.* Cambridge, MA: MIT Press, 2005.

Petrikin, Jonathan S., ed. *Environmental Justice: At Issue.* Farmington Hills, MI: Greenhaven Press, 1995.

Schweitzer, Lisa and Max Stephenson. "Right Answers, Wrong Questions: Environmental Justice as Urban Research." *Urban Studies*, 44 (2007).

U.S. Environmental Protection Agency. "Environmental Justice." http://www.epa.gov/environmentaljustice (Accessed September 2010).

Visgilio, Gerald Robert and Diana M. Whitelaw, eds. *Our Backyard: A Quest for Environmental Justice.* Lanham, MD: Rowman & Littlefield, 2003.

Aysin Dedekorkut
Griffith University

ENVIRONMENTAL JUSTICE (POLITICS)

Environmental justice has been defined as equal protection from environmental health hazards and equal access to governmental decision-making processes for people of all incomes and ethnic groups. More broadly, the concept of environmental justice includes redress of social inequalities related to the burden of environmental pollution or unequal access to resources leading to reductions in health and quality of life without equity in sharing the benefits of industrial activity. In the United States, affected communities have included groups of low economic and political status such as indigenous tribes and people of color. Recently, the concepts of sustainability and precaution have led to the inclusion of intergenerational ethics into environmental justice. The principles of justice or equity have figured prominently in public debates and the politics surrounding pollution, climate change, resource development, and food security.

In general, exposure assessment and risk management decisions are based on an analysis of the weight of scientific evidence that leads to conclusions about the potential risks to health and the ecosystem. However, policymakers must include both economic and political concerns as well as an evaluation of the uncertainty in the information and the possibility of social stigma in any decisions. Many times, in politically charged issues, the rigorous science base is lacking, and policymakers cannot assume that social or political concerns are not without merit. Choices and trade-offs will always be necessary when making any decision of significant consequences involving any complex ecological and social system because group values, perspectives, and long-term goals will vary. A committee of the National Academy of Sciences recommended that "In instances in which the science is incomplete with respect to environmental health and justice issues, the committee urges policymakers to exercise caution on behalf of the affected communities, particularly those that have the least access to medical, political and economic resources." Environmental justice and ethics can provide support for arguments that a sustainable environment, with its diversity and complexity, has intrinsic value that is important to the well-being of future generations.

As information is more widely distributed and democratic societies demand stakeholder participation in energy, resource development, and environmental conservation issues, the decision-making process has become more open and subject to widespread political debate. While the arguments are environmental, economic, and political in nature, the underlying issue in the debate is usually related to ethical values such as human rights, justice, benevolence, and beneficence. Outlined here are several discussions about four environmental justice areas that continue to generate political debate at local, regional, national, and international levels.

Health and Exposure to Chemical Pollution

Most issues in environmental justice are related to the health disparities within diverse groups of minority populations. These groups are either denied the use of resources or exposed to adverse impacts from chemical exposure. The basic disparity in these controversies relates to inequalities in the risk-benefit ratio experienced by socially or economically disadvantaged populations. Alaska Natives and Native Americans, for example, have been the focus of concerns about exposure to environmental hazards through occupational exposure, living conditions, or food systems. Uranium ore was discarded on the Navajo lands in New Mexico and Arizona during the 1950s, with resultant dust detected in the homes of Navajo miners. Mining issues related to radioactive and other metals still are being debated in Canada as the Canadian provincial and tribal governments consider uranium mining on Algonquin lands. In Alaska, gold and oil mining on both native and state lands are hotly debated due to the potential impacts on fish and wildlife resources. The political debates, while local in nature, are centered on who controls use of the land—the majority group or the minority inhabitants.

Political structures such as state or provincial governments overlap on conflict with tribal or federal governments. The legacy of colonialism where differences in cultural values, relationships between people with the land, and intergenerational ethics conflict remains a veiled context. History documents a process by which an advantaged group, having exhausted its sustainability options, dispossesses another group of its culture or lifestyle and traditional knowledge by economic, military, or government process. The potential outcome is a reduction in the innate ability of the group to adapt to its local environment and an increase in the vulnerability of future generations to survive in the real world as their natural wealth is destroyed.

Establishing causal relations between a reduction in community health and well-being with chemical exposures from the siting of an environmental hazard such as a mine, a waste site, or oil refinery in a minority or low-income community is politically complicated because of the modern definition of health and inherent uncertainty in the scientific process. "Irrefutable evidence" of causation is practically impossible to obtain in human populations because of expense and multiple confounders, but also because of ethical restraints on controlled experimentation on human populations. Developmental defects from chemical exposures to variable mixtures often do not become apparent until adulthood, creating a lag time that prevents nondebatable conclusions. The recent use of concepts such as "weight of evidence" and the "precautionary principle" to address the uncertainty and causation issues have not always been successful in the courts of government agencies.

In analyzing a suspected case of environmental justice, the U.S. National Academy of Sciences suggests that proximity and the characterization of exposure are key components. The following are some questions that might be addressed:

- Were sites located because of discriminatory motivations, cheap land, or lack of political power?
- Are the communities characterized by the same socioeconomic/ethnic indicators today as when the sites were originally developed?

Being able to quantify how the contaminants move through the environment and the actual human exposure (if possible) can add to the weight of the evidence. The characterization of exposure in the affected community requires an understanding of all potential pathways available to the contaminant that may lead to exposure. Besides direct pathways such as air, drinking water, and food sources, pathways related to differences in behavior, employment, and lifestyles should also be identified and characterized. For example, unlike urban populations, rural Alaska Natives have high intakes of subsistence foods (wild foods, country foods).

In some locations, the levels of persistent organic pollutants (POPs) or mercury have caused risk concerns, especially when the source is related to military legacy or resource development that provides little economic or other benefit. For urban populations, lead (Pb) is a primary environmental toxicant. Blood levels, in the past, have been consistently higher for poor and minority children in the central areas of cities. Many times, this research fails to demonstrate simple statistical support for causal certainty because of confounders such as parental bioaccumulation, risky behavior, differences in biomagnification, and biases related to census tract data. Currently, prevention intervention trials are being tested as a method to provide a link between exposures and health. Removal of the contaminant that leads to reduction in the illness is a key addition to the weight of evidence, but these studies are expensive and take many years. However, no matter how a particular health condition came to be, if it is an environmental hazard whose burdens are borne inequitably, then it must be mitigated. Over the years, with elevated research funding, there has been an increased awareness by governments of health disparities.

Health and Climate Change Impacts

There is scientific consensus that the rapid increase in fossil fuel combustion (consumption) associated with the Industrial Revolution, in conjunction with population increases, has contributed to changes in the direction and strength of climate patterns. The bulk of fossil fuel combustion and population expansion occurs in temperate and tropical regions. The biophysical manifestation of climate change occurs as an individual stressor in the north and includes losses of sea ice, snow cover, and retreating glaciers as well as decreases in forests and vegetation, distribution of fresh water, and instability in fisheries. The impacts of these changes are projected to be most prominent in high latitudes and, on the immediate time horizon, will affect northern people disproportionately. While some climate impacts may be advantageous, rural and indigenous people in the north are likely to gain less and lose more than people residing in the south. Over time, socioeconomic impacts such as increases in migration to the north and pressures to develop nonrenewable resources will develop. Traditional cultures and economic activities of the indigenous peoples of the Arctic will be affected by decreases in adaptive capacity and increases in vulnerability that threaten the sustainability of communities and their cultures that have existed for thousands of years.

Climate change impacts are a new issue in the field of environmental justice since the industrial and population drivers of negative impacts are located at a distance—usually thousands of miles away, even on different continents. Because the causes and impacts extend over international borders, the politics become global. These momentous environmental, social, and cultural repercussions experienced on the local scale will require strategic political activity that transcends national-scale solutions into international arenas. Recently, climate change debates have been presented not only in public health terms but also in human rights terms. For example, aside from nutritional and economic value, the procurement of subsistence foods can be deeply tied to cultural identity and values.

The issue of climate change, especially in the Arctic, also raises the important issue of intranational equity, which has received little attention. To date, the issues of equity and justice focus on the global scale and international negotiations occurring at political meetings such as the 2009 Copenhagen Climate Change Conference. The Copenhagen Summit and other meetings revolve around the question of equitable global distribution of greenhouse gas emissions as related to economic development around the world. In the United States, Canada, and the other Arctic countries, this issue of controlling climate change affects the resource-rich north and raises the question of unequal distribution of burdens of and benefits within developed countries. These impacts can be characterized as aspects of "internal colonization"—the exploitation of those living in distant and rural regions of developed countries. For example, many times these rural communities experience lower life expectancies at birth as well as higher rates of infectious disease and suicide, while their resources fuel the economies of the urban centers.

People around the world living in remote regions that depend on hunting and gathering bear not only the physical/nutritional burdens but also a cultural burden that is disproportionate to the benefits they receive from industrial production and consumption. Some cultures view plants and animals as the moral equivalent to humans and see themselves as an intimate part of this interconnected system. This holistic perspective will lead to psychological stress as their land and ecosystem are destroyed or lost as the climate changes. In the Arctic, environmental health is almost synonymous with public health and is linked to the vibrancy of cultural traditions. This psychological burden is described by Sheila Watt-Cloutier, vice president of Inuit Circumpolar Conference, in her testimony at POPs treaty negotiations:

> [I]magine for a moment if you will the emotions we now feel—shock, panic, rage, grief, despair—as we discover that the food which for generations has nourished us and keeps us whole physically and spiritually is now poisoning us. You go to the supermarket for food. We go out on the land to hunt, fish, trap and gather. The environment is our supermarket. . . . As we put our babies to our breasts we feed them a noxious chemical cocktail that foreshadows neurological disorders, cancer, kidney failure, reproductive dysfunction, etc. This is truly worrying.

The politics of local climate impact is framed explicitly as one of human rights and climate (environmental) justice. Thus, climate change becomes linked to human sustainability and adaptation. There is an inherent connectivity between sustainability and social equity leading to the concept of productive environmental justice, which hinges on both the global demographic and geographic source of harm. Politics affect the balance between beneficial environmental goods and services and risks to health. In this form of environmental justice, the "not in my backyard" distributional struggle of local environmental impacts collides with a newer collective consideration in which "my backyard" is essentially "everyone's backyard." Short- and long-term localized adaptation strategies will be needed to rectify these environmental injustices as well as additional research to understand the synergistic impacts of multiple environmental, social, and political stressors. Even with increased scientific understanding of the effects of climate change, both national and international political action will be needed to mitigate the current damage and to enable communities to adapt long term to the new conditions. The Arctic Council was created as an attempt to politically address these issues by coordinating the political efforts of northern people.

Regional Resource Development, Water, and Sustainable Well-Being

Well-being encompasses the provision of and access to environmental benefits such as safe water, good air quality, and a sustainable ecosystem rather than just the avoidance of harm or deprivation. Environmental justice issues are relative to the community rather than universal standards. These issues are also political—about who gets to ask the questions and how they are framed. How big an ecological footprint is too big, and for whose lifestyle is development sustainable? These issues also involve structural power, inequality, and cultural identity. The specific histories of resource development and its place in regional and national economics illustrate conflict and environmental justice issues, centered on current or future disproportion burdens and risk, including impacts on land use practices, in both modern and traditional fisheries, and water resources, related to energy production needed for distant industries and large population centers. Underlying some of these conflicts is the socialized neo-colonial construction of the land as empty. Tar sands development or natural gas production not only affects the distribution and quality of the water but also can affect wildlife resources and aesthetic value. Privatization of water resources is a political and social issue that encourages the accumulation of economic and natural capital in the control of a few in contrast to its dispersion throughout the population. This issue is related to the traditional values of stewardship and intergenerational sustainability.

In this context, whether the goal of different communities is environmental protection, remediation and mitigation of industrial health hazards, or rights to hunt or fish and use the land in traditional ways, it is clear that the land is deeply connected with the identity of the people and their cultural values and vision. People who identify with land view their responsibility as stewards for future generations. Not all resources are to be consumed now by mining, forestry, fishing, or hydroelectric and fossil fuel energy developments that risk making the land (and eventually the planet) unlivable.

In relation to resources and water, the environmental justice movement has tended to have a local political perspective focused on external impacts of resource development and industrial production. A broader view of the intersection of environmental quality and social inequality suggests that structural socioeconomic political power relationships often give rise to environmental inequality and injustice. The intercolonial or intracolonial economic model of the flow of resources (staples) to the processing and manufacturing of commodities industries in large population urban centers leads to economic control and priority setting by the urban center. Consequently, natural resource economic regions are stuck with investing capital in inflexible extraction technologies. These technologies leave waste, devastate the land, and destroy fish and water resources—a pattern actually found in both underdeveloped nations and rural regions of developed countries. This export structure based on primary resources, many times, is enabled by the power structure in both local and national governments.

The legacy of this structural inequality is regional disparities and regional dependence that give rise to social, occupational, and environmental risks that are associated with the local and specific context of the environmental issue: coal mining in Canada; gold mining in Alaska; offshore oil development in the United States, Russia, and Nigeria; diamond mining in Africa; and deforestation in South America. These localized environmental injustices are the consequences of specific constellations of political and economic power relations and expose community vulnerability as a product of environmental injustice. The connections between environmental well-being and health issues has become more obvious and may also help to counter the tendency of government agencies and risk management officials to "individualize" the risks, that is, "blame the victim." A vulnerability analysis can expose or make less obscure the relevant political and economic actors who, in the past, avoided criticism.

An example of this is the experience of the Lubicon Lake people with oil sands and the use of potable water. More than $13 billion in revenue was generated for oil companies and the Alberta government. The roads, oil wells, and processing led to a decline in people's land-based livelihood and created a state of dependency and poverty as well as an increase in health problems. All decisions regarding resource exploitation in the Lubicon Lake people's tradition were made elsewhere, with leases issued by the Alberta government. Future generations' water supply is at risk in that up to four barrels of water are needed to produce a barrel of oil.

Food Security and Environmental Justice

Current concerns about food systems and the modern agriculture industry overlap with many other issues. Conflicts between the sustainable agriculture movement and current industry organizational practices highlight many gaps in power between perpetrator and victim, production demands, technology, and sustainability. These issues include legacy effects of pesticides, allergic reactions to genetically modified foods, spread of antibiotic resistance, exposure hazards to workers, and damage associated with irrigation, disease, safety, and climate change impacts on livestock practices and declining wild stocks of fish. Basic questions related to the politics associated with these issues are the following:

- Is it economically viable?
- Is it socially just?
- Is it ecologically sound?
- Does it preserve cultural identity and values?

When these questions are answered in relation to food security, the need for reforms for greater justice is seen as well as the value of reform and environmental justice movements. As Earth's human population expands and overwhelms the planet's carrying capacity, the political aspects of environmental justice will increase in magnitude.

See Also: Environmental Justice (Business); Environmental Justice (Ethics and Philosophy); Intergenerational Justice; Precautionary Principle (Ethics and Philosophy).

Further Readings

Agyeman, J., P. Cole, R. Haluza-DeLay, and P. O'Riley, eds. *Speaking for Ourselves*. Vancouver, Canada: UBC Press, 2009.

Duffy, L. K. "Disease." In *Encyclopedia of Global Warming and Climate Change*, S. George Philander, ed. Thousand Oaks, CA: Sage, 2008.

Institute of Medicine. *Toward Environmental Justice*. Washington, DC: National Academies Press, 1999.

Maxwell, N. I. *Understanding Environmental Health*. Sudbury, MA: Jones and Bartlett, 2009.

Myers, N. J. and C. Raffensperger, eds. *Precautionary Tools for Reshaping Environmental Policy*. Cambridge, MA: MIT Press, 2006.

Wenz, P. S. *Environmental Justice*. Albany: State University of New York Press, 1988.

Lawrence K. Duffy
University of Alaska, Fairbanks

ETHANOL

Brazil and the United States are the two largest ethanol producers, creating ethanol out of sugarcane, sugar beet, corn, rice, wheat, or sorghum, which is pictured here.

Source: Randolph Femmer/U.S. Geological Survey Library of Images From the Environment

Ethanol (also known as bio-ethanol), a renewable source of energy, is made primarily by the fermentation of sugars produced by plants such as sugarcane, sugar beet, corn, sorghum, or cereals like rice and wheat. It is primarily used as a transport fuel either in pure form or as a gasoline additive. Brazil and the United States are the world's two largest ethanol producers. In Brazil, sugarcane is used to produce ethanol, whereas in the United States, ethanol is produced from corn (maize). Although Brazil was the first country to use ethanol as auto fuel, it is now being used in the United States, Canada, many European countries, and in some Asian countries like China and India, as a gasoline additive. Ethanol as a fuel first came to the limelight after the Arab oil embargo of 1973, when Brazil launched, in 1975, the National Alcohol Program known as PROALCOOL with the objective of partially displacing use of gasoline in transport.

Interest in ethanol was reinforced in the new millennium due to increasing volatility and rapid surge in crude oil prices. The high import dependence of the majority of the developed and developing countries on a handful of oil and gas producers in the Middle East and west Asia—owning the largest chunk of oil and gas reserves but plagued by geopolitical tensions—has made diversification of energy resources almost inevitable. Ethanol provides an apt alternative with its potential technical, social, economic, and environment benefits compared to gasoline. However, the commercial viability and competitiveness of the fuel is largely dependent on prolonged government nurturing and support. Furthermore, large-scale production of the fuel from the existing agricultural crop–based feedstocks might lead to a food–fuel trade-off, eventually leading to higher food prices and loss of biodiversity due to competing claims on limited land resources. Additionally, the net energy value impact of ethanol produced from the existing feedstocks on well-to-wheel greenhouse gas (GHG) emissions has often been contested by scientists around the world.

The interest in ethanol as an alternative fuel for transport, especially in oil-importing developing countries, is motivated by several factors:

- Diversification of energy sources and lower exposure to the price volatility of the international oil market; this is especially attractive for those oil-importing countries that have high delivered costs of petroleum (for instance, landlocked countries).
- The promise of contributing to rural development by creating jobs in feedstock production, manufacture, and the transport and distribution of feedstock and products.
- Usage of ethanol reduces harmful pollutants from vehicle exhaust. Ethanol has the greatest air-quality benefits where vehicle fleets are old, as is often the case in developing countries. It helps to reduce the exhaust emissions of carbon monoxide and hydrocarbons and particulate matter, especially in cold climates, and also contains no sulfur.
- It has technical advantages like ease of storage and high octane number (a measure of the ignition quality of a gas and indicator of the ability of the fuel to resist premature detonation in combustion chamber) of 120, much higher than that of gasoline, which ranges between 87 and 98. Ethanol, when mixed with gasoline, can also replace harmful lead additives that are used for raising the octane of gasoline.

Brazil uses pure ethanol in about 20 percent of its vehicles; in the rest of the vehicles, it uses at least 25 percent ethanol blend. The United States, Australia, and China use 10 percent ethanol blend. In India, 5 percent ethanol blend has been announced. The Brazilian PROALCOOL program started as a response to the Middle East oil embargo of 1973. At that time, Brazil's dependence on foreign oil made it even more vulnerable than the United States. Brazil's program has in fact been extremely successful, although its development has not come without hitches. The feedstock costs account for 58–65 percent of the cost of ethanol production in Brazil. Thus, the commercial viability of ethanol is critically dependent on the cost of cane production. However, there is no parallel to the center-south region of Brazil in terms of productivity, and the ethanol produced in the region is the cheapest in the world. This could be attributed to the following factors:

- Although cane cultivation is water intensive, the cane fields in this region are largely rain fed, in contrast to irrigated sugar production in countries such as Australia and India.
- Sugarcane and other activities do not need to compete for land because there is still a large amount of unused land in this region of Brazil for further expansion of cane production. Productivity in Brazil has also been receiving constant stimulus from decades of research and commercial cultivation. For instance, cane growers in Brazil use more than 500 commercial cane varieties that are resistant to many of the 40-odd crop diseases found in the country.
- Most distilleries in Brazil belong to sugar mill/distillery complexes that are capable of changing the production ratio of sugar to ethanol. This capability enables plant owners to take full advantage of fluctuations in the relative prices of sugar and ethanol as well as to benefit from the much higher price that can be fetched by converting molasses into ethanol.

- Flex-fuel vehicles that were introduced in Brazil in 2003 further increased the attractiveness of building hybrid sugar–ethanol complexes. More than 60 percent of all motor vehicles produced in Brazil are now "flexi," that is, they can run on any mixture of alcohol/gasoline, as well as on 100 percent alcohol. These engines can also operate with regular gasoline alone if there is short supply of biofuels.

Moreover, ethanol in Brazil provides significant reductions in GHGs compared to gasoline. This is due to the relatively energy-efficient nature of sugarcane production, use of bagasse (left over after the juice has been squeezed out of sugarcane stalks) as process energy, and the advanced state of sugar farming and processing. As set against this, GHG reduction from corn-produced ethanol has been small because corn farming and processing are far more energy intensive. Besides, the refining process for ethanol may also be based on usage of fossil fuels in considerable quantity. Thus, the saving in net energy value or balance (i.e., the difference between the energy in the fuel product, that is, output energy, and energy needed to produce the product through input energy) and associated GHG emission from blending of corn ethanol with gasoline, on a life-cycle basis, may turn out to be smaller compared to that of sugarcane ethanol.

The net energy value of ethanol produced from corn in United States has often been contested by scientists (like David Pimentel of Cornell University) who argue that ethanol uses more energy than it actually yields. However, since 2004, a number of studies have been undertaken in United States, which refuted these findings. Notable among these studies is the one undertaken by the Laboratory for Energy and the Environment at Massachusetts Institute of Technology in 2006. The study was called the Review of Corn Based Ethanol Energy Use and GHG. The study inferred that on average it takes 0.03 gallons of oil to produce 1 gallon of ethanol. Considering historically, the justification for increasing net energy value of ethanol produced from corn could be easily explained. In the 1980s, it was thought that the energy balance of ethanol produced from corn in the United States was neutral to negative, that is, the amount of energy that went into producing ethanol was equal to or greater than the energy contained in the ethanol. However, since then, the advances in the farming community, as well as technological advances in the production process of ethanol, have led to positive returns in the energy balance of ethanol. In fact, a modern ethanol plant produces more ethanol from a bushel of corn and uses less energy to do so.

In order to ensure that ethanol is produced in a sustainable manner and enhances saving in net energy value and GHGs, ethanol plants should use biomass and not fossil fuels; cultivation of annual feedstock crops should be avoided on land rich in carbon (above and below ground), such as peat soils used as permanent grassland; by-products should be utilized efficiently in order to maximize their energy and GHG benefits; nitrous oxide emissions should be kept to a minimum by means of efficient fertilization strategies; and the commercial nitrogen fertilizer utilized should be produced in plants that have nitrous oxide gas cleaning facilities.

It may also be noted that the crops used for ethanol production are also used as food crops or feedstock for cattle feed. An increase in the price of these crops on account of higher ethanol demand would essentially lead to a food–feed–fuel conflict. Thus, in the United States, for instance, because of higher usage of corn for producing ethanol, corn prices have doubled over the past three to four years, substantially increasing the feed costs for livestock and dairy farmers. Moreover, the acreage under corn has grown at the cost of the planted area for soybeans; further increases in the crop acreage may necessitate deforestation, with adverse effects on the environment, especially GHG emissions.

A critical issue that needs to be explored is whether Brazil's success in achieving self-sufficiency and a commercially competitive ethanol industry could be replicated in other developing countries. In this context, it should be kept in mind that Brazil's success was preceded by more than 20 years of government support. The country still continues to maintain a significant tax differential between gasohol (80 percent gasoline/20 percent ethanol) and hydrous ethanol. A large number of countries around the world are growing sugarcane, but none have been able to match Brazil's sugarcane cost structure. Thus, subsidies—indirect, direct, or both—would be needed to launch and/or maintain a biofuels industry in most developing countries. In the past decade or so, the key instruments that have been widely used to foster production and increase in consumption across various countries are mandatory blending targets, tax exemptions, and subsidies. In addition, governments have intervened in the production chain by supporting intermediate inputs (feedstock crops), subsidizing value-adding factors (labor, capital, and land), or granting incentives that target end products. Import tariffs have also played a significant role by protecting national industries from external competition. Although higher oil prices will make ethanol production and consumption relatively cheaper, an additional concern for governments is the lack of clarity on the extent and period of continued support that might be necessary before the ethanol industry becomes self-sustaining.

Most of the trade-offs and problems that arise out of ethanol produced from existing feedstock could be taken care of if ethanol were produced from lignocellulosic (cellulosic material containing lignin) materials such as grass, trees, and different types of waste products and residuals from crops, wood processing, and municipal solid wastes (known as second-generation biofuel). This is because lignocellulosic raw materials are geographically more evenly distributed than fossil fuels and minimize the potential conflict between land use for food (and feed) production and energy feedstock production. These raw materials are much cheaper than conventional agricultural feedstock and can be produced with lower input of fertilizers, pesticides, and energy. Furthermore, ethanol produced from lignocellulose enhances net energy value of ethanol substantially and generates low net GHG emissions and might also provide employment in rural areas. However, for the coming decade, the challenge still remains for policymakers and industry executives alike to nurture the continued expansion of the biofuel sector while abiding by ecologically sustainable production requirements. This is because large-scale production facilities of second-generation biofuels are yet to be available and the technology is yet to be commercially proven. What exists at present are only a few pilot plants that are making ethanol from cellulose in the United States. Additionally, the efficiency in generation of ethanol from cellulose needs to be improved if it is to compete with gasoline. Furthermore, a substantial clarity is necessary on the ecological implications of producing ethanol by using these materials.

See Also: Biofuels (Business); Biofuels (Cities); Biomass; Oil.

Further Readings

Borjesson, Pal. "Good or Bad Bioethanol From a Greenhouse Gas Perspective—What Determines This?" *Applied Energy*, 86 (2009).

Bourne, Joel K. "Biofuels: Boon or Boondoggle?" *National Geographic* (October 2007).

Energy Sector Management Assistance Programme (ESMAP). *Potential for Biofuels for Transport in Developing Countries*. Report 312:05. Washington, DC: World Bank, 2005.

Gerdal, B. Hahn-Ha, M. Galbe, M. F. Gorwa-Grauslund, G. Liden, et al. "Bio-Ethanol—The Fuel of Tomorrow From the Residues of Today." *Trends in Biotechnology*, 24 (2006).
Sorda, Giovanni, Martin Banse, and Claudia Kemfert. "An Overview of Biofuel Policies Across the World." *Energy Policy*, 38 (2010).

Kaushik Ranjan Bandyopadhyay
Asian Institute of Transport Development

ETHICAL SUSTAINABILITY AND DEVELOPMENT

The World Commission on Environment and Development (WCED) brought the issue of sustainability to the center stage of global discussions in 1987 when the United Nations sponsored a study to establish the relationship between economic development and the environment. Published by the Oxford University Press in 1987 as *Our Common Future* (also known as the Brundtland Report after the youngest person and first-ever woman to hold the office of prime minister in Norway, who chaired the commission), this report has been the foundation upon which several critical issues concerning the future of humankind have been anchored. The commission defined "sustainable development" as "development that meets the needs of the present without compromising the ability of future generations to meet their own needs." This can also be formally stated as the twin principles of intragenerational and intergenerational equity.

Apart from providing the conceptual and theoretical framework for the coordinated action among nations and institutions on issues on environment and development, the commission's recommendations also had a number of significant outcomes. The Earth Summit—the United Nations Conference on Environment and Development (UNCED), which was held in Rio de Janeiro, Brazil, in 1992—was as a result of one of the key recommendations of the WCED. The Earth Summit in turn marked the real beginning of international environmental protection initiatives and proposed a sustainable development agenda. This article on ethical sustainability and development explores the ethical issues associated with sustainability and development and the interconnections with critical issues facing humankind today. Sustainability has grown to become a global concern to governments at all levels, multinational corporations, nongovernmental and humanitarian organizations, and local communities, especially indigenous peoples.

Ethics as a Conceptual Frame

The word *ethics* is derived from the Greek word *ethos* meaning "character." It refers to one's ability to distinguish right from wrong. It is the sum total of the values, beliefs, and actions that shape the character of a person or society. It also consists of the fundamental issues of practical decision making, and its major concern includes the nature of the ultimate value and standards by which human actions can be judged. When these ethics define the relationship between the environment and man, it is termed *environmental ethics*. Environmental ethics, therefore, can be defined as a system of ethical values, human reasoning, and knowledge of nature that endeavors to forge a pattern of right conduct toward the environment so that the needs of the present generation are fulfilled without compromising the ability of future generations to meet their own needs. Environmental ethics is the discipline that studies the moral relationship of human beings to, and also the value and moral status of, the environment and its nonhuman contents. It is generally believed that sustainability ethics or ethical sustainability has grown out of environmental ethics.

Ethics may be approached from several viewpoints: a conceptual theoretical framework, which postulates the meaning of ethics from the viewpoint of a specific theory, based on a certain defined understanding and perspective. It may also be normative, that is, prescribing and explaining expected and required behavior in accordance with generally agreed-upon ethical intuitions, systems, and practices. The third approach may be empirical in that it places policy choices within historical and political contexts. The discussion on ethics may also transcend national boundaries, from local and national ethics to international ethics, with consequences for individual ethical behavior to that of institutions and the global society.

At Rio de Janeiro in 1992, the UN Conference on Environment and Development declared in Principle 4 of Agenda 21 that in order to achieve sustainable development, environmental protection shall constitute an integral part of the development process and cannot be considered in isolation from it. By this, the three concepts of ethics, sustainability, and development were linked in a way that requires humankind to understand and pay attention to how these three issues can contribute to discussions that affect life on Earth.

Examining the Components of Sustainability

Considering human life on Earth, food, water, and energy form the basic elements of sustainability considerations, which means that these three elements are important for ensuring sustainability. The importance of these elements can be further explored if one examines the key issues within these elements that have to be considered. The following sections consider these elements in detail:

- *Food*: Of importance are land tenure issues that determine availability, quality (fertility) farming systems and cultural practices, and food production and availability. Also critical are world trading systems that determine global prices of food, global food chains, and their practices, among others; are all issues of concern.
- *Energy*: Especially for forest-dependent rural communities in fragile ecosystems, the pattern of household energy use and sustenance of rural livelihoods are crucial, especially for women and children, on whom the burden of securing the household rural energy requirements rests.
- *Water*: Availability of water for domestic, agricultural, and industrial use; water quality and quantity and distribution issues; control and management; and patterns of use are all crucial issues to be considered. Pessimists have said that if there were to be a World War III, it would be over the control of resources, most likely over water.

Examining the components of sustainability also means addressing the issues and processes that affect sustainability considerations, and these issues include but are not limited to the following:

- Governance structures that affect the extent to which sustainability of especially natural resources is ensured. Ineffective governance structure in many countries of tropical Africa, for example, has been the cause of massive deforestation by the turn of the century.

Instances of military regimes spearheading the pillage of natural resources are also common in many parts of the developing world, especially Africa and Asia.

- Institutional frameworks in place and the strength of the public institutions. The strength of public institutions in a country determines the extent to which rules and regulations regarding resource use are adhered to.
- Decision-making processes at the national mesolevel and microlevels. Decision-making processes determine which members of society have a say in the way resources are used and for whose benefit. More often than not, the rich and advantaged in society have a more powerful voice than the poor, and as a result there arise inequities in the use of resources.
- Extent of collaboration with the private sector, nongovernmental organizations (NGOs), and community-based organizations. Development is now considered to be a collaborative effort, and a global partnership is required to ensure that development is meaningful to people.
- Bioresource consumption patterns. It has been postulated that bio-resource consumption patterns are important for the survival of the human race. Generally, affluent countries of the global north tend to consume more resources than the less affluent global south, and critics say that this pattern needs to be critically examined and addressed.
- Legal and institutional frameworks and their application and enforcement. The extent to which ethical issues can be resolved to ensure sustainable development may also depend on the extent to which legal and institutional frameworks are suited to delivering social justice and accountability in resource use.

Ethical Sustainability

The discussion on ethical sustainability means extending the frontiers of the Brundtland Commission definition of sustainability and looking at the United Nations definition, which has three dimensions, namely, environmental protection, economic development, and social equity or social justice. These are sometimes referred to as the three "Es." The third dimension of the definition of sustainability—social equity or social justice—is the domain where most of the issues of ethics feature. In this third dimension, issues of social justice, fairness, and a just society are some of the issues that are subject to several interpretations based on ethical leanings and ideologies.

The issues of ethical sustainability can be examined from different viewpoints and frameworks. These could be from the viewpoint of the use of the world's recourses, production systems, globalization, and technology transfer among the nations of the world. The examination can also be from the viewpoint of critical issues that border on the survival of especially poor people from developing countries. These issues may include mining, small-scale agricultural practices, climate change adaptation, arms trade, deforestation, food security, and renewable energy, among others. For example, 60 percent of food production in the world is done on degraded soils, and it is estimated that 60 percent of the world's population live near the poverty line, while the wealthy nations consume most of the world's resources. Wealthy nations have the capacity to make more choices than poor nations, and within countries, there are often wide gaps between the poor and the rich. Again, poor people are often forced to take short-term decisions and actions on resources because of their socioeconomic situation. A farmer on whose land commercial trees grow, for example, may decide to fell trees before they grow to commercial size because of the exigencies of the time. It is often said that when survival is at stake, issues of sustainability take second place. However, a combination of our consumption patterns affects the integrity of the planet Earth, which humankind depends on, at least for now.

Development Defined

Development can be defined as the process of positive growth and change in the living conditions of people. It may involve physical or nonphysical developments. It involves changes or improvements in the total stock of assets of an individual, household, or a country as a whole. Development can also be spatial or aspatial. Spatial development may involve physical developments that improve the ability of people to carry out their day-to-day productive, reproductive, and sociocultural activities such as road construction, housing development, health facility (hospitals, clinics, health posts) development, or reticulations development, among others. Aspatial development may involve capacity development, health education, and sensitization, which do not necessarily have physical manifestations in space.

Over the years, the measurement of development has been a subject of debate. From the measurement of development using economic growth and economic indicators, growth is now measured with a mix of indicators, including not only economic indicators but social indicators and other indicators such as the extent of freedom, human rights, and availability of democratic institutions, human security, and the freedom of speech. The Human Development Index of the United Nations Development Programme attempts to broaden the discussion on development by considering broad, cross-cutting indictors for most of the issues mentioned above.

On a global scale, concern about development led to the meeting in 2000 of heads of state and world leaders to set a target on development. This meeting led to the establishment of the Millennium Development Goals (MDGs). The heads of government agreed that the goals were to be achieved by 2015. The eight MDGs are as follows: eradicate extreme poverty and hunger; achieve universal primary education; promote gender equality and empower women; reduce child mortality; improve maternal health; combat HIV/AIDS, malaria, and other diseases; ensure environmental sustainability; and develop a global partnership for development. The seventh MDG, "ensure environmental sustainability," calls on countries to "reverse the losses of environmental resources" by 2015. Making this goal operational has proved to be a challenge for most countries, not least because of a lack of indicators of sustainable development.

Clearly, there is a strong link between sustainability and development. Issues of sustainability and development are related in a way that suggests that unless ethical principles are applied, development may not proceed on a sustainable basis. There is, therefore, the need to examine the issues of ethical sustainability and development with this conceptual frame. Undoubtedly, man is related to the environment, and he is solely dependent on nature. But in the past few years, overexploitation of natural resources has disturbed the environment. Increased population and higher consumption per head greatly affects the environment. The success of environmentally sound development depends on the proper understanding of social needs and opportunities and of environmental characteristics and the interrelationship between these. As the Commission on Culture and Development of UNESCO (1995) notes: "the principles and basic ideas of a global ethics furnish the minimal standards any political community should observe."

Ethical Sustainability and Development: Challenges and Opportunities

There have been a number of arguments for protecting the environment, the heat of it emerging in the 1970s, and these have generated investigations by environmental ethics.

It has been argued that it is morally wrong for human beings to pollute and destroy parts of the natural environment and to consume a huge proportion of the planet's natural resources. The arguments have been that these practices are unsustainable in the long term and are intrinsically harmful to humans. In these arguments, the issues of instrumental values and intrinsic values of the nonhuman aspect of nature have been of considerable importance. Instrumental values have been defined here to mean the value of things as means to further some other ends, while intrinsic values have been defined to mean the value of things as ends in themselves, regardless of whether they are also useful as means to other ends. Theorists have based most of their arguments on these two. Between the two values, objects that have instrumental values are considered right to be exploited by those objects that are intended to serve. However, because the intrinsically valuable is that which is good as an end in itself, it is commonly agreed that something's possession of intrinsic value generates a prima facie direct moral duty on the part of moral agents to protect it or at least refrain from damaging it.

Western traditional ethical perspectives have been argued to be anthropocentric, or human-centered, and in counter-position to the environmental ethics in that they assign intrinsic value to human beings alone such that the protection or promotion of human interests or well-being at the expense of nonhuman things turns out to be nearly always justified. A typical example is the argument put forth by Aristotle (*Politics*, Bk. 1, ch. 8) that maintains that "nature has made all things specifically for the sake of man" and that the value of nonhuman things in nature is merely instrumental. It is, however, counter-argued that anthropocentric positions find it problematic to articulate what is wrong with the cruel treatment of nonhuman animals, except to the extent that such treatment may lead to bad consequences for human beings.

Even though both sides, that is, the environmental ethics side and the anthropocentric side, justify their position, clearly neither can exist independently of the other. Most importantly, human development invariably affects the integrity of Earth and its ability to survive, and the argument should be how Earth's resources can be exploited for development by considering the certain limits and control of exploitation, so that exploitation and conservation can coexist for the sustainability of future generations. Put in other words, the resources in the world should be exploited with the mind-set of being responsible to the resource and other human beings (ethics) so as to ensure that development is sustainable. This is well echoed in Agenda 21 as well as a new crop of environmental ethics theorists who propose a new enlightened anthropocentrism also called "prudential" anthropocentrism, which emphasizes that all the moral duties humans have toward the environment are derived from their direct duties to its human inhabitants.

The practical purpose of ethical sustainability has been to provide moral grounds for social policies aimed at protecting Earth's environment and remedying environmental degradation in order to enhance development. In the 1972 publication *Limits to Growth* by D. Meadows et al., there is a quote that emphasizes the need to examine human values and development: "We affirm finally that any deliberate attempt to reach a rational and enduring state of equilibrium by planned measures, rather than by chance or catastrophe, must ultimately be founded on a basic change of values and goals at individual, national, and world levels."

Recognizing the Interconnections: Ethics, Sustainability, and Development

According to the Brundtland Report, sustainable development must find cohesion between economic, social, and environmental objectives. In the past decades, however, imbalances are recorded in these objectives in favor of the economic objectives. During the 2002 global review in Johannesburg, South Africa, it became obvious that a decade after passing Agenda 21, progress in implementing sustainable development has been extremely disappointing. It was established that since the 1992 Earth Summit, poverty and environmental degradation had deepened. As the world waited for action, the Johannesburg Summit set once again a number of action points with five priority areas: energy, health, water and sanitation, desertification, and biodiversity and improved ecosystem. This was against the backdrop that the world was also striving to achieve the MDGs in the midst of some gains being eroded by climate change and poverty. Even though neither Agenda 21 nor its review during the World Summit on Sustainable Development in Johannesburg 2002 were explicit on the ethical implications of implementation, the World Summit on Sustainable Development provided a forum for a number of opinions to be expressed on the issue of ethics and Agenda 21 during the parallel events, the Global Peoples Forum and partnerships in support of sustainable development. Among the key issues raised were the following:

- Partnership with civil society and the private sector reiterating the fact that there should be a global accord to fully integrate the efforts of civil society and the private sector in the overall efforts of the international community to achieve the goals of environmental protection and sustainable development.
- Environmental dimension of dialogue among civilizations that emphasizes that the ecological crisis facing humanity is deeply rooted in a complex web of economic, social, and cultural factors as well as belief systems, societal attitudes, and perception and that the emergence of the new environmental ethic for the 21st century should be based on a code of conduct and a code of moral duty for all human beings.
- Ethical and spiritual values stressing that concrete measures to harness the full potential of a new economy to make meaningful contributions in areas of information technology, biology, and biotechnology should take into account their ethical, spiritual, and social implications. This is in concert with the consensus of the 1972 team of researchers led by Dennis Meadows that any deliberate attempt to reach a rational and enduring state of equilibrium by planned measures, rather than by chance or catastrophe, must ultimately be founded on a basic change of values and goals at individual, national, and world levels.

Almost a decade after the Johannesburg summit, prime environmental and development issues such as the Deepwater Gulf oil spill in the United States and its aftermath still persist. The situation of damage to the environment and the livelihoods of people is worse in developing countries. In several instances, transnational companies, especially those involved in the extractive industries such as mining and forestry, still operate with significant adverse impacts on the environment and humans, threatening the sustainability of livelihoods. In many developing countries, fishermen still employ DDT when fishing in rivers, lakes, and other water bodies; commercial fishermen employ methods that threaten the sustainability of fish production; and commercial farmers use excessive pesticides and other chemicals in a bid to increase production. These actions endanger the total food chain from the environment to humans. Issues of oil exploration and inequities in the use of resources have brought strife in resource-rich local economies.

As R. A. Mathew and Paul Wapner describe in their paper "The Humanity of the Global Environmental Ethics," environmental (ethics) injustice has gone beyond the environment to the way humans treat each other. Several instances of excessive use of Earth's resources have been recorded. Parking lots are being accused of consuming huge amounts of energy, as much as 110–1,800 PJ, in their construction and maintenance. Recent studies have shown cases of colony collapse disorder of bees, with honeybees, bumblebees, and many other insects being slowly poisoned to death by persistent insecticides used to protect agricultural crops. These issues have birthed a "green revolution" within which every developmental activity factors in a level of environmental friendliness, preservation, and protection. The adoption of biofuel/bio-energy, the carbon market and clean development mechanisms, the use of solar-powered automobile engines, advocacy for the use of bicycles and metro mass transit systems instead of individual automobiles for routine work–home trips, and the emergence of a number of natural/herbal-based food supplements for treating ailments are but a few green initiatives. Researchers at Columbia University in New York City have demonstrated that a layer of plants and soil can cut the rate of heat absorption through the roof of a building in summer by 84 percent, redefining cooling plants and suggesting this new development is able to reduce significantly local warming.

In housing, the ancient neighborhood concepts of green belts and open spaces as ways to preserve some state of the natural environment are being strongly advocated again. Popular actors and film stars have advocated "green development" as a way of responding to the issues that confront the planet. The surge in "greenism" has been such that there is hardly any entity that deals with resources or production that is not associated with some green ideas, whether they are practical manifestations or not. In city development, Eco 2 cities are being developed to take into account the critical issues facing human settlements in an urbanized world. Care, however, must be taken in implementing these "green ideas," as some may have unintended effects. A typical example is bio-energy. Apart from what critics say about large tracts of land for the cultivation of food crops having been converted to production for biofuels, and the destruction and conversion of forests into bio-energy crops, there appear to be emerging concerns that carbon emissions emanating from biofuels could be more than those produced by fossil fuels.

Ethical sustainability lies at the core of the challenge of environment and human development, and the responses that are generated are important for resolving the dilemmas associated with them. Governments, business, environmental groups, households, and individuals all have crucial roles to play in designing appropriate responses. Recent events and challenges associated with human responses to environmental disasters indicate that there are frontiers of knowledge that are yet to be discovered and therefore need to be explored. Agreeing with the realists of the 1980s, it will be beneficial to stand for environmentalism to ensure sustainable development, but there is the need to work with business and governments to soften the impact of pollution and resource depletion, especially on fragile ecosystems and endangered species, as well as on the vulnerable around the world. Clearly, the emphasis should not rest on parochial interests of any group of people or governments, but there should be concerted efforts by all in the responses to the challenges. This calls for innovative solutions, flexibility, dialogue, respect for fundamental human rights, and upholding of the integrity of nature and man.

See Also: Environmental Justice (Business); Environmental Justice (Ethics and Philosophy); Sustainable Development (Business).

Further Readings

Aristotle. *Politics*. Internet Classics Archive. http://classics.mit.edu/Aristotle/politics.html (Accessed January 2011).

Edwards, A. *The Sustainability Revolution: Portrait of a Paradigm Shift*. Philadelphia, PA: New Society Publishers, 2005.

Environmental Research Web. http://www.environmentalresearchweb.org (Accessed September 2010).

Mathew, R. A. and Paul Wapner. "The Humanity of the Global Environmental Ethics." *The Journal of Environment & Development*, 18:2 (June 2009).

Meadows, D., et al. *Limits to Growth*. New York: Universe Books, 1972.

Daniel Kweku Baah Inkoom
Kwame Nkrumah University of
Science and Technology

F

FISHERIES

In 2006, almost six pounds of other marine animals were caught for every pound of shrimp (shown here). This "by-catch," or unintended capture of nontarget organisms, is a consequence of many fishing methods.

Source: Aldric D'Eon/Northeast Fisheries Science Center/ National Oceanic and Atmospheric Administration

Fish play an important role in maintaining healthy marine ecosystems, and they provide more than 950 million people with their primary source of protein. In addition, more than 200 million people, mostly from developing countries, actively depend on marine fisheries for their livelihoods. However, modern fishing vessels, often armed with advanced technologies and industrial-scale vessels, are capable of harvesting animals at rates that far exceed the natural replenishing ability of many species. A 2003 evaluation of global fish stocks concluded that large predatory fish, including tuna, cod, and halibut, have declined by 90 percent since preindustrial times. This continued depletion of stocks reduces the health and stability of fish populations and their aquatic ecosystems and also threatens to economically destabilize many developing and industrialized nations and undermine the food security of people around the world. Overfishing is not an environmental issue of the future; it is a current problem, acknowledged for much of the 20th century. However, a combination of sociopolitical pressures, scientific uncertainty, and lack of enforcement has resulted in the overexploitation of stocks and degradation of marine ecosystems.

Ecological Impacts

Many modern fishing techniques severely alter marine ecosystems. Bottom trawling and dredging change sea floor topography and destroy coral reefs and benthic invertebrate communities that provide important foundations for marine food webs. By-catch, the unintended capture of nontarget organisms, is an unavoidable consequence of many fishing methods because they are not specific to species, sex, or age class. By-catch tonnage can be substantial, sometimes greatly outweighing the target species. For example, in 2006, almost six pounds of other marine animals were caught for every one pound of shrimp landed. This poses a significant threat to many species, including sea turtles, marine mammals, birds, and other fish species, as well as to undesirable members of the target species.

Fishing can also alter the evolutionary trajectory of fish populations because humans exert selective pressures different from those of natural predators. Fisheries generally target larger individuals and increase the overall mortality rate of the population, which results in selection for maturation at younger ages and smaller sizes. This leads to a smaller-bodied, less fecund population that is more vulnerable to predation and environmental perturbations. In addition, reproductive-aged individuals are often removed by fishing before they have the chance to reproduce and make their genetic contribution to future generations.

Removing large amounts of biomass from an environment also has cascading effects that alter the stability of marine ecosystems and render them less resilient to changing conditions. If prey availability dwindles, predators may decrease in number or switch to eating different species. Or if predatory fish populations decrease, prey species may increase and overexploit their food base. These changes can occur unpredictably, making it difficult, if not impossible, for managers to respond appropriately. In order to preserve the food security, water quality, and resilience provided by a balanced aquatic ecosystem, specific management strategies are required to ensure the continued persistence of fish populations and, in turn, these important ecosystem services.

Policies, Management, and Solutions

On the high seas, responsibility for conservation and management of commercial fish stocks largely falls to regional fisheries management organizations (RFMOs). These are intergovernmental organizations such as the Northwest Atlantic Fisheries Organization (NAFO) and the Inter-American Tropical Tuna Commission (IATTC). RFMOs are created by treaty, tasked with setting catch limits, adopting regulations and limitations on the usage of fishing gear to reduce by-catch, and managing virtually every other aspect of high-seas fisheries.

International organizations like these and international fishery policies tie together the economic fortunes of developing and developed nations and, more importantly, can significantly reduce food security for inhabitants of poorer countries when poorly constructed. Wealthy nations in the European Union, for example, that have overfished their own waters spend $227 million annually to buy access to the oceans of developing nations and practice industrial fishing under the flag of the host nation. Countries utilizing these strategies argue that it is a fair trade to offer developing nations payment for access to supplementary fish stocks in their waters. In reality, these actions both directly reduce the available food source for the local population by removing large quantities of fish and indirectly reduce it through by-catch. In addition, fishery disputes between powerful maritime nations are likely to increase as stocks decline and countries increase their capacity to overfish. These disputes can occur for many reasons, including divergent goals (e.g., Japan versus Australia and New Zealand regarding southern bluefin tuna stocks) or territorial disputes (e.g., Spain versus Canada when Spanish ships crossed into Canadian territories to fish for turbot in 1995).

On a regional level, fisheries management policies are created to account for a combination of ecological, social, and economic factors that vary with local conditions and needs. Traditionally, fishery managers sought to achieve the maximum sustainable yield (MSY) a stock could produce; however, it has become clear that managing fisheries to simply maximize yield fails to take into account all stakeholder interests. Today, scientists use their knowledge of biological and environmental dynamics to present stock assessments, which predict how fishing affects the target species, to policymakers. Policymakers must then incorporate the concerns of stakeholders such as fishermen, conservationists, and consumers into their decisions. These conflicting pressures can lead policymakers to ignore or devalue scientific recommendations, resulting in a "ratchet effect" whereby managers continually increase the allowed effort. Once the allowed level is established, these restrictions are theoretically enforced by limiting gear, catch, fishing time, and area. Realistically, fishing quotas are regularly exceeded due to the limited enforcement capabilities of managers and the "race for fish" drive of fishermen motivated to maximize their personal short-term gain.

One potential solution to these problems is a rights-based management system, under which people or groups are allocated privileges to harvest a certain portion of a predetermined total allowable catch. Often, these privileges come in the form of individually transferable quotas, which allow owners to sell or trade fishing quotas. This reduces the pressure to overfish by increasing the owner's investment in the long-term future of the stock. Such management schemes have halted or reversed fishery collapse in a variety of systems, including the Alaskan halibut and New Zealand hoki. However, they remain controversial, in part because of concerns over the equity of rights allocation and the lack of support by some fishing communities.

Ecolabeling by organizations such as the Marine Stewardship Council is an emergent strategy designed to increase consumer awareness of overfishing and to incentivize sustainable management practices. However, the standards governing certification can be controversial and lack transparency, and some argue that consumer-based solutions are often ineffective. Ecolabeling can also be misleading if it only addresses a subset of ecological issues affected by a fishery. For example, cans labeled "Dolphin Safe" tuna may mask the fact that many turtles, birds, and other fish are killed by "dolphin safe" tuna fishing.

Finally, many scientists and policymakers advocate an end to fishery subsidies, which cost governments $20 billion to $50 billion annually and contribute to overfishing worldwide. Many subsidies lack accountability and are used to support fishermen who fish depleted areas. Advocates for subsidies argue that they sustain a culturally and economically valuable lifestyle, and policymakers often find it difficult to remove support from fishing communities. Though this may provide some short-term economic benefit, opponents contend that fishing subsidies are ultimately unsustainable as they maintain too large a standing workforce ready to exploit diminished fisheries and prolong the eventual decline of local fishing communities.

Each solution relies at least in part on stock assessments created by fishery scientists. These assessments have an inherent level of uncertainty that, if not properly accounted for, can produce policies that lead to overfishing. For example, the collapse of the Newfoundland cod fishery was partly due to the mishandling of scientific uncertainty by both scientists and managers. In response to such situations, many fisheries now take a precautionary approach to management, which requires implementing conservative strategies until there is solid evidence that another approach is more appropriate. To take it one step further, some have begun implementing adaptive management, which constantly adjusts management practices based on additional data.

Marine protected areas (MPAs), which restrict human activity in certain areas, have been introduced as a way to buffer marine systems from overexploitation. MPAs have been shown to have a positive effect on stock rebuilding, though their effect on yield remains controversial. Reserves have failed to help some species, either because of improper design or because the needs of these species are not met by MPAs. In addition, reserves created without the support of local communities and interest groups risk facing opposition and discontinuation. Therefore, reserves must be carefully designed and implemented such that both the surrounding people and marine species can benefit.

Many scientists and policymakers also advocate shifting focus away from maximum sustainable yield (MSY) to other goals such as maximum economic yield, which can allow for greater balance of the economic, social, and ecological consequences of fishing. Finally, the concept of ecosystem-based fishery management has gained popularity in recent years. This management approach prioritizes the ecosystem as a whole, rather than focusing on a single species or stock. However, much work remains in fully defining and implementing this approach.

The problem of overfishing is often exacerbated by other challenges to commercial fish stocks, including marine pollution and climate change. Strategies such as aquaculture have been presented as alternatives. Though these approaches may ameliorate one issue, they almost universally present managers with others. For example, while aquaculture may prevent by-catch and reduce pressure on wild stocks, it also introduces pollution though antibiotic leaching and escaped fish.

See Also: Aquaculture; Organic Foods; Tragedy of the Commons.

Further Readings

Clover, C. *The End of the Line*. Berkeley: University of California Press, 2004.

Food and Agriculture Organization (FAO) of the United Nations. *State of the World Fisheries and Aquaculture 2006*. Rome: FAO, 2007.

Garcia, S. and A. Rosenberg. "Food Security and Marine Capture Fisheries: Characteristics, Trends, Drivers and Future Perspectives." *Philosophical Transactions of the Royal Society. B: Biological Sciences*, 365 (2010).

Hilborn, R., et al. "State of the World's Fisheries." *Annual Review Environmental Resources*, 28 (2003).

Worm, B., et al. "Rebuilding Global Fisheries." *Science*, 325 (2009).

Veronica Yovovich
Kate Richerson
Yiwei Wang
University of California, Santa Cruz

GAIA HYPOTHESIS

The Gaia (Greek for "Earth goddess") hypothesis or theory was first proposed by James Lovelock in 1974. The hypothesis has been developed and refined since then, most notably by James Lovelock and Lynn Margulis, a biologist. It is a controversial ecological hypothesis that proposes that the various components of the Earth system, including the lithosphere, hydrosphere, and biosphere, are integrated into a system of interacting parts that overall maintain a homeostasis. The Earth is considered a living organism in which the biological and physical parts are inextricably linked together into a self-regulating system. The Earth system and the relative stability of its atmosphere over long periods of time is considered analogous to the human body and its maintenance of body temperature within a narrow range. According to the Gaia hypothesis, just as the body's temperature homeostat can only be understood in terms of its physiology, so can the Earth system be explained only in terms of a planetary physiology, that is, a systems view of the interactions of its various components.

There is a long tradition, predating Lovelock's invocation of Gaia, of belief in Earth as a superorganism. This quasi-mystical belief arose from a conjecture that the seemingly fine-tuned biological and physical systems characterizing Earth could not have arisen from chance alone. The relative stability of atmospheric composition over millions of years despite a state of disequilibrium signals the presence of control mechanisms. For instance, Earth's atmospheric composition remains unchanged although the amount of incoming solar radiation has increased considerably over the course of the evolution of life on the planet. The salinity of oceans has remained stable over geological time scales. The Gaia hypothesis in its most popular version postulates the existence of feedback mechanisms connecting biotic and abiotic components of the Earth system, which help to maintain equilibrium. An example of a feedback mechanism is the role that phytoplanktons are theorized to play in maintaining global temperature. According to the theory, the increased phytoplankton activity, in response to enhanced solar radiation or increased temperature, causes increases in dimethyl sulfide and ultimately the sulfate aerosols in the atmosphere. The increase in the number of these particles, which act as cloud condensation nuclei, enhances the albedo, thus reducing the incoming solar radiation and exerting a cooling effect. Therefore, the phytoplankton in the oceans would drive the negative feedback loop that helps to keep the global temperature in balance.

Gaia and Darwin

In order to combat the criticism that the Gaia theory is unscientific, Lovelock has proposed that it is an extension of Darwinian evolution. To demonstrate the emergence of planetary self-regulation from selection pressures operating on individuals, the Gaia theory has used a model called Daisyworld. Daisyworld refers to a hypothetical planet that is populated with only two types of daisies: black and white. Black daisies absorb heat and quickly grow to cover most of the planet. As the planetary temperature increases due to absorption of heat by black daisies, the white daisies that do well at higher temperatures start to increase in number and displace black daisies. Due to the higher albedo of the white daisies, the planet cools down. Thus, the planetary temperature is maintained within a certain range by the relative number of white and black daisies, which in turn is controlled by natural selection. However, in the long term, as the luminosity of the star warming the planet increases, the white daisies are unable to lower the temperature, and the planet turns into a hot desert.

Challenges to the Gaia Hypothesis

Various aspects of the Gaia hypothesis have been challenged. One of its central ideas is that the feedback loops between biosphere and atmosphere lead toward a global homeostasis. This has been deemed an oversimplification. It is pointed out that the feedback loops can be negative (stabilizing) or positive (destabilizing). Therefore, what is at issue is the claim that the feedback loops inevitably lead to an Earth system hospitable to life.

Another pivotal tenet of the Gaia hypothesis is that the environment modifies in accordance with the requirements of the organisms, thus creating favorable conditions. There is no denying that species do change their environment, and they adapt not to a fixed, predetermined set of conditions but to a constantly changing environment, partly a product of their own actions. However, the evident problem with this line of thinking is that since only organisms that are well adapted to the environment have survived, it is easy to ascribe the environmental fit to an intentional change to support life.

Yet another postulate of the Gaia hypothesis is that Earth maintains itself in a state conducive to life through self-regulation. The biologically mediated feedback loops between organisms and their environment are considered products of evolutionary forces acting on individuals. It is assumed that the selection of traits beneficial to individuals will also result in the creation of generally beneficial environmental conditions. A shortcoming of this assumption is that natural selection acts on traits that confer selective advantage on individuals carrying them, not on traits that are generally beneficial.

Evidence that contradicts the Gaia hypothesis also includes mass extinction events in the history of Earth. It is difficult to account for these extinctions in the context of a system that is self-regulatory and maintained in a stable state.

Criticism of Teleology of the Gaia Hypothesis

The strongest version of the Gaia hypothesis has been criticized for the claim that atmospheric conditions on Earth appear to have been created with specific purpose, that is, the emergence and perpetuation of life. The "Optimizing Gaia" life hypothesizes life, or biosphere, creates conditions optimum for itself. Another version of Gaia termed "Influential Gaia" has gained wider acceptance. This weaker form of Gaia proposes the existence of feedback loops between life and its surroundings and co-evolution of both, without making any claims for the achievement of overall homeostasis. Most recently, Lovelock has modified optimality, as an outcome of the interactive feedback loops between the biotic and abiotic components, to maintaining conditions generally favorable to life.

Gaia Hypothesis and Climate Change

Gaia theory proposes that climate change poses an imminent threat to the integrity of Earth as a living system. The ocean algae that provide a negative feedback to increasing temperature can be irreversibly harmed as a result of temperatures climbing above a certain range. The only biological feedback mechanism left will be the terrestrial forests that fix atmospheric carbon dioxide (CO_2). This can cause the global temperatures to escalate in an unprecedented fashion. Due to the increase in temperature and the resulting increase in stress on vital ecosystems, the feedback loops that have kept the planetary conditions remarkably stable over billions of years will be disrupted. According to Lovelock, Earth's climate is unlikely to change in the linear and largely incremental fashion, as predicted by the models. It is more likely that a tipping point will be reached that will cause drastic changes to not only the climate but the entire Earth system, rendering a majority of the planet uninhabitable. The loss of biodiversity due to climate change is another mechanism that will reduce the resilience of the Earth system as a whole.

The Gaia Hypothesis: Model and Metaphor

Evidence in support of the core propositions of the Gaia hypothesis has accumulated since it was proposed in the 1970s. The interconnectedness of the biotic and abiotic worlds has been proven with the discovery of various cycles that link the terrestrial world and the oceans. For instance, the atmospheric cycles of sulfur and iodine were shown as connected to the dimethyl sulfide and methyl iodide circulating in the oceans. The research demonstrating bioamplification of the rock-weathering effects by geochemical forces has also lent support to the claim of interconnectedness of the biotic and abiotic elements.

Over long spans of time, the level of oxygen in Earth's atmosphere has remained almost constant at 21 percent. Another line of reasoning that points to the existence of hitherto unknown carbon sinks is the relatively low level of increase in Earth's temperature despite unprecedented increase in atmospheric CO_2. To explain some of these characteristics of Earth's atmosphere, a number of scientists have come to rely on biogeochemistry (as opposed to just geochemistry).

On balance, the Gaia hypothesis has drawn attention to the interconnectedness of biological and geological processes on a planetary scale and their possible stabilizing effect on the planetary system. The idea of emergent properties, arising from simpler components of the Earth system and their interaction, and their significance to the maintenance of a life-supporting system has been one of the main contributions of Gaia hypothesis. Some of the other claims, especially those that viewed Earth as an organism, appear to have some utility as a metaphor, but not as a scientifically valid construct.

See Also: Adaptation; Carbon Emissions (Personal Carbon Footprint); Ecological Imperialism.

Further Readings

Charlson, R. J., J. E. Lovelock, M. O. Andreae, and S. G. Warren. "Oceanic Phytoplankton, Atmospheric Sulfur, Cloud Albedo and Climate." *Nature*, 326 (1987).

Kirchner, J. W. "The Gaia Hypothesis: Fact, Theory, and Wishful Thinking." *Climatic Change*, 52 (2002).

Lenton, T. M. "Gaia and Natural Selection." *Nature*, 394 (1998).

Lovelock, J. E. "Gaia: The Living Earth." *Nature*, 426 (2003).

Lovelock, J. E. and L. R. Kump. "Failure of Climate Regulation in a Geophysiological Model." *Nature*, 369 (1994).

Schneider, S. H. "A Goddess of Earth or the Imagination of a Man?" *Science*, 291 (2001).

Watson, Andrew. "Final Warning From a Skeptical Prophet." *Nature*, 458 (2009).

Neeraj Vedwan
Montclair State University

GENETICALLY ENGINEERED CROPS

Genetically engineered (GE) crops are the main product from advances in agricultural biotechnology over the past several decades. When an organism's genetic material (DNA) has been artificially altered, it is called genetically modified (GM), genetically engineered, or transgenic. Genetic engineering in agriculture initially focused on increasing production in widely planted crops through the insertion of genes with selected traits. The exponential growth of GE foods into a dominant feature around the globe has created uncertainty surrounding the perceived benefits and costs to society. Some of the pros and cons of GE crops regarding human health include increased nutritional values, reduced chemical applications, possible allergenicity and toxicity, and inadequate regulation and product labeling.

Soybeans, shown here, have been the primary genetically engineered crop, created by the insertion of genes with desired traits into regular plants to increase production.

Source: Wikimedia

The first generation of transgenic applications in the mid-1990s focused primarily on crops engineered for herbicide tolerance and disease and insect resistance. Herbicide tolerance (HT) occurs when a gene from a bacterium conveying resistance to some herbicides is introduced and results in using less herbicide for weed control. Disease resistance is engineered through the introduction of a gene from certain viruses that cause disease in plants, thus increasing crop yields. Insect resistance is achieved by incorporating the gene for toxin production from the bacterium *Bacillus thuringiensis* (Bt) into the food plant. Bt is currently used as a conventional insecticide in agriculture, but GE plants that permanently produce this toxin require lower quantities of insecticides when pest pressure is high. HT crops have been the dominant treatment, and HT soybeans and cotton are the most widely and rapidly adopted GE crops in the United States, followed by insect-resistant cotton and corn. Globally, soybean has been the principal GE crop covering 51 percent of the biotech area worldwide, followed by corn (31 percent), cotton (13 percent), and canola (5 percent). Other GE crops grown and in development internationally include alfalfa, sweet potatoes, rice, and plants designed to withstand extreme weather conditions.

Since GE crops were first introduced for commercial reproduction in 1996, their use has increased rapidly. The International Service for the Acquisition of Agri-Biotech Applications (ISAAA) reports that 7 million acres of transgenic crops were grown worldwide in 1996, and by 2007, a total of 282 million acres were planted. The number of countries electing to grow GE crops has also increased steadily from 6 in 1996 to 18 in 2003 and 25 in 2008. Though the use of transgenic crops is expected to level off in industrialized nations, it is gaining momentum in developing countries, which outnumbered industrialized nations 15 to 10 in 2008. Currently, three countries account for a little more than 80 percent of the total GE crops planted globally: the United States (50 percent), Argentina (16 percent), and Brazil (16 percent). Other countries (notably in the European Union) have been skeptical about adopting or importing GE crops and have played a major role in creating the discussion around GE products and their implications for human health.

Pros

GE crops have the potential to impact human well-being in both direct and indirect ways. The most obvious benefits are from the second generation of GM applications that provide value to the consumer through foods with increased levels of micronutrients, modified and healthier fats, and modified carbohydrates. Some examples of these GE foods include rice varieties with beta-carotene or higher levels of iron and zinc, corn enhanced with vitamin C, soybeans with improved amino acid composition, and potatoes with greater calcium content. Though improving human health through better nutrition is important for the general populace, it is seen as most valuable for those suffering from severe malnutrition. Women and children have higher requirements for vitamins and minerals that are not met in many developing countries. Globally, 3 million preschool-age children have visible eye damage from vitamin A deficiencies, leading to blindness and death in 250,000 to 500,000 cases annually. Clinical trials in developing countries showed that vitamin A supplementation through fortified food products reduced the mortality rates of preschool-age children on average by 23 percent and indicate the possible benefits that staple foods enhanced with GE technologies can incur for nutrition.

A widely touted attribute of GE crops is their ability to thrive with reduced sprayings of toxic chemicals, which is also favorable for human health. Studies of insect-resistant GE cotton productions in China, Australia, South Africa, and the United States showed reductions in insecticide between 40 and 60 percent compared to conventional cotton crops. These diminished chemical applications may be better for consumers ingesting the GE products and the farmers and other agricultural workers who are exposed to them. Additional advantages of decreased chemical applications are reduced costs for farmers and increased productivity from their land and labor, which translates into increased material wealth and food security. Again, this may be especially significant in developing nations where GE is seen as having the potential to address many issues pertinent to resource-poor or small-scale farmers. In an extreme example cited by ISAAA, farmers in India growing Bt cotton found that their yield increased by 31 percent, insecticide decreased by 39 percent, and profitability increased by 88 percent. The apparent cost-effectiveness of GE crop production is argued to confer health and accompanying socioeconomic benefits for farmers around the world.

Because there are many new and challenging facets to engineering foods for mass consumption, the research and development surrounding GE is being governed by federal and international agencies. In the United States, three federal agencies regulate different aspects of GE foods: the U.S. Environmental Protection Agency (EPA), the Food and Drug Administration (FDA), and the U.S. Department of Agriculture (USDA). These agencies monitor the environmental and human health risks under the 1986 Coordinated Biotechnology Framework. In the United States, risk assessment requires agencies to analyze and interpret the scientific data and make informed predictions about risks from the activity or product. Subsequent risk management requires the agency to make legal and policy judgments about how to implement regulatory options available to them. The international entities with authority over GE matters are the European Union (EU), the Food and Agriculture Organization (FAO), and the World Health Organization (WHO). More than 50 GE food products have been evaluated by the FDA and deemed safe as conventional foods, and WHO maintains that GE products on the market do not indicate any risk to human health. Proponents of GE crops argue that there is no convincing evidence yet of any new risk to human health and contend that there are ample measures in place to minimize risks and address public concerns.

Cons

Concerns over the impacts of GE crops on human health focus especially on allergenicity, toxicity, and unknown effects. There has been a significant rise in food allergies over the past decade, prompting questions of whether the introduction of GE products into the human food chain might cause unintended allergic reactions that are sometimes fatal. The FAO and the WHO developed a framework in 2000 to assess the potential allergenicity of GE foods containing genes from known allergens. Other risks are not so straightforward and include the possibility of gene transfer from biological sources with unknown allergenicity and the unintentional creation of novel allergens. Inadvertent transfer of allergens has been documented as early as 1996 in a study that revealed a Brazil nut allergen had been transferred to a soybean under development. Even industry studies have demonstrated allergenicity in pre-market GE crops, such as a 1996 Monsanto report showing a 28 percent increase in an allergen present in a soybean product. With limited methods available for addressing food allergies, and no strong animal model for predicting human allergens, uncertainties regarding health issues persist.

Genetic engineering can also alter both existing and unanticipated toxicological characteristics of foods. Some scientific data indicate that gene insertion can generate unexpected increases in levels of naturally occurring toxins, while the possibility that GE will produce new, rare, or previously unknown toxins remains a concern. In addition to relying on the chemical analysis of known nutrients and toxic compounds, regulatory agencies such as the EPA have tested GE foods in mice as an indirect assessment of toxicity to humans. However, these studies have limited applicability because they study a single chemical component and use high-dose exposures over a limited time period.

Additional unintended effects of GE processes could stem from gene transfer. Transfer from GE foods to the body's cells or bacteria in the stomach could adversely affect human health, particularly if antibiotic-resistant genes used in creating GE crops are transferred. The movement of genes from GE plants into conventional crops and related wild species as well as the mixing of crops derived from conventional and GM seeds may also have an indirect effect on food safety. This risk has already occurred in the United States. Corn approved only for animal feed (and deemed resistant to human digestion) appeared in household products for human consumption due to inadvertent contamination.

With these uncertainties surrounding GE crops and human health, the aforementioned regulating bodies have a formidable task in overseeing GE research and risk assessment. Many critics argue that these agencies have been too lenient. In 1992, the FDA adopted a policy whereby GE foods were presumed "generally recognized as safe." Similarly, the FAO and the WHO subscribed to the concept of substantial equivalence, which regards GE food products to be as safe as their conventional counterparts. Safety testing through the FDA has been documented by the nonprofit Center for Science in the Public Interest (CSPI) and others to be inadequate. Studies noted that biotechnology companies frequently have not released the requested information, there were undetected errors in technical data, and the FDA had a lack of necessary authority in the review and regulation of GE research and crops. Since 2003, official standards for food safety assessment have improved with the global consensus forwarded by the Codex Alimentarius Commission of FAO/WHO. These principles dictate a premarket assessment, performed on a case-by-case basis, which includes an evaluation of both direct and unintended effects. However, peer-reviewed studies found that despite these guidelines, risk assessment of GE foods has still not followed a defined prototype.

Labeling of GE products remains a contentious point in the debate. With upward of 60 percent of food in U.S. supermarkets containing GE ingredients, the main arguments for labeling include consumer awareness and tracking health problems should they occur. The FDA requires labeling of GE foods if the food has a significantly different nutritional property, includes an allergen that consumers would not expect to be present, or contains a toxicant beyond acceptable limits. Currently, mandatory labeling of all GE foods has been proposed—but not yet enacted—at the national, state, and local levels in the United States. Factors against mandatory labeling include the added expense to the producer and consumer and the potential for consumer preference of non-GE foods.

At the international level, not all countries have been keen to convert to GE crops. The EU has been noticeably cautious with GE products, which is likely due to numerous food scares unrelated to GE that brought food safety issues to the foreground in the late 1990s. Labeling is mandatory in the EU for products derived from modern biotechnology. Controversy has stemmed over the use of the Cartagena Protocol to regulate GM organisms from one country to another. Based on the precautionary principle, the treaty gives countries the right to ban imports of GE seeds and crops seen as a threat to their environment or to procure the information necessary to make informed decisions. By 2003, the Cartagena Protocol was approved by more than 130 countries and ratified by 50. Though the United States did not ratify the protocol, it must adhere to the provisions when shipping GE products to the countries that are signatories. Within the context of human health, GE crops have the capacity to provide potential benefits and risks that are shaped largely by national and international regulating bodies, the scientific community, and an informed public.

See Also: *Bacillus Thuringiensis* (Bt); Biotechnology; Genetically Modified Organisms (GMOs); Precautionary Principle (Ethics and Philosophy).

Further Readings

Bouis, Howarth E. "The Potential of Genetically Modified Food Crops to Improve Human Nutrition in Developing Countries." *Journal of Development Studies*, 43:1 (2007).

International Service for the Acquisition of Agri-Biotech Applications (ISAAA). "Global Status of Commercialized Biotech/GM Crops: 2008." http://www.isaaa.org/resources/publications/briefs/39/default.html (Accessed September 2010).

Magaña-Gómez, Javier A. and Ana M. Calderón de la Barca. "Risk Assessment of Genetically Modified Crops for Nutrition and Health." *Nutrition Reviews*, 67:1 (2009).

Taylor, Ian E. P. *Genetically Engineered Crops: Interim Policies, Uncertain Legislation.* London: Haworth Food and Agricultural Products Press, 2007.

Nicole Menard
Independent Scholar

GENETICALLY MODIFIED ORGANISMS (GMOs)

The combination of genes at the embryonic stage, creating chimeras, is still under contention. These chimera offspring still hold traits of the added embryonic cells.

Source: Transgenic Core Facility at the National Institute of Mental Health

The acronym *GMOs* stands for genetically modified organisms, a term that is often used interchangeably with the terms *GE* (*genetically engineered*) *organisms* and with *biotechnology*. The term *GMO* refers to animals, plants, or microorganisms that have been altered at the genetic level. The science that has paved the way for the development of GMOs is viewed with both enthusiasm and great skepticism, as are GMOs themselves. On one hand, GMOs are argued as providing solutions to global issues, including feeding the world population and reducing negative environmental impact. On the other hand, GMOs are argued to be an unnecessary development that could cause far more problems than they could potentially alleviate, as the following explains.

The development of GMOs has its very early roots in experiments undertaken by an Augustinian monk named Gregor Mendel in the mid-1800s. Mendel carried out meticulous experiments using pea plants in which he explored their hereditary traits. From these early experiments, Mendel argued that plants pass on hereditary traits from one generation to the next, that they were passed on according to a certain pattern involving each "parent plant" contributing half the genetic traits to the next generation of plants, and that this next generation would inherit different sets of traits.

It was not until the early 1900s when several scientists, including Sigurd Orla-Jensen (a Danish microbiologist), Patrick Geddes (an English biologist), and Károly Ereky (a Hungarian engineer), developed the early ideas around what we now understand as (new) biotechnology. It is from the work of these early pioneers that GMOs were eventually developed. Ereky is most notable here in that he created the term *biotechnology* in 1917 and provided a prescription for a new science of the same name—a scientific technology that is based upon a combination of biology and chemistry.

The discovery of the DNA (deoxyribonucleic acid) double helix by Francis Crick and James Watson in 1953 paved the way for GMO development. The DNA double helix is essentially the structure of genetic material, the discovery of which provided a basic premise for new biotechnological developments, including GMOs. Yet the development of GMOs really began only in 1973 when Herbert Boyer and Stanley Cohen invented genetic engineering—the altering of the genetic code through human intervention involving introducing new genetic material into a host, such as a plant or animal. Gene splicing, which is the recombining of DNA (rDNA), is the process that modifies an organism, with the resulting altered (host) organism being termed a GMO.

Countless examples exist of GMOs in the world today. The first GMO developed for commercial sale was a tomato, engineered to have an extended shelf life. However, this product was not sold widely and was soon withdrawn from the market when Calgene, its developer, ran into business-related problems. Many other GMOs have since been developed. Two food examples include *Bacillus thuringiensis* (Bt) corn, a corn that produces its own pesticides, which is made possible by the insertion of the bacteria toxin *Bacillus thuringiensis* into the corn. A second example is a type of rice, golden rice, which is designed to produce beta-carotene that can increase the vitamin A levels in those who consume the rice.

Other examples of GMOs can be found in the animal world. A range of animals have been genetically modified to produce human proteins for medicinal purposes, including sheep, cows, and goats. Fish have been genetically modified as novelty pets that glow in the dark. Other very contentious developments include the combination of animal with human genes at the embryonic stage. The products of this animal and human genetic mix are known as chimeras and are regarded by most as being strictly for research purposes.

Given the array of GMOs that have been developed or are under development, it is perhaps not surprising that pros and cons have been identified in relation to them. One argument in favor of GMOs is that they will help to feed the world by being able to provide higher-yielding crops for less cost on less land or on land that is not of high quality. Another benefit is seen in the ability of GMOs to help reduce negative environmental impacts by reducing the need for chemical application on plants and through making more resourceful use of land. Environmental benefits are also being argued as GMOs are thought to be useful for combating climate change through helping reduce water loss and through the development of drought-resistant GMO crops. Benefits to farmers and producers are derived through the decreased need for chemicals and the development of genetic traits in animals and produce that are resistant to various pests and diseases. It is also claimed that there are benefits to consumers through better-tasting food with increased nutritional value and foods that have a longer shelf life.

In addition to these claims for the benefits of GMOs, there is also a lengthy list of arguments against various GMOs. In research and field trials on animals, a variety of health problems have arisen, including enlarged ovaries in genetically modified cows, abnormal changes to internal organs of rats, and deformed heads in salmon. It is also argued that there are dangers to human health through the consumption of GMOs in food. Concerns in this area include a fear of increased antibiotic resistance due to antibiotic-resistant genes found in GMO plants and fears that foreign proteins in GMOs will increase the prevalence of various allergies. There is also some unease over the amount of chemicals used on GMO food plants. Essentially, these are concerns around the safety of GMOs that extend beyond animal and human health to environmental safety. In relation to the environment, examples of safety concerns are many. They range from the escape of GMO creatures such as the aforementioned salmon into the wild as threatening wild fish populations to the accidental dispersion or removal of genetically engineered material from animal and plant experimentation or trial sites. Environmental concerns are linked with threatening biodiversity. The spread of GMO crops such as canola into neighboring farmlands growing conventional canola has resulted in lawsuits in North America, where debates have emerged regarding whether farmers have purposely grown GMO crops without paying for the right to do so or whether it is due to accidental contamination from neighboring GMO crops spreading naturally. Farmers have complained about the unwanted spread of GMO crops, having to pay for the higher cost of GMO seed, and not being able to save seed for planting the following season. In a nutshell, critics argue that the proclaimed benefits of GMOs have not eventuated.

It is not surprising that a polarized debate has ensued around GMOs. Around the globe, resistance movements and protest actions, including direct sabotage of GMO crops, have emerged alongside the development of the technology. Nevertheless, the development of GMOs is occurring at a rapid pace.

See Also: *Bacillus Thuringiensis* (Bt); Biotechnology; Genetically Engineered Crops; Genetically Modified Products.

Further Readings

Bhargava, Pushpa M. "The Social, Moral, Ethical, Legal and Political Implications of Today's Biological Technologies: An Indian Point of View." *Biotechnology Journal*, 1 (2006).

Parekh, Sarad R. *The GMO Handbook: Genetically Modified Animals, Microbes, and Plants in Biotechnology*. New York: Humana Press, 2010.

Royal Society of Canada. "The Past, the Present and the Future." *Journal of Toxicology and the Environmental Health*, 64:1 (2001).

Sanderson, Colin J. *Understanding Genes and GMOs*. Singapore: World Scientific Publishing, 2007.

Stewart, C. Neal, Jr. *Genetically Modified Planet: Environmental Impacts of Genetically Engineered Plants*. New York: Oxford University Press, 2004.

Corrina Tucker
Independent Scholar

GENETICALLY MODIFIED PRODUCTS

Genetically modified (GM) products are produced by genetic manipulation, a technique through which the genetic materials from a living organism are harvested, modified, or recombined (i.e., manipulated with other genes) and replaced in the same or a different organism to obtain the desired traits. This technique of combining and manipulating genes from different organisms is also known as recombinant DNA technology or genetic engineering, and the resulting product is known as genetically modified or transgenic or genetically engineered. Genetic engineering or modification is carried out for various reasons, such as to treat or repair a genetic defect (gene therapy); to develop medicine or vaccines; to develop crops that are resistant to drought, cold, flood, or herbicides; to enhance the growth rate of animals or plants; to shorten the maturity or ripening time of fruits, vegetables, and grains; and to enable an organism to produce a substance, which normally it would not do (e.g., producing human insulin from a microorganism to treat diabetes), among others.

For the aforesaid reasons, several transgenic (i.e., GM) products—crops and animals, microorganisms, foods, medicine, vaccines, biologics, cosmetics, pesticides, and industrial feedstock—are produced for various uses. However, issues relating to genetic modification and use of GM products—such as concern for human health, concern for the environment, regulations and labeling, consumers' concern—continue to prevail. Although some consider genetic engineering and GM products a blessing for agriculture, food, and medicine development, others think it a threat to the natural world, harmful, unnecessary, and a failed and failing technology that benefits big businesses at the expense of others.

A thorough discussion about all GM products and their issues is beyond the scope of this article. Hence, this article presents an overview of GM products (i.e., transgenic crops, animals, and foods), discusses safety aspects of GM products, and highlights regulations and concerns relating to GM products, followed by a discussion on the public debate.

An Overview of GM Products

Genetic modification is successfully applied to grow plants, animals, microorganisms, pharmaceuticals, biologics, and chemicals, among others, to meet the increased global need for foods, feeds, livestock, medicine, and other products. A brief overview of the three major categories of GM or transgenic products (crops, animals, and foods) follows.

GM Crops

In 2006, of 97 percent global GM crops, the United States grew 53 percent, followed by Argentina (17 percent), Brazil (11 percent), Canada (6 percent), India (4 percent), China (3 percent), Paraguay (2 percent), and South Africa (1 percent). GM crops are developed to introduce agronomic and novel traits such as tolerance to herbicides, cold, drought, and salt; resistance to insects, pests, and diseases; increase in growth, yield, and nutrient content; and improvement in taste, quality, storage, shelf life, and transportation, among others. Some examples of popular GM crops are Roundup Ready soybeans (herbicide tolerant), *Bacillus thuringiensis* (Bt) corn and Bt cotton (pest resistant), GM tobacco and potato (cold resistant), and GM rice (rich in beta-carotene and iron), among others. Investigation and field testing of various novel traits to develop new GM crops are under way in many countries. Research and development of GM crops with novel traits is continuing, and a variety of GM crops with traits like enhanced nutritional profiles is expected to emerge in the coming years.

GM Animals

Genetic engineering is successfully applied to produce transgenic or GM animals. The purpose of applying genetic engineering to livestock is to produce animals and organisms to be used for biomedical research, pharmaceuticals in milk of GM animals (e.g., sheep), animals for xenotransplantation (i.e., process of transplanting animal tissues and organs into humans as a substitute for human organ donors; for example, transplanting heart and kidneys from GM pig to human), animals as models of human diseases, and animals and organisms to improve food supply and support human nutrition. Transgenic animals such as rats, mice, and rabbits are extensively used in biomedical research that helps understand human biology at the cellular and molecular level in health and disease. Researchers can often diagnose the root causes of diseases associated with gene defects by introducing or inactivating particular genes. Sheep, cattle, and goats are genetically modified to produce valuable proteins such as pharmaceuticals in their milk. For example, α-1-antitrypsin (AAT)—a protein used to treat emphysema (a lung disease) and cystic fibrosis (an inherited disease of the secretory glands)—is produced in the milk of sheep by incorporating the human gene that codes for synthesis of the protein (AAT) into the sheep. Another example is the effort to develop edible vaccines in tomatoes and potatoes, which will be easier to transport, store, and administer than the traditional injectable vaccines.

Some companies have developed transgenic animals to provide new organs for transplantation such as kidneys, livers, and hearts (xenotransplantation). Research is under way to breed transgenic pigs with genes that are compatible with humans in the hope that their organs will not be rejected by a patient's immune system. If successful, this research could transform the lives of many patients waiting for organ transplants. Further, transgenic animals are also used to evaluate the toxicity and safety of new medicines and vaccines. Because transgenic animal models can highlight specific characteristics such as mechanisms of tumor formation, they can clearly demonstrate the possible side effects of new medicines or vaccines. Furthermore, transgenic animals may serve as an alternative to nonhuman primate models in biomedical research. For example, GlaxoSmithKline, a multinational pharmaceutical company, is awaiting approval from the World Health Organization (WHO) to use a transgenic mouse model for neurovirulence testing of its oral polio vaccine. Thus, transgenic animals are vital in the discovery and development of new medicines and to treat several diseases. They also help scientists characterize the newly sequenced human genome. Without them, the ability of the pharmaceutical industry to discover new medicines or treatments seems to reduce significantly.

GM Foods

The first GM food introduced in the U.S. market during mid-1990 was the delayed-ripening tomato. Since then, a variety of GM foods has been produced and marketed, mainly in the United States, Argentina, Canada, the European Union, and South Africa. GM foods were developed because of some perceived advantages, either to the producer or to the consumers. The initial objectives of developing GM foods were to improve crop protection by incorporating agronomic traits such as herbicide resistance, insect resistance, and the ability to survive a range of stresses like drought, cold, and high salt and aluminum content of soil. However, in recent years, with the development of genetic engineering, besides incorporating agronomic traits, GM foods are produced to increase food supply, improve nutritional quality, enhance taste, and preserve quality during storage and transportation.

GM foods with improved agronomic traits currently available in the international markets are insect-resistant corn (maize) and potatoes; herbicide-tolerant corn, soybeans, canola, rape, chicory, and potatoes; and virus-resistant sweet potatoes, corn, and African cassava. All the genes used to incorporate these agronomic traits are derived from microorganisms. GM foods with improved nutritional qualities include vitamin A–enhanced rice (golden rice); iron-fortified rice (high-iron rice); protein-enhanced cassava, plantain, and potato; and antioxidant-enhanced (lycopene and lutein) tomatoes. In addition, some government authorities have approved several varieties of GM papaya, potato, rice, squash, sugar beet, and tomatoes that can withstand stress conditions such as drought, flood, and high aluminum and salt content of soil. Among animal foods, growth-enhanced Atlantic salmon containing a growth hormone gene from Chinook salmon is the first GM animal in the food market. Other growth-enhanced fish are grass carp, rainbow trout, tilapia, and catfish.

Further, to enhance the efficiency of cheese production, New Zealand has developed GM cows to produce milk with increased levels of casein protein. Other applications of genetic modification in animal species for food purposes such as to increase birth rate in sheep, increase egg production in poultry by creating two active ovaries, and breed "enviro-pig" (i.e., environmentally friendly pig, which excretes less phosphorus) are under way. Although the United Kingdom (UK) has approved GM yeast for beer making since 1993, such microorganisms were never intended to be commercialized. Attempts are also under way to produce other GM microorganisms; for example, starter fermentation cultures for various foods (bakery and brewing), lactic acid bacteria for cheese making, and probiotic bacteria for promoting health benefits, among others. Many enzymes used as food and feed processing aids—for example, α-amylase for bread making, glucose isomerase for fructose production, and chymosin for cheese making—are derived from GM microorganisms. In fact, some countries permit the GM microorganisms to produce dietary supplements such as vitamins (e.g., carotenoids used as food additives, colorant, or dietary supplements). Thus, over the past few decades, several GM products with specific traits have been developed and commercialized.

Safety of GM Products

Safety of GM products is a major issue in both developed and developing countries. Therefore, international organizations such as WHO, the Food and Agriculture Organization (FAO), and the Organisation for Economic Co-operation and Development (OECD) are involved in the safety assessment of GM products. In 2003, the combined effort of WHO, FAO, and OECD to build international consensus on safety assessment of GM products led to the development of principles for the safety assessment of GM products, which are incorporated in the Codex Alimentarius Commission (CAC). These principles demonstrate a premarket assessment of GM products performed on a case-by-case basis and a thorough evaluation of the intended and unintended effects from the inserted gene. Specifically, the Codex safety assessment of GM foods, GM plants, and GM microorganisms (CAC 2003) require investigation of potential toxicity (e.g., direct health effects) and allergenicity (e.g., tendency to provoke allergic reactions) of the GM foods or feeds, stability of the inserted gene, nutritional characteristics and effects associated with the specific genetic modification, horizontal gene transfer (e.g., natural genetic modification resulting from the gene exchange between different species), and any unintended effect that may result from the insertion of new or modified genes. Codex principles do not have a binding effect on national legislation. However, they are referred to specifically in the Agreement on the Application of Sanitary and Phytosanitary Measures (SPS Agreement) of the World Trade Organization and are used as a reference when any trade dispute arises.

Regulations of GM Products

GM products are not yet regulated in some countries. Countries that have legislation in place regulate them differently. The regulatory differences among countries vary primarily due to the differences in attitudes toward GM products. Regardless of attitude toward GM products, the regulatory processes of all countries mainly focus on the assessment of risk for consumers' health, risk for the environment, and issues related to control and trade (e.g., testing, labeling). The following paragraphs summarize the regulatory policies of some countries related to GM products.

In the United States, the federal government implements the regulatory process related to GM products. GM products are required to comply with federal laws and regulations. The 1986 Coordinated Framework is responsible for the regulation of biotechnology. Three other agencies—the Department of Agriculture (USDA), the Environmental Protection Agency (EPA), and the Food and Drug Administration (FDA)—are involved in developing, reviewing, and implementing the regulatory legislation of GM products. The USDA ensures that genetically modified organisms (GMOs) are safe to grow. The EPA ensures that GM pesticides, nonpesticidal toxic substances, including microorganisms, are safe for the environment, and the FDA ensures that GM food, GM drugs, GM biologics, and GM feed are safe to eat and use. In addition, the FDA has issued draft voluntary guidelines to industries for labeling of GM foods. Although new laws and regulations have evolved since 1986, policies of the Coordinated Framework continue to govern various agencies' action in the United States.

In the European Union (EU), the European Commission (EC) has issued the revised Directives (2001/18/EC) that regulate the release or use of GM products. According to this directive, the application for release of GM products must accompany a risk assessment that identifies and evaluates the potential negative effects (immediate or delayed) to human health and the environment. Recently, the EC has adopted a comprehensive proposal on GMO cultivation that considers member states' desire to produce GMOs. However, the EU's existing science-based authorization system regulates the GMO authorization. The labeling rule of the EC provides information to consumers, allowing them to make an informed choice. Any food or feed that contains or is produced from GMOs in a proportion higher than 0.9 percent of the food or feed ingredients must carry a label referring to the presence of GMOs.

In the UK, in line with international legislation, the Department for Environment, Food and Rural Affairs regulates the releases of GMOs. All assessments are approved via the GM Food and Feed Regulations, effective since April 2004. These regulations are led by the Food Standards Agency of the UK. In Australia, the Gene Technology Act (2000) regulates and enforces the safety and monitoring of GM products. With New Zealand, Australia has developed the Standard A18, which governs the safety and labeling aspects of GM products. According to the standard, all genetically modified foods need to be assessed for safety for human consumption before issuing approval for sale and use. Further, the standard requires all GM foods and ingredients to be labeled if they contain novel genes in the final products.

Asia has been in development of regulation of GM products since the early 1990s, and Asian governments are trying to find a regulation that will allow them to reap the benefits of biotechnology while avoiding trade issues with either the United States or the EU. Japan, Korea, China, Australia, Taiwan, India, Indonesia, Thailand, Malaysia, the Philippines, and Korea have granted some approval (e.g., food safety for imports, field testing, and commercialization) for GM products. Japan, China, Australia, India, and Indonesia have granted all three types of approvals. Several other countries in the region are also developing the regulatory process relating to import of GM products.

In Argentina, the National Advisory Commission on Biotechnology of Argentina regulates the release of GMOs into the environment through a number of requirements set for safety purposes. In addition to obtaining the marketing license, all GMOs must comply with the requirements set by the National Service of Health and Agrofood Quality. Mexico has banned the import of GM corn following repeated concern raised by farmers and consumers.

In Africa, only South Africa and Zimbabwe have legislation in place relating to the biosafety law of GMOs. Other African countries are in the process of developing biosafety policies and laws to comply with the requirements of the Cartagena Protocol on Biosafety—an international agreement that ensures the safe handling, transport, and use of living modified organisms (LMOs). At the regional level, other organizations such as Southern African Development Community, the Common Market for Eastern and Southern Africa, and the New Partnership for African Development are engaged in establishing biotechnology-related policy. So far, only South Africa has approved the commercial growing of GM crops.

Effective since 2003, Codex principles and the Cartagena Protocol on Biosafety regulate international GM food and environmental safety. Additional information about national and international legislation of GM products is available on the Organisation for Economic Co-operation and Development website (www.oecd.org). Thus, regulations and policy relating to GM products continue to evolve worldwide as genetic engineering advances.

Concern for GM Products

Increasing concern about GM products prevails since the introduction of the first GM food in the market. Although people generally accept the GM medicine or vaccines with improved treatment potential, they are often skeptical about eating GM foods. The primary concerns about GM products are related to human health and the environment. The major concerns for human health are that transfer of modified and or new genes between species may create new allergens that can cause allergic reaction in susceptible individuals (allergenicity), gene transfer from GM foods to the host's body if the transferred gene is harmful to human health (e.g., development of antibiotic resistance due to transfer of an antibiotic-resistant gene from a GM food to a person who consumes it), movement of genes from GM crops to the conventional crops or related species nearby, which may affect food safety and food security (out-crossing), and unintended effects on human health. The issues related to the environment include the ability of the GM organism to escape and introduce to the wild or non-GM population, potentially adverse effects on beneficial insects, induction of resistant insects, detrimental consequences on plant biodiversity and wildlife, decrease in crop rotation practices, and movement of herbicide-resistant genes to other plants. Although scientific studies and the FDA have ensured that GM products, including GM foods, are safe and nutritious, some people remain unconvinced, and public debate on GM products continues.

Public Debate

In several countries, the development and release of GM products has created public debate on issues relating to costs, benefits, safety (human and environment), and labeling and traceability of GM foods to address consumers' choice. Proponents of genetic engineering claim that development of GM products can help overcome poverty, ensure food security, prevent deficiency diseases (e.g., blindness caused by vitamin A deficiency), deliver affordable healthcare, and promote conservation of soil, water, and energy through use of bioherbicides and bioinsecticides. However, opponents of genetic engineering believe that tampering with the genes of living organisms is unethical and may exert unintended effects on human, animals, and the environment; encourages domination of world food production by a few companies; increases dependence on industrialized nations by developing countries; encourages biopiracy; and destroys biodiversity. This debate might continue in the broader context of other uses of genetic engineering.

Conclusion

The ability to introduce novel traits through genetic manipulation will continue to create GM products (plants and animals) for the purpose of producing food, therapeutics, industrial chemicals, and other high-value products. Despite the contribution of GM products to food supply and healthcare, issues relating to human health and environment and thus the public debate are expected to continue. Therefore, federal regulators responsible for reviewing the health, safety, and efficacy of transgenic organisms and their products will continue to face challenges to effectively address these issues.

See Also: *Bacillus Thuringiensis* (Bt); Biotechnology; Genetically Engineered Crops; Genetically Modified Organisms (GMOs); Roundup Ready Crops.

Further Readings

Baumüller, Heike. "Domestic Import Regulations for Genetically Modified Organisms and Their Compatibility with WTO Rules: Some Key Issues." International Centre for Trade and Sustainable Development (ICTSD). http://www.tradeknowledgenetwork.net/pdf/tkn_domestic_regs_sum.pdf (Accessed August 2010).

Kleter, Gijs A. and Esther J. Kok. "Safety Assessment of Biotechnology Used in Animal Production, Including Genetically Modified (GM) Feed and GM Animal—A Review." *Animal Science*, 28:2 (2010).

U.S. Department of Energy, Office of Science, Office of Biological and Environmental Research. "Genetically Modified Foods and Organisms." http://www.ornl.gov/sci/techresources/Human_Genome/elsi/gmfood.shtml (Accessed August 2010).

World Health Organization. "Modern Food Biotechnology, Human Health and Development: An Evidence-Based Study." http://www.who.int/foodsafety/publications/biotech/biotech_en.pdf (Accessed August 2010).

World Health Organization. "20 Questions on Genetically Modified Foods." http://www.who.int/foodsafety/publications/biotech/20questions/en (Accessed August 2010).

Meera Kaur
University of Manitoba

GREEN ALTRUISM

Altruism, also called helping behavior, refers to selfless acts of caring and kindness. Individuals who make decisions and take action toward sustainability are thought to be displaying altruism. Concerns arise over the motivation and reward driving environmental action, debating if there is such a thing as pure altruism or whether all decisions are ultimately egoistic, providing some personal payoff to the altruist.

Theories of kinship and reciprocity have been used to attempt to explain altruistic action from an evolutionary perspective. A recently introduced theory of competitive altruism explains that people show altruism because they are driven to seek status in social groups as a means of advancing their own genetics. In our green culture, there is debate over whether the choice to be green is purely altruistic to save the planet or ultimately self-serving and the current popular choice for achieving status and other rewards.

People make assumptions about each other every day based on their activities. Because of today's acceptance of green products as popular choices over nongreen products, it has become not only socially acceptable to go green but socially preferred. Environmentalists and others who make decisions that benefit the environment can be seen as either taking action for the public good or influenced by the marketing and social benefits promoted as part of our popular green culture.

There has long been debate about altruism in general, and whether there is such as thing as "true" altruism. Altruism has been defined as the intention to help others at a cost to oneself. Altruism is influenced by feelings of empathy, closeness in bond to the object of help, mood of the individual offering help, and the value of the helping or altruistic behavior. It is theorized that individuals decide to offer help based on evaluation of the rewards they receive from helping weighed against the costs of not helping. Motivation may ultimately reveal altruistic acts as egoistic, always providing some form of personal benefit to the altruist.

From an evolutionary point of view, altruism is a mystery. Survival of the individual can come under threat if the cost of helping another or the group as a whole causes personal sacrifice. And yet altruism is a behavior observed in social groups from humans to insects. Kin selection theory posits that altruism can be explained in a family social group that is connected genetically. Here, helping the group is necessary to ensure survival of the genetic code. Hamilton's Rule posits that the closer the blood relatedness, the more worthy of help. Therefore, altruism is inherently a form of egosim. This theory of altruism has recently been abandoned by long-term supporters, including evolutionist Edward O. Wilson. Wilson flips the scenario, believing that close kinship is the result or ultimate benefit of behaving with group-based motives, not the cause of it.

Whether relatedness is the cause or just one precondition of kin selection, applying the theory to human societies is providing some guidance for those trying to promote more sustainable behaviors. Understanding the motivations behind adopting certain behaviors can help inform the design of incentives-based systems such as money back for recycling and the framing of environmental messages. Audiences motivated by self-reward, however overtly, are more responsive to messages and programs that are built on an aspect of personal gain. Those audiences motivated by success of the group and society as a whole tend to be more responsive to messages and programs promoting sustainability as a societal benefit.

Another theory of altruism is based on the concept of reciprocity. Reciprocal altruism theorizes that an individual is willing to bear the personal cost of performing an altruistic action in anticipation of future payoff as the receiver of an altruistic action. Reciprocity is seen as a motivator for cooperation within a group. Reciprocity focused on a sustainable action may explain people's willingness to invest in green technologies that require higher start-up costs, having the expectation of ultimately reducing costs over time. Solar power and other types of alternative energy are expensive to adopt but have been proved to lower energy use and thus costs over the life of the system.

The popularity of going green has led to a theory of competitive altruism as a means of explaining the rise in purchasing green products. Competitive altruism labels pro-environmental behavior as a form of cooperative, pro-community behavior with a payoff in personal gain. People who are altruistic achieve personal gain in their communities because others observe their behavior, make positive assumptions about their personal qualities, and assign status to them as group members. Altruistic actions, then, serve as signals displaying the socially valued qualities of the individual. The more altruistic these individuals are, the higher they climb in status within the group.

In the mid-1990s, pervading stereotypes assigned to those who recycled or participated in conservation behaviors were negative, reflecting an assumption that they recycled because of a lack of resources. This translated into a lowering of their status socially. From an evolutionary psychology framework, it is in a person's best interest to achieve high social status in his or her group. This benefits an individual by increasing his or her power and resources in the group. Fifteen years ago, recycling lowered perceived status.

Today, going green is a popular decision. Celebrity endorsements and the higher price tags that come with green products broadcast the value of green for improving social status. The key to making going green work for status gain is in the public display of green products and decision making. Purchasing green products online, for example, without social witness, does not accomplish the desired effect of gaining status like driving a hybrid car. The car is a publicly visible choice that is on constant display, collecting and reinforcing the social status assigned to its driver.

When given a choice, people tend to purchase green alternatives in public, but not necessarily in private. If given a choice between two green products, people tend to select the more expensive product. This has been equated with the social signaling of status in terms of the choice of a higher-priced hybrid car displaying the individual's ability to waste more resources (money).

Individuals who are assigned status by a group subsequently become more likely to show altruism to the group and view the group more positively. This strengthens the group and reinforces group behavior. It has been observed that sustainable behaviors are often adopted in clumps. Solar panels are not spread evenly throughout a town but appear to be adopted in clusters of houses in a single neighborhood. Some social psychologists believe that this is the result of a localized social norm, where the behaviors adopted by certain leaders or high-status members of a group become widely adopted by members of the group. Triggering social norms is becoming a common strategy for encouraging broad-scale pro-environmental behavior.

This argument extends into the business sector, where green products that appear to be environmental solutions are merely labeled as such to answer consumer demands for a greener product. If the product does not in reality meet the environmental standards it promotes, it is considered a type of greenwashing—misleading the public to believe in the false environmental benefit of the product. Greenwashing appears to be a successful business strategy that feeds off the social desire for status.

It can be argued that the motivation for pro-environmental action is irrelevant as long as the choice ultimately moves society (and the individual) toward a more sustainable path. However, understanding motivation may be necessary to create long-term environmental care rather than promote a transient cultural impulse toward green as just another fad or popular choice.

See Also: Ecolabeling (Consumer Perspective); Individual Action Versus Collective Action; Social Action.

Further Readings

Dizikes, Peter. "E. O. Wilson Shifts His Position on Altruism in Nature." *Boston Globe* (November 10, 2008).

Griskevicius, V., et al. "Going Green to Be Been: Status, Reputation, and Conspicuous Conservation." *Journal of Personality and Social Psychology*, 98:3 (2010).

Hardy, Charlie L. and Mike Van Vugt. "Nice Guys Finish First: The Competitive Altruism Hypothesis." *Personality and Social Psychology Bulletin*, 32/10 (2006).

Sadalla, Edward K. and Jennifer L. Krull. "Self-Presentational Barriers to Resource Conservation." *Environment and Behavior*, 27/3 (1995).

Dyanna Innes Smith
Antioch University New England

GREEN ANARCHISM

Anarchism, much like socialism, is difficult to define because there are so many different ideologies embraced by those who identify as "anarchist." In its most general sense, anarchism is the belief that the best way to organize society is as a free association of all members with no overarching authority such as government or police. Most anarchists have as their goal a society that organizes its social, political, and economic structures as voluntary federations of decentralized, directly democratic, policymaking bodies. Still, even within this broad framework, there are those who go beyond the broad rejection of the state and argue for the abolishment of all social hierarchies, just as there are those anarchists who focus on the abolishment of capitalism or of violence; there are anarcho-capitalists and anarcho-socialists, pacifistic anarchists and revolutionary anarchists, and collectivist

Incidents like the World Trade Organization protests, pictured, are examples of insurrection by green anarchists, who criticize mainstream environmental movements for being shallow and ineffective.

Source: Steve Kaiser/Wikimedia

anarchists and individualistic anarchists. Thus, anarchism is a widely disputed label, encompassing a vast assortment of ideologies. Green anarchism, then, is but one manifestation of these broad notions of abolishment of centralized authority and hierarchies. What distinguishes green anarchism from other types of anarchism is its focus on environmental restoration and protection. Green anarchists argue that the state, capitalism, globalization, industrialization, domestication, patriarchy, science, and technology are all inherently destructive and exploitative forces that cannot be remedied through reform. They cite the degradation of nature and of humans by these forces as inexcusable and intolerable. They envision a world organized as small ecovillages, composed of no more than a few hundred people, arguing that this scale of human living is the proper framework to nurture both human fulfillment and ecological protection.

The roots of modern-day green anarchism reach back to Henry David Thoreau, an American anarchist and transcendentalist. In Thoreau's *Walden*, a meditation on simple living and self-sufficiency, green anarchists locate the initial strains of a movement to resist the advancement of industrialism and to elevate the protection of Earth and its natural resources to the forefront of all human activity. Today, John Zerzan is considered by many green anarchists to be the foremost contemporary advocate of green anarchism. Zerzan and other green anarchists argue, quite simply, that civilization is the enemy of nature. They object to what they see as a growing acceptance of the view of nature as an object to be manipulated and dominated, and they are further galvanized by statements that conceptualize nature as merely a source of raw materials to be exploited and depleted. For many green anarchists, this is where the opposition to technology arises—they see technology as the manifestation of these narrow views of nature, arguing that technological advances come at the profound expense of Earth's natural resources and ecosystems, and they see deep contradictions in the celebration of technology as something that can alleviate social, economic, and other societal problems, arguing instead that our insatiable thirst for technology and material possessions is at the root of the problems that technology purports to relieve.

Green anarchists are also intensely critical of mainstream environmental movements, which they characterize as shallow and ineffective. Green anarchists argue that actions such as swapping out incandescent light bulbs for compact fluorescent lamps (CFLs), driving hybrids and other vehicles that still rely on petroleum and other finite resources, and attempting to preserve nature, reduce pollution, and recycle materials are impotent activities that lack the deep critiques of our institutions and behaviors that are necessary to create a sustainable, equitable future. Green anarchists are motivated by a desire to foster a reconnection of humans with the Earth, and they advocate various forms of direct action to create pathways to this reconnection. While most green anarchists are opposed to violence and terrorism, particularly that which would result in injury or death to humans and/or animals, most green anarchists embrace the following:

- Sabotage
- Cultural subversion
- Insurrection

Examples of sabotage include the hundreds of millions of dollars of damage done to corporate offices, banks, luxury homes, ski resorts, genetic research facilities, car dealerships, sports utility vehicles, and timber mills carried out by groups such as Earth Liberation Front (ELF). One can see examples of insurrection in riots and protests such as those in Seattle in 1999 at the World Trade Organization meeting and in Genoa at the 2001 G8 Summit. Cultural subversion can take many forms, including co-opting billboards and posting flyers with green anarchist messages, capturing on film instances of corporate environmental degradation, and disseminating green anarchist "'zines" and newsletters to the wider public. Groups associated with green anarchists include the aforementioned ELF as well as Root Force and the Earth Liberation Army, among many others.

Green anarchists face intense criticism from many camps, both left- and right-wing, as well as from other anarchists. Many anarchists characterize the green anarchy movement as lacking a coherent vision and instead see green anarchists as needlessly theoretical; these critics contend that green anarchists reject both civilization and technology without offering a viable alternative, which leads to disappointment and inertia among those who would be swayed by a more coherent vision and to utter alienation of those who require a more nuanced message in order to embrace green anarchist philosophy. Left-wing criticisms of green anarchism, including those from environmentalists, socialists, and leftist progressives, include the following:

- Abolishment of technology would have devastating consequences; for example, those who depend on medical technology and medications to live would soon die out; further, technology is required to deal with issues such as nuclear waste. For example, the abolishment of nuclear energy would leave behind a legacy of environmental devastation that only technology can manage.
- Green anarchists refuse to declare which level of technology is acceptable and further decline to define what is meant by "civilization," and what it means to argue that civilization is "the enemy."
- Green anarchism's arguments regarding the inherent evilness of technology are misplaced; technology is not inherently bad—it is capitalism that is wasteful and destructive and needs abolishing.
- The primitive world that green anarchists envision would lead to the death of most of the world's population.
- Green anarchists are catastrophists who use their terrifying doom-and-gloom rhetoric to scare people into accepting their political ideas.

Right-wing and capitalistic criticisms of green anarchism echo many of the above criticisms and also include such criticisms as the following:

- Capitalism is not inherently evil and is the only viable method for structuring the world's economic system and meeting the needs of the world's population.

- Capitalism offers remedies such as the cap-and-trade system for thwarting ecological devastation.
- Green anarchism is coercive and limits personal freedoms, while capitalism is a contractual system based on rationality and freedom.

Thus, green anarchism is an intensely debated political philosophy, both within and outside the wider anarchist movement. Its advocacy for returning to a primitivist lifestyle, which green anarchists argue is the best way to protect and restore ecosystems, alienates far more people than it inspires, suggesting that green anarchism is a movement that faces immense difficulty in realizing its goals.

See Also: Anarchism; Capitalism; Earth Liberation Front (ELF)/Animal Liberation Front (ALF); Ecocapitalism; Ecosocialism.

Further Readings

Parson, Shawn. "Eco-Anarchism Rising: The Earth Liberation Front and Revolutionary Ecology." http://www.allacademic.com/meta/p176520_index.html (Accessed September 2010).

Stewart Home. "Green Anarchist." http://www.stewarthomesociety.org/ga/index.htm (Accessed September 2010).

Zerzan, John, ed. *Against Civilization: Readings and Reflections.* Los Angeles: Feral House, 2005.

Tani E. Bellestri
Independent Scholar

GREEN-COLLAR JOBS

Typically, America's workforce was divided into two different types of jobs. White-collar jobs were typically management, administrative, or financial in nature. This designation took its cue from the white dress shirt that managers, administrators, or financiers would wear. Hence "white-collar crime" includes embezzlement or some other financial or administrative crime. Blue-collar jobs were typically manufacturing in nature, coming from the collar of the blue coveralls that a person in a manufacturing position would wear. So, green-collar jobs deal with environment improvement or carbon-based fuel reduction since the forest or environment has a lot of green. This "collar" contrasts with a brown-collar job that may bring environmental destruction or increased carbon-based fuel usage. Due to the complexity in defining specifically how "green" the job is, a green-collar job should be seen in its shade along with its collar.

Defining the Green-Collar Job

One of the most difficult things in assessing green-collar jobs is determining exactly what is and what is not a green-collar job. Suppose an assembly-line worker for a major auto manufacturer makes a vehicle that runs on electricity only. However, the vehicle must be charged with electricity from a coal-fired (carbon-based "brown") plant. Is the assembly worker's job green? If a farmer grows corn that is made into biofuel because the price of gasoline is $5 per gallon, has the farmer become a "green-collar" employee just because he/she can gain more profit from turning food into fuel? If the price of corn then goes up and causes a tortilla shortage in Mexico (tortillas are corn based), are the workers on the farm really employed in green-collar jobs?

So, given these difficulties, the most commonly accepted definition for green-collar jobs comes from the United Nations Environment Programme in 2008: "work in agriculture, industry, services and administration that contributes to preserving or restoring the quality of the environment."

Carbon Credits and Renewable Energy Sources

Within the United States, green-collar job growth will center around a cap on carbon emissions by power-generation plants/industries and renewable energy sources. Recent congressional legislation has sought to put a limit on the amount of carbon-based (brown) power produced by the nation's electric grid. This legislation met with stiff opposition from carbon-based (coal and oil) plant operators and from power-intensive industries. Congressional leaders removed parts of the legislation that sought to put a cap on carbon emissions. Currently, brown energy produces the most power for the lowest costs. So, if a company or carbon-based plant exceeds its carbon cap, it must purchase a "credit" of renewable energy from a renewable energy source. This renewable energy source can be something like the power generated by a working windmill in California.

Categories of Green-Collar Jobs

Green-collar jobs in the energy supply sector focus on wind generation, biofuel production, solar power generation, hydropower, and geothermal energy. Unfortunately, investment in these technologies from private and public equity is moving away from the United States and Europe and into China. For instance, in the second quarter of 2010, China received $11.5 billion in clean tech asset financing, whereas U.S. companies received less than half that. In fact, for all of 2009, China received nearly $35 billion in clean energy financing compared to approximately $25 billion of clean energy financing in the United States and the United Kingdom combined (numbers 2 and 3 on clean energy financing). Still, the number of green-collar jobs in energy production should increase, since 2.3 million jobs have been created but still produce only 2 percent of the world's energy. Both China and Germany have seen massive upticks in green-collar job growth with more private equity money being put into renewable energy production. In the United States, the number of green-collar jobs is expected to grow to more than 40 million by 2030, up from 8.5 million currently.

In agriculture, green-collar jobs employ organic farming methods. Farming is one of the largest employment sectors in the world with 1.3 billion farmers worldwide. These organic methods also help in soil conservation and increase biodiversity. Green-collar farming also reduces wastewater runoff. In some areas of Detroit, Michigan, political action groups are looking to return vacant lots with urban blight into new urban farm areas. Additionally, new green-collar jobs can be created by using new Israeli methods of irrigation that use less water.

Green jobs in forestry are closely related to agriculture. Some of the farms that could be organic farms providing green-collar jobs are next to a forest. So here, where agriculture meets forest, using green methods in agriculture can reduce forest destruction and can aid in how the forest is affected by farm water runoff.

Defining transportation green-collar jobs can be quite tricky. If an electric vehicle recharges on electricity from a carbon-fired plant, are the assemblers of this vehicle employed in green-collar jobs? If an Amtrak train carries a lot of people, is the conductor a green-collar job? What about the driver of a carbon-based-fuel bus, since less carbon is used by the bus than by everyone on the bus driving their own cars? Certainly using a bus that is a hybrid vehicle would seem to make the bus driver's job more of a green collar.

The construction industry has green-collar jobs in improving buildings to become more energy efficient. Buildings will need improvements in insulation, lighting, and air conditioning and heating systems. Additionally, some green-collar construction jobs will build new buildings with energy efficiency in mind, including solar panels on roofs.

Industries can create green-collar jobs by effective use of materials. This includes employing production methods that attempt to recycle unused materials in production instead of simply discarding them. Workers involved in the recycling of these materials can be said to have some "green" to their collar. Industries can also create green-collar jobs by adopting cradle-to-cradle manufacturing, whereby a product at the end of its life span becomes raw material for a new product. Thus, manufacturing takes on biomimcry. In nature, biomimcry occurs when a dead tree decays and becomes nutrients for new plant growth. Communities can create green-collar jobs by simply encouraging communities to recycle and then employing refuse collectors to collect these recyclables along with the household owners' garbage.

Funding Green-Collar Jobs

In reality, though, green-collar jobs must pay for themselves if they wish to continue beyond initial funding from either governments or venture capitalists. The Environmental Protection Agency's website maintains a series of links to funding grants for creating green-collar jobs. Congress tried in 2007 to create the Green Jobs Act with millions of dollars in funding, but the bill didn't come to a full vote on the House floor.

Currently, brown energy (coal, oil, natural gas) costs less than 5 cents per kilowatt-hour to produce, while green energy costs more to produce. Solar energy costs approximately 20 cents per kilowatt-hour to generate, while wind-power generation cost is closer to brown-power generation at approximately 6 cents per kilowatt-hour. Hydroelectric plants can provide green-collar jobs, but only in certain areas around dams. Legislative bodies must enact laws enforcing strict building codes that require environmentally friendly methods of construction and retrofitting to promulgate green-collar job growth in the construction industry. Local communities must encourage recycling, and industries must be provided with economic incentives (including natural economic incentives by being paid for recycled material) to ensure that green-collar jobs in the recycling sector continue.

Green-Collar Job Education

In order to meet the demand for green-collar jobs in the energy sector, universities have developed engineering degrees in alternative and renewable energy systems. Job placement for graduates with these degrees is quite high. In the scientific field, one can receive a bachelor's degree to become a conservation scientist. Community action agencies, national agencies, and United Nations agencies must educate farmers on the economic benefits of becoming green farmers, including the recycling of plant seeds.

Green-Collar Jobs and the Bottom Line

As one can see, defining green-collar jobs can be quite hard to do, as a construction worker may retrofit a building one day and repair a coal-fired power plant the next. Similar examples occur in transportation and manufacturing. Education regarding the economic benefits of creating and maintaining green-collar jobs must occur at all levels of government. Earth-first initiatives must be enforced through enactment of aggressive legislation to protect the environment. With these actions, workers' collars can turn from brown to green.

See Also: Anthropocentrism Versus Biocentrism; Green Jobs (Cities); Green Jobs (Culture).

Further Readings

International Labour Organization (ILO). "Climate Change and the World of Work: ILO Director General Explains New Green Jobs Initiative." Press release. Geneva: ILO, September 24, 2007.
Pooley, Eric. "Commentary: America Sits Out the Race." *Business Week* (August 2, 2010).
United Nations Environment Programme (UNEP). "Green Jobs: Towards Decent Work in a Sustainable, Low Carbon World." New York: UNEP, 2008.

Charles R. Fenner Jr.
State University of New York, Canton

GREEN COMMUNITY-BASED LEARNING

Groups like this Outward Bound hiking group are examples of experiential learning and sustainability education at local and community levels.

Source: Wikimedia

Green community-based learning focuses on sustainability education that is rooted in what is local—a particular place's history, traditions, issues, culture, literature, art, and terrain. Using the area in which students are located to provide a context for environmental education, green community-based learning attempts to provide both content knowledge and awareness of the need for sustainable practices. Student work focuses on a community's environmental issues and needs, and community members serve as resources and partners in every aspect of teaching and learning. Concentrating on local issues has a powerful effect on instruction and environmental awareness. Because real-world problems that are meaningful to students are explored, students are engaged academically. Additionally, genuine citizenship is promoted, and students are prepared to live in a sustainable manner and respect any community in which they find themselves. Based upon many research-based classroom approaches, green community-based learning often incorporates techniques culled from community education, project-based learning, and experiential learning. Green community-based learning continues to grow in popularity and importance.

Sustainable Education

Awareness of environmental issues has grown in recent years, inspiring many students, teachers, and community members to learn more about, and to take actions to promote, sustainability. As this has occurred, an interest has developed in integrating sustainability education concepts into the K–12 curriculum in schools. Some states, school districts, and schools have developed thorough and rigorous standards and curricula that address sustainability education. Other administrators, teachers, and parents, however, desire to have the students with whom they work examine environmental issues in school but are provided with little material with which to begin. Fortunately, a variety of resources exist that encourage the inclusion of sustainability education concepts into the K–12 curriculum. A variety of approaches, including green community-based learning, can be used to integrate sustainability education into elementary, middle, and high school instruction. A number of colleges and universities also have added green community-based learning to their curricula. Additionally, some nongovernmental organizations (NGOs) as well as colleges and universities have developed ways to encourage green community-based learning outside the classroom.

Studying sustainability provides the tools necessary for life to continue. More specifically, sustainability education focuses on twin goals of allowing life to continue in a manner that supports human and animal health and promoting the possibility that future generations can be nourished by the Earth. In many ways, sustainability is a measure of the quality of one's life. Conditions such as air pollution, soil erosion, water contamination, and drought may decrease the quality of life of those in certain communities. In addition to quality-of-life considerations, sustainability education allows society to meet the needs of the current population without endangering future generations' ability to support themselves. Sustainability issues can be found on the local, national, and international levels, and all of these levels are interconnected. By helping participants learn how to make a difference at the local level, green community-based learning helps build a sense of connectedness and engagement between a place and its inhabitants.

Green community-based learning borrows from other instructional models, including community education, project-based learning, and experiential learning. In getting participants to interact with their community and environmental issues of consequence to that community, green community-based learning seeks to build a sense of the need for sustainable behavior while also empowering individuals and groups to understand their role in affecting change.

Community Education

Community education, sometimes referred to as community-based education, can be used with learners of any age, including those in traditional K–12 settings, students at colleges and universities, and adult learners. Community education is predicated on certain goals that are used as the basis for the development of related learning activities. These goals include empowerment, participation, inclusion, self-determination, and partnership. "Empowerment" refers to increasing the ability of individuals and groups to shape and influence issues facing them and their communities. Within an environmental context, this might include identifying a source of water pollution as diminishing a community's quality of life. "Participation" involves supporting members of the community to take part in the decision-making process involving issues confronting their community. Using the example of water pollution, community members might lobby policymakers to have their opinions heard. "Inclusion" ensures that all members of a community are given an opportunity to participate in the process, regardless of their backgrounds. Sometimes, this includes making special efforts to reach out to traditionally underrepresented groups such as African Americans or Latinos in order to ensure that all stakeholders are heard. "Self-determination" ensures that members of the community have the ability and right to make choices affecting them. This ensures that the members of a community are provided some sort of voice in the process. "Partnership" recognizes that many individuals, organizations, businesses, and agencies must work together to ensure resources are used effectively.

The process of community education is often regarded as equally significant to the content explored. Process frameworks have been developed to ensure that community education works in a manner that adheres to its goals. The National Coalition for Community Education promulgated a series of community education principles in 1991 to assist with this. These principles include the following:

- Self-determination, recognizing that local residents are in the best position to identify community needs and wants
- Self-help, based on the understanding that individuals are served best when their ability to help themselves is encouraged and enhanced
- Leadership development, because the leadership capacity of local residents is essential for ongoing self-help and community improvement
- Localization, so that services, programs, events, and other community involvement are made available to and utilize the skills of local residents
- Integrated delivery of services, encouraging cooperation and coordination between organizations and agencies so that the public good is served through careful marshalling of scarce resources
- Maximum use of resources, allowing the interconnection of physical, financial, and human resources of the community
- Inclusiveness, encouraging all persons' participation in the process, regardless of race, ethnicity, age, sex, income, religion, sexual orientation, or other factors
- Responsiveness, insofar that programs and services developed respond to continually changing needs and interests of a community
- Lifelong learning, in that even K–12 institutions appreciate that formal and informal learning occurs outside the parameters of formal education

With regard to green community-based learning, these principles provide a useful reminder of the importance of process. Initial professional development and continual fine-tuning of the process are often provided to support this.

Project-Based Learning

Often associated with instructional technology, project-based learning uses projects intended to provide participants with deep learning experiences. Project-based learning encourages participants to combine technology and inquiry to engage with issues and questions that they identify as meaningful and relevant to their own lives. When used in a formal academic setting, project-based learning traditionally uses classroom projects to assess student competence with subject matter instead of more traditional evaluation methods such as testing. Often confused with experiential education, project-based learning has roots in the educational progressivism movement.

Project-based learning consists of instruction that relates to student-generated questions. The questions are related to the students' everyday lives and thus bring a real-world connection to classroom projects. By combining complex tasks based on challenging questions or problems, project-based learning builds students' problem-solving, decision-making, investigation, and reflection skills. Because project-based learning emphasizes teacher facilitation rather than direction of student work, it drives students to encounter the core concepts and principles of sustainability and environmental studies via hands-on activities.

As does community education, project-based learning places a great focus on the process followed in the classroom. Rather than directing student learning, the teacher works in the role of a facilitator. As such, the teacher works with students to frame worthwhile questions, structure meaningful tasks, coach knowledge development and social skills, and assess what students have learned from the experience. Proponents of project-based learning often stress its usefulness in preparing students with collaborative and thinking skills required for the workplace. In order to gain maximum benefits from project-based learning, green projects should accomplish the following:

- Be organized around open-ended questions or challenges, focusing student work on significant sustainable or environmental issues, debates, questions, or problems that face their community
- Create a need to acquire essential content and skills, concentrating on the end-project that necessitates that students understand the reason to learn and are motivated to do so
- Require inquiry to learn and create knowledge, so that participants begin to see themselves as more than consumers of information
- Require critical thinking, problem solving, and various forms of communication, allowing those involved to use higher-order thinking skills, work as members of a team, and share information created with a wider audience
- Allow exploration of participant choice and voice, supporting the ability to engage in independent work and to take responsibility for the direction of work
- Utilize feedback and revision, permitting participants to use a peer-critique system to create higher-quality products and to build a better sense of self-evaluation
- Result in the sharing of a group-created product or performance in a public forum, demonstrating that work is intended to be shared with others, including some who may not support the groups' goals

Many advocates of environmental and sustainability education have utilized project-based learning to create instructional experiences for the groups with which they work.

Experiential Education

Experiential education combines hands-on experiences, often in the field or outside the school building, in a manner that builds upon, complements, and expands content covered in the classroom. Experiential learning requires a rigorous process and a strong working relationship between teacher and student, one where appropriate academic risks are taken and a level of trust must exist between the parties involved. Based largely on the works of

John Dewey, experiential education has long been attempted but seldom implemented with success. Influencing a host of other initiatives, including service learning and Outward Bound and thinkers such as Kurt Hahn and Paulo Freire, experiential education remains an important way to involve students with real-world experiences, democracy, and learners' creation of knowledge.

Experiential education may take on many manifestations in a learning setting. Some of these manifestations may include outdoor education, service learning, cooperative learning, active learning, and environmental education. Outdoor education uses organized instructional opportunities that occur outdoors, using environmental experiences as a learning tool. Service learning combines community service with stated learning goals, relying on prior educational understandings as the foundation to provide meaningful experiences throughout the service. Cooperative learning rearranges traditional heterogeneous ability grouping in order to support the diverse learning styles and needs a group may exhibit. Active learning places the responsibility for learning directly on the learners themselves and requires their past experiences to inform their process of learning. Environmental education seeks to inform learners about relationships between the natural environment and the interdependence between those relationships, something strengthened by being outdoors and learning through doing.

Because it takes so many forms, experiential education may look very different in practice from place to place. Certain principles permeate the model, however, and are believed to be crucial to maximizing the learning experience. These principles include the following:

- Journaling, which can be compiled in a simple composition or spiral notebook, that allows students to reflect upon what they have done, what they have discovered, and how they are responding to the process
- Cooperative planning, which includes both participants and those assessing them, so that the learners are provided a choice in the content to be explored
- Critical thinking strategies that are formally introduced, providing students with tools such as games, simulations, role plays, and stories in the classroom
- Discussions, so that learners may apply higher-order thinking skills so that they may synthesize information at a deeper level of understanding
- Peer-to-peer interaction throughout the experience, so that learners see themselves as active participants in the learning process
- Delineated goals and objectives for the experience, with repeated check-in points, that both allow teachers to facilitate problems and provide necessary support
- A culmination of the experience, which allows participants an opportunity to share findings from the field opportunities

Environment and sustainability can be used as the areas investigated as part of the experiential education sequence. With its focus upon participatory, community-based research, experiential education is ideal for identifying community needs and using diverse knowledge models to allow for participatory intervention.

Criticisms of Green Community-Based Education

Although green community-based education has been embraced by many, it is not without critics. Those who take issue with green community-based education often focus on several issues. First, certain disciplines, especially mathematics and the sciences, are primarily skill-based at the elementary level. Transforming the curriculum into overreaching projects does not, in the minds of some, allow for the necessary practice of particular skills, such as factoring quadratic equations. Rather than relying on students to devise their own problems in the course of a project, more traditional educators believe students require extensive practice that can only be provided in a more traditional setting. Second, pressures from legislation such as No Child Left Behind and the Elementary and Secondary Education Act have made schools much less willing to experiment with instructional strategies that are not correlated directly to assessments. Green community-based education tends to be centered on projects. In some settings, such as charter, private, or alternative schools, these projects may encompass the entire curriculum, although in other settings, these may include only a few hands-on activities. Regardless of the scope of the projects, however, their multidisciplinary approach makes them less appealing to school leaders who are focused upon achievement in specific subject areas, such as reading, mathematics, or science. Finally, while many may attempt green community-based education, implementation of these attempts is often lacking. Unless schools commit money, time, and support to teachers who are willing to initiate green community-based learning, such efforts will often prove unsuccessful.

Green community-based education provides a way to offer explorations into areas pertaining to environmental and sustainability issues. Centered on issues and questions generated by participants, it is an appropriate and motivating way for parents, teachers, or community leaders to introduce these topics. Green community-based education's focus on providing solutions to issues facing residents of a community is also empowering, especially for those who come from traditionally underserved populations. Certain pressures may make implementing green community-based education a challenge, but when properly implemented, it can support learning goals in most settings. Drawing as it does upon many established instructional strategies and models, green community-based education provides one way to build participant knowledge about and involvement with the environment.

See Also: Antiglobalization; Education, State Green Initiatives; Environmental Education Debate; Organic Farming; Social Action.

Further Readings

Bartlett, P. F. and G. W. Chase, eds. *Sustainability on Campus: Stories and Strategies for Change*. Cambridge, MA: MIT Press, 2004.

Edwards, A. R. *The Sustainability Revolution: Portrait of a Paradigm Shift*. Gabriola Island, British Columbia, Canada: New Society Publishers, 2005.

Timpson, W. M., B. Dunbar, G. Kimmel, B. Bruyere, et al. *147 Practical Tips for Teaching Sustainability: Connecting the Environment, the Economy, and Society*. Madison, WI: Atwood, 2006.

Treffinger, D. J., S. G. Isaksen, and B. Stead-Dorval. *Creative Problem Solving: An Introduction*, 4th ed. Waco, TX: Prufrock Press, 2006.

Wiland, H. and D. Bell. *Going to Green: A Standards-Based Environmental Education Curriculum for Schools, Colleges, and Communities*. White River Junction, VT: Chelsea Green, 2009.

Stephen T. Schroth
Jason A. Helfer
Diana L. Beck
Knox College

GREEN JOBS (CITIES)

At the beginning of the 21st century we are called to rethink—and reformulate—the manner in which we conduct our economic lives and our relationships with ecological systems that support life on Earth. The instability of the global economy demonstrated in the Great Recession of 2008 has resulted not only in a dramatic increase in unemployment and exacerbation of global economic polarization between the rich and the poor but also undermined public confidence in the fundamental stability and fairness of our economic systems. Furthermore, the dawning of a postindustrial age is met with the emergence of climate destabilization as an existential threat.

Derivative of these crises is increasing social dislocation and unease. The poor become poorer and more numerous at the expense of the increasing wealth of the very rich; those least able to protect themselves are most vulnerable to the vagaries of increasingly unstable economic and atmospheric climates. Humankind's capacity—not only capability, but also willingness—to constructively engage these challenges and better integrate ourselves into Earth's ecological framework will be reflected in future levels of societal sustainability.

Greening the Economy

According to the dominant economic canon, economic growth is essential to improve the lives and livelihoods of the global population. Economic growth begets industry and employment growth that provide to workers salaries to purchase the fruits of production. To the extent that low unemployment is essential for the health and well-being of any society, social, economic, and environmental policymakers have begun to tout "green jobs" as a key element not only in reducing unemployment levels but also in defining a more sustainable relationship with the environment, reducing the global economy's dependence on fossil fuels, regenerating urban infrastructure, and boosting economic growth.

The U.S. Department of Labor Bureau of Labor Statistics recognizes green jobs as those that contribute directly to the production of products or services that are beneficial to the environment or conserve natural resources or those that involve making manufacturing or service provision processes more efficient or environmentally friendly. Most green jobs will be located in our cities. The businesses and enterprises that support those jobs, the workforce that will be required to perform them, and the transportation networks that facilitate the movement of goods and workers are concentrated in our cities and metropolitan areas.

The economic sectors in which these jobs are most likely to be directly created include the following:

- *Energy supply alternatives*: renewable energy sources, including solar, wind, geothermal, and hydropower
- *Building efficiencies*: new construction and renovations; ecological design
- *Brownfield redevelopment*: urban land reclamation and infill development; urban planning
- *Transportation*: zero-emission vehicles, transit, and alternative transportation modes; transportation planning
- *Basic industry and recycling*: waste management, process design, quality assurance, environmental management
- *Food and agriculture*: organic farming, local and urban agriculture
- *Forestry*: urban forestry and landscaping, reforestation, ecology

However, many more jobs—many of them not necessarily recognized as green—will likely be created as the economy as a whole responds to the focus of stimulus spending and job creation in these sectors.

Policy Support for Green Jobs

Governmental policy plays a critical role in providing positive (as well as negative) incentives that promote green job creation. At all scales, tax policy can be an effective tool to encourage business to invest in processes and technologies that both create green jobs but also increase the efficiency with which goods and/or services are produced. Financing programs can help cities and states rehabilitate or replace aging, decrepit, or just inefficient infrastructure and built environment.

Federal and state programs can provide funding for educational institutions to develop and implement new curricula that enable workers sidelined by the contraction of the "old economy" to re-skill for newly created green jobs. Similarly, federal and state programs can provide grant and scholarship aid programs that enable those in need of retraining to pursue new credentials.

A proposal by Architecture 2030 to address the predicted crisis in the commercial real estate market (the "CRE Solution") advocates a government-led three-year program that encourages property owners to complete substantive efficiency renovation projects in return for targeted tax deductions. In addition to new private spending and increased government tax revenue induced by the tax deductions, the proposal estimates that 1.3 million new jobs will be created. A 2008 proposal by the Center for American Progress estimated that a federal investment of $100 billion according to its policy recommendations would create 2 million new jobs in an emerging clean-energy economy.

In the American Recovery and Reinvestment Act of 2009 (ARRA), the U.S. federal government allocated more than $90 billion in government investment and tax incentives "to lay the foundation for the clean energy economy of the future . . . [helping to] create a new generation of jobs . . . and improve the environment." Investment targets include energy efficiency (more than $2.7 billion through the Department of Energy's Energy Efficiency and Conservation Block Grant program), development of renewable energy sources ($4 billion), modernization of the electric transmission infrastructure ($10.5 billion), development of advanced vehicles and fuel technologies, green innovation and job training ($500 million in grant funding through the Employment and Training Administration), and clean energy equipment manufacturing (more than $1.6 billion). By the end of 2009, more than 116,000 new jobs (most, but not all, green) were attributable to these investments.

Cities and Green Job Creation

Much of what cities can expect to accomplish in generating green jobs clearly depends on policy forged at state and federal levels. However, within the constraints imposed by federal and state legislation, municipal governments can have a dramatic impact on the complexion of economic development in their communities through the regulatory frameworks they adopt that govern the production of their built environments. The U.S. Energy Information Administration estimates that by 2035, approximately 75 percent of the built environment will be new or renovated.

The adoption of green building and development ordinances and standards can have a long-term impact not only on the quality of the built environment but also on the complexion of the workforce. For example, the adoption of Leadership in Energy and Environmental Design (LEED) or similar standards for new construction and green retrofitting standards for building rehabilitation projects not only results in significantly more energy-efficient structures but also can lead to the direct and indirect creation of green jobs in the construction, manufacturing, and maintenance/operation sectors.

Cities can also encourage the development of residential and commercial green projects by streamlining permitting processes for qualifying projects. Cities can also more strongly encourage the development of green projects through zoning ordinances that specify density incentives or create green overlays to existing parcels. The Seattle Department of Planning and Development has created a green building program that provides a variety of incentives and support mechanisms for builders and developers at all scales. These include expedited permitting processes for qualifying projects, green development education programs and online resources, and a variety of financial incentives for energy conservation, material recycling and reuse, water conservation, and other green practices.

Observing that its metropolitan areas are the engines of U.S. economic growth, a 2008 report conducted for the U.S. Conference of Mayors projected that more than 4 million new green jobs will be created over the coming three decades. These jobs will be concentrated in renewable-power generation; residential and commercial retrofitting (energy-efficiency standards and implementation); renewable transportation fuels; and engineering, research, legal, and consulting. It further contended that the potential growth in green jobs would be the fastest-growing segment of the U.S. economy over the next several decades and would significantly increase its share of total employment. While recognizing the importance of national policy in the job creation and economic transformation process, the report states, " . . . if investment in green industries is to successfully transform the U.S. economy, it must happen at the metropolitan and local level."

Green Jobs and Social Transformation

If green jobs are to be economically transformative, they must also be socially transformative. The focus of economic development on the creation of green jobs has been criticized in some circles as a strategy that would neglect those who have lost their jobs in the sort of vanishing manufacturing jobs that created the post–World War II middle class. In response to these criticisms, $500 million of ARRA funding was allocated to retraining workers in the developing energy-efficiency and clean-energy industries.

In spite of the inevitable loss of jobs in the transition to a "low-carbon and sustainable society," the United Nations Environment Programme (UNEP) noted that a greater number of green jobs would ultimately be created. A "just transition" must emphasize job retraining and social transition assistance across a broad socioeconomic spectrum.

The European Union is basing the need to transform its economy to a more sustainable footing not only on addressing the ravages of the recent economic crisis but, as importantly, on the need to decarbonize its economies in light of the effects of climate change. Europe 2020, the European Commission's strategy for European economic and social development in the 21st century, identifies the dependency of sustainable and equitable development on the ability of its workforce to develop new skills to adapt to a greening economy. Recognizing that the share of low-skill jobs in total employment is decreasing, a core tenet of the Europe 2020 strategy is to ensure that future growth is fairly distributed geographically and economically. It observes, ". . . lifelong learning benefits mostly the more educated," and emphasizes "inclusive growth—a high-employment economy delivering economic, social and territorial cohesion." Members of the European Parliament have called on the European Commission to broaden the scope of the strategy by proposing a complementary and actionable green jobs strategy.

The Green Jobs Controversy

On its face, the push for development of green jobs seems to present a win-win scenario. New jobs would be created in technologies and industries that are more environmentally friendly. Government subsidization of these technologies, industries, and the green jobs that come with them is seen as "priming the pump" for the creation of a new, carbon-free economy that benefits business while it reduces unemployment and carbon emissions.

The controversy that has emerged over "green jobs" centers precisely on the role of government policy—and public expenditures—in creating those jobs. Advocates of government intervention argue that market action alone cannot be expected to achieve the transition to a clean-energy economy. Because its focus is exclusively on productivity—creating more output with less input—the free market is uniquely unqualified to address the employment crisis, at least in the short term.

For example, targeted investment of public revenues into alternative energy sources such as solar or wind energy is required to provide the incentives for these enterprises to compete on an equal cost footing with traditional energy sources. Without this investment, solar and wind energy prices would be too high to be competitive with coal- or gas-generated electricity. Recent dramatic increases in production of natural gas have driven down the market price of that fuel and drawn a significant amount of private investment capital away from clean-energy projects. Left to the free market alone, the large-scale wind and solar projects will not be pursued because the energy they produce will be more expensive than the energy produced by traditional fossil fuels.

Reports advocating green jobs policies appeal to the present confluence of climate destabilization through global warming and to the dramatic—and enduring—rise in global unemployment following the Great Recession of 2008. These reports and other advocates argue that the situation in which the global community finds itself is considerably different from that to which we had become accustomed. Instead of economic exploitation based on plentiful natural resources and tight labor markets, we are now faced with the declining availability of many natural resources (consider, for example, the concern stirred by the emergence of "peak oil" predictions) and an increasingly slack labor market.

The most vocal critics of green jobs policies vehemently argue that government green jobs policies are tantamount to picking winners and losers in the energy market. They reject the notion that "central planning" should not decide how public resources should be allocated. The "invisible hand" naturally "greens" both consumer goods and industrial process as a natural outgrowth of competition for consumer preferences.

Critics also claim that the most widely circulated and cited green jobs reports—from the UNEP, the Center for American Progress, and the U.S. Conference of Mayors—are based on assumptions and methodologies that overestimate the impact that green jobs policies will have on employment.

Moderate critics of green jobs policies caution that such investments are not likely to produce the sort of economic boost in the short term that would solve our current unemployment problems. They cite statistics that show that clean-energy jobs are unlikely to require the numbers of workers that the old, labor-intensive manufacturing industries did. Instead, government investments in the green sector are likely to produce a smaller number of jobs in the short run with larger numbers of jobs being created throughout a "greener economy" in the long run. Significant job growth, then, derives from a growing economy, not just a growing clean-energy sector.

A Green Future

The transition to a low-carbon, energy-efficient, sustainable economy will require the transformation of how we work and the impact of that work on our environment. Traditional carbon-intensive jobs will be eliminated; new green jobs will be created. However, while new technologies may provide the means toward a clean, green, sustainable future, supportive social, economic, and environmental policies must be implemented at all levels of governance—but especially at the municipal and metropolitan levels—to facilitate and manage the transformation.

See Also: Environmental Justice (Business); Green Altruism; Green-Collar Jobs; Green Jobs (Culture).

Further Readings

Architecture 2030. http://www.architecture2030.org (Accessed October 2010).

City of Seattle. "City Green Building." http://www.seattle.gov/dpd/greenbuilding (Accessed October 2010).

Corfee-Morlot, J., L. Kamal-Chaoui, M. G. Donovan, A. Robert, et al. "Cities, Climate Change, and Multilevel Governance." Organisation for Economic Co-operation and Development (OECD) Environment Working Paper No. 14. Paris: OECD, 2009.

"Europe 2020: A European Strategy for Smart, Sustainable and Inclusive Growth." Brussels, Belgium: European Commission, 2010.

Kamal-Chaoui, L. and A. Robert, eds. "Competitive Cities and Climate Change" Organisation for Economic Co-operation and Development (OECD) Regional Development Working Paper No. 2. Paris: OECD, 2009.

Lee, J., A. Bowden, and J. Ito. *Green Cities, Green Jobs*. San Francisco, CA: Apollo Alliance, 2007. http://www.greenforall.org/resources/green-cities-green-jobs-by-joanna-lee-angela (Accessed December, 2010).

Martinez-Fernandez, C., C. Hinojosa, and G. Miranda. "Green Jobs and Skills: The Local Labour Market Implications of Addressing Climate Change." Organisation for Economic Co-operation and Development (OECD) Working Paper. Paris: OECD, 2010.

Pollin, R., H. Garrett-Peltier, J. Heintz, and H. Scharber. "Green Recovery: Program to Create Good Jobs and Building a Low-Carbon Economy." Washington, DC: Center for American Progress, 2008.

Renner, M., S. Sweeney, and J. Kubit. "Green Jobs: Towards Decent Work in a Sustainable, Low-Carbon World." New York: United Nations Environment Programme, 2008.

United Nations Environment Programme (UNEP). "UNEP Background Paper on Green Jobs." New York: UNEP, 2008.

U.S. Bureau of Labor Statistics. "Green Jobs." http://www.bls.gov/green/home.htm (Accessed October 2010).

"U.S. Metro Economies: Current and Potential Green Jobs in the U.S. Economy." Washington, DC: United States Conference of Mayors, 2008.

Kent L. Hurst
Independent Scholar

GREEN JOBS (CULTURE)

The phrase *green jobs* marries two goals that are sometimes framed in opposition to each other: environmental sustainability and economic growth. Promoters of green jobs seek to eliminate the paradigm that environmental protection is only about limits—about limiting logging rights, taking land out of development, imposing restrictions on polluting industries, and so on—and choose instead to emphasize the ways in which environmentalism can be an economic driver that promotes new green industries, increases employment, and makes existing businesses more efficient. If governments provide the necessary incentives, the transition toward a lower-carbon economy will affect employment by both adding new green jobs and greening existing jobs. This article highlights the pros and cons of green jobs as well as their potential to truly transform the economy to put work in the service of environmental goals.

"Good" jobs—fair, well-paid, and offering reasonable hours—in the green industry include research (shown), engineering, and architecture, yet many labor-intensive green jobs offer low wages or hazardous conditions.

Source: Randy Wong/Sandia National Laboratory

Are Green Jobs Local?

In this age of globalization, jobs are often moved overseas to take advantage of cheaper labor, less regulation, or more favorable government incentives. To benefit a country's economy, green jobs must rely on the skills of the local labor force and must not be easily exportable.

There are many green jobs that cannot be outsourced. These include labor-intensive service and construction jobs such as installing solar panels, building dams, upgrading electric systems, or monitoring energy use. There will be more demand for such work in the energy, transportation, and construction sectors in the future as workers build new infrastructure in developing countries or repair and upgrade aging infrastructure in developed countries. The current state of American infrastructure, for example, is sorely in need of improvement and has earned an overall grade of "D" from the American Society of Civil Engineers. As the country makes the massive changes necessary to receive a grade of "A," there will be many opportunities to rebuild that take advantage of new technologies, save energy and water, and incorporate better transportation systems to lessen negative impacts on the environment.

Making all of these jobs green will require the help of government, trade unions, and industry leaders. These players will have to develop standards, implement effective zoning laws, and establish competitive best practices that will raise the bar on sustainability for every new project. Engineering and urban planning are two careers that can help to coordinate and review these activities.

An infrastructure renaissance also holds the potential to revive the manufacturing sector in the United States and other countries. Those that wish to become global leaders in the production of windmills, solar panels, and alternative fuels, for example, must implement policies that support these fledgling industries, as Germany, China, the Netherlands, and others are beginning to do now. Current manufacturing in the United States has proved to be very successful at creating jobs that supply vehicles, aircraft, and other equipment to the military. With a shift in national priorities, some of these industrial parks could be turned over to the production of equipment for greener purposes.

However, parts, equipment, and building materials can still be easily made overseas, and thus, manufacturing in the service of green projects is not immune to foreign competition. There continues to be concern in the United States that China and other countries, now the world leaders in the manufacture of renewable energy equipment, will dominate clean technology industries in the future. To create and hold on to jobs in this growing sector, the United States must take steps to become a producer rather than just a consumer of new green technologies.

Are Green Jobs Good Jobs?

To achieve both the social and the environmental goals of a transition to a greener economy, green jobs must be good jobs: fair, well-paid, and offering reasonable hours. They are just as much about career development, financial security, and quality of life as they are about the environment. Many green jobs that exist today do not meet these criteria: they may offer low wages, like agricultural work in the biofuel industry; hazardous conditions, like informal materials and waste recycling in underdeveloped countries; or long, low-paid hours, like factory assembly of solar panels or windmills in China. Green jobs must fulfill their economic promise by offering pathways out of poverty toward stable, desirable employment.

Education has a large role to play in training the workforce for desirable careers that require advanced training. Some green certification programs are beginning to fill this need. The Leadership in Energy and Environmental Design (LEED) program, which certifies architects and craftspeople who have gained a skill set for green building, is an excellent example of an educational standard that is leading the field toward sustainability. Other industries can emulate this program. In addition, trade schools, work programs, and institutions of higher education are beginning to weave environmental values into many of their program offerings. Science, technology, engineering, and math (STEM) education, particularly for younger learners, will be increasingly important to prepare for green jobs that require quantitative skills. If educators and policymakers can find better ways to improve today's substandard STEM education in the United States, then tomorrow there will be new generations of engineers, architects, environmental scientists and economists, water and natural resource managers, and other professionals who have the skills to solve pressing environmental problems.

The Impacts of Greening Existing Jobs

There are a few traditional environmental protection roles that will continue to be available, including those of park ranger, conservationist, and ecologist. But what promise does the greening of other jobs hold?

In agriculture, there is increasing interest in making current industrial systems less energy intensive and in promoting a more sustainable, local food system that supports small farmers. These jobs may be more numerous in the future, especially with the growing popularity of farmers markets and community-supported agriculture (CSA) efforts, but it may still be hard to earn a good living from this type of work.

Efficiencies gained by greening the energy, transportation, and construction sectors can have a ripple effect on other sectors of the economy. The use of renewable energy, which is currently responsible for 2.3 million jobs, can reduce the carbon footprint of a grocery store or a clothing boutique; greener building practices can reduce the footprint of the logging and materials industries. As infrastructure is rebuilt, it will be important to recognize which existing elements can be salvaged and which require radical transformation. In the building sector, substantial efficiencies can be achieved by retrofitting old structures instead of constructing new ones. Though a few cities have formal programs to employ retrofitters, most have at best informal networks of handymen and contractors; these can be employed and additional workers hired to flesh out more robust public works programs. In the transportation sector, which currently is responsible for 23 percent of worldwide greenhouse gas emissions, more significant changes are necessary. Opportunities to erase the heavy carbon footprint of the aviation industry are slim, relying mainly on advances in fuel economy and manufacture. Retrofitting cars by powering them with biofuels can reduce emissions, but efforts will be better spent to decrease their use by expanding rail, car sharing, and other public transit systems and by redesigning roadways to encourage walking and biking. Incentives and employment programs that are directed toward the latter goals will offer the greenest jobs in transportation.

There are some industries that will continue to have harsh environmental impacts despite efforts to bring them in line with goals of sustainability. It is debatable whether jobs in manufacturing can be called green; even if a worker produces a windmill that ultimately reduces carbon emissions, its creation uses large amounts of energy and raw materials and generates pollution and waste. The emerging field of industrial ecology, which seeks to make each step in the cradle-to-cradle life of material goods more sustainable, can ameliorate this problem. This life-cycle analysis work can provide green jobs for highly educated engineers and managers who are charged with making improvements. Work in other industries that operate at the beginning of the supply chain such as logging, mining, and other resource extraction will also continue to exact heavy tolls on ecosystems.

Green Jobs and Transforming Consumer Culture

Jobs in the clean energy and technology sectors; in the physical, biological, chemical, and environmental sciences; and in public policy offer opportunities to catapult business operations toward a lower-carbon future. But the transformative power of green jobs rests largely on the success of transforming the business culture to add new bottom lines: not just the bottom line of profitability, but also those of worker satisfaction and positive environmental impact.

Part of achieving these goals and truly moving to a green economy will necessitate a revision of the modern consumer culture that exists in the United States and other developed nations. The buying, selling, and discarding of a great volume of products is not green. Each of these items was once a raw material, processed in a factory and likely shipped across long distances in its lifetime. Currently, there are jobs whose function is to tame the impact of this cycle, but they are underutilized. Many cities employ people to manage recycling services that can handle paper, plastic, glass, and metal cans, but few offer convenient electronics recycling, food scrap recycling (composting) services, or services to promote the reuse of mixed material goods. In both developed and developing countries, items that are not formally recycled are often (when they are not thrown away) collected or redistributed informally for free or for small amounts of money. Greater attention to the life cycle of different products must spur the creation of well-paying jobs that close the recycling loop.

Business models must shift so that products are made out of durable, ecofriendly materials and are made to last. Designers who reject the planned obsolescence model—on which many businesses base their bottom line so as to use cheaper materials and sell greater quantities of flimsier products—will find that there is a market for these quality goods. Green goods and services capture $1.37 trillion currently, a number that is expected to grow in the future.

Finally, revising the very nature of work and the way people spend their time can ease the transition to a greener economy. In the United States and other countries, employees can be trapped in a work-and-spend lifestyle in which they take on long hours to pay for expensive consumer items. Moving to a lifestyle that involves more leisure time spent with family and friends and less emphasis on buying can go a long way toward reducing consumption.

The Future of Green Jobs

As human society prepares to reduce its enormous impact on the environment, structural changes must encourage work that makes human activity more efficient, less wasteful, safer, and happier. Though green industries will add a number of entirely new jobs to their economies, the overall effect of green initiatives will be to steer existing jobs toward more sustainable ways of doing business. Government can favor environmentally friendly businesses and their workers with tax credits, subsidies, and work programs. Nations can cooperate by sharing technology and ensuring that unsustainable practices are not more profitable than sustainable practices. Businesses can capitalize on the savings and benefits of greening their operations. And finally, societies can work to level the playing field between excessive consumerism and poverty. With all of these efforts, there may be a brighter future for green jobs.

See Also: Capitalism; Ecocapitalism; Ecolabeling; Environmental Justice (Business); Socially Responsible Investing; Sustainable Development (Cities).

Further Readings

Apollo Alliance and Green For All. *Green-Collar Jobs in America's Cities: Building Pathways Out of Poverty and Careers in the Clean Energy Economy*. San Francisco, CA: Apollo Alliance and Green For All, 2008.
Environmental Defense Fund. *Green Jobs Guidebook: Employment Opportunities in the New Clean Economy*. Washington, DC: Environmental Defense Fund, 2008.
Kannankutty, Nirmala. *2003 College Graduates in the U.S. Workforce: A Profile*. National Science Foundation, Division of Science Resources Statistics. Arlington, VA: NSF 06-304, December 2005. http://www.nsf.gov/statistics/infbrief/nsf06304/ (Accessed November 2010).
Renner, Michael. *Green Jobs: Working for People and the Environment*. Washington, DC: Worldwatch Institute, 2008.
United Nations Environment Programme. "Green Jobs: Towards Decent Work in a Sustainable, Low-Carbon World." (2008). http://www.unep.org/PDF/UNEPGreenJobs_report08.pdf (Accessed September 2010).

Sophie Turrell
Independent Scholar

GREEN PRICING

Green pricing is an optional service provided by some American electric utilities under which the customer pays an extra fee, in return for which the utility company invests further money in green power resources. The additional fee represents the utility company's cost, since most green power sources are not as cost-effective as traditional sources or have associated start-up costs. This is typically an option in states where competition among electric utilities is not allowed. In New England, New Jersey, Pennsylvania, and other deregulated markets, consumers can purchase green power directly on the retail market.

Premiums for green pricing programs vary from 0.2 cents per kilowatt-hour (Delaware Electric Cooperative) to 11.6 cents per kilowatt-hour (City of Tallahassee/Sterling Planet), with some companies simply accepting contributions of any size. Of course, with a pricing system like this, the lowest price is not necessarily the most desirable: the fee represents, after all, not a luxury charge for a premium service, but a promise of investment in green technology, and a fifth of a penny per kilowatt-hour is not going to accumulate as fast—and so arguably does not represent the same level of commitment to going green—as nearly 12 cents. Therein lies one problem with green pricing: if utility companies can offer such vastly different programs, could it not imply that many or most of them are simply placebos, meant to make the customer feel better, feel greener, while still consuming the same electricity off the same grid as everyone else and depositing micropayments into a piggy bank that will never be full enough to be spent?

One problem with promoting renewable energy sources is that, if consumed at the same rate as our present rate of energy consumption, the environmental consequences would still be dire, just not the same as the consequences of the use of fossil fuels. Hydroelectric power destroys river habitats; wind farms catch birds in their turbines; geothermal power production leaches toxins into the soil and air; and present-day biofuels redirect agricultural efforts away from staple crops, threatening commodity price shocks and the food security of the world's poor while increasing pressure on water resources. Widespread use of biofuel threatens deforestation and land-use problems: this is already an issue in parts of Africa, Central America, and southeast Asia, where farmers are clearing rainforests in order to plant oil palms, which have become a valuable crop because they produce almost eight times more vegetable oil than other vegetable-oil crops. Because palm oil is rarely used as cooking oil outside those areas—it lacks the neutral flavors of canola, sunflower, or soy—only the biofuel market was able to spike the demand for palm oil enough to lead to widespread deforestation. The fact that some of the growth in unsustainable palm oil plantations is Indonesia was funded by carbon credits programs—that some businesses' carbon credit purchases may actually lead to further carbon emissions rather than balancing out their own—demonstrates the unreliability and infancy of the global biofuel industry at this time. Furthermore, biofuels still contribute carbon emissions when they are used. In short, although renewable energy has become conflated with clean energy in the public imagination, they are not coequal. Arguably, a true long-term solution to the environmental problems caused by our present energy usage is to reduce our usage—regardless of the source of that energy.

Arguments against green pricing need not go so far as to villainize renewable energy. It is enough to note that not all green power sources are equally green, and that consumers signing up for green pricing services may not be informed as to what they are investing in, either in terms of the type of energy project or the specifics. Some utility companies may be more transparent than others; some customers may be more proactive in informing themselves than others; but again, opponents say, there is that lingering suspicion of a placebo effect. So long as utility companies are collecting fees earmarked for green energy, they can say they are doing their part. But is it enough?

Furthermore, not all "dirty" energy is equally dirty. Natural gas is argued to be significantly cleaner than it was even at the cusp of the 21st century and produces much less carbon dioxide (CO_2), sulfur dioxide, and nitrous oxides per joule of energy than oil or coal. Compared to oil or coal, it produces almost no particulates: 7 parts per million (ppm) versus coal's 2,744 ppm. And while it is a greenhouse gas itself, it has a residence lifetime in the atmosphere of only about 12 years—compared to CO_2's effects, which last for centuries. However, as natural gas opponents point out, it is composed mainly of methane, which has radiative forcing 20 times greater than CO_2: CO_2 is not the worst greenhouse gas, simply the most common, and should natural gas be used at a greater scale, its contribution to global climate change would increase in proportion rapidly.

There is also a reasonable financial objection to green pricing: green energy receives substantially greater public subsidies than other forms of energy. According to the Department of Energy's Energy Information Administration, electricity production subsidies and government support is approximately 44 cents per megawatt-hour for coal, $29.81 per megawatt-hour for refined coal, $24.34 per megawatt-hour for solar energy, and $23.37 per megawatt-hour for wind power. The federal government spends a considerable amount of money funding renewable energy projects and subsidizing renewable energy in order to make it price-competitive. As taxpayers, utility customers are already paying more for renewable energy than for nonrenewable energy. Why, then, should they pay a green pricing premium on top of that?

Proponents of green pricing would argue that the same logic would lead to one to conclude that charitable donations are unnecessary when government programs for the needy exist. And that logic would in a sense be correct—they are unnecessary, not undesirable. Green pricing programs, even when they are not transparent, involve the public to a greater degree in the greening of America's energy supply and act as a litmus test to demonstrate to utility companies the extent of the American public's commitment to green energy. So long as that commitment exists, this should be a good thing: utility companies can move forward, confident that the American public puts a high enough priority on going green that they are willing to pay a little extra to smooth the transition.

Commercial customers, though, have shown that they have no interest in participating. Businesses are run to generate profits; given the choice between two apparently equal services, it only makes sense that they will choose the cheaper one rather than fund some other business's projects. Only about one-third of electricity is consumed by residential customers, which is an obstacle to all opt-in green energy programs: for every household greening its habits by remembering to turn off the lights whenever a room is empty, there are two convenience stores keeping their lights on all night or a factory operating heavy equipment.

Focus group testing has shown that individuals' expectations of participation in green pricing are dependent on their expectation of other people participating. They will participate if everyone is doing it. This is an effect we see everywhere from the adoption of consumer technology (DVD players caught on more quickly than Laserdiscs in part because the prevalence of rental stores and services bolstered the impression that it was a popular technology) to new green habits (compact fluorescent light bulbs and low-flow showerheads seemed to catch on all at once, but not at the time of their introduction, as though their popularity reached a critical mass). Furthermore, those green habits that have been the biggest successes have represented savings to the consumer; asking consumers to pay for a public good is a considerably different venture.

By presenting green pricing as a premium service, the implied message is that green power is for the affluent, like the General Motors EV1 electric car that was marketed to the wealthy or the solar-powered homes presented as novelties. The start-up cost was the barrier to entry for most consumers who might have otherwise been interested in renewable energy for years; only recent innovations in technologies like thin-film solar cells and economies of scale as wind turbines are produced in greater numbers have mitigated this, 30 years after President Jimmy Carter installed solar panels on the White House roof. Even when consumers understand that renewable energy can be cost-effective in the long run, the price of transition is sufficient to disinterest them. Promulgating the idea that utility companies that already have access to renewable energy subsidies need to charge extra in order to build up the capital for significant green energy investment simply underscores this decades-old association.

Even a well-informed, proactive consumer who can well afford the additional cost may not be interested in green pricing because donating to such efforts does not confer any ownership or decision making. Green pricing, green consumption, is not green citizenship. There is no way to really influence how the utility company will pursue its green energy initiatives, no way to be involved in the decision-making process or the discussion of whether the motivating priority is energy independence, carbon neutrality, environmental responsibility, the creation of local green jobs, or some other motive. Not every green consumer will want to go green for the same reason; treating potential green consumers as homogeneous will inevitably exclude some of them. What is truly needed is an involved, invested citizenry that understands the differences among disparate potential energy sources and is able to come to a collective decision that serves their needs in an environmentally responsible manner.

See Also: Biofuels (Business); Carbon Tax; Commodification; Ecocapitalism; Environmental Justice (Business).

Further Readings

Boyle, Godfrey. *Renewable Energy.* New York: Oxford University Press, 2004.
Bryce, Robert. *Power Hungry: The Myths of Green Energy and the Real Fuels of the Future.* New York: PublicAffairs, 2010.
Jones, Van. *The Green Collar Economy: How One Solution Can Fix Our Biggest Problems.* New York: HarperOne, 2008.
Tertzakian, Peter and Keith Hollihan. *The End of Energy Obesity.* Hoboken, NJ: Wiley, 2009.

Bill Kte'pi
Independent Scholar

H

HEALTHCARE DELIVERY

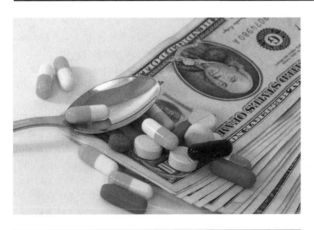

Many discussions about access to healthcare are really about access to money; some people forgo necessary healthcare in order to avoid the financial cost.

Source: Washington State Office of the Attorney General

Healthcare delivery refers to the ways in which healthcare services are supplied to individuals who need or desire them. Although the actual process by which healthcare is delivered varies among countries, it is safe to say that in most countries, delivering healthcare is a complex process that requires participation and cooperation among many different people and organizations. This complexity means that disputes about the best ways to deliver services are bound to arise, particularly in a country such as the United States that does not have a national system of healthcare but relies on a combination of private, public, and charitable programs. Although any system of healthcare delivery will have its own issues, much of this article focuses on comparisons between the U.S. method of healthcare delivery and the more nationalized or universal systems common in most of the rest of the industrialized world.

Issues and debates concerning healthcare delivery generally center around several related issues: How do different models of healthcare delivery affect the quality of care delivered? How do the different models affect the costs of providing healthcare, and who should have the right to make decisions about the healthcare of other people? These questions about methods of healthcare delivery have their root in two facts: health is an issue about which most people care intensely, and providing optimal healthcare can be a very expensive proposition in the modern world. Two additional facts contribute to the rising costs of healthcare that are behind many of the most heated debates about healthcare delivery. One is that usually people do not pay directly for their healthcare: instead it is covered by general tax revenues, by specific government programs, or by private insurance. In any case, the amount an individual pays to support the system may have no direct relationship to the costs of services that he or she uses (although he/she may be required to make a copayment or pay for a small percentage of those services) and hence he or she does not know the actual costs of the services received and has little incentive to try to limit those costs. A second fact is that providing goods and services to the healthcare market can be an extremely lucrative business and thus an incentive exists to create and provide those goods and services independent of whether they contribute substantially to the presumed goal of improving the recipient's health.

As noted in a 2000 report by the World Health Organization (WHO), the medical and financial aspects of healthcare delivery cannot be separated. This is because much of what modern healthcare offers is far too expensive for most individuals to pay out of pocket, and a single serious illness or accident could easily bankrupt the resources of a family if they had to pay all expenses themselves. Many discussions about access to healthcare are really about access to the money necessary to pay for care or the ability to require someone else to pay it. For instance, if the U.S. government determines that Medicare will pay for a particular service, that means essentially that U.S. taxpayers will pay for it, while if a private insurer is compelled to pay for a particular service, then everyone who pays premiums to that insurer is essentially paying for the service. If the government or insurer declines to pay for a medical service, an individual may still be free to pay for it out of pocket but the high cost of many procedures means that in the practical sense he or she does not have access to that service.

In the absence of any further system of regulation (for instance, national guidelines as to what services should be provided to treat particular medical conditions), the system of third-party payments can be expected to contribute to rising costs for healthcare without necessarily providing any improvements in health in return, based simply on the financial incentives such a system provides. This is not a criticism of the general concept of risk pooling, which is one of the fundamental reasons insurance companies and national or regional healthcare systems exist. Pooling the risk of providing care among many people means that if one person has high health costs, that will be balanced by many with lower costs, making the average cost within the reach of everyone. The advantage of this system is that it prevents the one person from being bankrupted by high medical bills; the disadvantage is that it may encourage some individuals in the system to overuse services and run up high bills since most or all of those costs will be passed on to others. Controlling costs given these contingencies requires some form of judgment beyond that of individual patients and providers about how healthcare funds should be spent, a process that has been accepted to some extent in most industrialized countries but has been strongly resisted in the United States, making any substantial reforms of the healthcare delivery system difficult.

The U.S. Healthcare Delivery System

Many observers have noted that the United States does not really have a healthcare system but instead has a multiplicity of systems. While in most countries healthcare spending is financed by taxes, in the United States the government plays a relatively minor role: among the 30 Organisation for Economic Co-operation and Development (OECD) countries, only in Mexico is a smaller proportion of the national healthcare budget provided by public expenditures. U.S. government spending on healthcare goes mainly to support care for the elderly (through the federal Medicare program) and the poor (through Medicaid and the State Children's Health Insurance Program, funded with a combination of state and federal revenues). Most Americans have health insurance through a private policy offered as a benefit of the employment of a family member; although they may have to pay something for this coverage, it is generally considered preferable to insurance purchased as an individual. An employer-based policy is called a group plan because, by condition of employment with a particular company, the employee is considered part of a group and thus is given the choice of one or more insurance programs offered through the group (but not available, at least not with the same price and conditions, to people who are not members of the group).

Americans who qualify for neither private, employer-based insurance nor one of the government programs may attempt to buy an insurance policy as an individual (such policies tend to be more expensive and to have more restrictions than group policies) or remain uninsured and either pay for medical expenses out of pocket or depend on charity care. Unlike most industrialized countries, in the United States, substantial numbers of people are not covered by any health insurance plan (45.7 million Americans were uninsured in 2007, constituting 17.2 percent of the nonelderly population). Research has shown that people lacking insurance suffer worse health and are at greater risk of mortality than those who are insured, and these concerns are a principal driver behind national health reforms that were enacted in 2010 in the United States. However, proposals to change the way healthcare is delivered and paid for in the United States have met with considerable opposition, and the future of these reforms is not certain.

Because the United States lacks a national system of healthcare, there is great variety in the ways healthcare is delivered in different parts of the country. For instance, the same services may often be delivered in multiple settings: basic diagnostic services may be provided in a physician's office, a school clinic, a neighborhood health clinic, or the emergency department of a hospital. A prescription may be filled in a neighborhood drug store, a hospital pharmacy, a branch of a national chain such as Walmart, or by mail. Surgery in some cases may be provided on an inpatient or outpatient basis, that is, for the same surgical procedure, a patient may or may not be admitted to the hospital. The same services may sometimes be provided by individuals with different qualifications as well: for instance, nurse practitioners and physician assistants can provide some of the services a physician may provide, and in some states, midwives are licensed to provide some obstetric services.

Although healthcare is sometimes conceptualized as a matter of individual choice as determined by consultation with one's healthcare provider, several factors influence where, how, and by whom healthcare services are delivered. Among these are laws, certification requirements by national governing bodies, and the willingness of insurers or other funding agencies to pay for particular services. Given the high cost of many medical services, the amount of care received may be more influenced by the willingness of an insurer to pay than by the medical judgment of the physician or the desires of the patient. This is a continuing issue that was highly publicized in the late 1990s by public outcry against "drive-through deliveries" in which a woman would give birth without being admitted to a hospital, a practice driven by the refusal of some insurers to pay for a hospital stay in conjunction with a routine delivery. The issue was resolved in late 1998 when a federal law was passed requiring health plans to pay for at least 48 hours of hospital care for mothers and infants following birth. A similar bill requiring insurers to pay for hospitalization following a mastectomy was also introduced but became a moot point when officials from a trade group for health maintenance organizations, the American Association of Health Plans, agreed that their members would pay for inpatient care after a mastectomy.

Comparing Models of Healthcare Delivery

It is well known that the United States spends more per capita on healthcare than any other country in the world. In 2007, the United States spent 16 percent of national income on health, versus an OECD average of 8.9 percent, and much more than the next higher spenders France (11 percent), Switzerland (10.8 percent), and Germany (10.4 percent). On average, Americans consumed $7,290 of health services in 2007, more than twice the OECD average of $2,984 (adjusted for differences in price levels in the different countries) and much higher than the next closest spenders of Norway ($4,763 per capita) and Switzerland ($4,417 per capita). Although government expenditures are relatively low in the United States as a percentage of total healthcare expenditures, absolute U.S. government expenditures on health were higher than in all but two OECD countries, Norway and Luxembourg. Even considering that rich countries spend more on health than do poor countries, U.S. expenditures are far higher than would be expected from per capita gross domestic products (GDP)—about $2,500 higher, in fact.

Several explanations have been offered for the higher costs in the United States. One is the aging of the population because, in general, as people get older, they have more health conditions and hence require more care at greater expense. However, while this may explain some of the growth in costs internationally, it cannot explain the higher growth in the United States because population aging is a phenomenon observed across industrialized countries. The proportion of elderly in the United States is not particularly high: only 12.6 percent of the population is over age 65, while in Europe the average is 16.7 percent and in Japan 21.5 percent, yet U.S. healthcare expenditures are much higher than in those countries. The poor health habits of Americans have also been proffered as an explanation, particularly the rise of obesity, which is associated with many chronic and expensive medical conditions, but this is not a sufficient explanation because rates of illness are not higher in the United States than in other OECD countries and the rates of some unhealthy behaviors such as smoking are actually lower in the United States.

A third proffered explanation is that costs are higher in the United States because Americans receive higher-quality care, a proposition that is difficult to address directly because there is no single agreed-upon way to measure quality of care across an entire healthcare system. However, it is difficult to argue that the U.S. model of healthcare delivery produces consistently higher-quality care, given a number of studies and comparisons that have resulted in the opposite conclusion. The WHO last produced a ranking of national health systems in 2000: France was the highest-ranked country, with many OECD countries also ranked highly (e.g., Italy was 2nd, Spain 7th, Austria 9th, and Japan 10th), while the United States ranked 37th, between Costa Rica and Slovenia. On internationally recognized measures such as life expectancy and infant mortality, the United States is far from a world leader. In life expectancy at birth, the United States ranks 49th (2010 estimate) with 78.11 years, far behind comparable countries such as Japan (82.12 years), Canada (81.23 years), and France (80.98 years). In infant mortality, the United States ranked 180th (lower rank is better) among 224 countries for whom information was available: the U.S. rate was 6.22 deaths per 1,000 live births, much higher than in comparable countries such as Sweden (2.75/1,000), Japan (2.79/1,000), and France (3.33/1,000).

In 2009, the OECD compared healthcare quality among its member nations, with particular emphasis on comparing the United States to the other OECD countries because of the different organization of healthcare delivery within the United States. The OECD found that several aspects of the U.S. system were commendable: care is delivered in a timely manner, a great deal of choice is offered to most people, and the United States offers new healthcare products to consumers more quickly than any other country. Looking at specific aspects of care, the United States excelled in cancer care, with substantially higher five-year relative survival rates for both breast cancer and colorectal cancer than the OECD average. For breast cancer, the United States was the OECD leader with a five-year survival age-standardized rate of 90.5 percent versus the OECD average of 81.1; in colorectal cancer, the United States ranked second with a five-year age-standardized survival rate of 65.5 percent (second only to Japan) versus the OECD average of 57.2 percent. However, the United States did less well in managing chronic disease, with the highest hospital admission rates for asthma and complications from diabetes of any OECD country, both conditions that modern medical standards say should normally be controlled and managed primarily through outpatient care. For asthma, the U.S. admission rate for people age 15 and over was 120 per 100,000 population (standardized by age and sex) versus the OECD average of 51/100,000; the diabetes complication admission rate was 57 per 100,000 versus the OECD average of 21/100,000.

The United States spends twice as much (on a percentage basis) as the OECD average on administration and three times as much as some countries. This can be explained in part by the administrative complexity of the U.S. system (a multitude of insurers, each with its own bureaucracy) and also by the failure to adopt a system of electronic health records (a reform that would be difficult given that the United States lacks a national system of care), which have become nearly universal in countries such as Australia, the Netherlands, the United Kingdom, and the Scandinavian countries. Prices for healthcare services are also higher in the United States than on average in other OECD countries. For instance, the average price of 181 pharmaceutical drugs in the United States was 30 percent higher than the average in any other OECD country, and further studies have indicated that while generic drugs are often cheaper in the United States than elsewhere, the price of branded pharmaceutical drugs tends to be much higher. The price of hospital services is also higher than in other OECD countries: some studies have estimated them to be as much as 40 or 50 percent higher. Physicians are also paid more in the United States than in other OECD countries, although this is due both to higher fees and to a higher volume of activity.

Somewhat surprisingly, the United States has lower consumption of some aspects of healthcare than is typical of other OECD countries. The United States has fewer hospital admissions per capita than the OECD average, and the average hospital stay for acute care is shorter. The United States also has fewer practicing physicians relative to the size of the population (2.4 per 1,000 population versus 3.1/1,000), and U.S. residents have about 30 percent fewer hospital visits than the OECD average. However, some surgical procedures and diagnostic tests are much more common in the United States than elsewhere. For instance, the United States ranks first among OECD countries in the number of CT scans (computerized tomography, also known as a CAT scan) performed with 227.8 per 1,000 population versus the OECD average of 110.7/1,000 and MRI (magnetic resonance imaging) with 34.3 per 1,000 population versus 22.8/1,000. The United States also ranks high in the number of knee replacements performed (183.1 per 100,000 population, versus 117.9/100,000 average for the OECD) and Caesarean sections (31.3 per 100 live births, versus 25.7/1000 average for the OECD). Although it is impossible to decide at long distance how often these procedures were necessary or enhanced the health of the recipient, the fact that there is also wide variation in the use of such procedures and the overall costs of care within the United States, as highlighted by *The Dartmouth Atlas of Health Care* and other studies, suggests that local customs and other factors may play a high role in choices usually assumed to be based on medical necessity.

There are many ways to organize healthcare delivery, and debate about the best ways to do so is expected to continue in the future. The very complexity of the issues and the lack of agreed-upon ways to measure even basic factors such as quality of care complicate the discussion. Costs can also be difficult to determine, particularly in the United States, where a complex system of accounting adds to the complexity of the issues. In addition, healthcare is an emotionally charged issue for many people, making it difficult to raise issues such as when care should be provided or denied and whether healthcare should be considered as a human right to which everyone has a claim or as a consumer good that is available only to those who have the ability to pay.

See Also: Asthma (Cities); Asthma (Health); Health Insurance Reform.

Further Readings

Center for American Progress. "The Health Care Delivery System: A Blueprint for Reform" (October 31, 2008). http://www.americanprogress.org/issues/2008/10/health_care_delivery .html (Accessed September 2010).

Committee of Health Insurance Status and Its Consequences, Institute of Medicine. *America's Uninsured Crisis: Consequences for Health and Health Care*. Washington, DC: National Academies Press, 2009.

Dartmouth Atlas of Health Care. http://www.dartmouthatlas.org (Accessed September 2010).

Glied, Sherry A. "Health Care Financing, Efficiency, and Equity." National Bureau of Economic Research (NBER) Working Paper No. 13881. Cambridge, MA: NBER, 2008.

Pearson, Mark. "Disparities in Health Insurance Expenditures Across OECD Countries: Why Does the United States Spend So Much More Than Other Countries?" Written statement to Senate Special Commission on Aging (September 30, 2009). http://www.oecdwash.org/ PDFILES/Pearson_Testimony_30Sept2009.pdf (Accessed September 2010).

World Health Organization (WHO). *The World Health Report 2000—Health Systems: Improving Performance*. Geneva: WHO, 2000.

Sarah Boslaugh
Washington University in St. Louis

Health Insurance Reform

The United States is one of the few industrialized countries in the world that does not have a national or universal system of health insurance. Instead, the healthcare Americans need is paid for by a patchwork of systems, including government programs, private insurance, out-of-pocket payments, and charity care. Although many are dissatisfied with the current system, which incurs far higher costs than the systems used in comparable countries while also leaving many Americans uninsured (45.7 million in 2007, constituting 17.2 percent of the nonelderly population) and producing a lower standard of health than is seen in many less prosperous countries, attempts to change the system have met with major challenges.

Two facts lie at the heart of the problem: most people place a large value on their health and react emotionally to anything that they feel might threaten it, and there is a great deal of money being made in the healthcare system as it currently exists so that organizations such as private insurers are willing to spend large amounts of money to contest any changes to the current system that might hurt their financial interests. On the other hand, many Americans have indicated that they place a high value on universal access to healthcare and do not feel the private insurance market can provide it. For instance, a *New York Times* poll in June 2009 indicated that 72 percent would like a government-sponsored healthcare plan similar to Medicare to be offered to all age groups to compete with the plans offered by private insurers, and over half indicated they would be willing to pay higher taxes to ensure that all Americans had health insurance. In addition, most felt the government would do a better job than private insurance companies in holding down healthcare costs (59 percent versus 26 percent) and in providing medical coverage (50 percent versus 34 percent).

There are several reasons why some form of national health insurance or national healthcare, which is the norm in most countries at a comparable level of economic development to the United States, has met with more resistance in this country. First of all, due to the way the health insurance industry developed in the United States, most nonelderly Americans receive their insurance from a private, employer-based plan. Most of these people are generally happy with their plan and do not want any changes that might disrupt or otherwise change the way they currently receive care. Second, extending health insurance coverage to people who do not currently have it would require additional expenditures of tax revenues (although how much is broadly open to interpretation; some argue that ultimately universal insurance would save money) that are not currently welcome as the federal government already has a large deficit, many states are in serious fiscal trouble, and many people are out of work. Third, the social safety net in general is much weaker in the United States than in most comparable countries and the idea of a national healthcare system strikes some Americans as a socialistic program or method of income redistribution that would impinge on their rights. Finally, to some, the requirement that everyone have health insurance coverage, which most economists believe is a necessity for any serious reforms, is seen as a violation of individual liberties and an unwarranted government intrusion into their lives.

Despite these objections, the health reform bill H.R. 3590, also known as the Patient Protection and Affordable Care Act (PPACA), became law on March 23, 2010, with some provisions taking effect immediately and many more timed to take effect months or years into the future. The bill is generally viewed as a compromise between those who wanted no health reform or who believed the federal government should be less involved in healthcare (excepting, generally, the very popular Medicare program that provides insurance to people over 65) and those who wanted the reforms to go further, for instance, moving to a single-payer system. An Associated Press poll in September 2010 indicated that twice as many people felt the PPACA did not go far enough in reforming the healthcare system (40 percent) and only 20 percent opposed it as improper government interference in healthcare.

Fundamental to the debates about health insurance reform are basic differences in how healthcare, and the funds to pay for it, are regarded. While most people value their health highly and may think of it as an absolute good, the fact is that it is also a commodity that must be paid for and therefore must compete with other potential uses for the same money. Many healthcare services are available to those who can pay so the question is: What should be provided to those who cannot pay? This gets down to a fundamental question of what healthcare really is. To some people, healthcare is a human right and for ethical reasons should be provided to everyone without regard to their ability to pay. To others, it is a public good that should be available to everyone and deliver benefits to everyone (the benefits in this case include maintaining a healthy population and the assurance that the care is available if needed). Finally, for some people, healthcare is more of a consumer good that should be available to those who can afford to pay for it. Currently, U.S. practice has aspects of all three philosophies, which often produces conflicting situations. For instance, highly expensive care being provided to some people because healthcare is an absolute good, some services (such as treatment in the emergency department of a hospital) provided to everyone because it is a public good, and many other services available or not depending on how much a person can pay because medicine is a consumer good.

The PPACA is an extremely complicated bill (more than 2,400 pages long), with many provisions that may be changed, repealed, or augmented even before they take effect (for instance, 21 states have already filed lawsuits challenging the mandatory insurance provision of the bill on the basis that it violates state sovereignty). The remainder of this entry concentrates on discussing some of the major provisions of the bill and the controversies surrounding them.

Immediate Provisions of the PPACA

Some previsions of the PPACA went into effect immediately. One such provision is that from the date of passage forward, existing Medicaid and state children's health insurance programs may be expanded but may not be cut back. Additionally, states are restricted from changing eligibility rules or adding paperwork requirements to make it more difficult for people to obtain coverage, rules that will be in place until 2014, when the state-based insurance exchanges are expected to be in operation. This is significant because both programs are dependent on state as well as federal funding and, in previous years, coverage offered has often changed from year to year depending in part on how much money individual states wanted to allocate to the plans. Given that many states are in a fiscal crisis in 2010 due to the recession, it is probable that without this provision these programs would have been cut back, leaving more people without insurance coverage. This is not a particularly controversial provision, although it has been criticized both by people who would prefer to eliminate or further restrict government-supported healthcare for the poor and by those who think simply maintaining existing programs is not sufficient because they already leave too many people without health insurance or with inadequate coverage.

More provisions took effect six months after the passage of the PPACA. As of September 23, 2010, insurance companies have been prohibited from imposing lifetime limits on benefits (i.e., limiting the total amount the company would have to pay to cover the expenses of one person over his or her entire lifetime) and will no longer be permitted to exclude children from policies due to preexisting health conditions. Companies must continue to offer insurance to children under age 26 through their parents' policies and are prohibited from dropping coverage of people due to technical mistakes on their applications. Many preventive services, including immunizations, mammograms, and colonoscopies, must be covered under insurance plans without requiring copayments. Consumers have the right to keep their own doctors when they join a new plan and to appeal to a third party if dissatisfied with insurance company decisions regarding reimbursements.

Most of these changes are also relatively noncontroversial, and the rule about keeping children under their parent's plan until age 26 is expected to lower the number of uninsured by about 72,000. One concern about several of these rules is costs, since health insurance is a business and any cost incurred by the company is ultimately expected to be passed along in the form of higher prices charged for policies. One unfortunate consequence has already occurred: several insurers no longer offer individual plans for children due to the requirement that they accept children with preexisting conditions. Requiring that policies be offered also does not mean that they are affordable because no restrictions are placed on how much insurers may charge for their policies, so adults or children with very expensive conditions may technically be able to purchase insurance (in the sense that plans are available to them), yet remain uninsured because the plans are too expensive for them to afford.

Other provisions that took effect in 2010 include establishment of a temporary high-risk pool on a national basis to provide health coverage to individuals with preexisting conditions and the creation of processes to review increases in healthcare premiums and require justification of such increases. In addition, a nonprofit institute will be established to conduct patient effectiveness research (i.e., to determine which treatments are effective and which are not, independent of the influence of pharmaceutical companies and device manufacturers). The most controversial provision among these is the institute to conduct patient effectiveness research: although the recommendations of the institute would not be legally binding, and similar types of recommendations are already being made by various professional bodies, some oppose effectiveness research on the grounds that it could be used to deny treatment or to ration care. Some also see it as threatening the doctor–patient relationship by inserting into it a third party (the guidelines) with no knowledge of the particular patient in question. However, advocates for patient effectiveness research argue that such guidelines based on research frequently benefit patients (for instance, by helping them avoid ineffective procedures or drugs), do not represent a threat to healthcare quality (such guidelines are in use in many countries that outperform the United States on many measures of healthcare quality), and will also be necessary in the future to control costs.

The next large group of insurance reforms will begin in 2013. The Consumer Operated and Oriented Plan (CO-OP) will be established to encourage the creation of member-run nonprofit health plans in all 50 states and Washington, D.C. Health insurance administration will be simplified by adopting a single set of operating rules for numerous procedures, including eligibility, verification, claims, payment and remittance, and enrollment and disenrollment, changes that must be implemented by 2014. Several changes will also be implemented in the tax code in 2013, including increasing the threshold for itemized deduction of unreimbursed medical expenses from 7.5 percent to 10 percent of adjusted gross income (waived for people age 65 and older through 2016). In addition, the Medicare Part A tax rate will rise from 1.45 percent to 2.35 percent for earnings over $200,000 for individual taxpayers and $250,000 for married couples filing jointly, and a 3.8 percent assessment will be imposed on unearned income for higher-income taxpayers. The latter changes are essentially tax increases, which are never popular, but the Medicare tax increase at least falls primarily on higher-income taxpayers who should be most able to afford to pay it.

The largest and most controversial changes will take place in 2014. U.S. citizens and legal residents will be required to have health insurance coverage, with a financial penalty for failing to do so. Companies with 200 or more employees will be required to offer health insurance coverage and automatically enroll employees (although the employees will be able to opt out), while companies with 50–199 employees will be penalized if they do not offer health insurance. State-based American Health Benefit Exchanges and Small Business Health Options Program Exchanges will be created to offer health insurance coverage to individuals and businesses with up to 100 employees. Various limits will be placed on health plans, including limiting ratings for individuals (age, geographic area, family composition, and tobacco use ratings will still be allowed, e.g., plans could charge higher premiums to smokers than nonsmokers) and limiting waiting periods to 90 days. Subsidies to cover purchase of insurance through the exchanges will be available for individuals and families with incomes between 133 and 400 percent of the federal poverty level (FPL), and all those not eligible for Medicare and with incomes at 133 percent of the FPL or below will be eligible for Medicaid.

The single most controversial provision of the PPACA is the requirement that by 2014 all individuals must have health insurance coverage or pay a fine. This requirement is considered necessary from an economic point of view because insurance is based on the concept of risk sharing so that people pay into the system (thus assuming the risks of others) even when healthy. If people were allowed to purchase insurance only after they became ill, the system would not be sustainable because costs would be too great to be supported by the amount of premiums coming in. Also, because individuals in the United States are essentially entitled to a certain level of care already (for instance, everyone must be treated in emergency rooms regardless of their ability to pay), they are in a sense insured because their care is subsidized by others who do pay their bills. Given this formulation, it seems fair to require everyone to pay something into the system. However, to some people and states this requirement seems an undue infringement on their liberty. In addition, some have protested that the bookkeeping and reporting requirements imposed on small businesses will be onerous and may discourage people from beginning or expanding businesses.

No one is certain how the healthcare exchanges will operate, but they have also been a source of anxiety as they are a major aspect of the move to universal coverage and yet are unlike anything that has been done in the United States. A few things are certain: there will be no national exchange; instead, the exchanges will be set up at the state level and administered by a government agency or nonprofit organization. The federal government will provide start-up money and in the early years the exchanges will be open only to people who do not work for large and medium-sized companies that already offer health benefits. Insurance plans will have to offer a standard package of benefits (with several levels of coverage available) and will therefore compete on the basis of price. However, the exchanges would not have the power to set premiums, although they could ask insurers to justify rate hikes and remove them from the exchange if the explanation is not satisfactory. The belief is that insurers will compete for business through the exchanges and will be motivated to offer plans at a good price that people will want to buy. However, since everyone will be required to purchase insurance and there will be no government plan to compete with the offerings of private insurers, there is nothing to guarantee that the plans offered will be either attractive or affordable.

Healthcare is a highly emotional issue and paying for it constitutes a substantial portion of government expenditures. The United States, unlike most industrialized countries, does not have a national system of health insurance, but legislation passed in March 2010 establishes a series of steps to move closer to the goal of universal coverage that is set to be achieved by 2014. This health plan has been attacked both as doing too much and as doing too little, but in some sense, we will have to wait until the various provisions go into effect to see how much it really accomplishes and how it affects the provision of healthcare services and, ultimately, the national level of health.

See Also: Corporate Social Responsibility; Healthcare Delivery; Precautionary Principle (Ethics and Philosophy).

Further Readings

Alonso-Zaldivar, Ricardo and Jennifer Agiesta. "AP Poll: Repeal? Many Wish Health Reform Went Further." *Huffington Post* (September 25, 2010). http://www.huffingtonpost .com/2010/09/25/ap-poll-repeal-many-wish-_n_739211.html (Accessed September 2010).

Henry J. Kaiser Family Foundation. "Health Reform Implementation Timeline." http://www .kff.org/healthreform/8060.cfm (Accessed September 2010).

"H.R. 3590 Patient Protection and Affordable Care Act." http://democrats.senate.gov/reform/ patient-protection-affordable-care-act-as-passed.pdf (Accessed September 2010).

Kotlikoff, Laurence J. *The Healthcare Fix: Universal Insurance for All Americans*. Cambridge, MA: MIT Press, 2007.

Nather, David. *The New Health Care System: Everything You Need to Know*. New York: Thomas Dunne Books, 2010.

Republican National Committee. "Health Care: 2008 Republican Platform." http://www.gop .com/2008Platform/HealthCare.htm (Accessed September 2010).

Sack, Kevin and Marjorie Connelly. "In Poll, Wide Support for Government-Run Health." *New York Times* (June 20, 2009). http://www.nytimes.com/2009/06/21/health/ policy/21poll.html (Accessed September 2010).

Tanner, Michael D. "Bad Medicine: A Guide to the Real Costs and Consequences of the New Health Care Law." White Paper, Cato Institute (July 12, 2010). http://www.cato.org/pub_ display.php?pub_id=11961 (Accessed September 2010).

The White House. "Health Care: Health Reform in Action." http://www.whitehouse.gov/ healthreform (Accessed September 2010).

Sarah Boslaugh
Washington University in St. Louis

HORIZONTAL POLICY INTEGRATION

Policy integration implies the incorporation of "new" objectives into existing sectoral policies. It is seen as a strategy for "greening" other policy domains by including environmental concerns. However, relatively autonomous policy sectors obstruct such integration attempts. Hence, policy integration, including different strategies to it, is a frequently underestimated challenge.

Policy integration may be referred to as the incorporation of specific policy objectives, which are extrinsic to a policy domain, into existing sectoral policies. For example, gender aspect may be integrated into research policy or agricultural policy may be greened by integrating environmental objectives such as biodiversity conservation and reductions of pesticides. Policy integration is frequently requested for making public policy more coherent. This claim is not new, and the idea of policy integration dates back to the 1970s' discussions on cross-cutting policies. The issue is currently largely being discussed under the term *environmental policy integration*. Still other objectives are being discussed in the context of policy integration, for example, rural development, gender, food safety, and freshwater conservation, which should be included into a strong sector's policies such as, for example, agricultural, energy, transport, or innovation policy.

Existing interpretations of policy integration are diverse and partly describe the phenomenon either as a process, a political result, or a mixture of both. The former comprises the management of cross-cutting issues that surpass established policy sectors and organizational responsibilities. The output of policy may be considered integrated where the policy elements are in accord with each other. Viewed as a conflicting organizational process, one may speak of policy integration when various public organizations work together and do not produce either redundancy or gaps in services.

Debates in Horizontal Policy Integration

In order for policy integration to be effective, the extrinsic objectives need to be formally included into other sectors' policies and they need to be actually implemented. Both aspects must be seen as serious challenges for policy integration. The literature distinguishes between two major political strategies for realizing integration: horizontal and vertical policy integration ($\rightarrow$ vertical policy integration).

Horizontal policy integration as a political strategy refers to attempts at being an advocate of extrinsic policy objectives to incorporate these into other sectors' policies at a given political level. Here, individual departments or ministries fulfill the task of horizontal or intersectoral coordination between a number of other departments.

Examples of horizontal policy integration are the early strategies of then newly established environmental agencies and ministries. In the 1990s, these organizations had been charged with the task of integrating environmental objectives into relatively strong policy sectors such as agriculture, innovation, energy, or transport policy at national levels.

Pros and Cons

The general idea of policy integration recently gained momentum due to two reasons. On the one hand, political issues are of a cross-cutting character. Environmental, energy, and growth policies have multiple dimensions and cut across different sectors of the economy. On the other hand, this change in the problem structure of political issues goes along with changes in policy approaches to them. Traditional sectoral and uncoordinated approaches, such as agricultural policy, have been criticized as being ineffective or not consistent with the holistic idea of sustainable development. In these cases, policy integration is expected to increase the efficacy and coherence of policies.

By formally including extrinsic objectives into sectoral policies, the horizontal approach has been employed successfully as a door opener prior to the World Summit in Rio de Janeiro in 1992. It acknowledges that formal integration into sectoral policies is the starting point, even though major policy impact still may be obstructed informally by policy sectors.

Although the concept of policy integration enjoys high political acceptance, realization in political practice encounters significant resistance due to two major obstructions. First, the integration of political objectives such as environmental goals, which are extrinsic to a given policy field such as agricultural or energy policy, disagrees with the logics of a sectorally designed government, which follows the concept of rational and effective public administration. In this concept, a number of separate government departments or ministries and other public agencies are charged with delivering public policy. Overlaps of responsibilities are avoided, and consequently each department develops a relatively autonomous policy sector, which delivers highly specialized, functional, and vertically organized policy through separate administrations. Second, policy integration goes against the economic interests of actors surrounding these relatively autonomous administrations. The extrinsic policy objectives often conflict with vested interests of their industrial clientele. For example, would agricultural stakeholders such as farmers' associations influence the ministry of agriculture and its agricultural policy much more effectively than would, for example, environmental groups, because they share specific material interests, which are touched upon by environmental claims?

The horizontal approach to integration does not pay due attention to the stage of policy implementation and its role in achieving effective integration. After the formal uptake of new objectives, policy implementation is administered and conducted by discrete sectoral organizations. During this stage, stalemates occur and such objectives consequently are often rendered ineffective. This orientation toward more formal integration aspects in the formulation of policies may be seen as the basic shortcoming of the horizontal approach.

The example of horizontal integration of environmental policy objectives, which had been applied prior to the World Summit in Rio de Janeiro in 1992, shows that a relatively weak policy sector had been charged with the task of "greening" the policies of stronger sectors. This exercise often resulted in the formal uptake of environmental objectives during the formulation of sectoral policies. However, during subsequent policy implementation, the full resistance of individual policy sectors operated against the environmental objectives through stalemates, resulting in a merely symbolic integration of environmental concerns into sectoral policies with very limited—if any—effect.

Hence, environmental policy integration as one form of policy integration means the inclusion of environmental objectives into policy as one among multiple and competing policy objectives. These objectives, in most cases, are conflicting with preestablished sectoral ones, which reflect the interests of powerful actors of the respective policy domain. Therefore, some scholars argue that the endeavor of sustainable development and consequently environmental integration must be seen as being superior to others and assume principled priority for these objectives. Here lies the difference between a rather realistic or rationalistic school of thought, which acknowledges the separation of policy into sectors and recognizes resulting sectoral and organizational interests, and an idealistic strand, drawing a picture of potentially coherent and rational policy.

See Also: Common Property Theory; Sustainable Development (Business); Vertical Policy Integration.

Further Readings

Briassoulis, Helen. *Policy Integration for Complex Environmental Problems—The Example of Mediterranean Desertification*. Aldershot, UK: Ashgate, 2005.

Giessen, Lukas, et al. "Politikintegration für ländliche Räume? Die (Nicht-) Koordination der Förderung [Rural Policy Integration? The (Non-) Coordination of Funding Programs]."

In *Land–Stadt Kooperation und Politikintegration für ländliche Räume* [Rural–Urban Cooperation and Rural Policy Integration], Sebastian Elbe, ed. Aachen, Germany: Shaker, 2008.

Jänicke, Martin and Helge Jörgens. "Neue Steuerungskonzepte in der Umweltpolitik [New Concepts for Political Steering in Environmental Policy]." *Zeitschrift für Umweltpolitik und Umweltrecht* [Journal of Environmental Policy and Environmental Law], 3 (2004).

Peters, B. Guy. "Managing Horizontal Government—The Politics of Coordination." *Public Administration*, 76:2 (1998).

Lukas Giessen
University of Göttingen

HUNTING

Waterfowl, like these geese and other animals that provide little sustenance, make up the largest proportion of animals killed by hunters, while deer make up only 5 percent.

Source: Stillwater National Wildlife Refuge Complex, U.S. Fish and Wildlife Service

Contemporary discourse regarding hunting is marked by passionate debate, despite the number of hunters in the United States declining over the past decade. Still, in spite of this decline, support for hunting has actually risen, and the numbers of those who disapprove of hunting have dropped. In the midst of these demographic shifts, and complicating this already firmly entrenched debate, is the rise of "green-" or "ecohunting," which is touted as providing hunters with the "thrill without the kill"; green hunters use tranquilizer guns to immobilize their prey, which is later released. Some laud this idea, while others question its "greenness." This debate over green hunting, and the wider hunting debate, touches on issues of ethics, environmentalism, politics, and U.S. culture in general. Hunting by aboriginal groups as an element of culture will not be addressed directly in this entry. It should be noted, however, that this type of hunting is also controversial.

Hunters: Stewards of Nature

Hunting advocates argue that hunting is an essential component of both the United States' cultural heritage and its economy and that hunting functions as an important wildlife management tool. They argue further that everyone benefits from the excise taxes that hunters pay on guns, ammunition, and outdoor equipment, claiming that through these taxes and hunting license fees, hunters contribute $200 million per year for wildlife conservation. They further claim the following:

- Hunters are among the foremost supporters of wildlife management and conservation.
- Hunters and hunting organizations provide direct assistance to wildlife managers and enforcement officers at all government levels.
- Hunting helps manage animal populations, resulting in less human/animal traffic accidents.
- Hunting contributes more than $30 billion to the economy annually and supports more than 1 million jobs.
- Hunters have provided thousands of pounds of game meat through donation programs such as Hunters Sharing the Harvest and Hunters for the Hungry.
- Through hunters' legislative efforts, funds have been channeled back into the conservation process, restoring populations of game species to record numbers.

Hunters: Cruel and Unusual

Hunting opponents argue that while hunting was once a vital part of human survival, hunting today is unnecessary, with most hunters engaging in the practice not for subsistence, but for recreation; they further argue that the following is true:

- Hunting contributes to the extinction of animal species.
- Hunting is cruel, particularly to animals that are injured but not killed and then suffer prolonged, painful deaths; research suggests that for every animal shot and captured, two more are wounded and not retrieved.
- Hunting disrupts wildlife migration and hibernation patterns.
- Hunting is not necessary for "wildlife management" and "conservation," which are euphemisms for programs that ensure that there are always enough animals for hunters to exploit. Because their revenue comes primarily from the sale of hunting licenses, wildlife agencies' major function is to propagate "game" species for hunters to shoot, resulting in the loss of biological diversity, genetic integrity, and ecological balance.
- Animals kill only their sickest and weakest cohorts, while hunters kill large, healthy animals that are needed to keep populations strong.

Hunting: A Closer Look

Hunting and poaching kills between 100 and 200 million animals in the United States annually. While hunters cite the need to manage overpopulations of certain species, particularly deer, hunting opponents point out that deer make up about 5 percent of all animals killed by hunters in the United States. Conversely, rabbits, squirrels, mourning doves, and waterfowl make up the largest proportion of animals killed by hunters—species that do not require population control and offer little sustenance. About 10 percent of the mammals and birds of North America are classified as "game animals"; none of the remaining 90 percent of North American species is overpopulated. Anti-hunting proponents argue that this is because animal populations naturally stabilize and meet the carrying capacity of their habitats.

Hunting opponents also point out that hunting accidents kill and injure numerous non-game animals as well as at least a thousand hikers and other hunters annually, though hunters contend that these injuries and deaths are negligible compared to those associated with, say, automobiles. Another hunting debate concerns "canned hunts," where hunters pay to kill captive native and exotic species on private reserves for the sole purpose of garnering a "trophy." Hunters cite these hunts as opportunities to commune with nature, while hunting opponents contend that canned hunts are excessively cruel and lack even the most basic notion of a "fair chase." "Ecohunting" has been promoted as a safe way to experience the thrill of hunting without harming the intended prey. Ecohunters pay for guided ecohunts, where they shoot any of a variety of animals from moving vehicles with tranquilizer guns. Ecohunting is touted as humane, relatively cheap, and good for the travel industry. However, a variety of criticisms have been raised regarding ecohunting, including the following:

- Immobilizing animals multiple times in short time frames leads to serious health and behavior problems.
- Insufficiently sedated animals charge ecohunters, often resulting in injury to the hunters and the animals being killed with traditional firearms.
- Insufficiently sedated animals run away and injure themselves.
- Animals fall in ways that can lead to asphyxiation or damage to internal organs.

These debates over ecohunting and traditional hunting are firmly rooted in notions of culture and ethics, with hunters maintaining, first, that it is their right to hunt and that hunting is a part of U.S. culture, and second, that hunting is beneficial to wildlife and ecosystems. Their opponents contend, instead, that hunting is not only cruel but that it actually creates many of the problems that hunters profess to remedy. They argue that while hunters have worked for decades to convince the U.S. public that hunting is good for society and good for wildlife, the truth is that hunting is merely a destructive, inhumane, recreational pursuit. Still, the fact that there are more supporters of hunting than there are those who hunt suggests that hunting is a part of the United States' cultural fabric that many feel is under attack.

See Also: Animal Ethics; Animal Welfare; Earth Liberation Front (ELF)/Animal Liberation Front (ALF); Ethical Sustainability and Development.

Further Readings

Bronner, Simon, J. *Killing Tradition: Inside Hunting and Animal Rights Controversies.* Lexington: University Press of Kentucky, 2008.
Dizard, Jan E. *Going Wild: Hunting, Animals Rights, and the Contested Meaning of Nature.* Amherst: University of Massachusetts Press, 1999.

Tani E. Bellestri
Independent Scholar

HYDROELECTRIC POWER

Hydroelectric power is currently the most used renewable energy in the world. There is a long tradition of using this technology when the terrain is appropriate, and it has been consistently used since the 19th century. Hydroelectric power has a number of advantages over other power technologies, be it thermal power (coal/gas/nuclear power) or other emerging renewable energies (particularly solar and wind power). Hydroelectric power provides a stable source of electricity, uses a renewable resource, has relatively low construction and operational costs, and does not generate significant greenhouse gases, making it an environmentally friendly technology. Nonetheless, there are negative impacts, including adverse effects in the geographical areas of hydroelectric facilities, theoretical risks related to constructing of hydroelectric structures, and a power supply that can be affected by rare meteorological events (such as droughts). The emergence of new hydro technologies (such as wave- and tidal-generated hydroelectric power) could mitigate many negative impacts, but these technologies are still not yet economically feasible.

Background

Hydroelectric power has been widely used since the 19th century. Conventional production methods generate power by using water flowing through a water turbine, powering a generator, which in turns produces electricity. The total power generated is determined by the speed and the height at which water flows through the turbine. As such, specific terrain is necessary to build hydroelectric facilities. Most hydroelectricity is generated by water retained by an artificial dam, but there are alternative generation methods, such as building a plant on a natural watercourse (such as a river). The power generated at a hydroelectric facility can range from 100w to 5kW (often called a pico-hydro facility, which can generate enough power for a few buildings) to upward of 20,000 MW.

Advantages of Hydroelectric Power

Hydroelectric power presents a number of advantages when compared to other power technologies. One of the main strengths of hydroelectricity is that the energy production is constant. This is especially significant when hydroelectric power is compared to other renewable energy technologies. For example, solar panels only produce energy when there is sunlight, and solar farms produce little power during cloudy conditions or at night. Another popular renewable energy, wind turbines, only produce electricity during optimal windy conditions, shutting down when the wind is either too weak or too strong. Hydroelectric power, on the other hand, can be used as a primary energy source for a community (rather than be relegated to a supporting role), since decision makers can adequately plan power input.

Hydroelectricity is a renewable energy because the water used to generate electricity is not consumed when the energy is generated. This makes public acceptance of the technology less controversial than with other nonrenewable methods. In addition, as the water used is constantly replenished naturally, there is less concern about running out of resources to power the facility (which is a major concern with other technologies, such as petroleum, natural gas, or uranium).

Furthermore, as hydroelectric power has been developed and constructed since the 19th century, it is a mature technology with a great body of extant knowledge and expertise. Construction costs can be readily defined and maintenance costs can be aptly planned. This also means these costs are comparable to some of the more popular combustible technologies (such as coal power plants), and production plants have extended life spans compared to thermal power plants. From a technical standpoint, hydroelectric power is one of the most responsive technologies—it is easy to start and to stop compared to other electric power-generating sources.

A significant benefit of hydroelectric power is the lack of chemical waste or pollution generated during the production process. While traditional technologies produce important amounts of greenhouse gases, hydroelectric power has a very limited greenhouse footprint. In turn, this implies that hydroelectric power can be included in a country's clean-energy portfolio when complying with world clean-energy targets. Other advantages include the use of hydroelectric dams and their reservoirs for flood control and agriculture (controlled flooding) as well as for aquaculture and water sports.

Disadvantages of Hydroelectricity

There are also negative impacts related to the implementation of this energy technology. One of the major weaknesses of hydroelectric power is the ecological impact hydroelectric structures can have on their local environment. For example, building a hydroelectric facility using a dam can necessitate the flooding of an entire valley. This, in turn, forces the relocation of local populations and affects the local fauna and ecological systems. Areas that could be used for living, recreation, and agriculture become unusable. Also, while using water from an existing watercourse is seen as having a limited impact on the environment (the power plant does not generate any greenhouse gases and does not consume the water during the energy production), recent studies have found that hydroelectric facilities can have an impact on local ecosystems (for example, preventing fish migrations or damaging river banks). In addition, some sites in tropical areas are believed to generate more greenhouse gases than conventional thermal generation plants.

From a structural perspective, there is a hypothetical risk that a dam retaining water could rupture and generate massive flooding in the surrounding areas. Dams are susceptible to structural failures during periods of sustained rain or in cases of shoddy construction, or can even be damaged by malicious intent (such as terrorism). The impact on surrounding populations could be catastrophic, as the flooding of surrounding communities can trigger considerable damages, even fatalities.

While energy generated by hydroelectricity facilities is generally constant, the risk exists that production could drop during adverse meteorological conditions (such as serious droughts). Dams can be used to regulate water flow and to attenuate some of this risk, but in periods of serious drought, reservoir levels can drop drastically, limiting the amount of electricity that be generated. As such, although the electricity generated is mostly constant, it is possible to have periods with low production output, making hydroelectric power slightly less reliable than traditional thermal power.

There are other minor drawbacks to hydroelectricity, including the fact that specific geographic characteristics are necessary to construct a hydroelectric facility. A new facility needs a strong water current, so not all locations are appropriate (compared to more traditional thermal technologies). Additionally, constructing a hydro facility can require a longer period to validate the project due to its impact on the ecological environment.

Future Outlook of Hydroelectric Power

New hydroelectric technologies are currently being developed. They are also known as ocean hydro technologies and include wave technology (where electricity is generated by large turbines installed underwater) and tidal technology (where electricity is generated by utilizing tidal movements). While these technologies attenuate some of the negative impacts of traditional hydroelectric power facilities (e.g., a limited impact on surrounding wildlife, more constant power production, and less risk on infrastructure), they do present two serious constraints. First, since these technologies are still in development, they present a considerable technology risk: there are very few large-scale models that demonstrate the viability of these technologies. Further, the cost to generate power can be two to three times higher than current hydroelectric technologies, having a serious consequence on the financial and economic feasibility of these projects. Overall, private companies working on these technologies expect market feasibility to take more than 10 to 15 years to achieve.

See Also: Alternative Energy; Solar Energy (Cities); Wind Power.

Further Readings

Karjalainen, Timo, et al. "Negotiating River Ecosystems: Impact Assessment and Conflict Mediation in the Cases of Hydro-Power Construction." *Environmental Impact Assessment Review*, 31:5 (September 2010).

Khan, J. et al. "Ocean Wave and Tidal Current Conversion Technologies and Their Interaction With Electrical Networks." Power and Energy Society General Meeting, "Conversion and Delivery of Electrical Energy in the 21st Century." *IEEE Conference 2008*, 20 (July 2008)

World Commission on Dams. "Dams and Development: A New Framework for Decision-Making." http://www.dams.org/report/contents.htm (Accessed September 2010).

Jean-Francois Denault
Independent Scholar

Hydropower

Nearly 90 percent of all renewable energy is hydropower, like this electrical plant, but the changes necessary to create hydropower can produce changes in river ecology.

Source: iStockphoto

Hydropower, the generation of electricity through the force of falling or free-flowing water, is one of the products of the multipurpose development of river systems. Those purposes often include flood damage reduction, navigation, drinking water supply, and irrigation, among others. Where feasible, hydropower generation has been included with these purposes and, outside the United States, can be a primary purpose for river control and development. Worldwide, hydropower is the most widely used source of renewable energy—nearly 90 percent of all renewable energy is hydropower—having an installed capacity in 2006 of 777 GWe (gigawatt electrical) supplying nearly 3,000 TWh (terawatt hour) of electricity, or roughly 24 percent of all the electricity in the world. Nations including Norway and Brazil generate more than 85 percent of their electricity from hydropower; in Canada and Venezuela, among others, more than 50 percent of electricity is generated by hydropower. Paraguay generates all the electricity it consumes inside its borders with hydropower, with substantial amounts of electricity remaining for export.

Advantages

Hydropower has a number of advantages compared to most other means for producing electricity on a large scale: (1) the hydrologic cycle means that hydropower is potentially completely renewable, though there are seasonal and even longer-term changes in this cycle; (2) hydropower generation converts roughly 90 percent of the energy in falling water into electric energy, an efficiency rate more than twice that of wind power and combined-cycle fossil fuel generation, and three times that of coal-fired steam generation; (3) life-cycle greenhouse gas (GHG) emissions associated with hydropower generation do exist, coming mostly from the reservoir behind the dam, but are much more than an order of magnitude lower than any fossil fuel electric generation, even with current carbon capture technology; and (4) hydropower machinery is relatively more durable since it is relatively more simple and runs at slow speed, and is more flexible in its power-generation scheduling since generation can be quickly brought online and offline to meet peak power system demands and help ensure load reliability and stability.

Disadvantages and Mitigation

Because hydropower requires at least some—and often very extensive—changes to river systems, there are also environmental costs related to using it for electricity generation. Among the most important costs are those derived from changing a river's natural environment with dams and reservoirs. This results in loss of open-river environment and can produce changes in riverine ecology from altered flow regimes. For example, dissolved oxygen (DO) levels in some reservoirs below the dams that impound water for hydropower can fall to levels too low to sustain aquatic life in some cases. This effect from storing water at depths much greater than that of the natural river can be partly offset with special aerating hydropower turbines and even with direct injection of oxygen into stream flow, but these measures are not implementable everywhere. In addition, regulating river flow behind dams for hydropower and other uses restricts the redistribution of sediment in the river, which can cause important habitat to be degraded and not naturally restored. Active mitigation measures to assist habitat restoration can be taken as well, though they, too, cannot be implemented in every case and often cannot be executed on the same large scale as the natural riverine sediment profiles. And as mentioned above, impounding water in reservoirs such as those used in conventional hydropower generation may increase emissions of some GHGs relative to the base or background emissions from the land before the reservoir was constructed. The emission of GHGs from freshwater reservoirs is an area of active research worldwide, with programs in many nations under way to characterize these widely variable but very small GHG amounts at different reservoirs.

Methods and Examples

Conventional hydropower generation converts the potential energy of elevated water (the difference between the water's source height and its outflow height is called "head") to electrical energy through generator turbines. Conventional hydropower is generated at Aswan Dam in Egypt, Bhakra Dam in India, Guri Dam in Venezuela, and Hoover and Grand Coulee Dams in the United States, among many others.

Other types of hydropower include pumped-storage generation wherein water is pumped from lower to higher reservoirs, then released back to its lower level through turbines. This allows these hydropower projects to meet periods of high electricity demand in a very timely manner. Examples include the Dniester Pumped Storage Power Station in Ukraine, Grand'Maison Dam in France, and Shintoyone Dam in Japan.

In addition, free-flow or run-of-the-river hydropower generates electricity from the natural flow of a river system without the need for large impoundment of water behind a dam. Although a smaller dam may be present, water in the flow of the river is generally diverted into turbines at lower levels without substantial water storage. Without storage, these systems are more dependent on river flow and seasonality and cannot usually alter generation to meet short-term demands. Because large reservoirs are not needed, run-of-the-river hydropower is sometimes preferred to conventional hydropower generation as having lower environmental costs. However, care must be taken to understand and engineer for diversions of water that do not adversely minimize aquatic habitat in the river through reduced volume or flow, elevated temperature, or other physical and chemical changes. Examples of run-of-the-river hydropower generation are Chief Joseph Dam in the United States, Ghazi Barotha Dam in Pakistan, and Beauharnois Hydroelectric Power Station in Canada.

Hydropower plants can be classified by their capacity for electrical generation. Although there is no international standard for the classifications, they extend from very large, or more than 10 GWe as at Three Gorges Dam in China, Itaipu Dam on the border between Brazil and Paraguay, and Guri Dam in Venezuela, down to micro hydro installations typically producing less than 100 KWe. Their low construction and maintenance costs and relative ease of installation have meant that micro hydro generation has often been used in developing nations with few other sources of energy. And increasingly, micro hydro installations are being proposed in developed nations, including Canada and the United States, where installation of new, large-scale conventional hydropower may not be feasible but sufficient hydropower potential exists to supply the electricity needs of small communities.

See Also: Alternative Energy; Hydroelectric Power; Solar Energy (Cities); Wind Power.

Further Readings

International Hydropower Association. http://www.hydropower.org (Accessed September 2010).

REN21. "Renewables Global Status Report 2006 Update." Paris: REN21 Secretariat, 2006.

U.S. Department of Energy, U.S. Energy Information Administration. "Country Energy Data and Analysis." http://tonto.eia.doe.gov/country/index.cfm?ref=bookshelf (Accessed September 2010).

Jeffrey R. Arnold
U.S. Army Engineer Institute for Water Resources

I

INDIVIDUAL ACTION VERSUS COLLECTIVE ACTION

The debate concerning individual action versus collective action has long been of interest to social science disciplines, including economics, political science, sociology, psychology, philosophy, and anthropology. This interest stems from the belief that certain goals can only be achieved with collective action. However, for a number of reasons that will be explained further, it is not always easy or even possible to ensure collective action. This entry explains why people, organizations, and nations engage in individual versus collective action, the outcomes of this choice, possible ways to encourage collective action, and finally, the role of individual versus collective action in greening the society toward a sustainable future.

Why Individual Action?

Rational choice theory posits that all action is fundamentally rational and that people calculate the likely costs and benefits of any action before deciding how to act. This assumption of economic rationality of the individual that supposes that rational individuals act to maximize their self-interest lies at the root of the choice between individual versus collective action. If the long-held belief that what is best for the individual is also best for the collective was true, individuals pursuing self-interest would not cause any problems in society. Economist Adam Smith argued that in a free, perfectly competitive market, rational decisions made by individuals intended for their own gain would produce the best outcome for the society as whole, as if an invisible hand were guiding them. Since then, this claim has been widely criticized due to the impossibility of fulfilling the assumptions of the perfectly competitive free market, thus resulting in the failure of the market. Among the notable market failures are the problems resulting from the lack of collective action: production of public/collective "goods," and "bads," termed *externalities*.

Collective action can be defined as action oriented toward achieving a common goal among a group. The dilemma of collective action is founded on the assumption that rational actors will not engage in collective action if they can benefit without participating and that in such cases they will choose to take a "free ride." Mancur Olson states in his 1965 book *The Logic of Collective Action* that due to the free-rider problem—the fact that individuals can benefit from action even if they do not participate—collective action is irrational. While his theory explains nonparticipation, it is criticized for failing to explain participation. If people believe they can benefit without paying the costs, why would anybody participate? However, the outcomes of Olson's "logic" of collective action, often termed the *problem* of collective action by others, merit some discussion first.

Consequences of Choosing Individual Over Collective Action

The collective action dilemma and the associated free-rider problem are manifested in the production and consumption of public and collective goods. As Garrett Hardin famously explained in his 1968 article "The Tragedy of the Commons," individuals pursuing their own interests bring about the collapse of natural resources and cause a tragedy rather than serve the public interest. Using the example of a common grazing land, Hardin showed that it is in the self-interest of the rational herdsman seeking to maximize his gain to add one more animal to his herd because his share of the costs of an additional animal in the form of damage done to the commons is smaller than his personal benefit. Even if he is inclined to self-restraint, the fear that the others are not prevents him. The result of focusing on short-term individual gain is competitive overexploitation of common-pool resources—resources held or used in common. In most cases of competitive overexploitation, individuals are not aware of the damage their acts are causing. But even if they are aware, their own responsibility seems infinitesimally dilute.

In terms of production of public goods, the same principles apply. Public goods have an important property that encourages free-riding: nonexcludability. Once the good is produced, it is available for all to utilize, for example, a lighthouse or a dam. This separates the benefits of the good from its production costs. It is then economically rational for people to try to make others pay for most or all of the costs of a public good that benefits everyone equally. People are not willing to contribute voluntarily to pollution reduction or the production of any kind of public good in optimal amounts. This results in the inefficiency of the market and the underproduction of the public good. Political scientists explain (non)participation in social action with the same "logic of collective action."

Game theory and versions of noncooperative games have also been used in attempts to explain why individual action is chosen over collective action even when the overall outcome is negative. The most famous noncooperative game, called the Prisoner's Dilemma, offers a simple representation of a two-person collective action problem and is also based on the assumption of the rational utility-maximizing individual. The game lays out a scenario in which two players who committed a crime are questioned separately. Each is given the option of going free if he confesses and implicates the other. The catch is that if they both confess, both get the full sentence; however, due to lack of evidence, if they both remain silent, both get a reduced sentence. The best outcome for both is achieved when both remain silent. However, neither trusts the other to remain silent, and with the lack of communication between them, they would both confess, focusing on their own self-interest. Their individual actions would result in a less-than-ideal outcome for them. The lesson here is that what is in the best interest of the group is different from what is in the best interest of the individual. Research shows that the divergence between individual and group interest increases as the size of the group increases. Furthermore, relative importance and visibility of each individual's contribution also decreases with increased group size.

As in the tragedy of the commons, individuals will think their (non)contribution is negligible and will not affect the outcome.

Applying Prisoner's Dilemma to a group setting (a game called *n*-person Prisoner's Dilemma) explains the tragedy of the commons. The best outcome for the group can only be achieved if all members contribute, for example, to the production of a public good. However, the best outcome for the individual is to exploit the contributors by reaping the benefits without paying the costs by being a free-rider. This not only gives him or her a better individual outcome but prevents possible exploitation by others in the group as in the tragedy of the commons situation.

Encouraging Collective Action

As the consequences of individual action are quite negative for the collective, the issue of how to induce self-interested individuals to engage in collective action received a lot of attention. Olson suggested that the free-rider problem can be overcome by either providing selective incentives to reward those engaging in collective action or monitoring and sanctioning systems to deny benefits to nonparticipants. These solutions are criticized because their administration also requires some form of collective action and would result in what is called the "second-order collective action problem." Hardin's solution to the tragedy of the commons was establishment of some form of property or use rights, which requires an exclusion system to be established. Examples of cooperative management of common resources exist, for example, in coastal fisheries, when the second-order collective action problem can be resolved.

In reality, more collective action is observed than Olson's logic of collective action, based on economic rationality predicts. Free-riding is the behavior expected of *homo economicus*, an individual guided by instrumental rationality, looking for the most efficient or cost-effective means to achieve a specific end. Yet *homo sociologicus*, a social individual who pays attention to social norms, may act differently. Social action and analytical philosophy literature provides some explanations as to why people participate in collective action besides the instrumentally rational goal of trying to influence social and political environment in order to change their circumstances. People who have a collective intention are rational in a broader sense than this economic, instrumental rationality of calculating individual costs and benefits. They engage in collective action because they identify with a group and act as its members or they want to express their views. They focus on more than a narrowly defined self-interest, care about responsibility, and have ethical standards that are not captured in economic terms. In these cases, participation may be viewed as a benefit rather than a cost that transforms the logic of collective action. Participants in political action realize that they are giving a free ride to nonparticipants but still are willing to participate because they care enough to share in the costs.

Contrary to Olson's model of decisions made in isolation or the noncommunication of the Prisoner's Dilemma, individuals do not live in isolation, and they talk to each other. They are influenced by the actions of others. Communication is said to give rise to a collective identity that motivates individuals to cooperate. Communication also elicits norms of cooperation and against free-riding. If the Prisoner's Dilemma is converted from a competitive to a cooperative game in which players communicate with each other, a different form of rationality takes precedence. Furthermore, even the Prisoner's Dilemma game produces different results in repeated iterations. That is to say, cost-benefit calculations include long-term gains and losses in cases of repeated social interactions. This suggests that when individual rationality leads to collective irrationality, a move from an individual to a collective rationality (such as Jürgen Habermas's communicative rationality) would be a more rational course of action.

Collective Action in Greening the Society

Many topics of green issues and debates are related to collective action problems because virtually all ecological resources are common-pool resources (airsheds, watersheds, land, oceans, atmosphere, biological cycles, the biosphere itself). As such, dealing with environmental pollution as well as depletion of natural resources requires collective action on the level of individuals, organizations with sometimes competing and conflicting interests and missions, and nation-states.

The *Stern Review on the Economics of Climate Change* described climate change as one of the greatest challenges of international cooperation that the world has faced. Climate change is an example of a global externality, and its mitigation faces the classic problems of provision of a global public good. A collective action for the reduction of global carbon emissions has not been taken yet due to the problem of collective action. Even if all nation-states were to be collectively better off if the climate change were mitigated, individual interests expressed by the perceived cost of emission reduction to each nation's economy is deterring participation. Nations are reluctant to make agreements that would increase the cost of production and lose any comparative advantages they might have. Debates come to a deadlock at the question of who will pay the costs as happened at the Copenhagen Climate Conference of 2009. Developing countries argue that developed countries caused the problem in the first place and are reluctant to pay the price now that it is their turn to develop. Developed countries only agree to cut back emissions significantly if developing countries also agree to cut so they would not lose comparative advantage for their products in the international markets. In this rhetoric of the north–south debate, few nation-states would go ahead individually and be taken advantage of for being the "good one." Individual self-interests of a nation require it to maintain low production costs. In the absence of global sanctions, there will be free-riders even if an agreement is reached.

Collective action research tells us that the way to overcome this gridlock should start with a broader vision. Frequent communication and repeated negotiations along with a green ideology/culture would overcome some of the difficulties posed by Olson's logic of collective action. This would require collective social action on the individual level to pressure organizations and nations to act. Individual participation in collective action to influence social and political environment can be a means of achieving a greener and more sustainable global agenda.

See Also: Common Property Theory; Fisheries; North–South Debate; Social Action; Tragedy of the Commons.

Further Readings

Hardin, Garrett. "The Tragedy of the Commons." *Science*, 162 (1968).
Hardin, Russell. *Collective Action*. Baltimore, MD: Johns Hopkins University Press, 1982.

Olson, Mancur. *The Logic of Collective Action: Public Goods and the Theory of Groups.* Cambridge, MA: Harvard University Press, 1965.

Ostrom, Elinor. *Governing the Commons: The Evolution of Institutions for Collective Action.* New York: Cambridge University Press, 1990.

Stern, Nicholas. *Stern Review on the Economics of Climate Change.* London: HM Treasury, 2006.

Aysin Dedekorkut
Griffith University

Intergenerational Justice

Intergenerational justice refers to the need for a just distribution of rewards and burdens between generations and fair and impartial treatment toward future generations.

Edith Brown Weiss identifies the following categories of activities as likely to undermine intergenerational justice:

- Wastes whose impacts cannot be confidently contained either spatially or over time
- Damage to soils such that they are incapable of supporting plant or animal life
- Tropical forest destruction sufficient to diminish significantly the overall diversity of species in the region and the sustainability of soils
- Air pollution or land transformations that induce significant climate change on a large scale
- Destruction of knowledge essential to understanding natural and social systems, such as residence decay times of nuclear wastes
- Destruction of cultural monuments that countries have acknowledged to be part of the common heritage of humankind
- Destruction of specific endowments established by the present generation for the benefit of future generations, such as libraries and gene banks

Unless substantial change occurs, the present generation is unlikely to pass on a healthy and diverse environment to future generations because of environmental problems, including loss of species, decline in arable land, tropical forests and water quality, and global warming.

But why care about the future? As cynics have said, "What has posterity ever done for me?" After all, the people of the far-off future are strangers; they are only potential people who do not yet exist and may not exist. They will be in no position to reward current people for what is done for them, to punish current society for a lack of care or responsibility, nor to demand compensation. It is not known what their needs, desires, or values will be.

Although future generations do not yet exist, it is reasonably certain they will exist and they will require clean air and water and other basic physical requirements for life. And although it is not known who the individuals of the future will be—they are not individually identifiable—they can have rights as a group or class of people, rather than individually, and we can have obligations and duties toward them. What is more, morality is not dependent on identity. Murder of an innocent person is morally wrong, whoever the victim is.

How can people who have not even been born yet demand rights? And if they cannot claim rights, do they have any? Future people may not be able to claim their rights today, but others can on their behalf, and various national and international laws protect the rights of future generations. Where future generations do not have formal legal representation, people are able to make claims on their behalf using reasoning based on moral principles, such as those outlined below.

Rationale

Desire for Self-Transcendence

It is part of being human to be able to relate to others and to care about the long-term well-being of the larger society, its values, institutions, and assets. It is this desire to be part of something that is larger than oneself and will endure beyond one's lifetime that motivates careers in public service, education, and scientific research as well as works of art and literature. Most people would be demoralized and saddened by the thought that Earth was to be destroyed in 200 years, even though they would be long dead.

The idea of contributing to and being part of an ongoing enterprise enables people to cope with the knowledge of their own mortality. It gives people a sense of purpose and identity. Ernest Partridge argues that it is only those who are alienated from the society around them, or who have some sort of narcissistic personality disorder, who do not have such feelings. These feelings enable people to transcend concerns about self, and people who do not have them are impoverished as a consequence.

Self-Interest

Morality can often be rationalized as being in one's own self-interest. It is far more pleasant and desirable to live in a moral community. Because humans can either make each others' lives miserable or help each other through cooperation, it makes sense to encourage mutual respect and moral obligations. A society in which citizens are concerned for the welfare of others is one in which individual welfare is best secured. In this view, there is an implicit social contract between members of a community that requires everyone to treat everyone else in a moral way. The question is: Who are members of this moral community? Does it go beyond the current generation to include all generations?

Philosopher John Rawls claims that most people would prefer a more egalitarian and just society if they were in the position of not knowing where in the society they were to be placed—at the top or the bottom, rich or poor. In a similar way, people would opt for intergenerational justice if put in a similar position of not knowing in which generation they are to be.

This "do unto others as you would have them do unto you" creed is exemplified by the scenario of the campsite. Most people will feel morally obliged to clean up a campsite they have been using so that it is at least in as good a condition for the next person as it was when they arrived. This is even though they do not know who the next campers will be or when they will come (time and identity are irrelevant). Part of the rationale behind honoring such an obligation is the knowledge that if everyone honors this obligation, then everyone benefits. The campers that are now leaving clean up the campsite in the hope that others will do so for them and with gratitude that others have done so before. When applied to generations, this creed is that each generation should leave sufficient natural resources and an unspoiled environment for the generations to follow.

Public Trust

The idea of a public trust or common heritage across generations means that environmental resources/values should not be destroyed merely because the majority of a current generation decides that it has better uses for them.

The idea that environmental resources are a common heritage of humankind was incorporated in the 1982 United Nations Treaty on the Law of the Sea. A similar doctrine is that of public trust, which is incorporated into U.S. environmental law. The doctrine of public trust has been reinforced by the courts. It affirms "a duty of the state to protect the people's common heritage of streams, lakes, etc., surrendering the right of protection only in rare cases when the abandonment of that right is consistent with the purposes of the trust. . . ."

Responsibility

Responsibility arises from power and ability to affect others and the knowledge that what we do may affect others. Increasingly, the activities of modern industrialized nations have impacts that are felt not only globally but well into the future. If people know that their actions may harm future generations, and they have a choice about whether to take those actions, then they are morally responsible for those actions. This is particularly pertinent to the environment because many environmental impacts such as radioactive waste disposal, global warming, and the spread of chemical toxins have long-term implications.

Because current generations can undermine the welfare of future generations, they have a measure of responsibility for that welfare. Inaction can also have consequences, and so inaction can be just as irresponsible as any action, particularly if it entails allowing existing trends to continue in the knowledge that these will be harmful. The fact that the consequences of our actions or inactions occur sometime in the future does not diminish our responsibility.

Because a healthy environment is a shared interest that benefits whole communities and is often threatened by the "cumulative effects of human enterprise," there is a collective responsibility to protect it. Individual actions can only offer limited solutions, and there is a need for government action and international cooperation.

Avoid Actions That Will Harm Future Generations

Some philosophers argue that the more distant future generations are from us, the less our obligations to them are because one cannot know what their needs and wants will be and what is good for them. Obligations to the current generation are termed *intra*generational. Others argue that even if it is not known what will be good for future generations, it is also not known what will be bad for them. Nevertheless, they are unlikely to want skin cancer, soil erosion, or frequent catastrophic weather events. Humans have fundamental needs that can be projected into the future, including healthy, uncontaminated ecosystems.

Therefore, we may not have positive obligations to provide for the future but negative obligations to avoid actions that will harm the future. We can fairly safely assume that future generations will want a safe and diverse environment. We cannot just assume that future generations will have better technological and scientific means to solve the problems we leave them. For this reason, we should endeavor to pass on the planet to future generations in no worse condition than past generations passed it on to us.

International Agreements

Intergenerational justice has been recognized in various international agreements, including the following:

- The Convention for the Protection of the World Cultural and Natural Heritage, 1972
- The United Nations Framework Convention on Climate Change, 1992
- The Convention on Biological Diversity, 1992
- The Rio Declaration on Environment and Development, 1992
- The Vienna Declaration and Programme of Action, 1993

These agreements led up to the UNESCO Declaration on the Responsibilities of the Present Generations Towards Future Generations, 1997. The text of the declaration was adapted from a Bill of Rights for Future Generations presented to the UN in 1993 by the Cousteau Society together with more than 9 million signatures of support from people in 106 countries.

Today, the principle of intergenerational justice is a principle of international law. A number of national laws and agreements also include intergenerational justice such as Australia's 1992 Intergovernmental Agreement on the Environment and the United States' 1969 National Environmental Protection Act (NEPA). Such sentiments go back as far as 1916 to the National Park Act in the United States, which charges the National Park Service with the duty of protecting the land "unimpaired for the enjoyment of future generations." In general, national parks in all countries have the same intergenerational goals.

What Should Be Sustained?

Even if it were agreed that we have an obligation to future generations, the nature of that obligation would be controversial. Do we merely need to protect those aspects of the environment necessary for survival and health such as a minimal standard of clean air and water? And what would that standard be? Which risks from hazardous and radioactive substances do we need to prevent?

The problem is that protecting the interests of the future may conflict with the interests of current generations. How do we balance our obligations to current generations with our obligations to future generations when these conflict? At one extreme is the "preservationist model," which requires that present generations not deplete any resources or destroy or alter any part of the environment. In this case, an industrialized lifestyle would not be possible and the present generations would make significant sacrifices living subsistence lifestyles in order to benefit future generations.

At the other extreme is the "opulence model," where present generations consume all they want and assume that future generations will be able to cope with the impoverished environment that remains because they will be technologically better off. Or, alternatively, advocates of this model assume that future generations will have the technological expertise to find new sources or substitutes for exhausted resources and extinct species. However, this model seems to be overly optimistic about the ability for wealth and technology to deal with environmental catastrophe and losses.

Many economists and businesspeople tend to argue that what is important is to maintain welfare and utility over time and that a community can use up natural resources and degrade the natural environment so long as they compensate for the loss with "human capital" (e.g., skills, knowledge, and technology) and "human-made capital" (e.g., buildings, machinery).

They point out that a depleted resource, for instance, oil, could be compensated for by other investments that generate the same income. If the money obtained from exploiting an exhaustible resource, such as oil, were invested so that it yields a continuous flow of income, this would be equivalent to holding the stock of oil constant. They therefore argue that not only is some substitution inevitable when it comes to the commercial exploitation of minerals but that it is consistent with intergenerational justice if the profits from the investment are reinvested so as to provide an ongoing equivalent income. This means that the Amazon forest could be removed so long as the proceeds from removing it were reinvested properly.

Such arguments provide a rationale for continuing to use nonrenewable resources at ever-increasing rates. Economists argue that although this might cause temporary shortages, those shortages will cause prices to rise, and this will provide the motivation to find new reserves, discover substitutes, and encourage more efficient use of remaining resources.

However, while the economic value of natural resources can be easily replaced, their functions are less easily replaced. Most people, even economists, agree that there are limits to the extent to which natural resources can be replaced without changing some biological processes and putting ecological sustainability at risk. They recognize that some environmental assets could not be "traded off" because they are essential for life-support systems and as yet they cannot be replaced.

In fact, the precautionary principle would prevent us from assuming that natural resources can be replaced without good evidence that they can. There are many types of environmental assets for which there are no substitutes: for example, the ozone layer, the climate-regulating functions of ocean phytoplankton, the watershed protection functions of tropical forests, the pollution-cleaning and nutrient-trap functions of wetlands. For those people who believe that animals and plants have an intrinsic value, there can be no substitute for them.

We cannot be certain whether we will be able to substitute for other environmental assets in the future and what the consequences of continually degrading nature will be. Scientists do not know enough about the functions of natural ecosystems and the possible consequences of depleting and degrading the environment. Therefore, it is not wise to assume that all will be well in the end because of some faith in economics and technological ingenuity. Many people do not agree that a loss of environmental quality cannot be substituted with a gain in human or human-made capital without loss of welfare. Therefore, they argue that future generations should not inherit a degraded environment, no matter how many extra sources of wealth are available to them. Environmental degradation can lead to irreversible losses such as the loss of species and habitats, which, once lost, cannot be re-created. Other losses are not irreversible but repair may take centuries—for example, the ozone layer and soil degradation.

Moreover, is it fair to replace natural resources and environmental assets—that are currently freely available to everyone—with human-made resources that have to be bought and in the future may only be accessible to people who can afford them? Poor people are often affected by unhealthy environments more than wealthier people. A substitution of wealth for natural resources does not mean that those who suffer are the same people as those who will benefit from the additional wealth.

Most environmentalists argue that the environment should not be degraded for future generations even if they are compensated with greater human-made capital. They claim that human welfare can only be maintained over generations if the environment is not degraded. They point out that we do not know what the safe limits of environmental degradation are; yet if those safe limits are crossed, the options for future generations would be severely limited. Retaining environmental quality for future generations means passing on the environment in as good a condition as we found it. It does not preclude some trade-offs and compromises, but it requires that those trade-offs do not endanger the overall quality of the environment so that environmental functions are reduced and ecosystems are unable to recover. When resources are depleted and species extinct, the options available to future generations are narrowed. According to Weiss, "each generation should be required to conserve the diversity of the natural and cultural resource base, so that it does not unduly restrict the options available to future generations in solving their problems and satisfying their own values." Overdevelopment reduces options and reduces diversity. Weiss points out that the principle of "conservation of access" implies that not only should current generations ensure equitable access to that which they have inherited from previous generations, but they should also ensure that future generations can also enjoy this access.

A minimal environment may be all that is needed for human survival, but people have come to expect a lot more than a subsistence lifestyle. Should that be denied to future generations? Justice would seem to require that future generations not only be able to subsist but that they have the same level of opportunities to thrive and be comfortable as current generations. Opportunities require more than mere survival-level environmental resources.

See Also: Environmental Justice (Ethics and Philosophy); Environmental Justice (Politics); Precautionary Principle (Ethics and Philosophy); Sustainable Development (Politics).

Further Readings

Beder, Sharon. *Environmental Principles and Policies*. Sydney: UNSW Press, 2006.

Hooft, Hendrik Ph. Visser 't. *Justice to Future Generations and the Environment*. Dordrecht, Netherlands: Kluwer Academic, 1999.

Partridge, Ernest. "The Gadfly Papers." http://gadfly.igc.org/papers/papers.htm (Accessed September 2010).

Shrader-Frechette, Kristin. *Environmental Justice: Creating Equality, Reclaiming Democracy*. Oxford, UK: Oxford University Press, 2002.

United Nations Educational, Scientific and Cultural Organization (UNESCO). "Declaration on the Responsibilities of the Present Generations Towards Future Generations." http://portal.unesco.org/en/ev.php-URL_ID=13178&URL_DO=DO_TOPIC&URL_SECTION=201.html (Accessed September 2010).

Weiss, Edith Brown. "Intergenerational Fairness and the Rights of Future Generations." *Intergenerational Justice Review*, 1:6 (2002).

Sharon Beder
University of Wollongong

INTERNATIONAL WHALING COMMISSION (IWC)

The International Whaling Commission (IWC) emerged from the 1946 International Convention for the Regulation of Whaling (ICRW). The convention's purpose is twofold: to provide for the appropriate conservation of whale stocks and to nurture the methodical development of the whaling industry. The IWC was created to review and revise as necessary the ICRW's "Schedule to the Convention," which lays out the guidelines for governing, on a global scale, the whaling industry. Some of the measures undertaken by the IWC include the following:

- Complete protection of certain species
- Designation of specific areas as whale sanctuaries
- Limitations on the numbers and size of whales that can be caught
- Prescriptions for open and closed seasons and areas for whaling
- Prohibition of the capture of suckling calves and female whales accompanied by calves
- Compilation of catch reports and other statistical and biological records

A zero-catch policy on whaling began in 1986 and has been debated ever since. Some countries want the ban lifted; others wish to ban commercial whaling while allowing a limited subsistence catch by indigenous peoples.

Source: National Oceanic and Atmospheric Administration

Membership in the IWC is open to any country that formally adheres to the 1946 convention; as of August 2010, there were 88 IWC members. Each member country is represented by a commissioner. The chair and vice chair of the IWC are elected from among the commissioners and usually serve for three years; the IWC also has a full-time, 17-member secretariat, based in Cambridge, England. Annual IWC meetings are held by invitation either in any member country or in the United Kingdom, where the secretariat is based. The IWC has several committees, including the Scientific Committee, which is composed of around 200 international whale biologists. The Scientific Committee provides an important function to the IWC, as any amendments made to the Schedule to the Convention are required to be based upon scientific findings. The broad goals of the Scientific Committee include the following:

- Encourage, recommend, and, if necessary, organize studies and investigations relating to whales and whaling
- Gather and analyze statistical data regarding the current condition and trend of the whale stocks and the effects of whaling activities on them
- Investigate, assess, and circulate information concerning methods of maintaining and increasing the populations of the whale stocks

The IWC relies on the information and advice of the Scientific Committee to review and revise the Schedule and to develop regulations for the control of whaling. In order for a regulation to be passed, a three-quarters majority vote by the commissioners is required. New regulations go into effect 90 days after a successful vote; any member country can lodge an objection to new regulations, which means that the regulation is not binding on that country. The regulations are implemented through the national legislation of the member states, which appoint inspectors to oversee their whaling operations and may also receive international, IWC-appointed observers.

By all accounts, the early years of the IWC were a complete failure. Quotas were set too high, there was rampant cheating, and key whale stocks continued to decline. This failure set the stage for the anti-whaling movement and eventually the moratorium on commercial whaling.

Conflict in the IWC

A global anti-whaling movement began in the 1970s, fueled by concerns over what was seen as the barbarity of the industry and of depleting whale stocks. The 1972 United Nations Conference on the Human Environment adopted a proposal recommending a 10-year moratorium on commercial whaling to allow whale stocks to recover, and the Convention on International Trade in Endangered Species issued reports in 1977 and 1981 identifying many species of whales facing extinction. Accompanying this growing concern with the whaling industry, the 1970s and early 1980s saw many nonwhaling and anti-whaling nations join the IWC; eventually, these countries gained a majority in the IWC, and in 1982, the IWC gained the necessary three-quarters majority vote to implement a moratorium on commercial whaling, declaring a zero-catch policy to begin in 1986. Arguing that the moratorium was not based on advice from the Scientific Committee, formal objections were lodged by Japan, Norway, Peru, and the Soviet Union, although Peru and Japan, under pressure from the United States, which threatened to reduce Japan's fishing quotas within U.S. waters, withdrew their objections.

The moratorium created great tension within the IWC; Norway lodged an objection to the zero-catch policy in 1992 and has been whaling commercially since 1994; Iceland withdrew from the IWC in 1992, rejoined in 2002, and began commercial whaling again in 2006. As the ICRW allows for scientific-research whaling and aboriginal-subsistence whaling, neither of which was affected by the moratorium, the United States and several other nations have continued aboriginal-subsistence whaling, while Japan has continued whaling under the auspices of scientific research. Japan is critical of U.S. aboriginal-subsistence whaling because several Japanese fishing communities that traditionally hunted whales before the moratorium were not allowed to continue doing so once the moratorium passed, in part due to U.S. objections. Environmental groups and anti-whaling countries have challenged Japan's continued scientific-research whaling, claiming that it is a front for commercial whaling, while the Japanese government argues that the refusal of anti-whaling nations to accept simple counts of whale populations as a measure of recovery of whale species justifies its ongoing studies on sex- and age-of-population distributions.

Japan, Norway, and Iceland continue to challenge the moratorium, and the 2010 IWC annual meeting was marked by tension as the IWC debated whether or not to lift the commercial whaling ban. The United States and several other anti-whaling countries proposed a compromise that would allow Japan, Norway, and Iceland to continue whaling but at substantially lower levels and under close scrutiny. More than 33,000 whales have been killed since the ban took effect in 1986, which critics say severely undermines the IWC. Under the compromise, trade in whales and whale products would be abolished; it is estimated that the compromise would result in 3,200 fewer whale deaths over the next 10 years. Japan, Norway, and Iceland contend that their activities do not threaten the survival of whale stocks and that the commercial moratorium should be lifted altogether, while the compromise's opponents wish to impose a ban on all commercial whaling but still allow for a limited catch by indigenous peoples for subsistence. Because sentiment over the moratorium and the compromise is so divided, reaching the necessary three-quarters majority vote for any side was not accomplished; after two days of negotiations, the IWC failed to reach consensus and instead postponed voting on the issue.

One of the controversial points in the compromise was its mandated creation of a whale sanctuary in the South Atlantic. There has been a prohibition against whaling in the South Atlantic since 1994, though it has not been honored. About 80 percent of the world's whales go to the South Atlantic each summer to feed, and creating a sanctuary there is necessary to protect the whale stocks, according to environmentalists and anti- and nonwhaling countries. Japanese authorities, however, insist that they have evidence that shows that the whale stock is sustainable in the Southern Ocean without a zero-catch policy there, something the European Union (EU), Australia, and the Latin American bloc are advocating. Japan defends its whaling in the Southern Ocean as being needed for scientific research and by insisting that the whale stocks there can be fished sustainably; opponents argue that whether the whaling is sustainable is not the issue because the area should be declared a sanctuary. Australia is suing Japan in the International Court of Justice to stop its Southern Ocean whaling.

Making an already tense situation even more ugly, Japan has been accused of offering bribes and prostitutes to IWC Commissioners from certain countries in exchange for their support in lifting the ban. London's *Sunday Times* filmed IWC Commissioners from certain countries admitting to the following:

- Voting with the whalers because of aid received from Japan
- Receiving cash payments for their travel and hotel bills at IWC meetings from Japanese officials
- Being offered the services of prostitutes when fisheries ministers and civil servants visited Japan for meetings

The countries that admitted taking bribes from Japan include St. Kitts and Nevis, the Marshall Islands, Kiribati, Grenada, Republic of Guinea, and the Ivory Coast.

While these bribes certainly do not paint Japan or the officials who took the bribes in a good light, the IWC is not without its critics, including Japan and others, who contend that the IWC is dysfunctional and that they display the following flaws:

- Disregard international law
- Ignore science-based policy and rule making
- Exclude whale-related issues from conversations regarding sustainable resource use
- Disrespect cultural diversity related to food and ethics
- Increase emotionalism concerning whales
- Engage in combative discourse that discourages cooperation
- Lack good-faith negotiations
- Pressure scientists, resulting in a lack of consensus on scientific advice from the Scientific Committee

At this time, the future of commercial whaling in general, and particularly in the Southern Ocean, remains uncertain. Conservationists as well as Australia and Latin American nations remain committed to keeping the moratorium in place, while the United States, New Zealand, and the EU, except for Denmark (which opposes whaling), support the compromise that would allow limited whaling. Japan, Norway, and Iceland continue to advocate lifting the ban completely. Further criticisms of the IWC have raised legitimate questions regarding its efficacy and policies.

See Also: Animal Ethics; Animal Welfare; Fisheries; Sustainable Development (Politics).

Further Readings

Friedheim, Robert L., ed. *Toward a Sustainable Whaling Regime.* Seattle: University of Washington Press, 2001.

International Whaling Commission. "Conference for the Normalization of the International Whaling Commission." http://iwcoffice.org/_documents/commission/future/IWC-M08-INFO2.pdf (Accessed September 2010).

International Whaling Commission. "IWC Information." http://iwcoffice.org/commission/iwcmain.htm (Accessed September 2010).

"Revealed: Japan's Bribes on Whaling." *Sunday Times* (June 13, 2010). http://www.timesonline.co.uk/tol/news/environment/article7149091.ece (Accessed December 2010).

Tani E. Bellestri
Independent Scholar

INTRINSIC VALUE VERSUS USE VALUE

Some environmental philosophies rely on the use of intrinsic value or the value of a natural organism or object for its own sake. Intrinsic value is often countered by an organism's or object's use value, defined as its usefulness to humans without regard for its intrinsic value. Key ethical debates in environmental philosophies have included the issue of whether intrinsic values extend beyond humans to the natural world and whether that extension should include only sentient beings. These determinations have important implications for the translation of environmental philosophies into actions.

An object's intrinsic value is the value of that object for its own sake, outside any economic valuation or use valuation. Intrinsic value can also mean the value of an object or concept that cannot be commoditized but is valued nonetheless, such as love, peace, justice, or the truth. An object's use value, sometimes termed its instrumental value, is its value based on its usefulness to humans as a means to an end rather than an end unto itself. This could be a monetary, cultural, or aesthetic value. An object may have both intrinsic value and use value. Determinations of intrinsic and use value are key concepts within environmental philosophy.

Environmental philosophies such as utilitarianism posit that it is a human duty to choose rules and courses of action based on promoting the greater good (pleasure) and preventing negative consequences (pain). The determination of intrinsic value and use value, a controversial

Environmental philosophers are undecided about at which point intrinsic value ends. Does it extend to only rational beings, all humans, all sentient life, or even inanimate natural objects?

Source: iStockphoto

practice that has resulted in much debate, is a key component of such moral decision making, as a moral agent must respect intrinsic value. Environmental philosophers agree on the existence of intrinsic value, but disagree over how it is to be determined and whether it should extend to only rational beings, all humans, all sentient life, or include inanimate natural objects. Debates also include whether all intrinsic values are equal or whether some are more valuable or higher than others.

Environmental philosophies debate whether intrinsic or use value is the higher or most important type of value. They also debate whether intrinsic value can be measured and how it should be measured, if possible. Some economists have attempted to determine the use value of objects that do not normally have one and assign them market prices through a method known as shadow pricing, for example, seeking to determine the price the public might pay to lessen or eliminate the environmental damage caused by pollution, as pollution is not a market commodity. Others argue that such shadow pricing reduces everything to a use value without consideration of intrinsic value.

Environmental philosophers also debate judgments on the nature of intrinsic value itself. Questions include whether it is an objective or subjective determination, whether it depends on internal factors only or can be determined based on a person or object's relations to other people or objects, and whether it applies to a particular person or object or to a state of affairs that a particular person or object is in. Another problem raised is the ability of humans to control natural desires in order to desire only that which is intrinsically good. Yet another problem is how to act in situations where actions may conflict, aiding the intrinsic value of one person or object while harming that of another.

The question of intrinsic value is also extended to other animate and inanimate components of the natural world. Do these objects have intrinsic value, or is intrinsic value restricted to rational beings, as the philosopher Immanuel Kant believed? Environmental philosopher Aldo Leopold and others have noted that historically, some people have not always extended intrinsic value even to other groups of humans, such as children, other races, or indigenous peoples. For example, African slaves were once considered property whose only value was their market price. Even those environmental philosophers who agree that animals or inanimate natural objects have intrinsic value do not always agree on whether intrinsic human values are of equal or greater importance or whether an object's intrinsic value is more important than its human use value.

Some environmental philosophers have sought to create environmental ethics that were in opposition to anthropocentrist philosophies such as utilitarianism, opposing their basis in the natural world's use value to humans as the dominant species. Anthropocentrism posits that humans have a greater intrinsic value and that natural resources are most valued according to their human usefulness. This philosophy can give rise to the opinion that humans have the right to utilize natural resources to best benefit their human use value. Critics of anthropocentrism note that these beliefs have resulted in the exploitation and degradation of the natural world and the mistreatment and extinction of animals as human use values often conflict with the natural resources' intrinsic values and rights. Such philosophies are therefore not sustainable in the long run, resulting in a future environmental crisis when the survival of the human species will hang in the balance.

Proponents of the environmental philosophies of biocentrism, ecocentrism, and ecofeminism offer alternatives based on the extension of intrinsic value regardless of any potential human use. These philosophies are centered on the belief that humans are not of superior intrinsic value and must therefore consider the intrinsic value of other components of the natural world when making environmental decisions. For biocentrism founder Paul Taylor and deep ecology founder Arne Naess, intrinsic value extended to other living objects, although to a lesser extent. Ecocentrism extended intrinsic values even further to encompass inanimate natural objects and stated that the intrinsic values of other objects were equal to the intrinsic values of humans, even if an object such as a plant does not have sentience to care about its interests.

Some people may thus argue that a natural object or life form should be conserved or protected based on its own intrinsic value regardless of whether it also has a use value, while others may argue that the same natural object or life form should be conserved or protected because of its use value to humans regardless of whether they agree that it has its own intrinsic value.

Environmental philosophers also debate the relationship between intrinsic values and intrinsic rights as well as human obligations to respect those rights. Although all human intrinsic rights are now recognized by most societies whether or not those humans are rational actors, not all philosophers believe those rights should be extended to animals or inanimate components of the natural world. Paul Taylor believed that natural objects such as plants had intrinsic rights based on their teleological life functions such as growth and reproduction, even if they were not consciously aware of such driving purposes.

Few environmental philosophers would argue, however, that a plant's intrinsic value confers upon it universal organic rights that must be respected. Instead, some argue that the interests of such organisms should merely be considered when debating actions. Environmental ethicist Holmes Rolston III believes that species as a whole as well as individual organisms within those species have teleological functions of survival that must be considered, extending questions of intrinsic and use value to whole ecosystems. Some environmental philosophers have noted that this accounts for the belief that while the individual intrinsic values of components of an ecosystem may be smaller or of lesser consequence, their collective total intrinsic value can be greater than the sum of these individual values.

As Aldo Leopold noted, determining the right action for an individual organism as well as for other natural components such as soil and water corresponds to the ecosystem as a whole, as threats to one organism or species can have negative impacts that threaten the whole. Leopold also allowed for the fact that sometimes the interests of an individual organism based on its individual intrinsic value can and should be subordinate to the collective interests of the whole ecosystem based on its collective intrinsic value. The duty of the moral agent would then be to evaluate all possible actions and consequences and choose the one that is best for the overall good or intrinsic value.

Another example of philosophical debates that have arisen over the issue of intrinsic values and rights as they conflict with human use values is the debate between the philosophies of animal liberation, as developed by Peter Singer, and animal rights, as developed by Tom Regan. Both philosophies recognize that animals as sentient creatures have an intrinsic value equal to that of humans and intrinsic rights that must be respected. Although Singer advocated vegetarianism, animal liberation could theoretically be extended to state that humans may morally consume animals as long as they are raised and slaughtered in a humane fashion that does not cause pain. Animal liberation, on the other hand, states that animal rights that are based on their intrinsic value preclude such human uses as food and animal research even when humane methods are utilized.

Critics argue that human noninterference in the teleological functions of organisms and species and determinations of intrinsic value fail to account for the fact that such interference is an integral part of the natural world. For example, animals rely on plants or other animals as food sources without regard for their intrinsic value. Should humans as reasoning organisms behave differently than their counterparts in the natural world? They argue that the focus of environmental philosophy and the activism it underpins should be on major environmental impacts such as species extinction, climate change, desertification, deforestation, and global pollution levels.

Proponents of these new environmental philosophies feel that older systems of thought based on anthropocentrism and the dominance of human use value must be replaced by new systems of thought based on ecocentrism and the extension of intrinsic value in order to achieve true change in the human relationship to the environment needed to achieve sustainability. In other words, revolutionary changes in action require revolutionary changes in thought, and true conservation requires overthrowing the older attitude of human dominion over nature based in part on the belief of the higher intrinsic value of humans.

See Also: Anthropocentrism Versus Biocentrism; Utilitarianism; Utilitarianism Versus Anthropocentrism.

Further Readings

Attfield, Robin and Andrew Belsey. *Philosophy and the Natural Environment*. New York: Cambridge University Press, 1994.

Baron, Jonathan. *Against Bioethics*. Cambridge, MA: MIT Press, 2006.

Beckerman, Wilfred and Joanna Pasek. *Justice, Posterity, and the Environment*. New York: Oxford University Press, 2001.

Beckmann, Suzanne C., William E. Kilbourne, et al. *Anthropocentrism, Value Systems, and Environmental Attitudes: A Multi-National Comparison*. (1997). http://e-archivo.uc3m.es/bitstream/10016/8451/1/anthropocentrism_pardo_1997.pdf (Accessed January 2011).

Branch, Michael P. and Scott Slovic. *The ISLE Reader: Ecocriticism, 1993–2003*. Athens: University of Georgia Press, 2003.

Callicott, J. Baird. *Beyond the Land Ethic: More Essays in Environmental Philosophy*. Albany: State University of New York Press, 1999.

Callicott, J. Baird. *In Defense of the Land Ethic: Essays in Environmental Philosophy*. Albany: State University of New York Press, 1989.

Dasgupta, Partha. *Human Well-Being and the Natural Environment*. New York: Oxford University Press, 2001.

Dobson, Andrew. *Green Political Thought*. London: Routledge, 2007.

Elliot, Robert. *Environmental Ethics*. New York: Oxford University Press, 1995.

Faber, Daniel. *The Struggle for Ecological Democracy: Environmental Justice Movements in the United States*. New York: Guilford Press, 1998.

Gillroy, John Martin and Joe Bowersox. *The Moral Austerity of Environmental Decision Making: Sustainability, Democracy, and Normative Argument in Policy and Law*. Durham, NC: Duke University Press, 2002.

Goldfarb, Theodore D. *Taking Sides: Clashing Views on Controversial Environmental Issues*. New York: McGraw-Hill/Duskin, 2001.

Gruen, Lori and Dale Jamieson. *Reflecting on Nature: Readings in Environmental Philosophy*. New York: Oxford University Press, 1994.

Jamieson, Dale. *Ethics and the Environment: An Introduction*. New York: Cambridge University Press, 2008.

Johnson, Lawrence E. *A Morally Deep World: An Essay on Moral Significance and Environmental Ethics*. New York: Cambridge University Press, 1991.

Lindahl Elliot, Nils. *Mediating Nature*. London: Routledge, 2006.

O'Neill, John, Alan Holland, and Andrew Light. *Environmental Values*. London: Routledge, 2008.

Postma, Dirk Willem. *Why Care for Nature? In Search of an Ethical Framework for Environmental Responsibility and Education*. Dordrecht, Netherlands: Springer, 2006.

Sarkar, Sahotra. *Biodiversity and Environmental Philosophy: An Introduction*. New York: Cambridge University Press, 2005.

Scriven, Tal. *Wrongness, Wisdom, and Wilderness: Toward a Libertarian Theory of Ethics and the Environment*. Albany: State University of New York Press, 1997.

Squatriti, Paolo. *Nature's Past: The Environment and Human History*. Ann Arbor: University of Michigan Press, 2007.

Stevis, Dimitris and Valerie J. Assetto. *The International Political Economy of the Environment: Critical Perspectives*. Boulder, CO: Lynne Rienner, 2001.

Traer, Robert. *Doing Environmental Ethics*. Boulder, CO: Westview, 2009.

Warren, Karen and Nisvan Erkal. *Ecofeminism: Women, Culture, Nature*. Bloomington: Indiana University Press, 1997.

Wenz, Peter S. *Environmental Ethics Today*. New York: Oxford University Press, 2001.

Marcella Bush Trevino
Barry University

IRRADIATION

Many of the conflicts over American food issues come down to questions of safety, but even when there is agreement over the existence of a danger, there can be conflict over its solution. Concerns about potential *Salmonella* contamination, for instance, are remedied in part by requiring that chicken be cooked to a temperature sufficient to kill contaminants (against the tendency to serve chicken still pink in many of the cultures of American ancestors) and warnings that accompany the serving of uncooked or undercooked eggs. Many argue that contamination should be prevented by reforms to the farming practices that increase its probability. Food irradiation is another area where much of the opposition to a practice designed to make food safer argues that it diverts attention from the conditions that made the food "dangerous" in the first place.

Food is exposed to ionizing radiation in order to eliminate insects, bacteria, viruses, and any other microorganisms that may be present. Ionizing radiation refers to radiation at the short-wavelength end of the electromagnetic spectrum, such as x-rays and high-frequency ultraviolet light, which consists of particles or waves energetic enough to detach electrons from atoms and ionize them. Ionizing radiation does not induce radioactivity; according to industry reports, irradiated food does not become radioactive, though some critics are not convinced.

Various radiations are used for food irradiation. The use of x-rays to preserve food was patented in 1918 and remains in use by some producers, in part because it is easy to obtain a uniform dose when using x-rays. They are less efficient, however, than more recent developments using gamma radiation—the radiation of photons in the gamma section of the electromagnetic spectrum, using radioisotopes like cobalt-60. Cobalt-60 irradiation is the most common method today, with the radioactive material carefully monitored and workers operating from behind thick concrete shields to protect them from gamma rays. Some setups keep the cobalt-60 underwater and lower the food being irradiated into the water using hermetic bells, which reduces the need for shielding.

Forty different countries use irradiation to increase the safety of their food, and international organizations like the United Nations Food and Agriculture Organization (FAO) have called for expanding use of irradiation. Shelf life is increased for some foods as well, as microorganisms causing spoilage are eliminated. Irradiation can also cause fruits to ripen more slowly and vegetables to delay their sprouting, which increases the time available to bring them to market.

The U.S. Department of Agriculture (USDA) has approved the use of irradiation at low levels as an alternative to pesticides because the exposure renders insects sterile, preventing the spread of pests like fruit flies and weevils. Furthermore, the importing of exotic fruit into the United States has been greatly expanded by using irradiation to simplify the quarantine procedures necessary for bringing plants into the country; this has allowed many fresh fruits to be imported for the first time. Previously, they would have perished before the quarantine period ended. For example, in recent years, fresh mangosteens from Thailand and Vietnam have been imported into the United States for the first time, thanks to the ability of suppliers to assure the USDA and U.S. Customs of the fruit's safety by irradiating it. The use of irradiation on guavas, mangos, star fruit (carambolas), and exotic citrus fruit has shortened quarantine times since 2008, enabling the fruits to be brought to market faster and in better condition than before. More fruits are expected to follow, as part of what could be an important new trade route for fruit farmers in southeast Asia and South America, providing fruits that Americans have previously sampled fresh only when abroad because they require growing conditions not found on American soil. Because of the large capital investment—at least $1 million—required to begin irradiating food, for now the producers doing so tend to be those selling products that can command a premium price or that, like the mangosteen, have no American competitors.

Various doses of irradiation are used, depending on the purpose of the irradiation. Because of concerns about radiation and possible undiscovered effects, some countries put upper limits on allowed doses or permit food irradiation only of specific food categories.

India, Mexico, New Zealand, and Australia, for instance, permit irradiation only of fresh fruit for fruit fly quarantine purposes. Other countries permit irradiation only of dried herbs, spices, and seasonings.

Low doses can be used to inhibit sprout formation in bulbs and tubers, requiring up to 0.15 kilograys of radiation. Delaying the ripening of fruit takes slightly more, about half a kilogray, while sterilizing insects and eliminating any foodborne parasites takes up to 1 kilogray.

At slightly higher doses, 2–7 kilograys, microbes that cause spoilage can be reduced in meat, poultry, and seafood. Spices can be cleansed of microorganisms at 10 kilograys. Although the Codex Alimentarius Standard on Irradiated Food does not put an upper dose limit on irradiation, most countries do not permit irradiation higher than these levels except in specific cases. Packaged meat can be fully sterilized at 25–70 kilograys, and the National Aeronautics and Space Administration (NASA)—which has special food and safety needs—sterilizes the frozen meat served to astronauts with doses of 44 kilograys of radiation. But these uses are unusual.

Regulation

Food irradiation in the United States is regulated by the Food and Drug Administration (FDA), which classifies the process as an additive. The Nuclear Regulatory Commission oversees issues pertaining to the safety of the processing facilities where irradiation is performed and the handling of materials like cobalt-60; the transportation of radioactive materials is regulated by the Department of Transportation. The USDA is also involved with fresh foods—fresh fruit and vegetables, meat and poultry products—and has been the government's agent in the bilateral agreements with foreign nations involving the importation of irradiated fruits to the United States. The most common irradiated food products sold in the United States are tropical fruits, spinach and other greens, spices, and ground meats. Irradiated food products must be labeled with the words "Treated with irradiation" or "Treated by radiation" and are labeled with a Radura symbol. Because of the public perception of irradiation as a health concern, stores that cater to organic and green customers tend not to stock irradiated products. Adoption of irradiation has been slow because of public perception, though it has diminished in recent years, perhaps in part because of the seemingly increasing number of public health scares associated with foodborne pathogens. In the 21st century, even as opposition to food irradiation has decreased, some of the push-back has originated with the local foods movement, which advocates moving away from the present model wherein food may travel thousands of miles from farm to plate when it could as easily have been grown mere blocks away. Food irradiation is one of the technological advances that make feasible and attractive the shipping of food from continent to continent and coast to coast, and its adoption seems to accelerate that trend.

The FDA is considering allowing the use of food irradiation without labeling so long as there is no material change in the food or the consequences of eating the food. The terms *electronic pasteurization* and *cold pasteurization* are also being considered as options for producers to use in their labeling in lieu of the word *radiation*. This has drawn considerable criticism. While the process of irradiation may be analogous to pasteurization in terms of its goals and positive effects, the processes themselves have nothing else in common, and there is little apparent reason to cloud the matter with the word *pasteurization* except for the fact that it is a word most consumers have positive associations with and that they understand to denote the product has undergone a process meant to make it safer.

Irradiated food, genetically modified organisms (GMOs), and the use of nanotechnology in food all involve a similar argument: on the one hand, ill effects have not yet been proved. On the other hand, the technologies are new enough that sufficient in-depth, long-term studies have not been conducted. There are valid points on both sides of this debate. There is a long history of harmful side effects of various substances and processes being discovered long after exposure. On the other hand, food irradiation is not being employed simply for aesthetic or cost-saving means: if there prove to be no harmful side effects, the gains are strong for both human health and the environment, as it helps to stop the spread of microbes, viruses, and insect pests that could be harmful both to humans and to the ecosystem. In an age increasingly alarmed by the possibility of pandemic or catastrophic consequences of the introduction of foreign pests, the benefits of food irradiation are compelling.

However, there are concerns about the indirect effects of food irradiation even if the process itself should prove safe. It may kill good bacteria as well as bad, a subject of increasing concern in recent decades as nutritionists and physicians emphasize the importance of good bacteria to human health and many worry that overreliance on antibiotics and other antibacterial regimes have gone too far. It may make some food products not dangerous but less nutritious, an argument some have offered in the case of pasteurized dairy products as well. It may have detrimental effects on the flavor or other aesthetic qualities of food, just as other sterilization techniques can. Perhaps, just as the widespread use of antibiotics has led to stronger antibiotic-resistant bacteria, the use of irradiation could lead to radiation-resistant strains of pathogens and microbial life. The common thread in this line of thinking is: the food may be safe, but that does not present a complete picture of the effects of this practice.

There are social effects to consider as well. If everyone boiled their water before drinking it, would water providers bother making sure of its safety? The adoption of food irradiation as a widespread practice almost invokes a sort of moral hazard: If the presence of germs on food will eventually be dealt with radioactively before the food is brought to market, is there less incentive on the part of producers to ensure safe practices and avoid those germs?

See Also: Antibiotic/Antibiotic Resistance; Ecolabeling; Genetically Modified Organisms (GMOs); Organic Farming; Pesticides.

Further Readings

Gibbs, Gary. *The Food That Would Last Forever: Understanding the Dangers of Food Irradiation.* Garden City Park, NY: Avery, 1993.

Urbain, W. M. *Food Irradiation.* Maryland Heights, MO: Academic Press, 1986.

U.S. Environmental Protection Agency. "Food Irradiation." http://www.epa.gov/rpdweb00/sources/food_irrad.html (Accessed January 2011).

Bill Kte'pi
Independent Scholar

Kyoto Protocol

The Kyoto Protocol, the international agreement connected to the United Nations Framework Convention on Climate Change (UNFCCC), sets binding targets for 37 industrialized countries and the European Union for reducing greenhouse gas (GHG) emissions. Highly controversial among the media in the United States, Europe, Asia, and other areas of the globe, the Kyoto Protocol is an environmental treaty that seeks to stabilize GHG emissions to a level that would reduce the risk of anthropogenic interference with Earth's climate system. Specifically, the Kyoto Protocol encourages Western developed nations to reduce their GHG emissions by slightly more than 5 percent from their 1990 levels. Creating a reporting process by which each nation monitors and manages its GHG emissions, the Kyoto Protocol contains penalties, considered weak by many, for those nations that are unsuccessful in reducing GHG emissions. The United States' decision to sign but not ratify the Kyoto Protocol was highly contentious, causing criticism of U.S. environmental policies and practices abroad while remaining unpopular domestically.

Background

The Kyoto Protocol was initially adopted on December 11, 1997, at a conference held under the auspices of the United Nations in Kyoto, Japan. The Kyoto Protocol came into full effect on February 16, 2005. As of September 2010, 187 states have signed and ratified it. Although the United States signed the Kyoto Protocol, it has never ratified it, and the administration of President George W. Bush affirmatively stated it was not seeking ratification by the U.S. Senate. As with most international treaties, while signing is a separate step from ratification, doing so usually indicates an intention to ratify the agreement. The United States' failure to ratify the Kyoto Protocol has subjected it to intense criticism. Ratification means that a developed country or one designated as having an "economy in transition" has agreed to cap emissions in accordance with the Kyoto Protocol. The United States based its opposition to the Kyoto Protocol in part on the exception granted China, the world's largest producer of carbon dioxide. Others believe the United States' failure to ratify the Kyoto Protocol was caused by unwillingness to dramatically reduce GHG emissions. Such reductions would have severely affected key sectors of the economy, including energy, transportation, manufacturing, and agriculture.

Greenhouse Gases

GHGs represent emissions into the atmosphere that absorb and emit radiation within the thermal infrared range. Primary GHGs include carbon dioxide, methane, nitrous oxide, ozone, and water vapor. The presence of GHG greatly affects the temperature on Earth. It has been estimated that without GHG, the average temperature of Earth would be approximately 59 degrees Fahrenheit (33 degrees Celsius), cooler than currently is the case. The burning of fossil fuels, common since the beginning of the Industrial Revolution, has substantially increased the presence of carbon dioxide and other GHGs in Earth's atmosphere.

The drafters of the Kyoto Protocol concentrated on individual nations' GHG emissions as a way to formulate a strategy to reduce the problem. Controversy arose, however, as those involved with the Kyoto Protocol focused on GHG emissions per capita rather than totals by nation. These different ways of looking at emissions result in different analyses of the problem as the rankings are very different. In 2005, for example, the following nations were the 10 largest producers of GHGs, with the following percentages representing each nation's share of global emissions:

- China: 17 percent
- United States: 16 percent
- European Union: 11 percent
- Indonesia: 6 percent
- India: 5 percent
- Russia: 5 percent
- Brazil: 4 percent
- Japan: 3 percent
- Canada: 2 percent
- Mexico: 2 percent

In terms of tons of GHGs produced per capita, however, the same 10 nations would rank as follows:

- United States: 24.1
- Canada: 23.2
- Russia: 14.9
- Indonesia: 12.9
- European Union: 10.6
- Japan: 10.6
- Brazil: 10.0
- Mexico: 6.4
- China: 5.8
- India: 2.1

These conflicting ways of looking at the production of GHGs proved problematic, as the drafters of the Kyoto Protocol decided to classify 44 nations as Annex I nations, including Australia, Austria, Belarus, Belgium, Bulgaria, Canada, Croatia, the Czech Republic, Denmark, Estonia, Finland, France, Germany, Greece, Hungary, Iceland, Ireland, Italy,

Japan, Latvia, Liechtenstein, Lithuania, Luxembourg, Monaco, the Netherlands, New Zealand, Norway, Poland, Portugal, Romania, Russia, Slovakia, Slovenia, Spain, Sweden, Switzerland, Turkey, Ukraine, the United Kingdom, and the United States. Under the Kyoto Protocol, the Annex I countries committed to make reductions in their emissions of four GHGs—carbon dioxide, methane, nitrous oxide, and sulfur hexafluoride—by 5.2 percent from 1990 levels. Because some of the largest producers of GHGs such as China and Indonesia were not included in the reductions, groups in the United States opposed the Kyoto Protocol. As a result of this opposition, neither the Clinton nor the Bush administrations sent the Kyoto Protocol to the U.S. Senate for ratification.

Mechanisms of the Kyoto Protocol

A key focus of the Kyoto Protocol is its implementation of "flexible mechanisms," also referred to as Kyoto mechanisms, which are aimed at lowering the overall costs of achieving emissions targets. These Kyoto mechanisms include such innovations as emissions trading, clean development mechanisms, and joint implementation. Countries that utilize flexible mechanisms have the potential to achieve emission reductions or to remove harmful substances, including carbon monoxide, either from their own airspace or that affect other countries, from the atmosphere through cost-effective measures. Implementers of these mechanisms, and proponents of the Kyoto Protocol itself, admit that the cost of limiting emissions varies from region to region. Despite these cost discrepancies, however, the focus and effect on the atmosphere is principally the same. Many of the intricacies of these mechanisms raised concerns about their integrity. Many advocates of environmental regulation were concerned that the mechanisms not operate so as to confer a right to emit for Annex I (industrialized and transitional) countries. This concern has led to systems to regulate while maintaining environmental integrity and equity. To ensure environmental integrity and equity, Annex I countries must have ratified the Kyoto Protocol, calculated their assigned amount as laid out in the Protocol, established a national system for estimating emissions and for removal of GHGs within their territory, created a national registry to record and track the creation and movement of Emission Reduction Units, and reported information on emissions and removals on an annual basis.

Differentiated Responsibilities

The UNFCCC's policy had been to encourage a common but differentiated responsibility for environmental regulation among signatories to the Kyoto Protocol. This policy is based upon the understanding that the current global emissions of GHGs have been caused by actions in developed countries. Per capita emissions in developing countries remain relatively low. While the Kyoto Protocol places responsibility on developing nations to control their emissions, they are not held to the same standard as developed countries, as this would prove disastrous to their economies. The Kyoto Protocol attempted to deal with this by recognizing that developing nations' share of global emissions will grow as their governments struggle to meet the needs of their populations for increased development. The common but differentiated responsibility is based on the measure of per capita emissions. Per capita emissions are a country's total emissions divided by its population. This places a greater onus on industrialized countries because on average their per capita emissions are 10 times greater than those of developing nations. As a result, the Kyoto Protocol called for cuts in 1990 levels of greenhouse emissions of 8 percent from the European Union, 7 percent from the United States, 6 percent from Canada and Japan, no cuts from Russia, and an 8 percent increase for Australia.

Results

Although the United States never ratified the Kyoto Protocol, it and the other industrialized nations made considerable efforts to comply with its basic principles. As a group, the Annex I nations have largely met their obligations under the Kyoto Protocol, reducing GHG emissions between 4 and 5 percent. Non–Annex I nations have tended not to reduce their emissions, with China's growing by more than 10 percent per year. The Kyoto Protocol has been criticized by some as having adopted a piecemeal approach that did little to reduce global greenhouse emissions overall. While many economists believe the commitments from Annex I nations are stronger than justified, the lack of quantitative emissions reductions from developing nations caused distress to many environmental advocates. Other critics of Kyoto believe that its commitments, even if achievable, will be insufficient to address the problem of climate change. Dissatisfaction with the Kyoto Protocol, and the expiration of its first phase in 2012, has led to discussions concerning a successor agreement.

In 2007, the heads of government of Brazil, Canada, China, France, Germany, India, Italy, Japan, Mexico, Russia, South Africa, the United Kingdom, and the United States signed the Washington Declaration, a nonbinding agreement to devise a successor to the Kyoto Protocol. The Washington Declaration envisioned a cap-and-trade system that would apply to both industrialized and developing nations. Although it was hoped to be implemented as early as 2009, negotiations are still continuing. The Intergovernmental Panel on Climate Change (IPCC), an agency of the United Nations, has also worked to evaluate the risk of climate change caused by human activity. IPCC reports, although criticized by some, will doubtless assist in shaping the successor to the Kyoto Protocol.

See Also: Alternative Energy; Antiglobalization; Carbon Emissions (Personal Carbon Footprint); Carbon Market; Carbon Tax.

Further Readings

Aldy, J. E. and R. N. Stavins, eds. *Architectures for Agreement: Addressing Global Climate Change in the Post-Kyoto World.* New York: Cambridge University Press, 2007.

Stewart, R. B. and J. B. Wiener. *Reconstructing Climate Policy: Beyond Kyoto.* La Vergne, TN: AEI Press, 2003.

Victor, D. G. *The Collapse of the Kyoto Protocol and the Struggle to Slow Global Warming.* Princeton, NJ: Princeton University Press, 2004.

Stephen T. Schroth
Jason A. Helfer
Jordan K. Lanfair
Knox College

M

MITIGATION

Mitigation can be locally regulated when urban land or water bodies are involved. If, however, the issue is air pollution, like this haze over Mexico City, mitigation must be a regional effort.

Source: Nancy Marley, Argonne National Laboratory

More than 50 percent of the world's population now resides in its cities and metropolitan areas. In the United States, that number exceeds 80 percent. And urban populations are growing rapidly. As a result of the attendant migration and development, humankind's relationship with and impact on nature is rapidly changing. Our increasingly urban lifestyles and consumption patterns are taking an increasing toll on the land—both geographically and generationally—and on ecosystems on which our survival depends. We have come to recognize that if we continue with "business as usual" development—to produce and consume in patently unsustainable fashion—we risk upsetting the delicate ecological equilibria that sustain human life on this planet.

Confronting Environmental Risk

The ecological networks that define life on Earth are enormously complex. In spite of technological advances and human ingenuity, there is inevitable uncertainty in our understanding of how these systems function and interact. The environmental consequences that derive from human development in the face of these uncertainties pose risks for the sustainability of our societies. We can respond to these risks either by adapting to their often-unforeseeable consequences or by proactively attempting to mitigate those consequences.

While adaptation implies reactive steps taken to protect against the consequences of risk, mitigation refers to proactive measures taken to reduce the severity or intensity of some future event or situation. There is a spectrum of possible risk mitigation that presumes that we can constructively manage risk through taking—or not taking—present action. At one end of the spectrum is action that results in completely eliminating a threat or the risk of its occurrence. The precautionary principle (in this case, that the risk of harm to humans or the environment of an action can be sufficient cause to refrain from taking that action) lies at this end of the spectrum. At the other end lies utter inaction through benign or willful ignorance or neglect. By choosing to not mitigate a risk, we make the de facto decision only to adapt to its consequences.

Most mitigative strategies occur somewhere in the middle of this continuum and are functions of our understanding of the risks, the severity of the likely consequences of those risks, our social or technological ability to take needed mitigative action, and our political will to do so. In modern society, whether or not to mitigate a risk is generally predicated on an objective analysis of the costs of mitigation weighed against the benefits of not doing so. Cost-benefit analysis assumes that all relevant aspects of the risk are known, understood, and quantifiable, and presumes to know where the threshold of acceptable risk lies.

Given the uncertainty inherent in our understanding of natural systems, it is often reasonable to presume that all relevant costs—especially environmental and social costs—associated with an action may not be known. Neither is it reasonable to assume that we entirely understand the long-term effects of any particular action. When it comes to human-nature interactions, risk management based solely on objective cost-benefit analysis provides an incomplete profile of the risks associated with a particular action.

This traditional, objective environmental risk management is often biased in favor of action that might otherwise not be taken if all costs were known. Humankind's willingness to confront the limits of its understanding of natural systems and seriously contemplate mitigation of environmental risk when appropriate reflects an evolved understanding of our place in Earth's ecological framework. To insist that human ingenuity and technology will ultimately triumph over the vagaries of natural processes is the essence of hubris in face of the great unknown.

A Question of Scale

The scale of action necessary to mitigate an environmental risk is related both to the geographical distribution of the source of the risk and also to the geographical and temporal distribution of its consequences. To the degree that the risk's causes and effects are local, so too can be policies and actions designed to mitigate it. Industrial pollution of urban land or water bodies may be mitigated by purely local regulation. If the issue is industrial air pollution, though, the impact is more likely to be regional and, consequently, mitigative action must address its expanded geographical impacts. In that case, effective mitigation will likely require the political and regulatory cooperation of multiple political jurisdictions.

Mitigation of global risks must be addressed at the international scale, reflecting the geographical distribution of risk causes and consequences. This is particularly problematic given the diversity of social, political, and economic interests that must be balanced at this scale. In order to be successful, a clearly recognized "global good" must motivate constructive mitigative action. But even when the international community recognizes this common good, there is often a problem in pursuing it.

An example of effective environmental risk mitigation at this geographic scale is the coordinated action taken by the international community to prevent the destruction of Earth's protective ozone layer (a global good). Based on conclusive scientific evidence that the ozone layer was deteriorating as the result of increasing concentrations of chlorofluorocarbons produced by human industry, aggressive international negotiations resulted in the Montreal Protocol (originally negotiated in 1987) banning their manufacture, use, and release. While there is still concern for the long-term effects of ozone layer depletion, scientists expect that the ozone layer will be largely repaired by the middle of the 21st century.

An example of a global environmental risk with potentially catastrophic consequences that has not been constructively engaged by the international community is the current debate over climate change and its likely consequences for Earth's climate. The economies of the developed world are dependent on the cheap and wide availability of carbon-based fuels. At the same time, however, the mining, processing, and combustion of these fuels— gasoline for our cars, coal for electricity generation, heavy oils to heat our homes—and the increasing demand for these fuels to power economic development in the developing world release such large quantities of carbon dioxide (and other greenhouse gases) that related warming of our atmosphere threatens to destabilize our global climate, the results of which would be catastrophic for life on Earth as we have come to know it.

In addition to the risk of continuing gradual warming, there is a significant risk that our greenhouse gas emissions, if not drastically reduced, will drive the atmospheric concentration of carbon dioxide across a climatic "tipping point" that will precipitate cataclysmic, abrupt, and irreversible ecological collapse. Aggressive global mitigation of this risk is essential to the sustainability of the developed world, but also to the very survival of many developing nations.

The intergenerational dimensions of climate change make its mitigation even more problematic. Investments in mitigative action are required now so that a persistent risk can be mitigated for future generations. According to economic analyses that dominate the mitigation debates, the financial return on these investments may be too delayed to attract the capital necessary to fund present mitigative initiatives.

Responding to Global Environmental Risk

How should we respond to global climate risk? To the extent that we are already seeing the effects of anthropogenic climate change (e.g., rising sea levels, changes in gross weather patterns, melting of the polar ice caps, and large-scale extinction of species) and know that further global change is "dialed in" by virtue of the greenhouse gases (GHGs) we have and continue to emit, we react by adapting to the world as we are coming to know it. If we live on the coasts, we build our structures on stilts (as in post-Katrina New Orleans). We build new reservoirs and pipelines to secure future freshwater supplies. We decide to relocate our communities in order to avoid the coming ravages of sea level rise, flood, drought, heat, cold, or pestilence.

However, while they may enable us to endure the challenges of a changed global climate, adaptation strategies do nothing to ameliorate the causes of global climate change. Neither do they do anything to reduce the risk, caused by continued GHG emissions, that further, more cataclysmic changes may befall us. In order to reduce the severity of future global change, we must mitigate those risks. That is, we must undertake measures that will reduce the volume of GHGs we release into the atmosphere, thus reducing the anthropogenic forcing of the processes that drive up the average temperature of our atmosphere and exacerbate the risk that more serious consequences related to that warming will transpire.

In the process of mitigation, we will also reduce the extent to which we must adapt to a changed climate in the future. If we do nothing (no mitigation), we will be forever adapting to a changing and increasingly foreign and unpredictable climate. However, if we mitigate (i.e., reduce our emissions of GHGs) sufficiently, we can expect that the extent to which we will have to adapt to future climatic changes will be reduced.

Environmental Risk Mitigation at the Local Scale

Much of the controversy over climate change mitigation at the urban scale derives from the geographical and intergenerational dimensions of the threat. GHGs are emitted by local (human) activity into the global atmospheric commons. All humans share the risk of human-caused climate destabilization, although it was caused and is being exacerbated primarily by a small subset of the world's population.

Unless motivated by legislative and financial support from legally binding international and national agreements that explicitly link local activities to global GHG concentrations (the goal of the Kyoto Protocol and subsequent rounds of international climate negotiations), it is often difficult for a city to recognize that local and global sustainability are often identical propositions. Traditional political jurisdictional limits can blind municipalities to local responsibility for global climate destabilization and the impact it will have on the many who live beyond their borders and will face its consequences in future generations. Further, municipalities face other sustainability challenges (e.g., local environmental pollution, poverty, and economic development) that, arguably, are of greater currency in their local political agendas. Local investments in climate change mitigation—the reduction of greenhouse gas emissions—provide little if any local return, either geographically or temporally.

The extent to which cities may be directly affected by climate change has a role to play in their considerations. Coastal cities, especially those that have endured catastrophic flooding in the past, may be more likely to be willing to undertake mitigative as well as adaptive action to reduce the risk of future flooding. Inland cities, historically susceptible to drought, may understand the potential impact of changing weather patterns to future freshwater availability and support local mitigative action. Unfortunately, adaptation is a simpler political sell because the investment of local funds is designed to provide local benefits, and the results of the investment are often more visible than GHG mitigation. A seawall or a freshwater pipeline are often much more attractive to local residents and businesses.

Bottom-Up Mitigation

While the European Union is making considerable progress in reducing its GHG emissions, the United States persists in resisting any legislatively mandated caps on or regulation of its emissions. Recognizing their increasing infrastructural and economic vulnerability to the effects of human-caused global warming and climate change, a group of northeastern states and cities sued the federal government to regulate carbon dioxide emissions under the Clean Air Act (*Massachusetts v. Environmental Protection Agency*, 2007). Their victory in this Supreme Court case brings regulation of GHG emissions a step closer in spite of inaction by the U.S. Congress, though the actual regulation of GHG emissions is likely to be contested for some time to come.

At the subnational scale, regional alliances have formed to pursue legally binding mitigation of GHG emissions in spite of resistance by the national polity. Members of these alliances—including states, provinces, and municipalities—enter into voluntary but legally binding commitments to reduce their GHG emissions. Prominent among these alliances are the Regional Greenhouse Gas Initiative (10 northeastern and Mid-Atlantic states), the Western Climate Initiative (seven states and four Canadian provinces), and the Midwest Greenhouse Gas Accord (six upper-Midwest states and the Canadian province of Manitoba).

The debate over climate change mitigation has been conducted largely at international and national scales, and global impacts of climate change documented by the Intergovernmental Panel on Climate Change are only painstakingly being translated into actionable guidance for local communities. The global economic impacts of the 2008 Great Recession will continue to render policy commitments of the type necessary for meaningful GHG reduction contested at all scales. The lack of international agreement on how to proceed after the Kyoto Protocol expires in 2012, and the refusal of the United States and China to pursue mandatory and verifiable GHG emission reduction targets will render climate change mitigation efforts made at subnational levels around the global all the more important.

See Also: Adaptation; Horizontal Policy Integration; Sustainable Development (Business); Vertical Policy Integration.

Further Readings

Howard, J. "Climate Change Mitigation and Adaptation in Developed Nations: A Critical Perspective on the Adaptation Turn in Urban Climate Planning." In *Planning for Climate Change: Strategies for Mitigation and Adaptation for Spatial Planners*, S. Davoudi, J. Crawford, and A. Mehmood, eds. London: Earthscan, 2009.

Intergovernmental Panel on Climate Change (IPCC). "Climate Change 2007: Synthesis Report. Contribution of Working Groups I, II, and III to the Fourth Assessment Report of the Intergovernmental Panel on Climate Change." R. Pachauri and A. Reisinger, eds. Geneva: IPCC, 2007.

Leary, N., K. B. Averyt, B. Hewitson, J. Marengo, et al. "Crossing Thresholds in Regional Climate Research: Synthesis of the IPCC Expert Meeting on Regional Impacts, Adaptation, Vulnerability, and Mitigation" (2007). Intergovernmental Panel on Climate Change. http://www.ipcc.ch/pdf/supporting-material/tgica_reg-meet-fiji-2007.pdf (Accessed February 2010).

Kent L. Hurst
Independent Scholar

N

Nanotechnology

Photovoltaic cells like these can be made using nanoparticles of materials, which is cheaper than traditional silicon solar cells and uses less semiconductor material.

Source: Randy Montoya/Sandia National Laboratory

Nanotechnology as a discipline deals with matter at an atomic or molecular level ranging from 1 to 100 nanometers (or billionth of a meter) in size and the consequent development of applications based on it. The Australian Office of Nanotechnology (AON) offers the following definition: "Nanotechnology is the precision-engineering of materials at the scale of 10–9 meters (one ten-thousandth the width of a human hair), at which point, new functionalities are obtained, resulting in products, devices and processes that will transform various industries." Nanoparticles are usually classified as either natural (such as those generated from volcanoes), incidental (generated in emissions from engine combustion), or engineered (manufactured on purpose). Nanoproducts have shown escalating sales in the recent past and are predicted to account for 14 percent of revenue generated from manufacturing by 2014, amounting to $2.6 trillion, matching the revenue generated from information technology and telecommunication industries put together. This is a huge increase compared to 2004, when these products accounted for less than 0.1 percent of the revenue generated from manufacturing. Furthermore, the industry is expected to provide employment to approximately 2 million workers by 2015. These promises have received the attention of several countries/regions around the world, including the European Union and the Asia-Pacific region.

It is hoped that nanotechnology will serve as an efficient source for green environmental technologies in both contemporary and future times. For instance, nanotechnology-enabled applications will help us reduce industrial contamination that results from the production of cosmetic and industrial products, utilize depleting natural resources like oil and coal in the most efficient manner, and also clean up industrial and other wastes.

Nanotechnology-Enabled Applications

Transforming Solar Energy

Nanostructured cadmium and copper indium diselenide can be easily applied to inexpensive materials like glass and plastic to make photovoltaic cells for trapping solar energy. This preparation is cheaper to manufacture compared to traditional silicon solar cells and also requires significantly less semiconductor material.

Organic Light-Emitting Diodes (OLEDs)

Other closely connected energy-saving devices are the OLEDs. These could be used for making skylights that mimic natural light and could be integrated with cabinets, curtains, tables, and so forth.

Safer and Stronger Lithium Batteries

Nanostructured electrodes laced with lithium cobalt oxide not only increases the power output of lithium rechargeable batteries by 50 percent but also makes them safer to use.

Better Storage for Emission-Free Fuels

Ecofriendly vehicles operate on hydrogen, but the hydrogen gas is highly flammable and presents considerable storage problems. Nanomaterials like metal hydrides react at near-room temperatures and at pressures that are only a few times higher than that of Earth's atmosphere, facilitating storage. Nanostructured materials also provide fast diffusion paths for hydrogen.

Cleaning Groundwater

New nanomaterials can bind pollutants and get them mopped up, a property that could be geared toward cleaning arsenic from groundwater (e.g., magnetic iron nanocrystals) in countries like Bangladesh, where high concentration of arsenic is responsible for civic and health problems. Nanoporous membranes filtering harmful pathogens from drinking water are even commercially available.

Assessing the Usefulness and Impacts of Nanotechnology

Developing technology and assessing its usefulness go hand in hand and, likewise, we need to gauge the usefulness of nanotechnologies at two different levels: (1) the production of such technologies (or green technologies) designed for creating a clean and green environment, and (2) the environmental impact of nanomaterials produced for other purposes. This is because the production of nanoapplications will lead to the appearance of nanoparticles in our environment. Therefore, research initiatives should be undertaken to understand the associated repercussions because we still do not have adequate knowledge about their environmental impacts. Additionally, we require scientific work to understand the degradation processes (e.g., by bacteria) for nanomaterials as some of these might resist degradation, persist longer in our environment, and possibly undergo changes with time, concomitantly changing their anticipated impact.

Potential Impacts of Nanomaterials

Air

Engineered nanoparticles are small in size and light in weight and thus may remain airborne for a longer period of time, increasing the likelihood of traveling long distances, crossing borders, and even interacting with other particles and gases in the air.

Water

There is lack of scientific data to show how nanoparticles aggregate, precipitate, and suspend in water, adversely affecting prediction about their interaction with the aquatic ecosystems.

Soil

Nanoparticles in soil may behave in unpredictable ways by supposedly binding chemically to soil particles, residing on the surface of the soil particles or in the pore space between the particles. The interaction between the nanoparticles and the soil particles may differ depending upon whether these bind together or are separated. Degradation of nanoparticles by naturally occurring soil microbes also remains an unanswered issue.

There might be some specific examples of impacts that appear alarming at this stage, for example, fullerenes or buckyballs (a class of carbon-based nanomaterials) used as antioxidants are sometimes shown to be impairing brain functions in fish. These buckyballs are fortunately least mobile but there may be other nanomaterials that behave differently and could be toxic and more mobile. Other associated risks may accrue from ingredients used to produce nanomaterials. Therefore, we also need greener methods for the production of nanomaterials.

Consider the use of nanotitanium dioxide for making sunscreens. A study conducted at the University of Toledo showed that nanotitanium dioxide reaches municipal sewage treatment systems after being washed off in showers and kills useful microbes. These microbes are helpful in removing ammonia from wastewater and reduce the amount of phosphorus in lakes. On the positive side, we have materials like carbon-nanotubes (long and hollow cylinders formed by a single layer of graphite) used for storing hydrogen in a cost-effective manner. Others, like nanosilver that acts as an antimicrobial agent used for burn injuries, have been shown to have damaging effects on aquatic organisms like the zebrafish.

It is best to carefully weigh the positive and the negative sides of nanotechnology against each other since research about each of these is still at a very nascent stage. It is hoped and has also been shown that nanotechnology will be able to contribute toward a healthy green environment by using up fewer naturally occurring raw materials and reducing harmful emissions, but nanotechnology may not necessarily live up to this promise, considering that we still require greener, efficient, and time-effective methods for producing nanomaterials. Any backlash against the nanoproducts tends to omit mentioning naturally occurring nanoparticles. Carbon-based nanotubes that some believe to be potentially dangerous also exist in natural systems like soot from candle flames.

Policy Measures

There are certain strategic measures that should be followed until we have substantial data to show the effects of nanotechnology on our ecosystems. These measures will ensure the best possible usage of technology in uncertain situations:

- Building cooperation between players in both the public and private sectors; strengthening cross-border cooperation
- Standardizing test protocols for facilitating generalization of research findings
- Increasing awareness about the risks and advantages of nanotechnology among the general public and environmental and national regulatory agencies
- Providing moral and financial support for boosting nanotechnology research efforts
- Evaluating the potential environmental and other health-related impacts of engineered nanomaterials that are already released and being used
- Evaluating risk-sharing methods with the private sector (e.g., the NanoBusiness Alliance)
- Utilizing knowledge about chemical issues related to health and environment and adapting it to address issues on nanotechnology

Joining Hands With Green Chemistry and Engineering

In recent times, there has been an increasing trend toward hopeful mergers between green chemistry and nanoscience for a greener environment. Green chemistry aims for the reduction or elimination of hazardous substances during the process of chemical production by focusing on a set of principles. The merger appears promising as green chemistry looks at preventing pollution at the source (or the initial stages of production itself) on a molecular scale.

A merger of nanotechnology and green chemistry aspires to the following:

- Design nanomaterials that yield new properties and performance but do not pose harm to the environment and human health
- Manufacture complex nanomaterials efficiently, minimizing the use of hazardous substances in the manufacturing process
- Assemble/interface nanomaterials using self-assembly or bottom-up approaches for the purpose of enhancing performance and reduction of wastes

The following are important guiding principles for green chemistry developed by the U.S. Environmental Protection Agency (EPA):

- *Prevention*: Prevention of waste is considered better than the treatment or cleaning up of waste after it is formed as a by-product of chemical production
- *Atom economy*: Designing of synthetic methods in order to maximize incorporation of all materials used in the chemical production process
- *Less hazardous chemical synthesis*: Designing of synthetic methodologies wherever feasible, for generating substances posing little or no toxicity to environment or health

- *Designing safer chemicals*: Designing of chemical products for the purpose of preserving efficacy of function while reducing toxicity
- *Safer solvents and auxiliaries*: Curtailing unnecessary utilization of auxiliary substances like solvents, separation agents, and so forth
- *Design for energy efficiency*: Recognizing energy requirements by keeping the environmental and economic impacts in mind and reducing the negative side effects; for instance, conducting synthetic methods at ambient temperature and pressure
- *Use of renewable feedstocks*: Using renewable rather than depleting raw material or feedstock wherever practicable both technically and economically
- *Reduce derivates*: Reducing unnecessary derivatization like blocking group, protection/deprotection, and temporary modification whenever possible
- *Catalysis*: Using catalytic reagents that are considered superior to stoichiometric reagents
- *Design for degradation*: Designing such chemical products that do not persist in the environment and/or break down into innocuous degradation products toward the end of their function
- *Real-time analysis for pollution prevention*: Developing analytical methodologies that allow for real-time, in-process monitoring and control prior to the formation of hazardous substances
- *Inherently safer chemistry for accident prevention*: Choosing such substances and the form of a substance that minimize the potential for chemical accidents, including releases, explosions, and fires

The following principles of green engineering developed in tandem with green chemistry were formulated in 2003 in Florida for developing energy-efficient products from biodegradable materials:

- Active engagement of communities and stakeholders in developing engineering solutions
- Improving, innovating, and inventing new technologies for sustainability or creating engineering solutions that go beyond current technologies
- Applying engineering solutions suitable for local geography, aspirations, and culture
- Preventing waste
- Minimizing the depletion of natural resources
- Ensuring that all energy outputs and inputs are inherently safe
- Deploying life-cycle thinking in engineering activities
- Protecting human health and conserving and improving natural ecosystems
- Holistically engineering products and processes, integrating environmental impacts and assessment tools, and using system analysis

The ultimate potential for intersections between nanotech and green chemistry/engineering may lie in the development of products (e.g., nanoelectronics, nanocomposites, and thermoelectrics) that eventually give way to greener products, processes, and applications servicing a green nanoscience. Specific examples include (1) functionalized metal nanoparticles, self-assembled onto biomolecular scaffolds (DNA, polypeptides) for developing nanoelectronic devices, (2) nanocomposites for making vehicle panels that help in increasing fuel efficiency and reduce carbon dioxide emissions. Increasing use of nanocomposites in place of steel panels can instrumentally reduce environmental impacts, and (3) nanostructured thermoelectric materials prepared by self-assembly process at relatively low temperatures for efficient cooling and heating or capture of waste heat.

The marriage between nanotechnology and the principles and practices of green chemistry/engineering cannot afford to ignore building an environmentally sustainable society. Therefore, sustainable development that stands for the ability to meet the needs of the current generation while preserving the ability of future generations to meet their needs acquires high priority for both current and future pursuits. Examples include (1) the use of DNA molecules for building nanoscale patterns on silicon chips and other surfaces, requiring less water and solvent compared to the traditional printing (lithography) method, (2) deploying nanoscale approaches to replace lead and other toxic ingredients in manufacturing electronics, (3) engineered nanoscale sensors for detection of pollutants at the level of parts per billion, and (4) technology for making inexpensive solar cells as well as improving the performance and lowering the cost of fuel cells for cars and trucks.

Future Challenges

There is still a long way to go before green nanoscience is a practical success; some of the research challenges are identified below:

- Diverse populations of nanostructures with well-defined structures and purity profiles are required in order to understand structure-activity relations.
- Assessment of atom economy and development of new synthetic and purification methods require improved understanding of product distributions, nanoparticle formation mechanism, and reaction stoichiometries.
- New transformations and reagents are needed that are more efficient, safer, and useful in a diverse range of reaction media.
- Improved analytical techniques are needed to permit routine analysis of nanoparticles in real-time to optimize production.
- More convenient purification methods need to be developed to provide access to pure nanomaterials without large solvent waste as a by-product.
- Alternatives to auxiliary substances should be developed in order to stabilize and control nanoparticle shape during synthesis and furthermore to incorporate the shape-controlling molecule as a functional part of the final product.
- Alternative solvents and reaction media are needed.
- Improved metrics are needed to compare the greenness of competing approaches that consider relative efficiencies and hazards across the life cycle.

See Also: Aquaculture; Carbon Capture Technology; Green Community-Based Learning; Nanotechnology and Food; Oil; Sustainable Development (Politics).

Further Readings

Anastas, P. T. and J. B. Zimmerman. "Design Through the Twelve Principles of Green Engineering." *Environmental Science & Technology*, 37:5 (2003).
Australian Office of Nanotechnology. http://www.industry.gov.au (Accessed September 2010).
Benn, T. M. and P. Westerhoff. "Nanoparticle Silver Released Into Water From Commercially Available Sock Fabrics." *Environmental Science & Technology*, 42:1 (2008).
Dionysiou, D. D. "Environmental Applications and Implications of Nanotechnology and Nanomaterials." *Journal of Environmental Engineering, ASCE*, 130:7 (2004).
Lux Research. *The Nanotech Report 2004*. http://www.luxresearchinc.com/tnr2004 (Accessed September 2010).
McKenzie, L. C. and J. E. Hutchison. "Green Nanoscience: An Integrated Approach to Greener Products, Processes, and Applications." *Chemistry Today*, 22:5 (2004).

Powell, M. C. and M. S. Kanarek. "Nanomaterial Health Effects—Part 1: Background and Current Knowledge." *Wisconsin Medical Journal*, 105:2 (2006).

Powell, M. C. and M. S. Kanarek. "Nanomaterial Health Effects—Part 2: Uncertainties and Recommendations for the Future." *Wisconsin Medical Journal*, 105:3 (2006).

Rickerby, D. G. and M. Morrison. "Nanotechnology and the Environment: A European Perspective." *Science and Technology of Advanced Materials*, 8:1–2 (2007).

Rocco, M. C. "Broader Societal Issues of Nanotechnology." *Journal of Nanoparticle Research*, 5:3–4 (2003).

Schmidt, K. F. *Green Nanotechnology: It's Easier Than You Think*. Washington, DC: Woodrow Wilson Center, 2007.

Neha Khetrapal
Indian Institute of Information Technology

Nanotechnology and Food

Nanotechnology or nanoscience is the study of devices and man-made materials at the molecular scale, typically under 100 nanometers in at least one dimension, and it has been a fast-growing area of research since the development of the scanning tunneling microscope and discovery of fullerenes (nanoscale carbon molecules with a number of useful special properties) in the early 1980s. In the 21st century, nanotechnology is proving to be a critical area of research and development, one that will affect all areas of engineering and the physical sciences, with implications for all industries. Endeavors range from attempting to manipulate matter at the atomic scale in order to build objects atom by atom to the development of various new materials. There have been numerous health and environmental concerns raised about nanotechnology and about its regulation as well as whether nanotechnology and nanomaterials represent an area distinct enough to warrant special regulation in addition to any other regulations to which they may be subject in accordance with their specific uses. For these reasons, the use of nanotechnology in food production is of special concern.

Nanotechnology can be used at various stages of food production, which are being actively studied and/or employed. Kraft Foods established the Nanotek Consortium in 2000 as a collaboration among 15 research labs and universities in the area of "interactive foods and beverages," from color- and flavor-changing foods to foods that can even alter their nutritive and allergenic content to suit the consumer. But many uses of nanotechnology in food production have to do with the raising of food, not with the final product.

For instance, nanocapsules can disperse pesticides, fertilizers, and other substances more efficiently on farms and could even be developed to dispense themselves reactively—nutrient capsules in soil could automatically break down when nutrient content in the soil dips below a certain level, for example. Nanocapsules could also deliver vaccines to livestock, and nanoparticles can be used to genetically engineer plants by delivering DNA. Nanosensors could detect pathogens, monitor crop growth and soil moisture, and other factors as needed. Nanotechnology can be used to determine the effects of enzymes more accurately than computer models.

In packaging food, fluorescent nanoparticles could be paired with antibodies to detect pathogens. Biodegradable nanosensors could monitor moisture, temperature, and chemical presences indicating spoilage or other problems, while many nanomaterials would serve as effective barrier materials to prevent oxygen absorption and spoilage. Nanoparticles of silver, magnesium, zinc, or other materials act as antimicrobial and antifungal surface coatings, and even the wrapping material for packaged food can be tailored with nanotechnology to be lighter, stronger, more heat resistant, less or more permeable as needed. Smart food wrappers can respond to ambient conditions, becoming more or less permeable to water vapor in order to regulate moisture. Most people have been taught the trick of adding a sliced apple to a sealed bag of cookies to make them softer, but the food industry is interested in teaching that trick to the bag itself—a bag that can keep cookies at exactly the texture you want them to remain, or make them more crisp, or make them chewy, by controlling its own microclimate. And a number of food, chemical, and food packaging companies are working on safer packaging materials—materials that will detect unsafe presences like *Salmonella* and *E. coli*, alerting workers long before consumers can purchase the food and bring it home. The same basic technology will help the federal government guard against attempts to poison the food supply. Already, nanomaterials are in use to make plastic bottles stronger and extend the shelf life of their contents, and several large breweries are investigating the possibility of using nanotechnology to transition from fragile, heavy glass bottles to sturdier, thinner plastic bottles that will actually keep their beer fresh much longer than glass does.

There is significant potential in revolutionizing the vitamin and supplement industry with nanotechnology. In comparison to the possibilities offered by nanoscience, present supplements are extremely inefficient and imprecise. Powders crushed to or formed at nanoscale can provide nutrients that are absorbed faster and more completely, and neutraceuticals can be nanoencapsulated for targeted delivery and better absorption. Nutrients can be added to any food without affecting the flavor or texture through use of coiled nanoparticles called nanocochleates, and nanodroplets of active vitamin molecules can be sprayed into the mouth for fast and efficient absorption with no noticeable flavor or mouthfeel other than the carrier fluid (water or a flavored liquid).

The potential uses of nanotechnology in food processing and in food additives are innumerable. Added nutrients can be made more bioavailable by encapsulation in nanocapsules—and can even be added to ingredients not currently nutrient fortified, like cooking oil. The idea of fast food chains being able to offer French fries fortified with the same nutrients as a bowl of cereal or salad is certainly a compelling one for the food industry, and it is notions like this that capture the imagination and research funding. Flavor enhancers, too, can be easily nanoencapsulated—mixed into ground beef before pressing into patties, they could boost the beefy and umami characteristics of a one-dollar hamburger. Nanoparticles could also selectively bind to pathogens to remove them from food, scrub the cholesterol from meats with plant-based steroids, and change the texture of foods. Moreover, existing food textures could be freed of the limitations that have been taken for granted: for example, ice creams and gelatin desserts could be developed that could be served hot without melting, and hot fudge could be made pliable and slow-flowing. Already, nanocapsules of fish oil are used to add omega-3 fatty acids to white bread sold in Australia, without leaving a fishy taste.

There are three areas of objection to the use of nanotechnology in the food industry: regulatory, environmental, and human health. The regulatory issue is that at this time, use of nanotechnology in the processing or manufacture of a given product is under the purview of the agency ordinarily responsible for that product—not a regulatory body dedicated to nanotechnology. The essence of the complaint is that nanotechnology presents its own special challenges and considerations and should be handled by a body conversant in them. For instance, substances do not always interact in the same way at nanoscale as they do at normal scale. Regulations need to reflect this and to regulate, for instance, silver and nanosilver (particles of silver at the nanoscale) separately—not only regulate them separately, but also test their safety separately. (In 2006, the Environmental Protection Agency declared that it intended to do exactly this, with silver specifically, because of its use as an antibiotic coating.) Further, the lack of a single regulatory body makes it more difficult for the public to have a voice and leads to the possibility of inconsistent nanotechnology regulations. In the long term, a single regulatory body is good for the industry as well—the sooner the better, if it is seen as an inevitability.

There is comparatively little objection to using nanotechnology in food packaging, except that it distracts from the root problem of recent foodborne illness scares: many practices of large agribusinesses are, even when legal and mindful of existing regulations, inherently unsafe and more likely to lead to outbreaks of *E. coli* and especially *Salmonella*. While there is a clear public good in decreasing the odds of contaminated food reaching the consumer, the greater good is in decreasing the odds of that food's becoming contaminated in the first place. The trend has been to shift responsibility for preventing consequences downstream, to the consumer whose job it has become to cook food to specific temperatures (temperatures higher than his great-grandmother would have needed to aim for), and to acknowledge the disclaimer accompanying uncooked or undercooked eggs in order to mitigate the risks undertaken by the food industry in its practices.

Proponents of moving forward with nanotechnology applications in food production point out the benefits to the developing world. Foods can be made more nutritious and safer, cheaply. Agricultural goods grown in the developing world—where they represent a more significant portion of a country's GDP and of its exports—can be made more productive, higher quality, safer, and given a longer shelf life, improving the lives of farmers. Food can even be tailored to help combat specific ailments that are particular problems in certain parts of the world, and drinking water can be purified more easily. The possibilities are as exciting as the glow-in-the-dark apples and carbonated mini-melons bound for American supermarkets. However, there are negative health considerations to take into account as well. While nanotech proponents point out that the present obesity epidemic is owed in large part to a dietary reliance on processed foods and that nanotechnology is the current best way to make processed foods healthier, less caloric, and more nutritious, this would again shift responsibility. The solution to Americans eating too much ice cream should be to convince them to eat less ice cream—not to trick them with a "healthier" ice cream laden with nanoadditives that prevent the body from digesting the fats and sugars. Further, those nutrient-fortified French fries mentioned above are still French fries, and when French fries can be marketed as a health food and a responsible meal selection for growing children, lines are blurred that will be difficult to unblur. Just as generations of children grew up eating sugar-filled breakfast cereal that was fruit flavored and advertised as "part of a nutritious breakfast," instead of simply eating fruit, so too might a future generation grow up on French fries and milkshakes, obtaining all their nutrients therefrom but surely—the argument goes—missing out on the big picture of nutrition. Surely it would still be better to get vitamins from fruits and vegetables, iron from greens and meat, protein from beans and fish, instead of depending on science to develop the perfect milkshake.

Furthermore, the toxicity issues of nanotechnology are from far understood. Nanoparticles are simply too new. A growing body of research suggests that, just as nanopowders are more easily absorbed by the body, so too are nanoparticles more reactive with the body, and substances that are inert at the macro scale are reactive at the nanoscale. Fullerenes, for instance, have been found to contribute to brain damage in largemouth bass, one of the species regulatory agencies use as a model for determining toxicological effects. We have only begun to explore nanotoxicity, and new materials and applications of those materials are developed all the time.

Finally, the use of nanomaterials leads to "nanopollution," which is also poorly understood. The environmental impacts of nanoparticles are still unfolding, and the bioaccumulation of nanoparticles in plant life needs considerable further research. Even when no nanomaterial is in the final food product, the use of nanotechnology in food production could pose environmental dangers, the risks of which have not been fully assessed.

See Also: Agribusiness; Ecolabeling; Genetically Modified Organisms (GMOs); Nanotechnology.

Further Readings

Chaudhry, Qasim, Laurence Castle, and Richard Watkins, eds. *Nanotechnologies in Food.* London: Royal Society of Chemistry, 2010.
David, Kenneth and Paul B. Thompson, eds. *What Can Nanotechnology Learn From Biotechnology?* New York: Academic Press, 2008.
Lawrence, Geoffrey, Kristen Lyons, and Tabatha Wallington, eds. *Food Security, Nutrition, and Sustainability.* New York: Earthscan, 2009.

Bill Kte'pi
Independent Scholar

Nongovernmental Organizations

From the highlands of Papua New Guinea to the halls of New York's United Nations, nongovernmental organizations (NGOs) with an environmental focus are shaping the establishment of global norms, public debate, and policy decisions. While many people value the contribution of "green" NGOs to society, others question their accountability, legitimacy, and effectiveness. Meanwhile, lively debates within environmental NGOs center on conditions of engagement with governments and the private sector, and the fundamental question of whether capitalism can be "greened." Some NGOs believe constructive partnerships will help major legal, economic, government, and corporate institutions forge a sustainable path. However, other NGOs are dubious of such collaboration; they insist these entities are incapable of changing, their interests too deeply entrenched in maintaining the status quo. This article outlines a number of areas where "green" NGOs operate, highlights multiple strengths viewed from a liberal democratic standpoint, and concludes with a critique.

Also referred to as peoples' groups, community groups, charities, grassroots organizations, civil society, and the "third sector," NGOs around the globe provide a range of services as well as catalyze policy and institutional change. NGOs span the spectrum of political and philosophical persuasions. Most aim to put pressure on traditional power structures of government and business, perceiving their investment in unsustainable practices and policies as critical barriers to change. Mirroring the diversity of NGOs' raison d'être, their tactics and approaches vary.

Environment-focused NGOs are often referred to as "ENGOs." The "old wave" of ENGOs has been characterized as capitalizing on conservation (often to the detriment of communities who live there) and centering debate around ethnocentric versus ecocentric approaches to conservation. The "new wave," however, is incorporating social considerations into its mandate, taking a position that sustaining livelihoods is critical to achieving conservation objectives. Thus, this article uses the broad "NGO" term, though it could also employ "SNGOs" to emphasize their broader sustainability objectives.

NGOs engage in a variety of activities, including advocacy, lobbying, research, legal investigations, outreach, and publicly reporting out from international negotiations. Many NGOs also partner with academic institutions, businesses, and governments to deliver services and programs and transform manufacturing operations. In addition, NGOs are involved with on-the-ground conservation and restoration work as well as monitoring environmental and human health indicators in communities with active natural resource extraction and other types of industrial activity. NGOs' reframing of environmental concerns as human health issues is helping bridge the North–South divide between economically wealthy regions that can "afford" to conserve their environment and economically impoverished areas that remain heavily reliant on natural resource extraction for subsistence.

Traditionally, NGOs have relied on financial support from membership fees and the occasional large, philanthropic donor. Because governments have also seen NGOs as useful vehicles to deliver services no longer provided by the state, many NGOs have come to depend on significant contributions from government for their operating budgets. As both philanthropic and government dollars dry up in the current global recession, many NGOs are not only looking to creative fund-raising methods but also thinking more strategically about leveraging resources, expertise, and influence through collaboration and networks. Such coalitions also help to muster critical mass attention around both local initiatives and international governance discussions with outcomes that affect peoples' day-to-day lives. NGO participation in the Conference of the Parties on the United Nations Framework Convention on Climate Change (UNFCCC) provides an example of how, through their active, often coordinated presence at key international environmental negotiations, NGOs help disseminate state positions and identify strategic opportunities to exert public pressure around specific issues of concern. In addition, through their use of independent media and social networking tools, NGOs are pioneering mechanisms to mobilize sustainability-inspired active citizenship.

Capitalism as the Problem Versus Greening Capitalism

Grassroots organizations such as Greenpeace and Rainforest Action Network tend to draw on membership dues and donors for support and often target specific businesses for high-profile, shame-based campaigns. Their tactics include advocacy and education as well as nonviolent civil disobedience and direct action. NGOs in this camp are suspicious of "green" corporate intentions, arguing that companies largely derive their profits at the expense of human health and the Earth's ability to maintain ecosystem services. Alarmed at the growing number of "green" NGO-business partnerships, they argue that mainstream environmentalism has been bought by corporate interests. NGOs see such partnerships as a way for companies to "greenwash" their businesses, that is, show an environmentally friendly face while proceeding with practices that undermine the health of people and the planet.

Other NGOs work inside government and corporate structures to promote change from within. These NGOs help institutions respond to growing public pressure to be much more responsive to all stakeholders, including communities where they operate, their employees, and the Earth. NGOs help companies meet their commitments to corporate social responsibility through mechanisms such as extended producer responsibility and product life-cycle analysis. For example, the World Wildlife Fund (WWF) has partnered with Walmart to help the company green its supply chains, and the Suzuki Foundation explicitly states that it works with individuals, governments, and business to conserve the environment.

Strengths

Many people with liberal democratic values generally welcome the contribution of "green" NGOs to society. When measured by their contribution to advancing social, ecological, and environmental justice, NGOs provide a number of valuable services. In this article, social, environmental, and ecological justice are distinguished as follows: (1) social justice assumes that each individual (and human entity such as government and businesses) has a responsibility to work for the greater good of society, (2) environmental justice is justice among humans on environmental issues, and (3) ecological justice is justice toward the natural world. The general term *justice* from here onward encompasses all three.

NGOs raise public awareness of and respond to the most pressing issues facing us today such as escalating food prices, deteriorating soil quality, rapid biodiversity loss, climate change, and disease. An indication of NGO effectiveness in this realm is the growing trend to consider them as key stakeholders in government—and increasingly corporate—engagement at local, national, regional, and international levels. Not only do many NGO supporters see a thriving NGO sector as an integral part of a healthy, liberal democracy, but there is a mounting appreciation for NGOs that fill gaps left by governments in social and environmental service provision.

Some NGOs also conduct their own monitoring, analysis, and documentation of environmental destruction and its impacts on human health. For example, NGOs such as Mining Watch Canada, Tar Sands Watch, Oxfam, and Baikal Environmental Wave challenge official research that often indicates little cause for concern. From Canada's oil sands and Russia's Lake Baikal to the highlands of Papua New Guinea and Guatemala, the above NGOs suggest that government reports and industry-funded research often purposefully exclude local perspectives and evidence of harm to people and the environment. These organizations insist that the perspective and traditional knowledge of local, often indigenous communities be considered and valued in all decision-making processes that affect them.

Finally, when viewed through the conceptual lens of new grassroots movements, NGO networks provide a useful framework for many environmentally inspired social movements such as 350.org, and those associated with "take back your time," "voluntary simplicity," "degrowth," and zero waste. Such possibility-oriented movements also offer a response to critics who accuse civil society of failing to articulate a positive alternative future.

Critique of NGOs

Because "green" NGOs operate across a spectrum of political persuasions and fill a variety of roles with different measurements for success (e.g., service delivery, advocacy, research), it follows that the grounds for criticism also reflect this diversity. For example, whether (particular) NGOs are perceived as "too radical" or "not radical enough" is a subjective verdict based on one's political stance and personal conviction of NGOs' role—or lack of role—in society. "Green" NGOs face both external and internal criticism. Some critiques apply to NGOs in general, while other assessments are specific to those engaged in environmental issues. A handful of general and specific opinions are offered here.

As some NGOs gain political strength, they encounter growing concerns from outside stakeholders about their accountability, legitimacy, and effectiveness. Critics of the civil society sector assert that while some NGOs' size, influence, resources, and responsibility have grown—in some cases to rival that of governmental and intergovernmental agencies (e.g., United Nations) they work alongside—NGOs lack the accountability mechanisms of democratically elected governments. Other critics argue that along with growth comes a preoccupation with greater expansion, increased bureaucracy, and the risk that NGOs are reduced to contractors for the state and business, rather than retain a strong degree of independence. The trend toward offloading government responsibility to provide public goods and services to nonstate actors further highlights two concerns: that of NGO accountability to the public and that, by filling holes left by government, NGOs erode strong public institutions by legitimizing the retreat of government from the public sector. The proliferation of small grassroots organizations has also come under fire, with critics suggesting efficient action is limited by multiple entities jostling for scarce donations and projects.

Some organizations, like the American Enterprise Institute, question the legitimacy of NGO participation in global public policy discussions. The institute posits that NGOs operating abroad weaken state sovereignty and democracy. This perspective is shared by countries like Russia and China, which have been known to suggest that foreign-based NGOs (environmental orientation is no exception) serve as a conduit for foreign states (e.g., the United States and European Union) to meddle in these countries' domestic political affairs.

Other critics of civil society propose its contribution as a "third sector" to promoting sustainability has been exaggerated, that its effectiveness remains contingent on robust, legitimate political and legal institutions whose existence does depend on cooperation, trust, and solidarity.

In a variety of campaigns, contradictions emerge between local and foreign advocates. For example, in the case of Sakhalin II, the campaign's lead organization—Sakhalin Environment Watch (SEW)—called for the project to adhere to best practices. However, some foreign campaigners such as Friends of the Earth called for an end to all oil development projects due to their impacts on climate change. Another common point of contention between local and foreign campaigners is in the framing of issues. In the case of Sakhalin, international campaigns tended to focus on gray whales as the primary topic of concern, effectively minimizing issues of greater importance to the Sakhalin population such as oil pollution and the project's impact on fisheries. While Sakhalin II offers only one example, the legitimacy of foreign entities "speaking for the locals" and emphasizing "conservation over livelihoods" is a key topic of debate among and outside the NGO community. In response, many large, international, environmentally oriented NGOs such as the International Union for Conservation of Nature (IUCN) have turned their attention to the social-environment pillars, as described above.

Just as civil society calls on corporations and governments to be more accountable to a range of stakeholders, pressure is mounting on NGOs themselves to adopt parallel practices.

In her work "Mechanisms for NGO Accountability," Lisa Jordan suggests that NGOs work in tandem with academics to develop a means to be more accountable "upwardly" to donors, "internally" to NGO staff, and "downwardly" to the broader public and recipients of NGO services. Because trust is essential to NGOs' maintaining their social license to operate, their ability to tackle these issues will determine the sustainability of the sector itself.

See Also: Environmental Justice (Ethics and Philosophy); Environmental Justice (Politics); North–South Debate; Social Action.

Further Readings

Braun, Bruce. *The Intemperate Rainforest: Nature, Culture and Power on Canada's West Coast.* Minneapolis: University of Minnesota Press, 2002.

Jordan, Lisa. "Mechanisms for NGO Accountability." GPPi Research Paper Series No. 3, Berlin, Germany (2005). http://www.worldbank.org.kh/pecsa/resources/mechanismfor_ ngo_accountability.pdf (Accessed September 2010).

Lepsoe, Stephanie. *Wasted Opportunities? NGOs in Waste Management Bridge the Social-Environmental Divide.* Geneva: Centre for Applied Studies in International Negotiation, 2006.

Maathai, Wangari. *The Canopy of Hope: My Life Campaigning for Africa, Women, and the Environment.* Brooklyn, NY: Lantern Books, 2002.

Stephanie Lepsoe
Independent Scholar

NORTH AMERICAN FREE TRADE AGREEMENT (NAFTA)

The North American Free Trade Agreement (NAFTA) between the United States, Canada, and Mexico took effect on January 1, 1994. An environmental side agreement was added to NAFTA because of concerns that its main provisions did not do enough to ensure environmental protection and sustainable business practices. The side agreement allows private individuals, nongovernmental organizations (NGOs), corporations, and governments to challenge lax environmental oversight as well as to challenge government environmental legislation or regulations that interfere with NAFTA free-trade provisions. Environmentalists have debated NAFTA's environmental costs and benefits during its planning and implementation and remain divided over the issue.

Key NAFTA measures included a gradual phasing out of tariffs and other barriers to international trade and the liberalization of international investment regulations. NAFTA also included several key side agreements, including an agreement with regard to environmental issues critics felt were not sufficiently covered by the main body of NAFTA. Among the numerous controversies that arose before and since the implementation of NAFTA was the debate over whether NAFTA would have a positive or negative effect on the environment. Proponents of NAFTA viewed it as a key development in international law because it addresses the interconnections between trade and environmental policies as well as economic growth.

Some of the products and industries with possible environmental benefits and consequences that are affected by tariff eliminations under NAFTA include agricultural products, energy, and transportation. Agricultural products that are raised with the use of harmful chemicals, inadequate safety protections for workers, or inhumane methods of confinement and slaughter may find their way onto a greater number of international tables. U.S. and Canadian firms have the ability to invest in alternative-energy industries in Mexico, facilitating their expansion. The establishment of international transportation routes for trucks and other vehicles may increase traffic and its subsequent pollution. These are just a few examples of the environmental costs and benefits associated with NAFTA, leading to division over its support.

Environmental groups have been divided over the implementation of NAFTA since its earliest proposal and remain so to this day. Well-known environmental groups that have spoken out in support of NAFTA include the National Wildlife Federation, the World Wildlife Fund, the National Audubon Society, the Environmental Defense Council, the Natural Resources Defense Council, Conservation International, Defenders of Wildlife, and the Nature Conservancy. Well-known environmental groups that have spoken out in opposition to NAFTA include Greenpeace, the Sierra Club, the United States Public Interest Research Group (PIRG), and the Student Environmental Action Coalition.

Initial criticisms of NAFTA's lack of sufficient environmental protections led to the inclusion of a special side agreement on environmental issues, one of several side agreements added to NAFTA to address issues of special concern and aid in its passage. This agreement satisfied some early critics but not all. Some still argue that the side agreement, known as the North American Agreement on Environmental Cooperation, has limited provisions that are inadequate and ineffective. Goals of the agreement include supporting open discussion of environmental concerns, promoting economic preservation and sustainability, and balancing environmental and economic benefits. The Commission for Environmental Cooperation (CEC) was created to oversee the side agreement. Supporters of NAFTA point to the various methods incorporated to ensure that each country enforces its environmental laws and regulations.

The individual countries party to NAFTA may hold each other accountable for the enforcement of national environmental laws through a series of escalating steps, including consultations between the two countries, involvement of the commission, involvement of an arbitration panel, negotiation of an action plan, imposed fines, and the suspension of NAFTA benefits. Private individuals or NGOs may also file a complaint with the commission if they observe a pattern of environmental violations and lack of national government enforcement. Critics also note that the environmental side agreement does not establish any international economic legislation, but instead only calls for the continuation of each country's individual regulations.

Critics have also challenged NAFTA on the basis that national environmental laws that run counter to NAFTA provisions are left open to challenge and may be ruled unenforceable with regard to such provisions. Similar arguments were made earlier against other international trade agreements such as the General Agreement on Tariffs and Trade (GATT). For example, the United States may not be able to impose an embargo against Mexican agricultural products that were grown using pesticides banned in the

United States. This is possible because of the NAFTA provision contained in chapter 11 that member nations must consider all North American capital as domestic, regardless of its international origin. Thus, a company from the United States could sue the Mexican or Canadian governments and vice versa. The provision also relates to state, provincial, and local governments.

The threats to national sovereignty in the area of environmental protection were realized in the form of a series of international lawsuits based on corporate claims that national environmental laws violated NAFTA provisions. The most well-known example involved the U.S.-based Ethyl Corporation's 1997 lawsuit against the Canadian government's ban of the chemical gasoline additive methylcyclopentadienyl manganese tricarbonyl (MMT), which it manufactured. The case eventually went to arbitration, with the result that Canada was forced to overturn its ban despite the fact that MMT was a known carcinogen (cancer-causing agent). Canada also paid $13 million in damages plus legal fees. Such high-profile cases proved that corporate interests had taken precedence over national sovereignty and environmental interests under the framework of NAFTA.

NAFTA opponents argue that the outcome of such lawsuits reaffirms earlier fears that governments will be unable to hold international corporations accountable for environmentally irresponsible practices and that such corporations will no longer obey environmental laws and regulations if not held accountable. Critics claim that the industrialization of Mexico would increase at a faster rate as U.S. businesses transferred their operations to Mexico to take advantage of lax enforcement of environmental laws, resulting in further environmental degradation in Mexico. They point to scientific studies revealing increased rates of pollution in Mexico to bolster their claims. Such environmental consequences were added to already existing fears that free trade would motivate corporations to shift jobs to Mexico, where workers could be paid less and production was less expensive, thus damaging the U.S. and Canadian economies.

Many environmentalists have raised concerns over NAFTA with regard to Mexico and its legacy of environmental degradation and lax enforcement of environmental legislation. Although Mexico's 1988 Código Ecológico (General Ecology Law) is a comprehensive piece of environmental legislation with similarities to U.S. environmental legislation, its provisions and the other government laws are rarely enforced. Critics of NAFTA argue that widespread Mexican public distrust of the government and legal system, along with rampant political corruption, have resulted in a culture of acceptable noncompliance. They also note that the Mexican government lacks the financial resources, technological development, and infrastructure such as hazardous waste disposal facilities to effectively enforce its environmental legislation.

Environmental proponents of NAFTA countered such arguments by highlighting the ability of economic growth through increased trade to increase the international and internal pressure to strengthen environmental legislation, regulations, and enforcement. Such arguments rely on an environmental theory known as the Kuznets curve, which posits that there is a direct correlation between pollution levels and per capita gross domestic product (GDP) when based on national income. By this theory, Mexico's pollution rates would decrease per capita GDP as its level of national income increased. Critics note that so far statistical data has not shown that this relationship between trade, wealth generation, and environmental benefits is working in Mexico.

The environmental debates over NAFTA have implications beyond North America in the era of increased economic globalization. The environmental controversy surrounding NAFTA has been extended to debates over whether NAFTA should be extended throughout the Americas, including North, Central, and South America and the Caribbean with the exception of Cuba. This proposed extension is known as the Free Trade Area of the Americas (FTAA). Environmentalists already concerned with the negative impacts associated with NAFTA oppose its extension while those who support NAFTA's environmental benefits support its expansion.

See Also: Nongovernmental Organizations; Socially Responsible Investing; Sustainable Development (Business).

Further Readings

Audley, John J. *Green Politics and Global Trade: NAFTA and the Future of Environmental Politics.* Washington, DC: Georgetown University Press, 1997.

Baer, M. Delal and Sidney Weintraub. *The NAFTA Debate: Grappling With Unconventional Trade Issues.* Boulder, CO: Lynne Rienner, 1994.

Deere-Birkbeck, Carolyn and Daniel C. Esty. *Greening the Americas: NAFTA's Lessons for Hemispheric Trade.* Cambridge, MA: MIT Press, 2002.

Dreiling, Michael. *Solidarity and Contention: The Politics of Security and Sustainability in the NAFTA Conflict.* New York: Garland, 2001.

Faber, Daniel. *The Struggle for Ecological Democracy: Environmental Justice Movements in the United States.* New York: Guilford Press, 1998.

Hufbauer, Gary Clyde and Jeffrey J. Schott. *NAFTA Revisited: Achievements and Challenges.* Washington, DC: Institute for International Economics, 2005.

Hufbauer, Gary Clyde and Jeffrey J. Schott. *North American Free Trade: Issues and Recommendations.* Washington, DC: Institute for International Economics, 1992.

Johnson, Pierre-Marc and Andre Beaulieu. *The Environment and NAFTA: Understanding and Implementing the New Continental Law.* Washington, DC: Island Press, 1996.

MacArthur, John R. *The Selling of Free Trade: NAFTA, Washington, and the Subversion of American Democracy.* New York: Farrar, Straus & Giroux, 2000.

Mander, Jerry and Edward Goldsmith. *The Case Against the Global Economy: And for a Turn Toward the Local.* San Francisco, CA: Sierra Club Books, 1996.

Mayer, Frederick. *Interpreting NAFTA: The Science and Art of Political Analysis.* New York: Columbia University Press, 1998.

Moran, R. T. and J. Abbott. *NAFTA: Managing the Cultural Differences.* Houston, TX: Gulf Publishing, 1994.

Starr, Kevin. *Coast of Dreams: California on the Edge, 1990–2003.* New York: Knopf, 2004.

Suchlicki, Jaime. *Mexico: From Montezuma to NAFTA, Chiapas, and Beyond.* Washington, DC: Brassey's, 1996.

Marcella Bush Trevino
Barry University

NORTH–SOUTH DEBATE

International climate change discussions often include a debate between developed countries (referred to as the North) and developing countries (referred to as the South). In general, the North takes the position that all countries should protect their environments and enact policies to reduce the growth of greenhouse gas (GHG) emissions. The South argues that it should not be asked to undertake measures that could slow its economic growth and development. The South also argues that, because the North is responsible for a majority of the world's GHG emissions and because the North benefited historically from polluting during its own industrial growth, the developed countries should be primarily responsible for reducing global GHG emissions.

The North

The term *the North* refers to the G8 countries and other economically and technologically developed countries that, with the exceptions of Australia and New Zealand, are located in the northern hemisphere. The North does not include any countries in Africa or South America. There are two important attributes shared by much of the North that are relevant to this debate.

First, many countries in the North have contributed to environmental degradation since the Industrial Revolution. Collectively, earlier development by the North led to environmental degradation from mining and burning coal, deforestation, and other industrial practices. These practices helped the North to grow its economies and advance its technologies so that it is developed today. This is especially relevant in a debate about climate change because some GHGs, such as carbon dioxide, can stay in the atmosphere for hundreds of years. The term *historical emissions* refers to GHG emissions produced by a country in the past that are still present in the atmosphere. When historical emissions are accounted for, the North is responsible for as much as 90 percent of the anthropogenic carbon dioxide emissions.

Second, most countries in the North have much higher per capita emissions (average emissions from a country, per person, per year) than the per capita emissions in the South. In some case, countries in the North have per capita emissions that are 5 to 10 times that of some countries in the South. This point comes up as an argument made by the South.

The South

The South includes a broad range of countries, from small developing countries that are among the least responsible for climate change but will be among those most affected by climate change to China (the world's largest current emitter of GHGs), India, and Indonesia. In discussions at the United Nations, the South usually speaks as a group of 130 countries called the G77 and China. The South is developing economically and technologically. According to a 2010 United Nations report, 1.5 billion people lack electricity, and billions of people lack access to cars, airplanes, and other polluting forms of transportation.

There are several ways in which the South's emissions are much lower compared to the North's emissions. Collectively, the South is responsible for relatively little of the historical GHG emissions. Although the South's emissions are increasing to meet a growing demand for electricity and transportation, and some large developing countries are currently major GHG emitters, factoring in historical emissions means that the South is still responsible for only a minority (around 10 percent) of the elevated atmospheric levels of GHGs. Unlike the North, the South has relatively low per capita GHG emissions, with China's emissions around 6.1 tons of CO_2 per person per year, and India's less than 2 tons of CO_2 per capita compared to the European Union's average of 7.9 and the United States' average of 17.2 (2009 figures). Furthermore, some of the emissions from the South are produced in order to manufacture consumer items for the North, which can be seen as a type of emissions outsourcing.

History of the Debate

The North–South debate dates back at least to the 1992 Rio Summit (known as the Earth Summit), which was the first major international summit to address climate change. The summit led to several important agreements to pursue sustainable development and to prevent catastrophic climate change, including the United Nations Framework Convention on Climate Change (UNFCCC) and Agenda 21.

Developed countries wanted wide agreement about the principles of sustainable development, and developing countries wanted to ensure that the concept of sustainable development would not be used by the North to limit the South's economic growth or restrict the South's rights to pursue development policies. The principle of common but differentiated responsibilities, as it was written in Principle 7 of the Rio Declaration, helped convince developing countries that agreeing to the goal of sustainable development would not hinder Southern governments' abilities to pursue development. The principle agrees that states should cooperate in protecting the environment, but with a view that different states have made different contributions to environmental degradation. Because the developed countries are especially responsible for that degradation and because the North has more technology and financial resources, it should primarily be responsible for environmental protection. The principle of common but differentiated responsibilities is acknowledged in the UNFCCC and is frequently cited by developing countries in speeches regarding climate change or development.

The North's Argument

Generally, the North makes two main arguments when encouraging the South to practice policies that protect the environment. The first argument is that the way the North did things in the past is not a good model for how the South should do things today. The second argument is that if people are going to prevent catastrophic interference with the climate system (the purpose of the UNFCCC), then global GHG emissions will need to peak and then decrease, and that is impossible unless all major emitters, including countries in the South, reduce their emissions.

The first argument comes in two forms. Sometimes the North argues that pollution in the past happened at a time before a global awareness of the importance of sustainable development, and before an understanding of the negative effects of that development. It would not be the same, the North argues, for the South to continue developing in a way that is detrimental to the environment today because people have a better understanding of the negative consequences of that development on the environment. The second part of the argument focuses on the potential for the South to adopt new, cleaner technologies that were not available during the Industrial Revolution. Countries that lack technological development can "leapfrog" past older, more polluting technologies and adopt cleaner, more efficient technologies from the beginning of their development.

The North's second main argument more frequently applies to the countries in the South that are major GHG emitters. The North argues that, although there is not a consensus about how or when global emissions need to peak, or how quickly and by how much emissions should be reduced, there is essentially a consensus that global levels of GHG emissions need to peak eventually and then global emissions need to decline within this century. The North argues that without cooperation from all the major emitters, it will be impossible to ensure that global GHG emissions are reduced. If the developing countries with the highest emissions continue to increase those emissions, it could offset the North's efforts to reduce its own emissions. This argument is directed toward only some of the South, and it may actually resonate with the smallest and most vulnerable developing countries.

The South's Argument

Generally, the South makes three main arguments for why the North should primarily be responsible for reducing GHG emissions and why the countries of the South should be allowed to pursue economic development as their governments see fit. The first argument is to invoke the principle of common but differentiated responsibilities. The second argument is to discuss the importance of development to reach targets such as the Millennium Development Goals (MDGs). And finally, sometimes the South points out that it is just trying to achieve the same standard of living as the North.

The argument invoking common but differentiated responsibilities is raised frequently. The South points out that the North is historically responsible for a majority of the world's environmental degradation and GHG emissions and that it was that very polluting development that fueled the North's growth and allowed the North to be prosperous today. The North should use some of its money to help undo the effects of its pollution. This argument is made by most of the South.

The South's second main argument in the North–South debate is to frame international efforts to reduce GHG emissions and otherwise protect the environment as primarily an issue of the South's right to develop. The South points to the MDGs—internationally recognized goals for development in the South including reducing poverty, decreasing infant mortality, and increasing primary and secondary education. Pursuing policies that help countries meet these goals, the South argues, is a priority, and it can participate in reducing GHG emissions as long as it does not interfere with a country's sovereign right to development. Any policy that adds an extra cost to governments of the South, it argues, reduces their ability to meet the MDGs and cannot reasonably be expected to be enacted without funding from the North.

The South's final argument for why it cannot be asked to reduce its emissions can be summed up as "But we are just trying to be like you." This argument is generally used by the major emitters in the South. Major emitters cite their countries' relatively low per capita emissions and point out that it is unfair for the North to ask the South to reduce its emissions when the North has much higher per capita emissions and enjoys a higher standard of living.

One final point about the nature of the South side of the debate is important. Although most countries that make up the South do not want their development to be regulated or their emissions to be reduced, there is a growing group of countries within the South that do recognize the importance of the major emitters in the South reducing GHG emissions. This group is led by those countries, such as small island developing states, that are among the least responsible for and the most vulnerable to the effects of climate change.

A Resolution

There is, in theory, a global consensus that countries should work to prevent catastrophic climate change. The ultimate objective of the UNFCCC, which has been ratified by 193 states, is to achieve the "stabilization of greenhouse gas concentrations in the atmosphere at a level that would prevent dangerous anthropogenic interference with the climate system." While there may be differing opinions about what level of GHG concentration could cause dangerous interference with the climate, the North and South widely agree that global GHG emissions need to peak within the next few decades or sooner and to be cut substantially by the middle of this century. In fact, the South often calls for global emissions to peak sooner and to be cut more substantially than the North. The North and South will continue to debate who should be responsible for making those cuts. The debate will play out during international climate change negotiations, including the UNFCCC negotiations for a post-2012 commitment period (which marks the end of the commitment period covered by the Kyoto Protocol to the UNFCCC). Barring major technological advances in reducing GHG emissions, the North and South will need to cooperate in reducing emissions or else neither side will be able to achieve the ultimate goal of the UNFCCC and prevent dangerous climate change.

See Also: Horizontal Policy Integration; Kyoto Protocol; Nongovernmental Organizations; Sustainable Development (Politics).

Further Readings

Lohmann, Larry. "Carbon Trading: A Critical Conversation on Climate Change, Privatisation and Power." *Dag Hammarskjöld Centre Development Dialogue* (Uppsala, Sweden), 48 (September 2006).

Olivier, J. G. J. and J. A. H. W. Peters. "No Growth in Total Global CO_2 Emissions in 2009." Bilthoven: Netherlands Environmental Assessment Agency, June 2010.

Secretary-General's Advisory Group on Energy and Climate Change. "Energy for a Sustainable Future." United Nations (April 28, 2010). http://www.un.org/wcm/webdav/site/climatechange/shared/Documents/AGECC%20summary%20report%5B1%5D.pdf (Accessed January 2011).

United Nations. "United Nations Framework Convention on Climate Change." New York: United Nations, 1992.

Jesse Cameron-Glickenhaus
Independent Scholar

Not in My Backyard (NIMBY)

"Not in my backyard" refers to usage of land that people don't want near them, including homeless shelters, jails, truck garages, landfills, and incinerators, like this one in Chicago.

Source: Great Lakes National Program Office, U.S. Environmental Protection Agency

The term *NIMBY*, derived from the phrase "not in my backyard," refers to oppositional group activities related to locally unwanted land uses (LULUs), which run the gamut of undesirable activities, including mental or other outpatient health facilities; social service providers such as homeless shelters; correctional facilities such as jails, drug treatment centers, and halfway houses; infrastructure facilities such as roads, highways, railroads, and other public utilities; sanitation facilities, including garbage truck garages, trash incinerators, and landfills; and energy infrastructure, including both renewable and nonrenewable sources and processing stations as well as associated by-product waste stations.

NIMBY is often used in a derisive sense, as oppositional activities can be perceived to be detrimental to the functioning of cities and municipalities. Indeed, in some cases, services and functions considered "vital" have been disrupted by oppositional groups and tactics. As such, NIMBY activity can seriously disturb, delay, or, in some instances, completely thwart public and private efforts at land use change and development. However, there are other interpretations of the meaning and value of NIMBY behavior. Some see it as an exercise of the democratic process, as a means for interrupting dominant narratives of use, function, and value.

Nevertheless, the typical treatment of NIMBY-type groups and activities in the literature is that of a syndrome, an unfortunate and inconvenient problem to be dealt with and overcome. The NIMBY framework implies that opposition is strictly an unpleasant obstruction rather than an integral part of the land use debate. Thus, NIMBY groups and the individuals involved in them are stereotyped and devalued as selfish, parochial, irrational, protectionist, and reactionary; in short, they are framed as groups that generate conflict and hinder progress toward significant and supposedly universally desirable social, economic, and environmental goals.

Proximity to proposed facilities and/or land uses is the most frequently cited inspiration for NIMBY activity, and several authors offer profiles of the "typical" NIMBY activist: someone who is older, well educated, relatively wealthy, and likely to attend meetings. However, proximity is not the only concern mobilized by opposition groups. Other variables that can affect oppositional activities are the type and spatial scale of the facility/use being proposed, the character of the place and the people in that particular location, different perceptions or associations locals have with proposed uses, and fear of being branded as NIMBY and therefore losing legitimacy in the community and among decision makers.

More positive characterizations of the NIMBY phenomenon claim that NIMBY-like activities exemplify the democratic process. Researchers in this vein see NIMBY activists as pursing political and environmental justice, demonstrating grassroots public opposition despite persistent negative stereotyping and sometimes formidable odds against meaningful influence. In this way, political activism, like that of NIMBYs, can empower and engage marginalized communities and can help refocus debate, perhaps drawing collective attention outward to remind involved stakeholders of the larger contextual issues surrounding a particular situation.

Indeed, there are vocal critics of the NIMBY concept and its wide application to complex and multifaceted land use and facility siting situations. Some go so far as to call the theory "invalid," claiming it can obscure the real issues when it is applied improperly. Also, although the slow growth of the renewable energy industry has been at times blamed on NIMBY-type opposition, some claim that such groups do not have much influence on policy in actuality. Some note that although opposition efforts have garnered much attention, their efforts are generally effective only in delaying outcomes rather than stopping them altogether. Critics instead claim that institutional factors figure more importantly in the slow adoption of renewable energy technology and infrastructure, with NIMBYs being used as convenient scapegoats for ineffectual and sluggish implementation plans.

Framing local opposition groups as NIMBYs has two inherent assumptions: (1) the solution advanced by power holders is the only (and correct) one, and (2) local groups oppose the most efficient and rational solution based on self-interest, disregarding the public good. Yet even governments and other decision makers have motives and interests of their own; a particular solution is chosen from among any number of alternative options precisely because it holds some institutional, political, or economic value to the one(s) making the choice. This means that groups may not be opposing the "highest and best use," as advocated by policymakers or public servants, but may instead be simply offering a contrasting point of view.

In fact, there is wide agreement in some circles that local residents can and should be accepted as experts in their own right, experts of their own localities and particular situations. Some authors explain the broader meaning of facility-siting conflicts as an issue of control and definition of space, in that local opposition and political participation may stem from dissatisfaction with decision-making processes and the loss of control over proximate land uses. From this perspective, outside technocrats are seen as having valuable, specific knowledge, but they should nevertheless listen and take heed of the concerns and advice offered by local resident experts.

The NIMBY literature, however, is very focused on mitigating, redirecting, and managing opposition rather than examining the context of it, its attributes, and why local residents would in fact engage in it. The NIMBY framework rests on the idea that the rational-technical decision-making process may impose certain costs on local residents but only because those costs produce widely distributed benefits to the rest of society. NIMBY groups, then, are seen as simply putting themselves first, ignoring the wider benefit produced by the facility in question. At its core, the NIMBY concept pits "experts" and rational policymakers against narrow-minded self-interest groups, ultimately framing public participation in planning and land use debates as a "problem" that must be managed rather than as a resource to employ. However, greater, and perhaps fairer, debate about the localized and common interests served by various land use and facility-siting options may reveal a more nuanced context in which those decisions must be made.

See Also: Anarchism; Environmental Justice (Ethics and Philosophy); Environmental Justice (Politics); Nuclear Power (Health).

Further Readings

Dear, M. "Understanding and Overcoming the NIMBY Syndrome." *Journal of the American Planning Association,* 58 (1992).

Gibson, T. "NIMBY and the Civic Good." *City & Community,* 4 (2005).

Lake, R. W. "Rethinking NIMBY." *Journal of the American Planning Association,* 59 (1993).

Schively, C. "Understanding the NIMBY and LULU Phenomena: Reassessing Our Knowledge Base and Informing Future Research." *Journal of Planning Literature,* 21 (2007).

Takahashi, L. M. and S. L. Gaber. "Controversial Facility Siting in the Urban Environment—Resident and Planner Perceptions in the United States." *Environment and Behavior,* 30 (1998).

Wolsink, M. "Wind Power and the NIMBY-Myth: Institutional Capacity and the Limited Significance of Public Support." *Renewable Energy,* 21 (2000).

Colleen C. Hiner
University of California, Davis

Nuclear Power (Energy)

In 1953, President Dwight D. Eisenhower made his "Atoms for Peace" speech, declaring that nuclear energy could be used to provide energy to "power-starved areas of the world." In 1955, Arco, Idaho, became the first town in the United States to be powered by nuclear energy, and today there are more than 100 nuclear power plants in the United States. Still, the United States remains deeply divided over nuclear power, with some arguing that it is a viable method of generating cheap, clean energy and others contending that nuclear power plants are expensive, dangerous, and leave behind a legacy of nuclear waste that we still do not know how to manage. With the prospect of "peak oil" looming over the world and growing concerns regarding greenhouse gases and environmental devastation related to mining, refining, and burning fossil fuels, the nuclear power debate is incredibly polarizing.

The largest detriment facing nuclear power is the disposal of radioactive waste, which is stored here in large containers.

Source: Randy Montoya/Sandia National Laboratory

Nuclear Power: How Does It Work?

Nuclear energy is energy that is stored in the nucleus of an atom, and a nuclear power plant is a steam power plant that relies on a radioactive element like uranium for its fuel. The uranium is formed into ceramic pellets and then stacked end to end in 12-foot metal fuel rods; hundreds of these fuel rods are bundled together to form a fuel assembly—a nuclear reactor contains many such assemblies. The individual atoms that comprise the uranium are split apart by a process called fission, which is accomplished by firing neutrons at the uranium; this causes some of the uranium atoms to split apart. Each time this happens, more neutrons emerge from the split atom and strike other atoms; this process of energy release is called a chain reaction. Fission releases massive amounts of energy, and that energy is used to heat water until it turns into steam, which in turn pushes turbines, which then force coils of wire to interact with a magnetic field, generating an electric current.

This entire process produces no soot, smog, or greenhouse gases and does not contribute to global warming. The World Nuclear Association (WNA) estimates that if the world's nuclear facilities were replaced with coal-fired power plants, 2.6 billion tons of carbon dioxide would be added to the atmosphere annually. Still, there is a by-product of nuclear fission, and that is nuclear waste—the spent nuclear fuel from the reactor. Nuclear waste is highly radioactive and must be stored in steel-lined concrete pools or in dry concrete caskets. The storage of nuclear waste is one of the most contentious issues in the debate over nuclear energy; the United States accumulates more than 50,000 tons of nuclear waste annually, and with talk of building more nuclear reactors, that number will only increase. One example of how prickly the issue of nuclear waste has become in the United States is the debate over the Yucca Mountain storage facility for nuclear waste, which has been under development for decades, yet still has not begun accepting nuclear waste for storage (as of December 2010).

Nuclear Power: Safe and Viable

Nuclear power is the largest source of emission-free energy in the United States, providing 70 percent of the nation's carbon-free energy; one out of every five homes and businesses in the United States gets its electricity from a nuclear power plant. Further, 16 percent of the world's electricity is supplied by nuclear power, produced by 440 nuclear reactors in 31 countries; the United States has more reactors than any other country, with 104. France, with 59 nuclear reactors, generates 80 percent of its electricity from nuclear power—a higher percentage than any other country in the world. Further, in relation to nuclear waste, many of France's nuclear power plants reprocess spent fuel rods back into usable fuel, which defuses some of the arguments regarding nuclear waste.

Nuclear power plants do not contribute to global warming, and nuclear energy proponents also claim that they do not emit pollutants that contribute to haze, smog, or to human health problems. Nuclear fission, unlike the burning of fossil fuels, produces neither sulfur dioxide nor nitrogen oxides—the pollutants that cause acid rain. Further, nuclear fission is the source of energy that provides the most power for the smallest amount of fuel; one pound of uranium can supply the same energy as 3 million pounds of coal. This low fuel requirement is part of the reason why nuclear power is so cheap.

Another reason that advocates support nuclear power is that it reduces the United States' dependence on fossil fuels in general and, more specifically, on foreign oil. In fact, even many environmentalists have joined the ranks of those supporting the expansion of nuclear power in the United States, noting that the threat of global warming, considered by many to be the most pressing environmental threat today, makes nuclear power more attractive than many once thought. While no new nuclear reactors have been built in the United States since 1973, the Department of Energy recently approved a nearly $10 billion loan for the construction of two nuclear reactors in Georgia. While many environmentalists reacted negatively to this news, many others welcomed it. Individuals and groups supporting the proliferation of nuclear power include the following:

- Stewart Brand, founder of The Whole Earth Catalog
- The African-American Environmental Association
- James Lovelock, prominent environmentalist and Gaia theorist
- Patrick Moore, cofounder of Greenpeace
- Fred Krupp, director of the Environmental Defense Fund
- The Pew Center for Climate Change
- Stephen Tindale, former Greenpeace activist

It should be noted that among environmentalists who support nuclear power, some have grudgingly accepted it, as many bills that approve of the use of nuclear power also include measures for increasing funding for alternative energy sources, such as solar and wind power. Further, among those environmentalists who "support" nuclear power, while some view it as viable and worth pursuing, others have simply decided not to oppose it as strongly as they did previously, noting that concerns over global warming have provoked them to accept, reluctantly, the proliferation of nuclear power plants. Many environmentalists recognize that there is no single approach that can accomplish the goal of reducing the negative environmental effects of energy generation and energy use, and they see nuclear power as but one of many methods for scaling back environmental degradation.

Supporters of the proliferation of nuclear power in the United States also argue that nuclear power offers the following economic benefits:

- Each nuclear power plant employs approximately 500 workers from the local community and creates an additional 500 additional jobs in the local area.
- Each new U.S. nuclear power plant creates as many as 1,800 construction jobs and adds, on average, 700 permanent positions to support continued operations.
- Every dollar spent by a nuclear power plant generates about $1.13 for the plant's local economy.
- The economic activity of a nuclear power plant generates around $20 million in state and local tax revenues.
- Economic activity from an average nuclear power plant generates more than $400 million in total output and about $60 million in total labor income for the local economy.

In addition to environmental, economic, and political considerations, nuclear plant safety is also a point of debate among foes and advocates. Nuclear power supporters argue that nuclear power plants are among the best-defended industrial facilities in the United States, and they further note that nuclear plants today have state-of-the-art features in place to prevent accidents. The government's Nuclear Regulatory Commission (NRC) has a set safety goal for every reactor in the United States that mandates that the chance of an accident that results in radioactivity being released to the environment must be no more than one in a million. Further, nuclear power champions argue that the most damning information about nuclear waste is actually misleading; they contend that within 40 years, spent nuclear fuel has less than 1 one-thousandth of the radioactivity it had when it was removed from the reactor. Further, they argue that 95 percent of the fuel's potential energy is still contained within the fuel after the first cycle; recycling spent fuel, then, not only decreases how much nuclear waste is produced but also allows nuclear reactors to get even more energy from fuel than has previously been allowed.

Is Nuclear Power Dangerous and Unsustainable?

Opponents of nuclear power argue that despite the "clean" energy that nuclear power produces, there are numerous problems with nuclear power that render it both dangerous and impractical. For example, while new nuclear plants are considered safer than older plants, the consequences of a major accident remain the same: extensive and enduring radiation pollution. They point to the 1986 Chernobyl explosion, which killed 31 people and caused hundreds of thousands of cases of delayed illnesses, as evidence of the devastating implications of embracing nuclear power. Pennsylvania's 1979 Three Mile Island incident also illustrates just how potentially dangerous nuclear reactors are, though nearly all researchers who have investigated the Three Mile Island partial-core meltdown say that there have been no discernible adverse health effects related to the meltdown. In fact, the Three Mile Island incident was the last nuclear energy emergency in the United States, though this gives little comfort to those who point out that until the 2010 Deepwater Horizon oil spill; 1979 was also the last time there was a major oil disaster (the Ixtoc spill). This shows, at least for nuclear power opponents, that just because it has been several decades since something last happened is no guarantee that it will not happen again. For their part, supporters of nuclear power point out that comparing oil spills to nuclear accidents is irrational because Chernobyl and Three Mile Island forced a profound rethinking of nuclear reactor design and safety measures, resulting in modern nuclear plants that are much safer than their predecessors. Conversely, petroleum companies are not as focused on creating safer methods for recovering oil as they are on capturing harder-to-get-oil.

The March 11, 2011, earthquake and tsunami that devastated Japan provided a chilling rejoinder to the notion that decades without an incident implies that the threat is gone. Battered by the earthquake and tsunami, the integrity of several of Japan's 51 nuclear reactors was compromised, particularly at the Fukushima I and II power plants, each located in the Fukushima Prefecture, about seven miles apart from each other. Both Fukushima I and II were automatically shut down following the earthquake, though at both power plants, tsunami waves destroyed the cooling backup systems, which led to several explosions at Fukushima I, resulting in radioactive leakage into both the air and surrounding sea. A state of emergency was declared at both plants, and more than 200,000 people were evacuated from nearby areas. On March 18, Japan's nuclear safety agency rated the Fukushima accident a Level 5 (based on a seven-level ranking system), two steps below Level 7's "Major Accident" designation. There is intense debate about whether or not Level 5 is a severe-enough ranking for the Fukushima incident, which thus far has suffered partial-core meltdowns in several of its reactors.

As of April 4, 2011, engineers were no closer to regaining control of Fukushima I, which by now was known to be leaking radioactive water into the Pacific Ocean. It could be years before the full extent of the damage caused by the Fukushima accident is known, and concerned citizens in Japan, as well as in nearby countries (including the United States), are struggling to know how to interpret information regarding radiation levels and human health. For nuclear energy opponents in general, and particularly for those who have been alarmed by the proliferation of nuclear reactors in earthquake-prone Japan, this most recent incident is proof that nuclear energy is a risk not worth taking. Several years ago, seismologist Katsuhiko Ishibashi even coined a term for the danger associated with building nuclear reactors in a country known to be an earthquake and tsunami zone: *genpatsu-shinsai*, which combines the Japanese words for "nuclear power" and "quake disaster." While true genpatsu-shinsai, which implies a full meltdown and the deaths of millions of people, might have been averted this time, that fact gives little comfort to those who recognize it was luck as much as it was ingenuity that has thus far held this most recent nuclear incident in check.

Critics of nuclear power also contend that it is a misnomer to refer to nuclear power as "clean energy" because manufacturing the nuclear power plants, mining and enriching uranium, and transporting the processed fuel all typically rely on a carbon-dioxide-emitting fuel source. However, even when this entire process is taken into account, nuclear energy still warms the planet less than does capturing, refining, and burning coal or natural gas.

Even in the absence of a major nuclear incident in the United States in the past 30-plus years, environmentalists and other concerned citizens are troubled by the "near misses" that have occurred such as the 2002 discovery of a pineapple-sized hole in the six-inch-thick steel cap bolted to the top of Ohio's Davis-Besse Nuclear Power Station reactor. If the corrosion had penetrated less than half an inch deeper, radioactive steam would have flooded the containment dome and the long-dreaded Three Mile Island repeat may have occurred. This incident drew the ire of the Union of Concerned Scientists (UCS), which notes that there have been 47 incidents over the past three decades in which the NRC failed to address safety issues at nuclear facilities until the problems became so severe that the plants had to be closed for repairs. The UCS also notes that there were cases in which the NRC allowed reactors with known safety problems to continue operating for months, and sometimes years, without forcing owners to repair the problems. Critics use evidence such as this to argue that the nuclear power regulatory system is severely compromised.

Also of grave concern to nuclear power opponents is that 27 nuclear reactors in the United States are known to leak radioactive tritium, which is linked to cancer if inhaled or ingested through the throat or skin. Most of the United States' nuclear reactors are old and in need of renovation, and nuclear energy opponents are dismayed by the movement to build new reactors while so many old ones are in disrepair. Critics are appalled by lax safety measures that have led to failures to inspect thousands of miles of buried pipes, for instance, and they oppose renewing operating licenses for facilities that are need of updating and refurbishing.

Environmentalists and other nuclear energy opponents are also incredibly critical of the fact that after more than 50 years of nuclear energy in the United States, there still is not a long-term plan for storing high-level nuclear waste. The Yucca Mountain storage facility remains idle and, for now, nuclear waste is stored at various sites across the United States in water-filled basins about the size of swimming pools. Most of the pools are close to full and, according to a 2002 report by the National Academy of Sciences, vulnerable to terrorist attack (nuclear power opponents also argue that nuclear power plants themselves make attractive targets for terrorists). Though the spent fuel is supposed to stay in the pools for only six months, it is often kept in these pools for years before being transferred to dry-cask storage containers. This type of storage is viable for about a century, yet the used fuel rods will continue to be radioactive for 9,900 years after the cask system expires. While supporters of nuclear power point out that reprocessing spent fuel helps to alleviate this problem, high costs and earlier bans against reprocessing have kept this method from being used in the United States; further, reprocessing does not eliminate nuclear waste but merely decreases it.

Reprocessing fuel has other drawbacks, say critics. France's reprocessing operations have accumulated such a list of problems that in the United States they would be shut down under the Clean Water Act. France's reprocessing plants dump so much radioactive liquid into the English Channel—100 million gallons annually—that the sediment at the foot of effluent pipes is considered radioactive and nearby beaches have been closed to the public due to radioactive contamination. Elevated rates of leukemia have been detected in neighboring populations, and radioactivity from the channel has been detected as far away as Canadian Arctic waters. Ten European governments have demanded that France stop these practices but to no avail. In addition to pollution problems, the reprocessing of nuclear waste isolates plutonium, which can be used to make nuclear bombs; France currently has enough isolated plutonium to make 10,000 nuclear bombs. In 1974, India made its first nuclear bomb with plutonium skimmed off reprocessed nuclear waste.

Water use is also of grave concern to those who oppose nuclear power. Nuclear power plants use massive amounts of water, about 2 billion gallons a day, both for creating steam to drive turbines and for cooling the reactor towers; drawing and releasing this water has adverse effects on aquatic ecosystems, particularly because the water often returns to its source up to 25 degrees warmer; reactors' intake pipes kill fish, and their outflow distorts ecosystems to favor warm-water species. Further, uranium mining is dangerous and contaminates the environment. During the uranium mining boom in the U.S. Southwest between the 1940s and 1980s, more than 15,000 men worked the mines, often with no protection; many of these men were later diagnosed with cancer and other respiratory diseases. Further, when these mines closed, uranium tailings were left along the Colorado River, leaching at least 15,000 gallons of toxic chemicals a day into water destined for human consumption. Today, most of the uranium used in the United States comes from Canada, Australia, and Russia. Of course, this leads to another concern for those opposed to nuclear energy: it is not renewable, as it relies on a finite resource for its fuel.

Some also find cause for alarm in the fact that although no one has yet proved conclusively that proximity to a nuclear reactor can cause cancer, the question still has not been conclusively answered. Some studies link various kinds of cancer with proximity to nuclear reactors, though others refute these conclusions. It is known that nuclear plants routinely release small amounts of radioactive gases, so nearby residents are exposed to small yet reoccurring doses of radiation; the best that experts can say about this is that the levels probably will not increase the risk of cancer. Elevated levels of radioactive tritium, which gets into water and is easily ingested, have also been detected downstream from nuclear plants. And so, while researchers have not reached consensus on whether or not cancer and nuclear power are related, the scientific consensus most certainly holds that there is no amount of radiation that is good for you.

Finally, critics of nuclear power disagree with the claim that nuclear facilities are safe and well guarded. For example, in 2007, the NRC received evidence that guards were sleeping on duty at Pennsylvania's Peach Bottom nuclear facility; the NRC ignored this claim until a guard videotaped the naps and a New York TV station aired the footage. There is also evidence of guards being improperly armed and trained, and Wackenhut, a company that provides guards for nuclear facilities, was found guilty of tipping off guards in advance about security drills. Critics again fault the NRC for lax oversight and point out that the NRC, which is funded by nuclear industry fees, may be prone to buckling to conflicts of interest.

With the NRC currently reviewing about 20 applications for new nuclear reactors in the United States, the nuclear power debate is likely to get more heated, and those who are concerned with the environment will be forced to decide which is a worse legacy: global warming or nuclear waste. Further, in contrast to questions regarding nuclear power and human health, we know definitively that around 24,000 Americans die every year from diseases related to coal plant emissions, which leads to another of the more complex points of the debate: one source of energy is known to kill people every day, and another, only if something goes terribly wrong.

See Also: Alternative Energy; Carbon Market; Coal, Clean Technology; Nuclear Power (Health); Oil; Yucca Mountain.

Further Readings

Caldicott, Helen. *Nuclear Power Is Not the Answer.* New York: New Press, 2007.

The Guardian. "Fukushima Nuclear Power Plant Update: Get All the Data." http://www .guardian.co.uk/news/datablog/2011/mar/18/japan-nuclear-power-plant-updates (Accessed April 2011).

Smil, Vaclav. *Energy at the Crossroads: Global Perspectives and Uncertainties.* Cambridge, MA: MIT Press, 2005.

Tucker, William. *Terrestrial Energy: How Nuclear Energy Will Lead the Green Revolution and End America's Energy Odyssey.* Washington, DC: Bartleby Press, 2008.

Tani E. Bellestri
Independent Scholar

NUCLEAR POWER (HEALTH)

Few can dispute the fact that the world will continue to need more energy in the coming decades: even taking into account economic recessions/depressions, conservation measures, and environmental disasters or concerns, most government forecasts predict increases in world energy consumption. Significant debates exist, however, around how this energy will be produced. Nuclear power—though potentially on the cusp of a renaissance—is one of the more controversial forms of electricity production. This controversy stems from many sources: its history is deeply intertwined with the history of nuclear weapons; the risks of catastrophic accidents—though low in terms of frequency—are potentially enormous in scale; and significant uncertainties about the construction and maintenance of nuclear plants remain.

Yet nuclear power production, which has held mostly steady but has not substantially grown in the past two decades, is potentially set to experience growth once again because of a number of current and pending crises: the threat of climate change, increasing demands for electricity, and aging power supply grids that "prefer" consistent forms of baseload power (like nuclear, coal, and natural gas). As a result of these emerging concerns, the arguments for supporting or rejecting nuclear power—which held fairly steady over the past several decades—are now shifting. The four major points of debate we might have about nuclear power typically have to do with cost, safety, waste, and proliferation. Understanding these four areas and the uncertainties that are inherent in each can help us to know what is at stake with moving nuclear power forward, or not.

Cost

There are a number of ways to compare the cost of nuclear power with other forms of electricity generation—so many ways, in fact, that it can be difficult to perform common-sense comparisons. Usually, however, we look at electricity cost in terms of cost per kilo-watt-hour (kWh). For example, in the United States, coal costs consumers anywhere from $0.02 to $0.04 per kWh which keeps electricity bills relatively low. Electricity from solar power, on the other hand, currently costs anywhere from $0.14 to $0.60 per kWh, depending on the maturity and availability of the technology. Nuclear power, when it comes from power plants that are already up and running, is often quite competitive with coal, costing from $0.02 to $0.06 kWh on average. When it is coming from a new plant, however, nuclear power can cost consumers substantially more—$0.15 cents per kWh by some estimates—because consumers have to pay (over time) for the tremendous start-up costs of building nuclear power plants. These numbers are frequently subject to varying energy politics and methodologies and can be hard to pin down.

Another major cost related to nuclear power has to do with the tremendous capital needed to build nuclear power plants. Because the United States has not built any power plants to speak of since the 1980s, it does not have the infrastructure it needs to build more plants quickly or easily. Some large parts for power plants (such as containment vessels) may have to be manufactured in countries such as Japan and imported. Furthermore, because of the global economic recession, getting financing for nuclear power plants (which can cost $12 billion on the low end) is increasingly difficult.

Another matter of concern for those interested in calculating the costs of nuclear power has to do with "externalities." For example, some environmentalists might argue that fuel cost comparisons (say, between uranium and coal) do not include the costs of long-term disposal of nuclear waste for future generations or the environmental costs of coal pollution or of uranium mining. The pollution created by coal when it burns, some argue, should be included in the cost of the electricity produced from that coal. If there were to be a carbon tax, then burning coal could become much more expensive, making uranium seem relatively cheap. However, if the costs of storing nuclear waste were included in the costs of nuclear-produced electricity, those prices could increase dramatically.

Finally, building nuclear power plants in the United States is very expensive because, historically, individual corporations have been encouraged to build and compete for nuclear contracts. As a result, there are a number of reactor designs—all with their own particular engineering and regulatory needs—that could be used. This variability increases the amount of "red tape" and construction costs needed to build new plants. Contrast this situation with that of France, which endorsed nuclear power as its main source of electricity generation several decades ago and has streamlined its design, construction, and regulatory regimes so as to maximize efficiency and minimize cost.

One final word about cost: critics of nuclear power point out that there are other "hidden" costs to nuclear power that industry advocates conceal. These costs have to do with risk and insurance. For example, if there were a serious nuclear accident in the United States, no utility company or private corporation would be able to cover the costs of cleanup, reparations, lawsuits, and so on. There is no private insurance that could cover such a disaster. As a result, the government adopted the Price-Anderson Nuclear Industries Indemnity Act in 1957, which effectively indemnifies the nuclear industry in the case of a catastrophe. The federal government has also subsidized—in various ways at various times—the building and maintenance of reactors. Again, given these realities, it can be very difficult to determine the actual cost of nuclear-produced power.

Safety

Although cost may be of primary concern for policymakers and utilities that are actually invested in building more nuclear power plants, the primary concerns for those who do not want more power plants built relate to safety and waste. We will examine safety first.

Generally speaking, most nuclear scientists and engineers believe strongly in the safety of nuclear power plants, citing their "defense-in-depth" designs and their endlessly redundant safety features. Historically, there has been only one truly catastrophic failure of a commercial nuclear power plant, which occurred in Chernobyl, Ukraine, in 1986. Given the immense production of nuclear power from plants worldwide, this could be said to be a relatively strong safety record.

Advocates of nuclear power could also point to the fact that, although nuclear accidents are potentially massive and catastrophic, other forms of electricity production are equally dangerous; the risks they pose, however, are often slow and invisible, and therefore less likely to cause public fear and outrage. Coal-fired power plants, for example, are extremely polluting technologies, putting public health and safety at risk by contributing to climate change; decreasing air quality; polluting streams, habitats, and drinking water (via mountaintop removal); and even leading to coal ash spills, mine collapses, and other human and environmental disasters. But because coal burning is ubiquitous and fairly mundane as a technology, it is rarely feared with the same level of intensity as nuclear power production.

Finally, nuclear advocates argue that nuclear power—once plants are up and running—is virtually "carbon-free," has a relatively clean safety record, and provides readily available, consistent forms of "baseload" power using small amounts of fuel.

Yet many of these points are open for interpretation, and detractors of nuclear power provide compelling counterarguments. For example, although there have been no other disasters on the scale of Chernobyl to date, there have been multiple small accidents—some taking human lives—at research reactors in the United States and elsewhere. In 1979, a near-accident at the Three Mile Island plant in the state of Pennsylvania forced the evacuation of thousands of residents. Though most technical experts agree that significant radiation was not released as a result of the disaster, and that a crisis was in fact averted, critics of nuclear power point out that determining what we mean by "significant" amounts of radiation is full of uncertainty and interpretation. They see Three Mile Island as proof that nuclear accidents are indeed possible. They go on to point out that the accident demonstrated that human error, combined with even small technical failures, can pose huge risks. For these critics, Three Mile Island puts the lie to the "techno-optimism" of scientists and engineers who believe nuclear plants are nearly accident-proof.

Some are not willing to accept the risks of nuclear power because the scale of nuclear accidents is potentially so tremendous. The Chernobyl accident may be the world's most severe nuclear disaster, but its reach was truly tragic and horrifying. The radiation from that plant killed and sickened thousands (more if one considers subsequent illness and birth defects); stretched over much of western and eastern Europe, adversely affecting plant and animal life; and remains to this day in the areas around Chernobyl, which have been closed to human habitation.

The safety of nuclear reactors and, particularly, their ability to withstand unforeseen natural disasters and other risks were again brought into question as a result of the 2011 crisis at the Fukushima Daiichi plants in Japan. On March 11, 2011, at 2:45 pm, a 9.0 earthquake occurred off the east coast of the island of Honshu, Japan. Almost directly due west of the earthquake's epicenter lie the six reactors of the Fukushima Daiichi plant; shortly after the earthquake hit, a devastating tsunami washed over the coast of Honshu, causing extensive loss of life and damage in the Fukushima Prefecture and beyond.

Three of the six reactors at Fukushima Daiichi had been shut down for maintenance in the fall of 2010, meaning they were already relatively cool. But three of the reactors were still producing power, and the backup power supply systems for all six reactors were almost immediately disabled by the force of the tsunami. A series of explosions, fires, and releases of radioactive gases ensued at the three "hot" reactors. Of significant concern was the integrity of the spent fuel pools, which are housed directly above the Daiichi reactors. A number of these pools may have been compromised by the earthquake and ensuing hydrogen explosion, posing a serious radiation risk.

At the time of this writing, the outcome of the Fukushima nuclear crisis is still unclear. The full extent of the damage to the reactors and the spent fuel pools may not be known for some time. Currently, the International Atomic Energy Agency is rating the severity of the disaster somewhere between that of the Three Mile Island accident and Chernobyl. However, many nuclear scientists and engineers believe the Daiichi reactors withstood the two natural disasters relatively well—most of the containment structures at the six reactors seem intact despite the pummeling they took from the quake and tsunami. Nonetheless, the utility that owns the plants—Tokyo Electric Power Company, or TEPCO—will be working for some time to contain radiation leaks at the plant, including overflow of seawater that is being used to cool the plants and is feared to be leaking back into the sea, causing contamination. Those who are critical of nuclear power see this as yet more evidence that nuclear power is an unnecessarily risky and dangerous power source.

For now, there is significant debate about whether the radiation leaks out of Fukushima Daiichi will cause long-term health impacts in Japan and, if so, what those impacts will be. Further, the extent of the environmental impacts of the radiation releases is not yet known. The scientific community does seem to agree that the accident will have negligible-to-minor health and environmental impacts internationally: Most agree that radiation that reaches other continents will be widely dispersed and have little impact on health or environment. Whether this holds true in the long term remains to be seen.

Waste

Debates around waste disposal are significant to nuclear power concerns because of the nature of uranium (and sometimes plutonium), the fuels that power nuclear reactions. Unlike other fuel sources, spent fuel from nuclear power plants is highly radioactive and can remain so for tens of thousands of years. Developing waste solutions for spent nuclear fuel and other forms of nuclear waste is central to moving forward with nuclear power plant construction in the United States.

Some countries deal with their nuclear waste by reprocessing it. Essentially, this means that the spent nuclear fuel (fuel that is left over after being used to produce electricity) is recycled. The benefits of reprocessing are that fuel lasts much longer because it can be reused; it also means that the nuclear power plant itself produces less waste over the long term. Not all of the high-level radioactive waste (HLRW) from reprocessing can be reused: at the end of the reprocessing stage, there will always be some extremely potent HLRW that must be disposed of. Countries such as France and Japan reprocess their fuel or send it abroad to be reprocessed elsewhere, and thus they produce much less HLRW than the United States, which does not reprocess fuel.

All countries with nuclear reactors must have some form of nuclear waste storage. In the United States, most low-level radioactive waste (LLRW) is either stored in casks or other containers "on-site," meaning at or near the location where the waste is produced. HLRW is often also stored on-site in lined pools or other containment vessels. Nuclear scientists and engineers generally believe these temporary storage methods do not pose public health risks in the immediate or mid-range future. Most agree, however, that long-term storage solutions are needed. Other countries such as Germany have invested in geologic depositories for federal disposal of waste.

One such proposed long-term solution has been the Yucca Mountain repository in Nevada. Following the 1982 Nuclear Waste Policy Act (NWPA), the federal government embraced the idea of building one enormous repository for the country's nuclear waste. The Yucca Mountain site was chosen (for both political and technical reasons) in 1987 but has been a source of notable controversy ever since, plagued by resistance from Nevadans, technical disputes, and the waxing and waning of federal funding as different presidents and political parties have assumed office over the years. The site was initially scheduled to begin accepting waste in 1998. Due to intense opposition, the site has yet to open—and may never open—though construction is well under way.

Critics contend that even if Yucca Mountain were to begin accepting waste someday, it would not be able to house all of the nation's nuclear waste—the amount is just too great. One possible alternative, proposed by Secretary of Energy Steven Chu and others, is to build multiple regional repositories. This could reduce the distance nuclear waste would have to be transported for disposal and could increase the chances of a repository's political success if regional stakeholders were involved in siting processes (rather than having it forced on them, as was the case in Nevada). The future of such sites, however, remains incredibly uncertain, and the waste problem persists as one of the great stumbling blocks to public acceptance of nuclear power plant construction in the United States.

Proliferation

Concerns about nuclear proliferation are tightly bound with complex geopolitical relationships and highly technical details that can seem impenetrable to most of us. In essence, those who worry about nuclear proliferation fear not only that nuclear weapon stockpiles worldwide are unnecessarily large and poorly protected but also that waste from nuclear power plants is massive and poorly secured and regulated. This waste, critics fear, could be used by rogue nations, terrorists, or others to make "dirty bombs" or other weapons with devastating consequences. Furthermore, they fear that nations such as Iran have used nuclear power plant construction as cover for secret development of nuclear weapons.

The reason the United States has not historically reprocessed its spent fuel the way countries such as France do is that in the 1970s, concerns over nuclear proliferation were substantial. President Jimmy Carter ordered a halt to reprocessing plans in the hopes of setting an example of restraint for other countries. Furthermore, some feared that the spent fuel left after reprocessing would be ideal for bomb-making and could lead to terrorist attacks. Although this decision was later reversed by President Ronald Reagan, reprocessing has not been adopted in the United States, primarily because it is not cost-effective—it is much cheaper to simply store waste on-site—and there has been no movement from federal leadership on this topic.

For some, the building of nuclear power plants is still twinned with the legacies of nuclear weapons. They worry that as nuclear power plants proliferate, so too will the risks of accidents, the use of nuclear materials as weapons, and the centralization of electricity production.

Conclusion

In sum, the world we live in is already a nuclear world. Nuclear power plants are an essential part of our electricity production strategy in the United States, and it is hard to imagine that they will be phased out anytime soon: the demands for electricity are too great, and most experts believe that even alternative forms of energy (such as wind and solar) could not produce enough baseload power to displace nuclear. Furthermore, nuclear power does not produce carbon, and many believe that coal-fired power plants now pose an even greater risk than nuclear. It is possible that our emphasis should be placed on phasing out coal first, before attempting to phase out nuclear. Most likely, our future energy strategies will include some combination of all of these energy sources. But the question of whether to build more nuclear power plants in the future, and to what extent compared with other sources of power, remains entirely open.

See Also: Carbon Tax; Nuclear Power (Energy); Solar Energy (Cities); Wind Power; Yucca Mountain.

Further Readings

Allison, Wade. "Viewpoint: We Should Stop Running Away From Radiation." http://www .bbc.co.uk/news/world-12860842 (Accessed April 2011).

Caldicott, Helen. *Nuclear Power Is Not the Answer*. New York: New Press, 2006.

Cravens, Gwyneth. *Power to Save the World: The Truth About Nuclear Energy*. New York: Alfred A. Knopf, 2007.

Elliott, David, ed. *Nuclear or Not? Does Nuclear Power Have a Place in a Sustainable Energy Future?* New York: Palgrave Macmillan, 2007.

Fisher, Jenna. "Radioactive Leak Plugged, Officials Now Eye Hydrogen Buildup: Japan Nuclear Timeline." http://www.csmonitor.com/World/Asia-Pacific/2011/0406/Radio active-leak-plugged-officials-now-eye-hydrogen-buildup-Japan-nuclear-timeline (Accessed April 2011).

The Future of Nuclear Power: An Interdisciplinary MIT Study. Cambridge, MA: MIT Press, 2003 (updated 2009).

Herbst, Alan M. and George W. Hopley. *Nuclear Energy Now: Why the Time Has Come for the World's Most Misunderstood Energy Source*. Hoboken, NJ: John Wiley & Sons, 2007.

Perrow, Charles. *Normal Accidents: Living With High-Risk Technologies*. Princeton, NJ: Princeton University Press, 1999.

Silverman, Craig. "Misinformation Clouds Much Japan Coverage." http://www.cjr.org/ the_observatory/misinformation_clouds_much_jap.php?page=1 (Accessed April 2011).

Siracusa, Joseph M. *Nuclear Weapons: A Very Short Introduction*. Oxford, UK: Oxford University Press, 2008.

Jen Schneider
Colorado School of Mines

Oil

The word *oil* is on the cusp of shifting meaning again. Originally referring to olive oil and extended to other hydrophobic liquids, including other vegetable oils, essential oils, and mineral oils, the word *oil* by itself quickly came to refer to petroleum. In the 21st century, as alternative fuels explored in the previous century are being adopted with greater frequency, that usage will eventually stop being clear, and "crude oil" will become as necessary a disambiguation as "vegetable oil." One of the interesting aspects of the oil debate is the certainty that someday the world will be without oil. It may not be clear how much oil remains or at what rate it will be consumed, but it is certain that oil is finite and cannot be replaced: the processes that create fossil fuels are, as the name implies, so slow that they transpire on a geologic timescale. The reality of this fact is sometimes overlooked because so many predictions about oil depletion have proved false, but they were false only in their time statements, not their facts. The oil will run out. Despite this, the arguments in favor of its continued use are compelling.

Petroleum is a mixture of hydrocarbons and other organic compounds found beneath the surface of the Earth and recovered primarily through oil drilling or, more expensively, extracting the oil from naturally occurring mixtures of sand, clay, and viscous petroleum called oil sands. Today's oil supply was formed from the remains of prehistoric plankton and algae, organic material that settled to the bottom of lakes and seas, mixed with mud, and was buried under layers of sediment and, as a result of heat and pressure, turned into liquid and gaseous hydrocarbons over a period of millions of years. Algae seem to be the primary source of oil; prehistoric plant life undergoing similar processes became coal. There is a wide range of petroleum products: while fuels are the most common, petroleum is also used to make lubricating oils, asphalt and tar, paraffin wax, and a wide variety of petrochemicals (chemical compounds made from the hydrocarbons left in petroleum after using the rest to make fuel), which are in turn used to make products like plastics, adhesives, paints, inks, cleaning supplies, fragrances, cosmetics, and flavorings and food additives.

Oil reservoirs develop when oil has formed in a place where there is porous, permeable reservoir rock in which it can accumulate and a cap rock or some other obstacle that prevents it from rising to the surface. Such reservoirs usually contain three layers: a layer of natural gas on top of a layer of oil floating on a layer of water. These reservoirs extend horizontally over a large area and are exploited as oil fields, home to multiple oil wells both active (drawing oil up) and exploratory (looking for oil). In some cases, the natural pressure in a reservoir is enough to force the oil to the surface once a route has been drilled—this was the case with many of the reservoirs responsible for the early oil boom and many of those in the Middle East. Once the natural pressure eases, mechanical pumps must be used, and as the oil becomes more difficult to pump over time, injections of water, gases, or chemicals are used to flush out the reservoir. There are today more than 40,000 oil fields around the world, both on land and offshore.

An enormous petroleum industry involving both private industry and government has developed around the processes of oil exploration, extraction, refining, transportation, and sales, with oil fuels including gasoline (a petroleum-derived liquid) accounting for most of the industry's products. Since the mid-1950s, oil has been the world's leading source of energy, due to its abundance, energy density, and the ease with which it can be transported. Today, oil makes up about 40 percent of the energy consumed in North America; 44 percent in South America; 41 percent in Africa; 32 percent in Europe and Asia; and 53 percent in the Middle East. In the United States, much of that consumption is through the use of vehicular fuels—worldwide, petroleum products account for 90 percent of vehicular fuels. Only about 2 percent of the United States' electricity generation, by contrast, is provided by oil. About 80 percent of the world's easily accessible oil reserves are in the Middle East, 62.5 percent of them in Saudi Arabia, the United Arab Emirates, Iraq, Qatar, and Kuwait. Other than Saudi Arabia, the top oil-producing countries are Russia and the United States, which nevertheless imports more than two-thirds of the oil it consumes. Oil consumption in the United States is, in a word, extraordinary: the country is responsible for more than a fifth of global oil consumption despite having only a 20th of its population, and has a per capita consumption more than double that of the European Union, which is just as highly developed and industrialized. Further, American dependence on foreign oil has grown steadily, nearly doubling from 35 percent in 1973. The United States is not only the number one importer of oil but also the only top-10 producer that is also a top-20 importer. This rate of consumption, and the magnification effect it has on the consequences of oil price volatility, has been a driving force in U.S. politics domestically and internationally since the Cold War.

In addition to the Strategic Petroleum Reserve, the largest emergency oil reserve in the world with a supply of about 720 million barrels, the United States has proven oil reserves of about 21 billion barrels. A Department of the Interior estimate of the undiscovered recoverable oil resources throughout the United States is 134 billion barrels, though it is impossible to estimate the expense of exploration and recovery.

The Environmental Impact of Oil

Oil is a toxin. Whatever practical uses it may have, it is toxic to almost every form of life on Earth. Crude oil consists of numerous toxic and carcinogenic organic compounds, including benzene (which remains present in gasoline), which has been linked to leukemia, Hodgkin's lymphoma, and other immune system and blood diseases, even when the exposure to benzene is diluted enough to be measured in parts per billion. A single quart of motor oil is enough to make 250,000 gallons of ocean water toxic to aquatic life.

Marine oil spills from tankers, offshore platforms, drilling rings, and wells are a frequent concern, and despite advances in technology that should make them easier to prevent or quickly halt, 60 percent of the 10 worst oil spills in history occurred in the past

20 years prior to this writing, from 1990 to 2010. Marine oil spills do considerable damage to the wildlife of the seas and shores: most affected birds die without human intervention and laborious cleaning by hand because the oil permeates their plumage, making it more difficult for them to swim, float, or fly and quickly causing organ damage and hormonal imbalances as the birds ingest the oil while trying to clean themselves. Marine mammals become more vulnerable to the temperature, less able to stay warm, and may be poisoned by consuming oily water. The oil floating on top of the water blocks some of the sunlight that would otherwise feed marine plants by enabling photosynthesis, and it provides food to oil-consuming bacteria that displace other populations, upsetting the microbial balance of the ecosystem. Cleanup efforts themselves have potential environmental consequences. Dispersants, used to clean up oil spills the way detergents are used to clean up kitchen grease, break up globules of oil and cause them to disperse, the theory being that smaller droplets can degrade more easily. But many of these smaller droplets wind up in deeper water, sinking to where they can damage undersea life like coral. And seafood that survives oil spills can become contaminated, leading the seafood industry to train inspectors to detect oil taint in an attempt to stop tainted seafood from coming to market. Barring chemical tests, it remains to be seen how effective this will be.

When oil overtook coal as the world's leading energy source, the industrial world was already conditioned to view air pollution as a fact of life, so the toxic, polluting exhaust caused by burning oil for industry or in a car's internal combustion engine was unremarkable—no worse than what had come before. The combustion of oil is usually not complete, meaning that incompletely burned compounds present in the oil are released into the air along with the water vapor and carbon dioxide—notably, methanol and carbon monoxide. The soot that is released contributes to cystic fibrosis, cancer, chronic obstructive pulmonary disease, and respiratory problems.

Petroleum combustion heats the nitrogen in the surrounding air to the point of oxidation, creating nitrous oxides, while the sulfur in the oil is released to become sulfur dioxide. Sulfur dioxide and nitrous dioxides combine with precipitation to form rain with unusually high levels of hydrogen ions: acid rain. Though oil is not the only anthropogenic cause of acid rain—the ammonia of livestock production and the gases emitted by coal-burning power plants are major factors—it is certainly a significant one. Acid rain kills insects and small aquatic life, interferes with the hatching of fish eggs (and can kill adult fish in extreme cases), and has eliminated the brook trout from many of America's streams and creeks. Soil microbes are killed off, dramatically changing the ecosystem of the soil and thus its chemistry, which affects the trees and other plant life dependent on minerals in that soil. High-altitude forests are especially vulnerable to acid rain because of the clouds and fog that form around them and are more acidic than the rain itself. Acid rain also damages man-made structures, eroding stone with large amounts of calcium carbonate by creating gypsum, a brittle substance that flakes off rocks like limestone or marble. Copper and bronze oxidize faster under the effects of acid rain, one of the contributing factors necessitating the restoration of the Statue of Liberty in the 1980s.

The three main anthropogenic factors responsible for recent and ongoing climate change are the increased concentration of aerosols in the atmosphere, the increased concentration of greenhouse gases, and deforestation and other changes to the surface of the Earth; oil drilling and combustion contribute significantly to the first two of these. Vehicle fuels alone account for 14 percent of the global greenhouse gas emissions, which alone is significant; however, the oil industry is also implicated in the 11.3 percent contributed by fossil fuel recovery, processing, and distribution; and some of the 16.8 percent emitted by industrial processes originates with petroleum combustion. The minimum contribution of oil to greenhouse gas emissions, then, is about a quarter.

When vehicle fuels are mentioned, typically the gasoline and diesel used for cars and trucks come to mind. But aviation fuel has a significant and growing impact on total greenhouse gas emissions. A European Union study found that greenhouse gas emissions from aviation in the states comprising the current EU increased 87 percent from 1990 to 2006, for instance, despite substantial gains in aircraft fuel efficiency in that period. Petroleum-burning aircraft emit carbon dioxide (CO_2), nitric oxide, nitrogen dioxide, sulfur oxides, hydroxyl, tetra-ethyl lead, incompletely burned hydrocarbons, and significant amounts of soot and sulfate particulates, and civil aircraft have a total contribution of about 2 percent to global CO_2 emissions—but it is thought that many of the non-CO_2 emissions may have substantially greater effect when the aircraft is passing through or near the stratosphere, partially depleting ozone.

Environmental Impacts of the Oil Shale Industry

Oil shale is fine-grained rock containing large amounts of kerogen, the material that releases crude oil or natural gas when exposed to sufficient heat. When kerogen is processed, liquid hydrocarbons can be extracted but at a considerably greater cost and effort than recovery of crude oil from reserves. As oil reserves have been depleted and the cost of crude oil has risen, the cost differential between oil shale and oil reserves has shrunk, increasing interest in shale. As crude oil continues to become scarcer, turning to the 3 trillion or so barrels of oil shale seems inevitable. The environmental impact, however, is more severe than from oil reserve extraction.

Surface mining of oil shale requires significant land use and should not be done near highly populated areas due to the dust, pollution, and noise. The impact of mining is essentially the same as that of open pit mines, and reclaiming the land when the mining is completed is a costly and slow process. Disposing of mining waste, including ash and spent shale, requires additional land—and because of processing, the waste takes up more volume than the original oil shale, up to 25 percent more, so it cannot simply be returned to the space from which it was taken. Waste material includes toxic and carcinogenic substances, including polycyclic aromatic hydrocarbons, sulfates, and heavy metals, which can contaminate groundwater or leach toxic compounds into the soil.

Mining occasionally requires lowering the groundwater level below the level of the oil shale, which can affect surrounding land, disrupting the surface and depriving agricultural or forested land of water. The amount of water used in oil shale extraction and processing is significant, and the wastewater carries tar, phenols, and other toxins.

Carbon dioxide emissions are greater from shale oil and gas than from oil and gas processed from crude oil, and the decomposition of the kerogen in processing releases still more carbon dioxide, along with methane, in addition to any emissions that may result from the power necessary to process it.

Environmental Impacts of Offshore Drilling

Offshore drilling refers to the underwater exploration and recovery of oil and gas resources, especially in the ocean. Major offshore oil fields lie in the Gulf of Mexico, the North Sea, and off the coasts of West Africa, Newfoundland and Nova Scotia, and Brazil. Offshore production is more difficult because of the more challenging environment and technology and the significant logistics and human resources difficulties because of the remote location. Offshore facilities become small communities where workers live for shifts of two weeks or more.

In addition to oil spills, offshore drilling affects the environment through the drilling fluid, metal cuttings, and other chemicals released into the ocean over the course of operation—90,000 tons of fluid and cuttings over the lifetime of a rig, according to one study.

Offshore drilling has been the focus of most of the debate over oil in the United States in recent years. A 1990 executive order by President George H. W. Bush banned new offshore wells, but this order was lifted by his son, President George W. Bush, in 2008 in response to the previous five years of oil price increases. A moratorium on drilling on the Outer Continental Shelf (OCS)—divided into four regions: the Gulf of Mexico OCS, the Atlantic OCS, the Alaska OCS, and the Pacific OCS—is due to expire in 2012 and may or may not be renewed. The debate touches on issues both of the environment and of energy independence. The major offshore oil disaster in the Gulf of Mexico in 2010 considerably intensified the debate over offshore drilling.

Other than reducing usage of oil by an extraordinary amount, the only way to reduce the United States' dependence on foreign oil is to produce more U.S. oil, and offshore reserves represent the largest untapped domestic source, say proponents of broadening offshore drilling. This could free the United States from vulnerability to oil price shocks and to sanctions by oil-producing countries hostile to the United States, particularly those in the Middle East. There are already 5,500 offshore leases not in use and 68 billion barrels of oil estimated to lie in parts of the Gulf of Mexico and Alaska covered by the existing leases. If the oil industry is not finished exploring this land, it is not clear that they have the resources or initiative to conduct explorations on additional leases. Furthermore, the Department of Energy's Energy Information Administration has estimated that access to the currently restricted areas will increase domestic oil production by only 1.6 percent by 2030—a drop in the figurative bucket presuming consumption (total consumption, at that, not per capita) does not drastically drop in that time.

The strongest argument against offshore drilling, however, is that it is an effort that distracts from the long-term reality of the eventual depletion of oil. Expanding offshore drilling, even removing all restraints from drilling and oil exploration entirely, is, in a sufficiently long view, a short-term solution to U.S. energy needs. The resources expended on assisting the established oil industry should, by this argument, be diverted to putting serious resources in the hands of those who are developing alternative energy sources, especially renewable ones.

The Peak Oil Debate

"Peak oil" is the time at which global petroleum extraction rates have reached a maximum, after which point the rate declines. The peak theory was developed by Shell geoscientist Marion King Hubbert, who used the mathematics behind it to predict that U.S. oil production would peak between 1965 and 1970 and global production in the first decade of the 21st century. He seems to have been proved correct about U.S. production. It is contested whether global—and thus total—oil production has yet hit its peak. During the oil price shocks of 2006–08, the idea that peak oil had arrived was very popular, but some have argued that the perception of oil as a finite resource has been exaggerated by the Organization of the Petroleum Exporting Countries (OPEC) in order to keep prices high; this argument does not deny the finitude of oil, but considers it abundant and plentiful enough that peak oil is nowhere in the near future, nor the end of oil anywhere on the horizon.

The truth is that most of the actors engaged in oil at the international level—private industry producers, governments, multigovernment organizations like OPEC—have reason to obfuscate levels of production and amount of reserves, and our estimates about reserves not yet exploited are only that—estimates. There may be a great deal more oil remaining than we believe, but it is just as possible that there is a great deal less. Certain areas have been very poorly explored, either because of the difficulty and expense of doing so (as with many large oil shale reserves) or because they are "off-limits" (as with the Outer Continental Shelf and the Alaskan Wildlife Reserve). Arguments in favor of lifting those limits presume that vast quantities of oil exist to be tapped there. Industry and regulatory estimates are probably close—but estimates are still only estimates, not guarantees.

The cornucopian view, more often labeled as such by its opponents than by its advocates, argues that peak oil theory focuses too much on the impact of supply on oil production, losing sight of the role played by demand, and that there is no end in sight to the oil supply. Some argue that the end will never come—that oil is so abundant that to speak of its end is to speak of other far-off inevitabilities like the dimming of the sun or the statistical inevitability of an extinction-event-level asteroid colliding with the Earth, things we can say with confidence will happen but that will happen so far in the future that there may be no human civilization around to worry about them. Others believe that the end is merely far off, some hundreds of years from now—far enough off that current policy changes cannot possibly be used in preparation. Still others—a minority—believe in the abiogenic origin of petroleum, which is to say, that petroleum is not a fossil fuel at all but is a continually renewed abiotic (made from nonliving materials) product. The evidence does not support this point, but neither can it be proved in a laboratory that oil was made from algae.

There is opposition to the peak oil theory from factions other than the cornucopian camp. Some have challenged not the notion that peak oil has arrived or will arrive shortly but rather that it will "peak" at all. The Cambridge Energy Research Associates have proposed that instead of following a bell curve as Hubbert's model predicts, oil production will reach an "undulating plateau" subject to ups and downs lasting several years each while gradually rising overall, before gradually falling overall as reserves eventually deplete. Should this prove to be the case, it could be a very long time before there is hard proof that the "peak" has been passed; short-term fluctuations would make it harder to know how to read the significance of a production decline.

See Also: Biomass; Carbon Tax; Oil Sands; Sustainable Development (Business).

Further Readings

Goodstein, David. *Out of Gas: The End of the Age of Oil.* New York: W. W. Norton, 2005.

Jones, Van. *The Green Collar Economy.* New York: HarperOne, 2008.

Kunstler, James H. *The Long Emergency: Surviving the End of the Oil Age, Climate Change, and Other Converging Catastrophes.* New York: Atlantic Monthly Press, 2005.

Ruppert, Michael C. *Crossing the Rubicon: The Decline of the American Empire at the End of the Age of Oil.* Philadelphia, PA: New Society Publishers, 2004.

Bill Kte'pi
Independent Scholar

OIL SANDS

Bitumen—a dense, sticky form of petroleum—can now be extracted at a lesser cost from the sand, clay, and water of oil sands, but it generates as much as 45 percent more greenhouse gas than conventional oil.

Source: iStockphoto

Oil sands, also known as bituminous sands, tar sands, or extra heavy oil, are a type of crude oil deposit found in many countries, but particularly in Canada and Venezuela. In the United States, oil sands are mainly concentrated in eastern Utah, primarily on public land, though there are currently no oil sands projects in the United States. Oil sands are composed of a mix of sand, clay, water, and a particularly dense, sticky form of petroleum known as bitumen. Until recently, oil sands were not considered when evaluating the world's oil reserves because extracting the bitumen from the sands was prohibitively expensive; now, with new technology, coupled with higher oil prices, bitumen can be profitably extracted and upgraded to a usable product. Beyond the general concerns about continuing to rely on finite resources for energy and the damage that producing and burning oil does to the environment, making a usable product from oil sands generates far more greenhouse gases (GHGs) per barrel than does conventional oil—as much as 45 percent more. While champions of oil sands point to them as a resource to be refined and utilized, opponents are deeply concerned about the damage that extracting and refining the bitumen causes to the environment and to human health.

The United States imports more oil from Canada than from any other nation—more than 2 million barrels per day. Nearly all of this oil comes from Canada's oil sands, located in Alberta. Thus far, Venezuela has lacked the capital and technology to extract and refine its oil sands. Canada, however, is aggressively extracting and refining the Alberta oil sands, much to the dismay of those concerned with the environmental and human health impacts of extracting bitumen from the oil sands and then refining and burning it. Oil sands projects are the fastest-growing source of GHG pollution in Canada, and beyond the damage caused to the environment by GHGs, oil sands extraction not only uses copious amounts of water—between two and five barrels of water for every barrel of oil produced—but that water ends up in "tailings lagoons." It is estimated, based on the industry's own data, that these lagoons currently leak more than a billion gallons of contaminated water into the environment annually. Oil sand extraction and refining also requires massive amounts of energy—nearly a billion cubic feet of natural gas per day. In addition to the water and energy required to extract and refine oil sands, parts of Canada's Boreal Forest, the world's largest terrestrial carbon storehouse and site of the world's largest forest wetland ecosystem, have been slated for tar sands projects. Other, more general environmental concerns related to oil sands projects are related to the necessity of clearing trees and brush away as well as removing topsoil, sand, clay, and gravel in order to be able to mine the oil sands.

Beyond these devastating environmental impacts, oil sands projects also have grave implications for human health, related to both the extraction process and the refining process. The contaminated water that seeps from the tailings lagoons is directly affecting communities downstream from Alberta's oil sands projects; citizens of those communities have alarmingly elevated cancer levels, and their subsistence economies are threatened because of polluted fisheries. Because tar sands oil is burdened with more pollutants than conventional oil, the refineries that process tar sands oil release carbon dioxide, heavy metals, and sulfurs, exposing communities near these refineries to dangerous levels of these toxins.

Studies of communities downstream from the Alberta oil sands projects show high levels of carcinogens and toxic substances in fish, water, and sediment. While Alberta's own studies, conducted earlier than these more damning studies, dismissed such health concerns, residents and other concerned parties cite elevated cancer levels and have complained about the oily scum left behind in their glasses after consuming a glass of water. Researchers who have studied sediment near and around the oil sands projects have also discovered dramatic increases in the sediment of a certain type of hydrocarbon that is a natural carcinogen; in some cases, the levels of these carcinogens were as much as four times greater than the recommended limits in the United States; as of yet, Canada has no guidelines for this particular carcinogen. While the study concluded that treated drinking water in these communities is safe, high levels of mercury, arsenic, and the aforementioned hydrocarbon have been found in local fish. This is particularly upsetting for those residents who rely on fishing as part of their subsistence economy, especially members of Native communities, who depend on fish for a significant portion of their diets.

For their part, supporters of oil sands projects challenge the findings of studies that point to negative environmental and health consequences. They argue that the negative impacts are overstated or that they cannot be linked conclusively to the oil sands projects. When John O'Connor, a family physician, raised concerns in 2007 about possible links between Alberta's oil sands and cancer rates in nearby community Fort Chipewyan, he was discredited, and the Alberta College of Physicians and Surgeons filed charges against him for "causing undue alarm" among Fort Chipewyan residents. O'Connor was later vindicated when the Alberta Cancer Board conducted a study and found higher-than-expected rates of several rare types of cancer among Fort Chipewyan residents; O'Connor was cleared of the charge of "causing undue alarm." Still, the study stated that the increases could be merely due to chance and suggested continued monitoring of residents for the next 5 to 10 years.

Some who support the oil sands projects are those who disbelieve, in general, that human activity of any kind affects the environment in negative, irreparable ways. Others who support oils sands projects may believe in anthropogenic climate change and other undue effects of human activities but do not believe that the negative environmental effects of oil sands projects have been proved conclusively. These supporters, and others, point to the strides that Alberta is making to minimize the consequences of oil sand extraction and refinement. These measures include the following:

- Alberta's $2 billion investment in carbon capture and storage (CCS) technology; two of these CCS projects are specific to oil sands mining and refinement.
- In 2007, Alberta became the first jurisdiction in North America to legislate greenhouse gas (GHG) reductions for large industrial facilities.
- GHG emissions per barrel of oil from the oil sands have been reduced by an average of one-third since 1990; some facilities have achieved reductions as high as 45 percent.

- The amount of water permitted to be taken from the Athabasca River for all oil sands projects is equivalent to less than 3 percent of its average annual flow. During periods of low river flow, water consumption is limited to 1.3 percent of annual average flow, which means that, at times, industrial users will be restricted to less than half of their normal requirement.

Finally, some are more comfortable with importing oil from Canada, even from oil sands, than from Saudi Arabia, Venezuela, Nigeria, Angola, Iraq, Algeria, or Brazil, which are other top oil importers for the United States. Whether it is for political or other reasons, supporters of the Alberta oil sands projects cite discomfort with relying on what they characterize as unstable, volatile, and corrupt regimes. Setting aside whether these are fair assessments, the problem with this argument is that Canada does not produce enough oil to satiate the United States' appetite, so imports from these other nations will continue, whether or not the United States continues to import oil from Canada's oil sands.

In the end, like many of the dilemmas we are faced with when contemplating our future, there are no easy answers. It seems likely that extracting and refining bitumen from oil sands presents a host of problems for both humans and for the environment; yet, without a viable plan for utilizing alternative energy sources, many believe that the best we can do is to minimize the damage caused by these endeavors. Further, Alberta's oil sands projects employ more than 100,000 people who would be without work if these projects were to cease, and the oil sands projects provide the Canadian government with needed revenue. While many of Alberta's citizens, like concerned citizens elsewhere, are adamant in their demands that oil sands projects be halted, and while environmentalists decry the devastation of the environment that they believe is linked to oil sands projects, without a major shift in how we structure our lives, it appears that we will continue to rely on oil sands and other forms of unsustainable energy sources.

See Also: Carbon Capture Technology; Carbon Sequestration; Corporate Social Responsibility; Oil; Sustainable Development (Business).

Further Readings

Dirty Oil Sands: A Threat to the New Energy Economy. "The Dirt on Oil Sands." http://dirtyoilsands.org/thedirt (Accessed September 2010).

"Scenes From the Tar Wars." *Mother Jones* (May 2005). http://motherjones.com/politics/2008/05/scenes-tar-wars (Accessed September 2010).

Sweeney, Alistair. *Black Bonanza: Canada's Oil Sands and the Race to Secure North America's Energy Future.* Hoboken, NJ: Wiley, 2010.

Tani E. Bellestri
Independent Scholar

ORGANIC CONSUMERISM

Organic consumerism is arguably the only reason why there is any widespread discussion of organic agriculture and food, as it has established not only an organic industry in the food system, but has also created a constituency for the ideas around organic farming and food. This is not to suggest that it is not debated among those who support organic food, let alone those who oppose it.

Organic consumerism is a subsection of a wider movement of ethical consumption whereby people try to use their power to exercise choice in the marketplace to influence the production of goods and, in turn, wider society. This is obviously a power that is best used by those who have considerable influence as consumers, generally those in the wealthy West, but it is not confined to solely such groups. The roots of ethical consumerism lie in the 19th century but can be seen in acts of protest such as the Boston Tea Party, where the symbolic power of certain goods singled them out for acts of protest. During the 19th century, as the movement to end slavery raged, it was possible for consumers to buy sugar that was not produced by slaves. During the same period as the boycott, antislavery groups encouraged people not to purchase goods produced by slaves. This established a dynamic of using purchasing power to reward actions seen as "ethical" by consumers and punish those that are not. Although conducted at an individual level, it is only effective and discernable through the aggregation of many decisions.

Early Efforts

Organic consumerism developed since the 1970s as proponents sought new ways to advance their principles. The movement started in Europe in the 1930s, quickly diffusing through the European empires and through publications around the world. Opposition to farming technologies that degraded the soil was based on the observation that farming literally depleted the soil, as in the Dust Bowl in the United States. It was also lodged in the soil as a complex ecosystem on which civilization and human health was dependent, the health of which was either being ignored or being abused. The answer was a new form of agriculture based on Earth-friendly technologies such as composting, returning fertility to the soil, and discovering nature's patterns in order to grow crops and raise animals in a way that did not damage the soil. In those first few years, farming in this manner was a matter of personal experimentation, the most ordered of which was run by Lady Eve Balfour in England in 1939. Balfour's Haughley Experiment (attempted over the course of nearly 30 years) to run an experimental farm comparing organic and nonorganic farming systems was a remarkable exercise in science, supported by voluntary efforts and funds.

Toward the end of the 1960s, the Soil Association, which with an international membership had been backing the experiment since the 1940s, began to lose patience. This involved two factions: those who were more scientifically literate doubted that such a large-scale experiment would produce answers, and those who doubted that scientific evidence was the best course to advance the organic argument looked to practical example. While evidence of the damage done by chemical-dependent agriculture abounded, the organic movement's own scientific experiments did not produce the evidence it needed. The experiment ended in a swirl of debt and acrimony, leaving the organic movement in need of a new way to advance its cause.

Early efforts to label and certify goods as organic in the 1920s began in Germany, and talk of doing so was followed by experiments in the United Kingdom in the 1940s and 1950s. The appropriately named Whole Food Society started to sell organic produce in the 1950s from a store in London, but found it difficult to source supplies. The organizers found that they had to locate goods globally, such as dried fruit from India, France, and California, jams from Switzerland, and vegetables from France. Without an adequate supply of organic goods grown to agreed standards, it would not be possible to advance the organic movement through consumerism. By the late 1960s, the attempt to retail organic foods had led to a revival of the idea of organic standards, just as debate about how to advance organic farming reached a crisis as outlined above.

The early 1970s saw an increase in the supply of food as the counterculture and back-to-the-landers of the late 1960s looked for new ways to shop for food. Historian Warren Belasco describes how a wave of whole food or health food shops opened during this period to supply this demand. Behind these new stores, a new infrastructure of processing and manufacturing arose. Belasco uses the example of Celestial Seasonings, which started producing herbal teas that were caffeine- and additive-free, as well as organic. By 1978, it employed 200 people and had a profit of $9 million and was sold in 1984 to multinational Kraft Foods, which was looking for a way into what was a new market for it. In his book *Organic Inc: Natural Foods and How They Grew*, Sam Fromartz uses the example of soya processor White Wave to describe a similar journey from the counterculture to mainstream and supermarkets. Some early organic companies grew and literally sold out, others remained small and focused on supplying health food stores. Those that were bought by multinationals demonstrated the opportunities developing to take organics and "alternative" foods mainstream.

The Ethical Dimension of Organics

Organic foods became part of a broader trend in consumer goods that displayed an ethical dimension. By the early 1980s, this area was being led by ethical cosmetics through The Body Shop, led by Anita Roddick, and the broader Fair Trade movement, which seeks a fair exchange for goods such as coffee from developing nations. For some, this was the realization of countercultural ideas from the 1960s, and for others, it resembled the neoliberalism of Margaret Thatcher and Ronald Reagan. While many involved in marketing ethical or natural products could define their own terms, organic produce, in order to be considered such, had to be produced to already legally defined standards. It was the expansion of these standards through the 1980s in the United States and Europe that laid the final groundwork necessary for organic food to grow during the 1990s.

Contemporary multiple food retailers, or supermarket chains, have a wide range of quality assurance standards, most of which are not visible to the consuming public. The exact range and specifications vary between companies and the markets they are addressing. Standards may cover the size of the product, the conditions for the workers picking the crop, or packaging specifications. For the European-based retailers, these standards were developed after experiences with food poisoning, food adulteration, and scandals over labor practices like child labor. Although based in Europe, these standards are applied wherever production takes place. Fieldworkers growing fine beans in Kenya for French supermarkets will have their workplace run in accordance with the regulations. By the early 1990s, the organic movement presented retailers with a scheme that they did not have to operate and for which consumers would pay a premium, so it was perhaps not surprising that it became widely adopted by supermarket chains.

During the late 1980s, organic foods became fashionable. As celebrities like Sting campaigned to protect the Amazonian rainforest, many people sought to protect the environment through a new range of ecologically sound products. The consumerism and market choice of the period took on a green tinge, and even Margaret Thatcher addressed the perils to humankind from climate change.

The international recession of the early 1990s saw the sudden growth in demand for organic food decline, but the potential had been established, and as the recession receded, sales of organic goods began to grow again. From the mid-1990s and for more than a decade, sales of organic products grew strongly across most of the Western world.

An interesting case study of the growth of organic food sales is the United Kingdom (UK), which played an important role in shaping attitudes to food safety in Europe more widely. By the early 1990s, the UK had seen its first boom in the sales of organic food and then its decline during the recession. Legislation underpinning organic standards had been developed in line with the requirements of the European Community. Organic food received an inadvertent publicity boost in 1996 with the announcement of a link between mad cow disease, a fatal brain disease in cattle, and a human disease named Creutzfeldt-Jakob disease (CJD), leading to a new variant of the disease, nvCJD. Until that time, British cattle and their products had been traded globally, an industry that ceased as British cattle products were banned. With no treatment and no diagnostics for nvCJD, the British population could only wait to find out their fate. The only group that could demonstrate that their products were free of the disease were the organic farmers, who had previously banned use of the contaminated feed that spread the disease. For many, this was a dramatic affirmation of the arguments that the organic movement had been making for many years.

Following this crisis came the first attempts to introduce genetically modified (GM) plants into Europe. Sensitized by the outbreak of nvCJD, many European consumers considered the safety of any new food-related technology paramount. Although the U.S. Food and Drug Administration (FDA) found GM products to be substantially equivalent to their non-GM equivalents in 1992, European consumers were not convinced. Many European supermarket chains found that they were being threatened with a consumer boycott for stocking some products that contained GM ingredients. In 1999, Carrefour, the world's second-largest supermarket chain based in France, announced that it would seek to remove GM products from its supply chains. Throughout this controversy, the organic movement continued to publicize that genetically engineered products were banned under its standards of production.

The organic movement was not content to offer a safe haven from GM crops for consumers; it wanted them banned in order to safeguard the integrity of organic crops. Studies demonstrated that pollen from GM crops could fertilize organic crops and be found on organic produce if grown too closely to organic fields. The 200 meters suggested as a separation zone was argued to be inadequate as pollen could travel up to 180 kilometers and remain viable for nine days. The organic movement began to argue that it was not possible to grow both GM and organic crops, and that there was no case for GM crops. In order to make their point that even the scientific testing of crops as being conducted by the British government was not safe, the organic movement, in alliance with radical environmentalists, began to destroy test crops. The executive director of Greenpeace UK, Peter Melchett, was arrested in 1999 for deliberately damaging a GM crop. While Melchett was resting in jail in England, Jose Bove was arrested in France for dismantling a McDonald's as a protest against the corporation's use of beef produced using growth hormones and in defense of European food culture. These acts of protest that courted legal sanction and damaged property were backed up by rising sales of organic food. As sociologist Hugh Campbell has argued, organic food managed in these circumstances to become the opposite of GM crops, forcing many actors in the food chain to make a choice.

For many in North America, this account of how organic foods have become part of the political resistance to food technologies and big business will appear in sharp contrast to their experience. Journalist Michael Pollan and sociologist Julie Guthman have argued that corporate interests have undermined what is distinctive about organic agriculture. Guthman has argued that the logistics of business in California have forced organic production to become less distinct from nonorganic practices or "conventionalized." This has been joined with poor regulation of the organic standards–setting bodies and inaccurate association of organic food with food scares to undermine the credibility of organic products for many. Yet farmers markets and community-supported agriculture remain vibrant aspects of organics in the United States.

These differing experiences of organic consumerism in practice inform the critics of organic food both within and outside the organic movement. For authors such as Guthman, the difference between organics and the rest of agriculture has largely been eclipsed in the United States. This is echoed by other scholars who argue that organics are just another part of the selling of counterculture. It has been argued that organic foods benefit from an unwarranted amount of social status and consumption of them is only to emphasize how "cool" consumers are, and that the only claims being made about organic food are health claims, rather than environmental or social claims. Certainly, claims of the health benefits of organic food are not as easy to prove as claims that it lacks pesticide and antibiotic residues. The fact that organic farming secures environmental benefits and that it can create more employment in rural areas has been confirmed in scientific studies.

In his book *Market Rebels: How Activists Make or Break Radical Innovations,* Hayagreeva Rao argues that activists can make or break radical innovations. He identifies the alliance between the organic food movements and anti-biotechnology activists as having "created a formidable challenge for food companies purveying standardized and homogenized products to customers." His argument is that activists create "hot" issues that can be for or against a product, and then mobilize "cool" protests to press their case. Other scholars point out that consumers increasingly control or cocreate the brands of their favorite products.

Conclusion

In many ways, organic consumerism is an old story of expressing political and social preferences through choosing where to spend money. Organic products can be seen as a symbol of what purchasers oppose and what they aspire to see changed in the world—the possibility of a better future. The contemporary panoply of marketing and branding is layered on top of this. The tension between these two positions remains: Who retains control of the meaning of the actions that people undertake when buying organic food? This returns the discussion to whether consumption is solely a self-regarding action or whether it is involved in a more complex set of calculations about the individual in society. Significantly, the difference may be in the context in which that consumption takes place.

As in this discussion, the European and U.S. experiences have been compared, which was addressed by Julie Guthman. She argues that the only way to stop the pressures that lead to organic food becoming conventionalized is through state action. In Europe, the state acts in just that way through schemes that compensate farmers for protecting the environment and help farmers convert to organic farming. The context in which European consumers purchase their organic foods is also different, in that although dominated by supermarkets, the corporations are sensitive to their customers' concerns.

See Also: Agribusiness; Agriculture; Genetically Engineered Crops; Genetically Modified Organisms (GMOs); Irradiation; Organic Foods.

Further Readings

Belasco, Warren James. *Appetite for Change: How the Counterculture Took on the Food Industry*. Ithaca, NY: Cornell University Press, 2006.

Campbell, Hugh, Michael Mayerfeld Bell, and Margaret Finney. *Country Boys: Masculinity and Rural Life*. University Park: Pennsylvania State University Press, 2006.

Fromartz, Samuel. *Organic, Inc.: Natural Foods and How They Grew*. Boston, MA: Mariner Books, 2007.

Guthman, Julie. *Agrarian Dreams: The Paradox of Organic Farming in California*. Berkeley: University of California Press, 2004.

Pollan, Michael. *The Omnivore's Dilemma: A Natural History of Four Meals*. New York: Penguin, 2007.

Rao, Hayagreeva. *Market Rebels: How Activists Make or Break Radical Innovations*. Princeton, NJ: Princeton University Press, 2008.

Matt Reed
Countryside and Community Research Institute

ORGANIC FARMING

The actual size of the global organic market is difficult to estimate; however, in recent years it has become a multibillion-dollar industry. It is now regarded as one of the biggest growth markets in the food industry, and this has led to organic foods being viewed as mainstream rather than as a niche market. In developed countries, the public appears willing to pay additional costs associated with organic food production. Many one time conventional growers are in the process of transitioning their agriculture to organic in order to seek more lucrative markets for their products. But despite this tremendous growth, the entire organic food market still only constitutes a small percentage of overall farm production.

Alternating crops in a field, as shown here with soybeans and corn, can increase biodiversity and sunlight capture, all without the use of chemical pesticides or fertilizer.

Source: iStockphoto

Consumer demand for organic products, however, is helping to frame several global debates. These debates pertain to the sustainability of future food production and the world's natural ecosystems and resources such as its biodiversity. Though timely and relevant, the debates between conventional and organic farming advocates are also contentious. By purchasing organic foods, consumers can express their concerns economically about conventional farming methods and the environmental and health impacts of synthetic fertilizers and pesticides.

At the anticipated rate of growth, the total worldwide human population is expected to reach between 9 and 10 billion by 2050. World food demands must be met, and with current conventional farming methods, there is a general consensus that this can occur; however, there is little consensus on how this can be achieved by sustainable means that will also conserve current world biodiversity. The term *sustainability* in this context implies maintenance of high food yields consistently over time and agricultural practices that are fully measured and have acceptable environmental impacts.

Even though conventional farming methods have dramatically increased overall food production, they have also had detrimental environmental impacts. The environmental impacts of agriculture come from the conversion of natural ecosystems to agriculture, from agricultural nutrients that pollute aquatic and terrestrial habitats and groundwater, and from conventional farming practices and pesticides that reduce biodiversity. Currently, about half the world's usable lands are used for pastoral or intensive agriculture. Usable land can be broadly defined as land that is not tundra, desert, rock, or boreal. In order to meet food demands, more land will need to be converted for agricultural use. According to some predictions, over the next 50 years, nearly a billion hectares of natural habitat, primarily in the developing world, may need to be converted to meet demand. This agricultural expansion could threaten entire ecosystems and worldwide biodiversity on an unprecedented scale. In addition, pesticides, nitrogen, and phosphorus application rates may triple in order to double current food production.

Sustainable farming systems such as organic farming are now seen by many as a potential solution and in Europe are receiving substantial support in the form of subsidy payments. Advocates for conventional farming, however, assert that alternative forms of agriculture such as organic methods are incapable of producing the same amount of food as intensive conventional methods, given the same amount of land. They assert that if organic agriculture requires more land to produce the same amount of food, then this would offset any environmental benefit. They dispute organic growers' claims that sufficient quantities of organic nitrogen fertilizers can be supplied through the growing of leguminous alternating crops to continually meet the increasing food needs of the future. Critics also assert that organic food is more costly due to lower yields and inefficient use of land instead of due to the higher human-labor intensity involved in raising organic crops.

Organic Farming Benefits

Society receives many benefits from ecosystems such as food, fiber, fuel, and materials for shelter. Ecosystems also provide a range of benefits that are difficult to quantify and have rarely been priced, such as water and air purification and waste decomposition. Organic farming methods attempt to work in harmony with and utilize the natural functions of these ecosystems.

Organic sustainability refers to a farmland's or ranch's ability to be in continuous production with minimal external input. Sustainable methods used in organic farming can increase species diversity and soil matter content, while simultaneously decreasing pollution and soil loss. Typical conventional farm wastes are related to soil degradation such as soil erosion, organic matter depletion, and nutrient leaching. For conventional growers to maintain continuous crop production on depleted soils, fertilization, irrigation, and energy are continually required, and because the soil cannot hold these additives, they often run off, polluting waterways with nitrates and phosphates. Organic growers seek to eliminate inputs such as synthetic fertilizers by naturally replacing nutrients. Consequently, soil is conserved and little runoff occurs because minerals needed for plant and animal growth are continuously recycled.

Greater diversity can enhance stability within a system and minimize pest and pathogen problems. Diversity refers not only to the numbers of species living above and below ground but also to the genetic diversity and age structure within each population. The decrease in biodiversity observed on conventional farms has generally been attributed to farming methods such as monocropping, tillage, and the use of herbicides and insecticides. However, this type of farming, once viewed as ideal because it was thought to be easier for the farmer, results in lowered biodiversity and has led to many farmlands becoming more vulnerable to pests, pathogens, and natural disasters. Production of only a single product such as wheat, corn, or soybeans is risky, particularly if growing conditions are not ideal or pests and pathogens become resistant to expensive pesticides. To increase biodiversity, an organic farmer uses many techniques described below.

Effective management of water is another resource utilized by organic farming practices. Properly managing soil's natural water-holding abilities can result in the lowering of surface runoff, surface evaporation, drought incidence, and flood incidence. Also, minimizing or eliminating tillage, growing high-residue crops and cover crops, and adding compost or manure to the soil maintains ground cover and builds organic matter. Just raising the amount of organic matter within soils by a small percent can dramatically increase the available water content.

Organic Farming Methods

The aims of an organic farming system are to maintain soil fertility, avoid pollution, use natural means to produce crops, consider animal welfare, and participate in minimizing detrimental environment impacts. The two main principles that distinguish organic farming are that soluble mineral inputs are prohibited and synthetic herbicides and pesticides are rejected. Also, organic farmers can use only regulated natural pesticides and fertilizers. Other principal methods used by organic farmers are seed selection, crop rotation, livestock and green manure, cover crops, composting, mineral-rich rock powders, biological control of pests, and mechanical (by hand) cultivation and weeding.

Seed selection is the initial strategy used by organic farmers prior to producing a crop. Because organic farmers' economic success is not based on producing a crop priced by weight, they can instead focus on nutrition and taste. This allows them to select crop varieties that may appeal more to health-conscious consumers. Connected to this health-conscious attitude is the philosophy of feeding the soil so that the soil can then feed the plants. Many of the strategies employed by organic farmers are directed at enhancing soil quality. Crop rotation is the practice of growing dissimilar crops with different nutritional needs in the same area in sequential seasons. This technique can help balance soil fertility by avoiding excessive depletion of soil nutrients and improve soil structure by alternating deep and shallow-rooted plants. Also, alternating crops helps to keep any one specific crop's pest and pathogen numbers low by forcing them to endure long periods without a favored-food source. Strip intercrops (alternating row crops) can increase biodiversity within the field and increase sunlight capture for plants. Strip intercropping of corn and soybeans or cotton and alfalfa are two examples.

To naturally enhance soil nutrients, a traditional component of crop rotation is the replenishment of nitrogen through the use of green manure. Green manure begins as a cover crop that helps hold soil moisture, maintain soil integrity, and prevent erosion. Typically, these cover crops are leguminous, meaning they have the ability to fix atmospheric nitrogen (the primary limiting agricultural nutrient) through a symbiotic relationship formed with nitrogen-fixing bacteria that colonize their root systems. Typical leguminous crops are clover, peas, alfalfa, beans, lentils, peanuts, and vetch. The growth of leguminous crops acts as a natural fertilizer, and if the crop is not harvested, it can be plowed under, adding organic fertilizer to the soil. To prevent any nutrient loss, catch crops (fast-growing crops capable of utilizing excess nutrients) can be grown between seasonal crops or between rows.

Composting is another technique utilized by organic farmers to input nutrients into the soil. Composting utilizes the natural tendency of microorganisms in the soil to break down organic material. The necessary ingredients for composting are carbon, nitrogen, oxygen, and water. Once the material is composted, it can be mixed and added to soil as an amendment to increase fertility. Another advantage to composting if it is done correctly is that it can also destroy unwanted weeds, seeds, and many pathogens. Compost tea or compost steeped in water can also be sprayed on topsoil and non-edible plant parts during critical growth periods when additional nutrients are needed. Additional approved organic fertilizers may be in dry or liquid forms. Dry organic fertilizers typically are rock phosphate and blood meal. Liquid fertilizers such as fish emulsion or kelp are less concentrated but more readily absorbed. Use of organic fertilizers is known to result in higher levels of earthworms and other beneficial organisms being present in fields and increasing biodiversity. The presence of these organisms leads to more efficient utilization of nutrients, aeration of the soil, greater water retention, and over time, greater soil vitality.

Organic farming methods can greatly reduce the impact of pests and pathogens; however, they can still be a problem for growers. The use of biological control agents involves the use of living organisms or their metabolic by-products to control pests or disease-causing pathogens. Control typically relies on predation, parasitism, and herbivory. The strategy of a biological control is to limit and manage the damaging effects of crop pests and pathogens while not eliminating them as with pesticides, which tends to encourage resistance over time. Organic farmers must use only approved biological control agents as their use is regulated.

Some additional methods are mixed farming, undersowing, and on-farm feeding of livestock and manure use. Crop borders, windbreaks, and hedgerow management along with the creation of noncrop habitats can enhance biodiversity and encourage beneficial organisms. Additional plantings of shrubs and trees can increase ecosystem diversity still further. Organic farmers use all these techniques and more in varying degrees to increase biodiversity and enhance the natural sustainability of the land.

Regulation

Even though organic farming in the United States and Europe is currently regulated, there is still considerable confusion surrounding the term. Most consumers have heard of the term *organic* and are aware that it is chemical free; however, most are unfamiliar with organic farming standards and practices.

All growers planning to market their products as "organic" must become certified. For the grower, the certification and regulation process can be tedious and time-consuming, but it was developed to assure consumers that food products labeled as "organic" were grown according to specific conditions. Prior to the certification process, there were several well-published accounts of false marketing by disreputable growers. Organic farming is subject to national and international law. In the United States, to ensure product integrity, the Organic Foods Production Act was passed in 1990, which mandated creation of the National Organic Program (NOP) and passage of uniform standards. In Europe, organic farming is controlled and marketed according to EC Regulation 2092/91.

In order to meet NOP regulations in the United States, organic producers are prohibited from using synthetic fertilizers and pesticides and must demonstrate their products are free of genetically modified organisms (GMOs), although for the latter, some trace amounts are allowed. Land upon which organic food is to be produced must be free of these products for at least three full years preceding harvest of the initial organic crop. Farms or specific fields not yet meeting this requirement may be considered in transition. Organic livestock producers have additional responsibilities and must agree to manage their livestock in ways deemed as not cruel while taking into account animal behaviors and providing pasture and outdoor access.

Additional information required from organic growers includes details about soil fertility, seed varieties, weed and pest management practices, and storage and handling routines. The farming histories of all fields are required, along with farm maps. The grower also must develop strategies to prevent contamination with prohibited substances and commingling with nonorganic products. Equipment and storage areas employed in organic production must either be dedicated to organic use or properly cleaned between conventional and organic use. All organic standards and regulations need to be closely adhered to as site inspections can occur. At times, growers have lost their organic certification for various reasons, resulting in a negative impact because the grower also loses the organic price premium.

Marketing Trends of Organic Foods

Public interest in organic food has grown in part due to consumer concerns over the broad use of genetically modified organisms, health and environmental effects of pesticides, and well-publicized food safety breaches. These concerns and the purchasing power of consumers have raised important questions for governments, growers, distributors, retailers, industry planners, and marketers. In an attempt to better understand consumers' attitudes regarding organic foods and to better understand who buys organic foods and why, numerous surveys have been conducted.

The resulting information reveals there are two main groups of organic consumers: those who purchase organic foods on a regular basis and those who purchase occasionally. In general, organic consumers tend to have children. It may be that families initially try organic food with the arrival of a newborn by selecting to purchase organic baby food. Also, even though younger individuals tend to have more favorable attitudes toward organic food, older consumers are more likely to purchase it. One explanation is that due to the price premiums on organic food, it may simply be more affordable for older consumers.

As for motives behind the purchase of organic foods for both regular and occasional consumers, health was viewed as the most important reason. Consumers are buying organic foods because of their desire to avoid chemicals used in conventional food production. The use of pesticides is perceived to be associated with long-term and unknown effects on health of humans and may be environmentally harmful. Concerns about food safety have also been identified as a reason for consumers selecting organic food. Recent food scares such as *Salmonella*, BSE (mad cow disease), and *E. coli* 0157 outbreaks have all contributed to increasing concerns about conventional food production methods.

Even though current consumer opinions about organic foods are generally favorable, reasons why consumers may not purchase organic foods include cost, lack of availability, and/or inconvenience. Some consumers are unwilling to accept blemishes or imperfections present in organic produce. For some consumers, a persisting level of skepticism continues to surround organic food labels, leading to questions concerning the genuineness of organic products. For whatever reason, many consumers are currently satisfied with conventional food sources.

Whether an individual is an advocate for conventional, organic, or a blended form of farming practices, there is little doubt that interest in organic farming is helping to focus a needed examination of the future global food supply. Consumers are using their most powerful weapon—their money—to encourage growers and all sectors of the multifaceted food industry to examine public concerns regarding food safety, environmental, biodiversity, and sustainability issues.

See Also: Agribusiness; Agriculture; Animal Welfare; Certified Products; Organic Consumerism; Organic Trend; Sustainable Development (Business).

Further Readings

Badgley, Catherine, Jeremy Moghtader, Eileen Quintero, Emily Zakem, et al. "Organic Agriculture and the Global Food Supply." *Renewable Agriculture and Food Systems*, 22 (2007).

Durham, Leslie A. *Good Growing: Why Organic Farming Works*. Lincoln: University of Nebraska Press, 2005.

Friedman, Laura S., ed. *Organic Food and Farming*. Westport, CT: Greenhaven Press, 2010.

Hole, David G., Allen J. Perkins, Jeremy D. Wilson, Ian H. Alexander, et al. "Does Organic Farming Benefit Biodiversity?" *Biological Conservation*, 122 (2005).

Hughner, Renee S., Pierre McDonagh, Andrea Prothero, Clifford J. Shultz II, et al. "Who Are Organic Food Consumers? A Compilation and Review of Why People Purchase Organic Food." *Journal of Consumer Behavior*, 6 (2007).

National Organic Program (NOP). http://www.ams.usda.gov/AMSv1.0/ams.fetchTemplate Data.do?template=TemplateA&navID=NationalOrganicProgram&leftNav=NationalOrga nicProgram&page=NOPNationalOrganicProgramHome&acct=AMSPW (Accessed July 2010).

National Sustainable Agriculture Information. http://attra.ncat.org/attra-pub/organcert.html (Accessed July 2010).

Newton, John. *Profitable Organic Farming*. Oxford, UK: Blackwell Science, 1995.

Tilman, David, Kenneth G. Cassman, Pamela A. Matson, Rosamund Naylor, et al. "Agricultural Sustainability and Intensive Production Practices." *Nature*, 418 (2002).

Mary Ruth Griffin
Independent Scholar

Organic Foods

Organic food often appears in journalism and marketing as a simple product—food grown without the use of artificial fertilizers, pesticides, growth hormones, fungicides, irradiation, and genetically modified organisms. For animal products, the beast in question needs to have spent most of its life outdoors, not be restricted in its movements, not be subject to antibiotics or hormones to boost its size rather than treat disease, and be fed almost entirely on organic produce. Many describe this as food that is more "natural" or more "traditional" in its production; it certainly measures a gap between organic farming practices and those of most nonorganic farming. As the following discussion demonstrates, one of the clear differences between organic food and its nonorganic counterparts is a foregrounding of the importance of how food is produced.

Beyond the production system, organic food has for most of its proponents two important advantages for the human or animal consuming it. The pioneers of organic farming viewed organic food as having more life-enhancing qualities than its nonorganic equivalents. At a time when little was understood about vitamins or trace elements, the claims were that organic food was better because it had additional properties—either opaque to science or, in some instances, mystical, so beyond the reach of science. The additional benefits of organic food, however defined, would be transmitted to the humans eating them, with positive impacts on their health. During the 1950s, this began to change as organic food became defined by the absence of certain qualities that were thought to be carried on nonorganic food. After interventions such as Rachel Carson's *Silent Spring*, consumers became increasingly concerned that there were residues of pesticides, in particular, on the food they were eating. Pesticides were increasingly being linked to damage to ecosystems and to the health of animals, and concerns grew that persistent exposure to low levels of pesticides was affecting human health. It was these latter concerns that were important in raising the public profile of organic food as various "scares" about high levels of pesticides were brought to the attention of consumers, raising questions about the effects of lower levels of the chemicals as well as potential "cocktails" when different chemicals combined with unknown consequences.

Concerns about pesticide residues were generally confined to fruits and vegetables, and it is perhaps in the area of the production of animals and animal products that the greatest differences can be seen between organic and nonorganic food. The pioneers of organic farming in the 1930s claimed that their animals were more resistant to animal disease because of how well they were fed and kept; they therefore opposed the use of vaccines as unnecessary and potentially dangerous to human health. It was during the 1960s, as intensive animal production was developed, that the organic movement began to reshape its concerns about meat and animal products.

In addition to concerns about the use of pesticides to protect animals from insects and infestations, there was particular worry about the use of antibiotics in animals. In 1967, the British government was concerned that residues of antibiotics were being carried into humans via meat and that resistance to antibiotics could develop in animals and then pass to humans. At the same time, the first growth hormones were being used commercially in farm animals, a development also opposed by the organic movement. The new combination of technologies for animal production—antibiotics, concentrated animal feeds, close confinement, and growth promoters—were all opposed by the organic movement. This opposition was not only to the production of flesh but also of milk and eggs using similar systems. The organic farming systems developed to oppose what had become often-common technologies in the production of animals.

At the beginning of the organic movement, the emphasis had been on the production of basic foodstuffs that required no or very little processing, with most that was necessary taking place on the farm. As the scale and scope of organic production increased, a number of questions were raised about the extent of processing possible for something to still be called "organic." The most immediate of these was around the pasteurization of milk. Many in the early organic movement in the 1930s saw pasteurization as being unnecessary if the animals producing the milk were healthy and believed that many of the health-giving aspects of organic milk were to be found in its raw or natural state. Over time, this argument receded as organic milk was sold in stores and distributed more widely, but it informs the attitudes taken about processed foods by many in the organic movement. Organic products should not be subject to processes that destroy the qualities of the ingredients, and neither should there be additives that might be prejudicial to human health. Although seemingly restrictive, this does allow for the combination of organic ingredients into complex processed projects such as bread, cola, confectionary, and prepared meals.

The seeming simplicity of organic food's being based on the avoidance of certain technologies belies a complex reality of realization. When is a pesticide "artificial" and when it is "natural," what proportion of a animal's rations need be organic for it to be organic, or is it possible to have ultra-heat-treated (UHT) organic milk? Even when these things are agreed, how is it possible for a consumer to know that the product has actually been produced in accordance with these strictures? An early certification scheme was experimented with by organic groups in Germany in the 1920s, was made moot again in Britain in the aftermath of World War II, and was being used by some groups in California in the 1950s. The effort to galvanize some form of assurance scheme started in the United Kingdom (UK) in the 1960s and was almost simultaneously adopted in California and the UK in 1973. This scheme was a combination of peer accreditation and a public list that allowed those in compliance to use a symbol denoting their produce as being organic. A farmer or producer would agree to abide by a set of restrictions, as part of which they would record their activities and someone appointed by the certification body would inspect this on an annual basis. Generally, these standards of production instructed farmers and growers as to what they were not permitted to do, allowing them to create a farming system suited to their circumstances and allowing for the variety of opinions within organic farming.

The first standards arrived in the UK and California in 1972 as in Britain the Soil Association and in the United States the Californian Certified Organic Farmers (CCOF) put their schemes in place. In the UK, a competition was held among members to provide a symbol to denote the presence of certification and since that time, these forms of certification have been often called "symbol schemes." The first legislation backing organic certification appeared in California in 1979 with the California Organic Food Act. Consumer legislation became the way in which organic food standards are observed, as to be "organic" became a legal definition and those claiming it without proof are liable to prosecution. In turn, this led to a proliferation of terms that imply organic status without the rigors of certification, for example, *pesticide free*, *chemical free*, or variations on *natural* production.

Globally, the most significant legislation started its long route to enactment in the European Community at the suggestion of Denmark in 1981. For organic goods to be freely traded within the European Community, there needed to be a common definition of "organic" in place for member states to enforce. Once the European Commission (EC) in 1987 required states to put in place such legislation, this accelerated the development of many national organic movements and provided templates for nations outside the EC to follow. The legislation started to come into place in 1992, along with support for conversion to organic production and a suite of research programs to foster the development of the market. European organic standards and the market for organic goods began to play a central role in fostering export-orientated production in Australia and New Zealand.

Federal regulations in the United States came into force much later and with a far greater degree of contention than in Europe. Many opposed to organics either directly or through seeking to weaken organic standards saw the creation of the National Organic Program (NOP) as an opportunity to intervene beyond the reach of state or local organic initiatives. In response to public consultation in 1998, the U.S. Department of Agriculture (USDA) received more than 250,000 replies from those opposed to attempts to include genetically modified crops, sewage sludge, and animal confinement under "organic" standards. After being passed in 2002, the federal organic standards have been consistently blighted by poor enforcement.

One of the most common models for enforcing organic regulations is that there are baseline national or federal standards; individual bodies/organizations can apply to the governing agency to offer certification services, and certification is granted to products that either meet or exceed those standards. The explicit goal of many of those framing these policies was to create a market in standards. In many jurisdictions this has been thwarted, as existing organic bodies establish not-for-profit organizations to provide those services. The UK saw the Soil Association set up a certification body that now conducts those services for more than 70 percent of produce sold in the UK, making the organic movement dominant in establishing standards. In contrast, in 2008 the NOP suspended temporarily 15 of the 30 entities authorized to inspect U.S. organic organizations because of failures in their activities.

Organic food is far from simple, a point that many who market organic food belabor as they argue it does not have a narrative that can be easily sold to consumers, while many in the organic movement argue that it is this complexity that is the point of organic food. Broadly, rather than traditional or natural, it can be argued that organic food is a social product, and "organic" is defined by the processes of certification and labeling that allow it to carry the label "organic." The specific meaning of that label will vary among societies, jurisdictions, and supply chains. In turn, this affects the meaning of "organic" in particular areas, with an increasing gap opening between organics in the United States and the EU. While in the United States "organic" is embroiled in controversies over food safety, corporate dominance, and the veracity of the products' organic status, in the EU debates are focused on whether organics are achieving the environmental standards often claimed for them. Organic food remains a central part of the campaign for a more sustainable food supply, but it remains a topic of debate and controversy.

See Also: Agriculture; Organic Consumerism; Organic Trend.

Further Readings

Belasco, Warren James. *Appetite for Change: How the Counterculture Took on the Food Industry*. Ithaca, NY: Cornell University Press, 2006.

Guthman, Julie. *Agrarian Dreams: The Paradox of Organic Farming in California*. Berkeley: University of California Press, 2004.

Reed, Matthew. *Rebels for the Soil. The Rise of the Global Organic Food and Farming Movement*. London: Earthscan, 2010.

Matt Reed
Countryside and Community Research Institute

ORGANIC TREND

The trend in organic food and farming can be understood in two ways. First is the objectives, patterns, and regularities in the trajectory of how organic food and farming have developed. We could consider how the quantity of organic food sold or acres of land certified as organic has developed around the planet. Equally, we can view "trend" in a slightly more pejorative manner as a fashion in food, garments, and cosmetics. The latter treatment of the term *trend* is often viewed as being a subject that is flippant and shallow. As this discussion demonstrates, there are links between the supposedly serious business of the development of the organic food and farming sector, and the business of "fashion."

The organic sector has been growing very quickly and is associated with high social status, but it is still a very small percentage of the total groceries sold.

Source: iStockphoto

The figures and statistics about organics need to be treated with caution, not because of deliberate falsification, but because of the nature of what is being discussed. For many years, the organic industry was a small-scale activity, and few people were interested in products grown without the use of many of the technologies of contemporary agriculture—not only potential consumers, but producers as well. Therefore, the number of producers was small and often widely scattered, and the volume and value of production were very low. Collecting statistics was often difficult and incomplete, hampered by a lack of interest or at times capacity from those who were interested. But with a small sector, growth can appear to be exponential, and how that growth is expressed can make the picture very different.

For the purpose of this entry, organic production will refer to goods that are certified as such and produced organically deliberately rather than by those who lack access to technologies that might give them a choice. Certification also means that the products are legally organic under the legislation that allows them to be bought and sold. On this basis, it can be said that farms did not produce organic food in the United States before 1940, that organic food was largely confined to Germany and the United Kingdom (UK) before that time, and that this was restricted to probably a handful of farms. During the next 40 years, there are very few figures about the production of organic products, as people who could buy it largely did so from the scattered farmers who grew it.

Some of the first statistics concerning organic food sales were recorded in the UK during the 1980s. In 1985, a UK government report estimated all organic sales at £1 million. By 1988, the Henley Centre, a forecasting organization, put the figure at £110 million for vegetables sold in supermarkets alone. These figures are confused because there was no central method of collecting them, and at the time, demand outstripped supply, and at some point in 1989, 95 percent of organic vegetables sold in the UK had been imported. The amount of land managed organically in the period between 1985 and 1989 rose by 700 percent, but it was not sufficient to supply the demand for organic produce. It is also important to note that these figures were for retail sales and not for the value that the farmer received.

Figures for organic sales, the amount of land managed, and the number of certified producers improved throughout the 1990s. Estimated total global sales of organic food in 1997 were $11 billion and rose to $30 billion by 2005. A similar pattern in the amount of land managed under organic systems is apparent—from 7.5 million hectares in 2000 to 31.5 million hectares in 2006. These figures are regularly quoted by those promoting the organic industry and movement as compelling evidence of an extraordinary rate of growth in the sales of organic produce.

It is certain that organic products rose at a very rapid rate before the global recession of 2008, and in some cases may have continued to do so. Placing the figures in the context of total sales of food or groceries, the U.S. chain Walmart had a total turnover of $285 billion in 2005, or approximately 10 times the total global sales of organic food. To place these sales in a national context, organic food sales were £1 billion in the UK in 2007; out of a total grocery market of £128 billion, organic foods accounted for 0.78 percent of total grocery sales in 2007 in the UK.

Considering the figures for organic sales, it is apparent that it has been a very dynamic sector, growing very quickly from a very low base. If we consider it as a fraction of total sales, then it is apparent that it is a very small percentage of the total value of groceries sold. These figures are probably far smaller than most people would think, in part because they have formed their impression from the media coverage of organic foods and the arguments surrounding it. This is where a consideration of the organic trend, as in fashion, becomes pertinent.

One of the criticisms of organic food has been driven in part by it being based on fashion: the creation of social status without any foundation in evidence. This has led to its being dismissed as the food of "yuppies" or the overanxious middle classes. At the same time, a phalanx of celebrities from rock stars such as Bruce Springsteen to actors such as Daryl Hannah and Christian Slater have promoted organic food. In recent years, this has become more pronounced as cosmetics and fabrics sourced from organic plants have been adopted by the fashion industry. This has meant that organics have become part of the physical products of the fashion industry.

Many celebrities and musicians have become aware that they can exercise considerable influence by promoting causes and bringing them to public attention. This is in many ways the voluntary form of the celebrity endorsement that is a mainstay of the advertising industry. Endorsements such as this are premised on the celebrity not only being someone that the public will recognize but with whom they will associate the positive aspects of the individual with the product. By choosing organic products, turning their farms over to organic production, or launching product lines, celebrities are promoting organics. This is not an unalloyed "positive" for the organic movement as it can mean that it becomes associated with the wealthy and as such exclusive.

Conclusion

As this discussion has shown, there has been strong and consistent growth in the amount of land globally certified as organic, along with the value of sales of organic goods.

Organic products remain a small percentage of the total value of food expenditures even in countries where they are well established. This growth has in part been driven by the cachet associated with organic as it is the choice of those with high social status as well as the desire by consumers to do "their part" for the environment. Equally, although the evidence is limited, it would appear that the growth trend in organic dips during economic recessions. It can therefore be concluded that organic products are a growing force in the food, cosmetic, and textile markets, but that their relation to the fashion industry and those of high social status amplifies this growth.

See Also: Agriculture; Organic Consumerism; Organic Foods.

Further Readings

Guthman, Julie. *Agrarian Dreams: The Paradox of Organic Farming in California*. Berkeley: University of California Press, 2004.

Participant Media and Karl Weber, eds. *Food Inc.: A Participant Guide—How Industrial Food Is Making Us Sicker, Fatter, and Poorer—And What You Can Do About It*. New York: PublicAffairs, 2009.

Reed, Matthew. *Rebels for the Soil. The Rise of the Global Organic Food and Farming Movement*. London: Earthscan, 2010.

Matt Reed
Countryside and Community Research Institute

PESTICIDES

Pesticide use is prevalent in the contemporary agricultural landscape and has ignited controversies about the safety implications for humans and the environment. Expanding international trade underscores that decisions on pesticide management made in one country can have global repercussions. Some of the costs and benefits associated with agricultural pesticides include increased crop yields, improved products, adverse health effects, and environmental contamination.

The Environmental Protection Agency (EPA) defines a pesticide as any substance used to mitigate pests such as insects, microorganisms, fungi, and weeds. Modern conventional agriculture employs a wide range of synthetic chemicals that inevitably leave residues in the produce and the environment. Residues enter the food chain through on-farm pesticide use, post-harvest use (which accounts for the largest part), pesticide use on imported food, and canceled pesticides that persist in the environment.

In 1991, the U.S. Department of Agriculture (USDA) began collecting data on pesticide residues in food through the Pesticide Data Program (PDP). The EPA uses PDP data to prepare pesticide exposure assessments and set acceptable limits in food as required by the 1996 Food Quality Protection Act (FQPA). This act mandated the EPA to reassess the risk of current and previously registered pesticides, with a focus on children's health. The EPA also provides licenses and registration for pesticide products once it is determined that they will not cause any unreasonable risks to human health or the environment. For example, the acceptable daily intake (ADI) value indicates the amount of a substance that can be ingested daily over a lifetime by consumers without any appreciable health risk. The Food and Drug Administration (FDA) and the USDA then conduct surveys to determine whether pesticide residues in or on foods are below these tolerance limits. The EPA records that U.S. agricultural trade has more than doubled since 2001, emphasizing the critical importance of ensuring that foods that may contain pesticide residues meet high safety standards.

Pros of Pesticide Use

Pesticide applications in agriculture persist because of the perceived benefits to both producers and consumers. Pesticide use arguably permits a greater yield from farms, and industry estimates place increased crop productivity between 20 percent and 50 percent.

Farm work is less arduous and labor intensive with pesticides, allowing farmers to maximize the benefits of other inputs such as high-quality seeds, fertilizers, and water resources. Pesticides also contribute to the production of a stable and predictable supply of high-quality and affordable food and ensure better storage and distribution. In the developing world, protecting food or industrial crops such as cotton can also prevent financial ruin by pest infestations and contribute to improved food security.

Stronger regulatory standards against pesticide residues in the conventional food chain have reduced consumer risks and the possible advantage of organic foods (which cannot be grown with pesticides). While studies have indeed shown less pesticide residue on organic produce, the products from conventional agriculture fall well within the acceptable level for human consumption set by the EPA, which takes a conservative approach. In the United States, an FDA survey measured the highest average intake of 38 different pesticides in the total diets of various population subgroups and found that for 34 of the 38 pesticides, exposures were at less than 1 percent of the ADI level. The remaining four were all less than 5 percent of the ADI value. Arguments for organic agriculture also ignore that this approach has a larger footprint on land per unit of production. The total land devoted to agriculture in the United States has declined by 25 percent since 1950 due, in part, to higher yields afforded by pesticide use.

Cons of Pesticide Use

It is generally acknowledged that acute, massive exposure to pesticides can cause adverse health effects and most, if not all, commercially purchased food contains trace amounts of agricultural pesticides. Whether chronic exposure to trace amounts of pesticides in foods poses significant health risks is still uncertain. Human health impacts linked to pesticide exposure for the general population include birth defects, childhood leukemia and other cancers, Parkinson's disease, developmental and neurological disorders, and reproductive and hormonal system disruptions. However, these effects depend on how toxic the pesticide is and how much of it is consumed.

Infants and children may be especially sensitive to pesticides because their internal organs are still developing and maturing; their enzymatic, metabolic, and immune systems may provide less natural protection than those of an adult; and they ingest more food and drink per pound of body weight than adults, which may increase their exposure. Organophosphate insecticides (OPs) are among the most widely used pesticides in the United States and have long been known to be particularly toxic for children.

Agricultural workers have also suffered from their exposure to pesticides. Farmworkers represent the highest rate of chemical-related illness of any occupational group in the United States and report approximately 300,000 pesticide-related illnesses each year. Their exposure to pesticides ranges from chronic to acute and includes symptoms from dizziness and nausea to death. The EPA has put more stringent precautions in place to protect workers and alleviate hazards as a result of research associating worker exposure to Parkinson's disease, leukemia, and other cancers. Adverse reproductive effects, such as abnormalities in sperm, have also been detected in workers and suggest risks for congenital malformations in their children.

Trace amounts of pesticides also affect the environment by contaminating the groundwater and soil, affecting nontarget species, and drifting into unintended areas. This was exemplified in the 1960s by the DDT crisis and Rachel Carson's book *Silent Spring*, which made the unintended impacts of pesticides a salient issue. Chemicals and compounds already in the soil from previous pesticide use can expose organic produce to pesticides as well. If organic crops are not grown under cover, they are also vulnerable to pesticide drift from neighboring farms and global transport of chemicals. Thus, organic farmers, who are prohibited from pesticide use to comply with their certification, may find that these residues still exist on their produce. Still, studies have shown not only that organic produce has less (one-third to one-half) pesticide residue than conventional crops but also that it is much less likely to have multiple residues. Nutritional research has also linked organic produce to increased levels of vitamin C, iron, magnesium, and phosphorous compared to the same foods under a conventional agricultural system and presents an argument for continued alternatives to the industrial agricultural model with intensive pesticide use.

See Also: Agriculture; Organic Farming; Organic Foods; Pesticides and Fertilizers (Home).

Further Readings

Crinnion, Walter. "Organic Foods Contain Higher Levels of Certain Nutrients, Lower Levels of Pesticides, and May Provide Health Benefits for the Consumer." *Alternative Medicine Review*, 15:1 (2010).

Magkos, Faidon, Fotini Arvaniti, and Antonis Zampelas. "Organic Food: Buying More Safety or Just Peace of Mind? A Critical Review of the Literature." *Critical Reviews in Food Science & Nutrition*, 46:1 (2006).

McCauley, L., W. Anger, M. Keifer, R. Langley, et al. "Studying Health Outcomes in Farmworker Populations Exposed to Pesticides." *Environmental Health Perspectives*, 114:6 (2006).

Nicole Menard
Independent Scholar

PESTICIDES AND FERTILIZERS (HOME)

While pesticides and fertilizers have come under scrutiny in an agricultural context, smaller home applications present similar and relevant concerns. Personal lawns and gardens may be small but the total area in urban environments is significant. Individuals have the power to make informed choices, control which products they use, and affect their environment. Debates on conventional and alternative methods circulate around effective pest removal and nutrient enrichment strategies, ecological contamination, health risks, and assumptions regarding "natural" products. Pesticides and fertilizers for home use aim to control pest species in the home and yard and enhance nutrients in gardens, flower beds, and lawns. Pesticides are substances used to control pests such as weeds, insects, fungi, microorganisms, and rodents, and can be grouped into categories of organic and inorganic. Organic pesticides contain carbon, while inorganic pesticides contain carbon found only in the form of carbonate or cyanide. These pesticides are derivatives made from arsenic, mercury, fluorine, sulfur, copper, and cyanide. Organic pesticides can be divided into three groups: synthetic pesticides (developed in laboratories and manufactured), natural pesticides (derived from animal, microbial, or vegetal origin), and microorganisms (e.g., *Bacillus thuringiensis*).

Pesticides are used control weeds, insects, and other pests, while fertilizer provides essential nutrients; both are used to enhance gardens, flower beds, and lawns.

Source: iStockphoto

Inorganic pesticides are mostly created from minerals like boric acid, sulfur, salts, and copper.

A fertilizer is any material that supplies one or more of the essential nutrients to plants. Fertilizer usually adds supplies of nitrogen (N), phosphorus (P), and potassium (K) to soil where these naturally occurring nutrients have been depleted. There are single-ingredient fertilizers—such as ammonium nitrate and urea for nitrogen—and combination fertilizers. Combination varieties are labeled such as 10-10-10 or 5-10-10, where the numbers indicate, in order, the percentage of nitrogen (N), phosphorous (P), and potassium (K) in the fertilizer. Fertilizers can also be classified into categories of inorganic or organic, referring more to the distinction of natural and man-made rather than carbon. Inorganic fertilizers are derived from nonliving sources and include the commercial fertilizers that are predominantly used. Synthetic fertilizers are typically manufactured into salt compounds, which dissolve quickly into their chemical elements and release nitrogen, potassium, and so forth. Other nitrogen sources such as ammonia, ammonium sulfate, and urea are by-products from the oil and natural gas industry. Organic fertilizers are derived from living or once-living material, including animal wastes, crop residues, compost, and numerous other by-products of living organisms. The variety of available materials and methods presents many options and outcomes for consumers to evaluate. Discussed below are some of the advantages and disadvantages of both synthetic and natural pesticides and fertilizers for home use.

Pros of Pesticides and Fertilizers

Pesticides are often used in the yard but can also be employed in the home to remove germs, insects, or rodents that carry disease. Pesticides are frequently used to kill insects or inactivate pests that destroy garden produce or yard vegetation and to control mosquitoes, which transmit deadly diseases like West Nile virus, yellow fever, and malaria. They are also used to remove harmful microbes that can damage food or cause health problems. Many common household products are considered pesticides, such as cleaners to disinfect and sanitize or products to remove mold and mildew. A 2006 study in the *Journal of Exposure Science and Environmental Epidemiology* reported that 75 percent of households in the United States and the United Kingdom use pesticides, a tribute to how effective and commonplace they have become.

Due to the concerns surrounding conventional pesticides and human and environmental health, there are alternative strategies that vary in effectiveness, safety, and ease. As discussed above, organic pesticides come in several forms, including synthetic, but the natural varieties provide many options. These biologically based pesticides are becoming increasingly popular and often are safer than traditional pesticides. Biopesticides are pesticides derived from natural materials such as animals, plants, bacteria, and certain minerals. Microbial pesticides consist of a microorganism (e.g., fungi) as the active ingredient and can control many different kinds of pests in a relatively specific way. Biochemical pesticides are naturally occurring substances (e.g., pheromones, scented plant extracts) that control pests by nontoxic mechanisms and are regulated by the U.S. Environmental Protection Agency (EPA).

Other solutions to pest management at home may include a program called Integrated Pest Management (IPM), where the driving force behind decisions of when, where, and what to spray is a thorough knowledge of the life cycles of pests and their interaction with the environment. This understanding of population dynamics, available predators, and other variables can reduce pesticide use to a minimum. IPM follows the steps of setting a threshold, monitoring and identifying pests, prevention, and control. The control stage starts out with targeted chemicals (e.g., pheromones) or activities (weeding and trapping) and uses pesticides as the last resort. There are also plenty of home remedies for pesticides that are affordable and safe. Some examples include canola oil, baking soda, salt, garlic, and red pepper, which all have applications for pesticide use. However, these may require more time to create and apply or may seem less effective.

For fertilizers, it is most important that the right nutrients be present in the form that plants can utilize. Conventional fertilizers are considered easy to use because they are concentrated, prepackaged, and affordable. They also provide certainty regarding the content of nutrients through their labeling. This gives the user more control and the ability to apply nutrients accurately and according to the requirements of the soil. In addition to the composition of N, P, and K, the remainder of the material in the fertilizer may contain minor nutrients and filler material that may allow the even application of nutrients across the area to be fertilized. Some synthetic fertilizers are now coated to facilitate a slower rate of nutrient release, which addresses a main criticism that quick-release varieties contribute to excess nutrients and soil and water contamination.

Organic fertilizers add valuable matter to the soil through natural products like compost, animal waste, bone meal, and lawn clippings that can often be obtained for little or no cost. They generally have slow-release action that occurs through decomposition by microbes so fewer nutrients are leached from the soil. Organic fertilizers are often preferred by vegetable and flower growers because they act a little more slowly and, since most are insoluble in water, they do not damage germinating seeds and young plants. Store-bought organic fertilizers often come in bulk, which can reduce packaging. Commercially packaged varieties now have labels with optimum NPK ratios as well (e.g., alfalfa horse feed pellets have 2-1-2) that can help ensure safe and balanced application.

Cons of Pesticides and Fertilizers

Despite the effectiveness of synthetic pesticides and fertilizers, there are some disadvantages, including local ecological degradation and negative health effects on families. Though pesticide use in the home and garden can intend to eliminate a particular pest, these applications can be ineffective or produce unintended consequences. Insecticides that do minimize one species of insect may give another type a chance to thrive or overpopulate, throwing off the ecological balance of the area. Nonselective insecticides have harmful effects when they kill beneficial insects as well as the ones that can damage the grass or garden. Soil benefits from the activities of many insects that enrich the soil and keep it from getting compacted.

The impact of pesticides on human health, especially in children, has been a great source of concern. A National Cancer Institute study in the United States indicates that children are as much as six times more likely to get childhood leukemia when pesticides are used in the home and garden. A study published in the *American Journal of Public Health* found elevated levels of cancer in children when pesticides were used in their homes and yards, with particularly high correlation in homes where dichlorvos pest strips were used. The weed killer 2,4-D has been associated with non-Hodgkin's lymphoma in studies in the United States, Canada, and Sweden and linked to other health issues, including birth defects, reproductive problems, and immune system damage.

Pesticides and fertilizers have also been shown to enter the soil, water, and air where they are applied and persist in the environment. Several research studies have linked pesticides with inhibiting symbiotic nitrogen fixation processes. According to a citizen activist group in the U.S. Midwest, 13 of the 18 most commonly used lawn pesticides have been found in groundwater or surface water in the Great Lakes Basin. Quick release or overapplication of nitrogen fertilizers can cause excess nitrates not taken up by plants to leach downward with percolating water and enter the groundwater supply. When phosphorus in fertilizer does not get taken up by the plant, it is quickly bound to soil particles that can erode into runoff and surface water. This fertilizer runoff is partly to blame for the proliferation of lake weeds and algae. While agricultural runoff of fertilizer is a major contributor, homeowners who use fertilizers high in phosphorus do exacerbate the damage. One pound of phosphorus in a lake can result in 300–500 pounds of algae. The overabundance of decaying algae depletes the water oxygen supply and can kill fish and desirable vegetation. Buying synthetic fertilizers may also contribute to dependence on fossil fuels by way of using their by-products.

Organic fertilizers possess some additional limitations. The content of available nitrogen and other nutrients is often unknown and variable unless a store-bought variety with a label removes this uncertainty, which is more expensive than a synthetic fertilizer. Organic fertilizers often have a lower percentage of nutrients and can require more applications and work. While safe if used properly, the overapplication of organic fertilizer can be just as detrimental to groundwater as synthetic fertilizers and can result in nitrate leaching, salt toxicities, or excessive vegetative growth. Similarly, organic pesticides are not without scrutiny. A study comparing natural organic and nonorganic synthetic pesticides raised valid concerns regarding effectiveness and safety. Rotenone and pyrethrin are two common plant-based pesticides, and imidan is a "soft" synthetic pesticide designed to have a brief lifetime after application. Up to seven applications of the rotenone-pyrethrin mixture were required to obtain the level of protection provided by two applications of imidan. These increased applications of rotenone and pyrethrin are unlikely to be any better for the environment, especially when rotenone is extremely toxic to fish and other aquatic life and this mixture is known to be quite potent. It is unclear which system is more harmful because less is known about how long these organic pesticides persist in the environment or the full extent of their effects (clearly stated on labels).

Further confounding these issues is a common perception that what is "natural" is automatically safe. Other terms that can have misguided assumptions include *chemical* pesticides and fertilizers, which conjure up the most potent, toxic chemicals available. Instead, some chemicals are naturally occurring, and both natural and synthetic fertilizers are composed of chemicals or chemical elements. "Organic" may seem natural but organic pesticides include man-made products, while synthetic fertilizers can be produced from mined minerals and are allowed in organic farming. These ideas and terms can challenge the consumer when navigating which products to use and why.

An argument that has gained momentum over the years opposes having a traditional lawn in the first place and eliminating the need for accompanying pesticides and fertilizers. In place of pristine and meticulous grass, homeowners may consider plants that are native to the area and adapted to local conditions, using ground covers, or creating rain gardens. This is one of many options to consider when evaluating the effectiveness and cost of products, impacts on the environment, risks to health, and views toward pesticides and fertilizers used at home.

See Also: Biological Control of Pests; Organic Consumerism; Organic Farming; Organic Trend; Pesticides.

Further Readings

Environmental Protection Agency. "Pesticides: Controlling Pests." http://www.epa.gov/pesticides/controlling (Accessed September 2010).

National Coalition for Pesticide-Free Lawns. http://www.beyondpesticides.org/pesticidefreelawns/index.htm (Accessed September 2010).

Staley, J. "Varying Responses of Insect Herbivores to Altered Plant Chemistry Under Organic and Conventional Treatments." *Proceedings: Biological Sciences*, 277:1682 (2010).

Tang, S., G. Tang, and R. Cheke. "Optimum Timing for Integrated Pest Management: Modelling Rates of Pesticide Application and Natural Enemy Releases." *Journal of Theoretical Biology*, 264:2 (2010).

Ussery, H. "Why Natural Insect Control Works Better." *Mother Earth News*, 240 (2010).

Nicole Menard
Independent Scholar

PRECAUTIONARY PRINCIPLE (ETHICS AND PHILOSOPHY)

The 1992 Rio Declaration on Environment and Development's definition of the precautionary principle states that "lack of full scientific certainty shall not be used as a reason for postponing cost-effective measures to prevent environmental degradation." This does not preclude other reasons for postponing or avoiding such measures, even economic or political reasons. In this case, the precautionary principle does not require environmental protection to be given priority.

The precautionary principle says that if the environmental consequences of human action may be serious, efforts should be made to avoid or lessen them. Lack of scientific certainty should not be an excuse for inaction. If the environmental consequences of an action could be serious, we should be cautious. It is based on the classical virtue of "prudence" and embraces the folk wisdom of "better safe than sorry" or "look before you leap."

In the case of the precautionary principle, it is not only a matter of considering consequences for the individual or the action-taker, but the broader consequences for the planet and for future generations. It is antithetical to a "wait-and-see" approach, where policymakers wait until they have more information before acting.

UNESCO's World Commission on the Ethics of Scientific Knowledge and Technology (COMEST) defines the precautionary principle as the following:

> When human activities may lead to morally unacceptable harm that is scientifically plausible but uncertain, actions shall be taken to avoid or diminish that harm. *Morally unacceptable harm* refers to harm to humans or the environment that is threatening to human life or health, or serious and effectively irreversible, or inequitable to present or future generations, or imposed without adequate consideration of the human rights of those affected.
>
> The judgment of *plausibility* should be grounded in scientific analysis. Analysis should be ongoing so that chosen actions are subject to review.
>
> *Uncertainty* may apply to, but need not be limited to, causality or the bounds of the possible harm.
>
> *Actions* are interventions that are undertaken before harm occurs that seek to avoid or diminish the harm. Actions should be chosen that are proportional to the seriousness of the potential harm, with consideration of their positive and negative consequences, and with an assessment of the moral implications of both action and inaction. The choice of action should be the result of a participatory process.

While environmental regulations are often anticipatory and preventive, they are not necessarily precautionary. They generally aim to prevent known risks rather than anticipate and prevent uncertain potential harm. It is only when the risk is uncertain because either the probability of damage is uncertain and/or the extent of damage is uncertain that the precautionary principle applies.

In the 1970s, the U.S. Environment Protection Agency (EPA) imposed limits on lead in gasoline based on increasing scientific evidence that it was causing problems but without proof that it had actually harmed particular people. The oil industry unsuccessfully opposed the regulations in the courts, and the case became a landmark in U.S. environmental law for the application of the precautionary principle.

The precautionary principle achieved widespread recognition after it was incorporated into the Rio Declaration. In 1993, the Treaty of Maastricht required European Community countries and the European Commission (EC) to base environmental policy on the precautionary principle. In 1999, the Council of the European Commission urged the commission to ensure that future legislation and policies were guided by the precautionary principle.

Many international environmental lawyers would argue that the precautionary principle is moving toward the status of customary international law, having already been adopted in numerous treaties and other key international documents. It has also been incorporated into national laws in several countries, including Germany, Belgium, and Sweden, and it has influenced several court judgments. In France, it has even been included in the nation's constitution as part of an environmental charter. This gives the principle priority over other legislation.

The legal system in English-speaking countries is less conducive to the incorporation of broad principles and tends to be based on specific rules and regulations. In the United Kingdom, for example, the precautionary principle is not included in statutory law. Nor has the precautionary principle made much headway in United Kingdom (UK) courts.

In the United States, the term *precautionary approach* is preferred. However, although U.S. environmental and health laws do not refer to the precautionary principle or approach by name, some of the earlier environmental legislation nevertheless adopted it. This has changed in recent years as politicians, under pressure from corporate donors, have demanded all environmental legislation be limited to situations where scientific proof of harm is available.

Shifting Burden of Proof

Traditionally, it was up to consumers, environmentalists, or government authorities to make a convincing scientific case that activities or products were harmful before they could be regulated. The thinking was that regulations constrained economic activity, and useful products and developments would be banned and technological innovation impeded unless scientific proof of harm was required before they were regulated. Excessive caution was said to lead to paralysis and stagnation.

These days, certain activities require developers to prepare environmental impact statements or assessments, and some products such as pharmaceutical drugs, pesticides, and food additives must gain approval before they can be marketed. In these cases, it is initially assumed that the activities in question or the products may be hazardous or environmentally damaging and the burden of proof has been shifted to the manufacturer or developer to produce scientific evidence that the activities or products are safe in order to gain approval. Although it is said the burden of "proof" has been shifted, proof is not actually required, just a convincing case—supported by scientific evidence—that the product or activity is safe.

This shifting of the burden of proof from one party to another, for example, from the regulatory authority to the developer, is only one element of the precautionary principle. However, the fact that those proposing an activity or product have to show that it is safe before it is approved—rather than the government needing to show it is unsafe before it can be restricted—is an important aspect of the precautionary principle.

In practice, the burden of proof has generally been shifted only for new products and activities where there is a long history of harm arising from like products and activities. Existing products are generally "presumed safe." This bias is partly based on the assumption that it is cheaper and more politically acceptable to prevent new products from being manufactured than it is to ban existing products, and it is easier to prevent new developments than to dismantle exiting ones. Similarly, synthetic substances may require licenses but natural substances are assumed safe, even if they are added in unnatural quantities to the environment.

Those proposing new environmental regulations still often have the burden of making a water-tight scientific case that the proposed regulations are necessary to protect human health or the environment. This gives opponents the opportunity to undermine the justification for such regulations by emphasizing the uncertainties in their scientific evidence.

What the precautionary principle does is ease the standard of proof so that scientific evidence of possible harm rather than scientific certainty is sufficient to prompt regulatory action. The assumption that an activity or product is safe until proved harmful shifts so that it can be considered harmful before that proof is available. It is no longer sufficient to raise doubts about whether the harm will happen to prevent an activity or product from being regulated.

When Does the Precautionary Principle Apply?

The precautionary principle involves two decisions. The first decision is whether the precautionary principle should be applied, and the second is which precautionary measures should be taken. The first decision involves assessing whether the potential harm is great enough, and the evidence for it strong enough, to trigger the precautionary principle.

All human activity has some impact on the environment. The question is which impacts are acceptable and which impacts need to be prevented or mitigated? Clearly this is a political question that requires broad community participation rather than a scientific question, given that the scientific evidence is inconclusive and the question of acceptability is a value judgment.

Most definitions of the precautionary principle restrict precautionary measures to situations in which the potential harm is "serious and irreversible" or "unacceptable" or "transgenerational" or "global" or "significant" as in "significant reduction in biological diversity." But most of these terms cannot be quantified scientifically or economically. For this reason, many advocate that the judgment should be made by a wide cross-section of the community and not just a few experts.

Uncertainty may not only relate to the probability of a serious event occurring. It may also relate to how serious the consequences might be. For example, there is a general scientific consensus that global warming will occur if greenhouse gas emissions are not reduced, but the consequences of this are uncertain. There is no scientific consensus about the scope or rapidity of sea level rise or its consequences. Nor is there consensus about the impacts in particular parts of the world. Nevertheless, because there is strong evidence that the consequences of global warming could be serious, the precautionary principle applies.

Opponents of the precautionary principle argue that it is unscientific because it can be triggered by irrational concerns. However, the precautionary principle should not be applied without scientific evidence of harm. The Canadian government points out that "sound scientific information and its evaluation must be the basis" for applying the precautionary principle and in deciding whether scientific evidence is sound, "decision makers should give particular weight . . . to peer-reviewed science."

But where, between the extremes of speculation and the unattainable full scientific certainty, is the point where there is sufficient knowledge to act? How much evidence does there need to be before the precautionary principle is triggered. If no evidence were required, then any nonscientific speculation or irrational fear would be enough to require precautionary measures and the principle would become impractical. On the other hand, scientific proof would render the precautionary principle unnecessary.

Most definitions of the precautionary principle try to define the level of evidence in terms of "reasonable grounds for concern" or "reasonable scientific plausibility" or "scientific credibility" or require decisions to be made "on the basis of available pertinent information."

Measures to Be Taken

Critics also argue that the precautionary principle aims at an unrealistic goal of zero risk. However, the precautionary principle does not aim to reduce risk to zero but rather to avoid or mitigate likely harm. The measures to be taken in response to the precautionary principle being triggered are not dictated by the precautionary principle. The precautionary principle is merely a guide as to when precaution needs to be exercised and which criteria should be used to evaluate measures adopted. There is no requirement on the part of the precautionary principle to ban anything, although decision makers may decide that a ban may be appropriate in certain circumstances.

Many economists and governments argue that action to avoid or mitigate potential harm should not be taken if acting on precaution means forgoing substantial economic benefits that outweigh the cost of environmental degradation that might otherwise occur. They assume in this that costs and benefits can be measured despite the uncertainty surrounding them and that the damage can be compensated for by the economic benefits.

Economists also argue that in some circumstances it may be preferable to postpone acting on a problem and incur the costs of fixing it later rather than now because of the following:

- Future costs are perceived to be less burdensome than current costs
- If good scientific research accompanies the delay, the extra information might enable the problem to be solved in a cheaper and more effective way

However, it can also be argued that postponing action is not the best decision because it may in fact cost considerably more to solve a problem in the future than it does to solve it now. For example, if the damage done in the ensuing time is irreversible, then it may not be able to be solved at all. Moreover, it is not fair to pass on environmental risks to future generations and assume that they will have the knowledge and/or technology to deal with them.

A stronger version of the precautionary principle dictates that positive action must be taken to avoid or mitigate the potential harm. In this view, if the harm is judged unacceptable or serious and irreversible, then inaction is not precautionary and is not compatible with the precautionary principle. Merely monitoring impacts or undertaking further research is a way of delaying intervention until more is known (i.e., wait and see) and therefore is not a precautionary approach.

The Wingspread Statement on the precautionary principle, like the UNESCO definition above, clearly mandates precautionary measures and is therefore a strong version of the precautionary principle:

> When an activity raises threats of harm to human health or the environment, precautionary measures should be taken even if cause and effect relationships are not fully established scientifically. . . .

The strong approach assumes that serious or irreversible environmental harm should not be tolerated and that other, less environmentally damaging ways can be found to achieve the economic benefits that the proposed action would have brought. Nevertheless, even in the stronger version of the precautionary principle, the action that should be taken is not determined by the principle. In only a few rare cases is the precautionary principle defined in a way that dictates measures. For example, the Oslo Commission of 1989 agreed that the dumping of industrial wastes, "except for inert materials of natural origin" into the North Sea should cease and only be allowed where it could be shown that there were not practical alternatives and it would cause no harm to the marine environment.

Measures can either "constrain the possibility of the harm" or "contain the harm" should it occur, by limiting its scope or controlling it. According to the EC, measures taken in response to the precautionary principle should be proportional to the scale of potential harm, nondiscriminatory, consistent with measures taken in similar circumstances, beneficial, and provisional so as to take account of new knowledge and assessments over time.

See Also: Intergenerational Justice; Kyoto Protocol; Precautionary Principle (Uncertainty).

Further Readings

Andorno, Roberto. "The Precautionary Principle: A New Legal Standard for the Technological Age." *Journal of International Biotechnology Law*, 1 (2004).

Beder, Sharon. *Environmental Principles and Policies*. Sydney: UNSW Press, 2006.

Commission of the European Communities. "Communication from the Commission on the Precautionary Principle." February 2, 2000. http://ec.europa.eu/dgs/health_consumer/library/pub/pub07_en.pdf (Accessed January 2011).

de Sadeleer, Nicolas. *Environmental Principles: From Political Slogans to Legal Rules*. Oxford, UK: Oxford University Press, 2002.

Environment Canada. "A Canadian Perspective on the Precautionary Approach/Principle." September 2001.

World Commission on the Ethics of Scientific Knowledge and Technology (COMEST). "The Precautionary Principle." Paris: United Nations Educational, Scientific and Cultural Organization (UNESCO), 2005.

Sharon Beder
University of Wollongong

Precautionary Principle (Uncertainty)

The precautionary principle has many definitions and applications, all of which include avoiding undesirable consequences by not taking a proposed action until it is proved safe. It is a means of avoiding risk when the outcome of an action is uncertain and has been a major focus of the environmental movement of the late 20th and early 21st centuries. The precautionary principle has applications in multiple sectors including, for example, agriculture, fisheries, forestry, and energy.

The modern term *precautionary principle* is often traced back to direct translation of the German word *vorsorgeprinzip*, prevalent in popular culture in 1930s Germany. But the first real discussion of the precautionary principle in the global community occurred at the United Nations Conference on Environment and Development in 1992, also referred to as the Rio Earth Summit. At this conference a declaration was made, the Rio Declaration of Environment and Development, which reaffirmed and expanded a 1972 declaration about the human environment. The declaration included 27 principles, of which Principle 15 specifically called for the use of the precautionary principle to protect the environment, as follows:

Principle 15: In order to protect the environment, the precautionary approach shall be widely applied by States according to their capabilities. Where there are threats of serious or irreversible damage, lack of full scientific certainty shall not be used as a reason for postponing cost-effective measures to prevent environmental degradation.

With this statement, the United Nations Conference on Environment and Development concluded that the member states of the United Nations shall use caution when considering actions that affect the environment, applying the precautionary principle.

Uncertainty and Risk

The definition of uncertainty varies based on the discipline within which is it used. Economists, engineers, statisticians, physicists, psychologists, and actuarial scientists all have slightly different interpretations of uncertainty. In general, uncertainty is used in reference to attempts to predict the effects or outcomes of a proposed action or decision. Uncertainty occurs when there is a lack of certainty about possible outcomes, particularly when more than one possible outcome occurs or simultaneous outcomes occur. Risk is a type of uncertainty in which one or more of the possible outcomes have negative effects (loss).

Regulations regarding human and environmental health are established as a means of risk management, among other driving factors. When the probability of an undesirable outcome can be quantified, some acceptable level of occurrence of undesired events may be established. Guidelines may then be established for environmental and health issues such as chemical exposure based on the established acceptable level of undesired outcomes.

For example, the U.S. Environmental Protection Agency (EPA) has methods for calculating the excess cancer risks for exposure to various chemicals of concern. These calculation methods can account for various exposure doses, durations, and pathways and provide an overall number representing the excess cancer risk due to exposure to a given chemical under the circumstances calculated. These calculations guide the EPA in establishing acceptable levels of exposure to given chemicals. Generally, the EPA considers excess cancer risks of less than one in a million (0.000001) to be acceptable. For many contaminants, this acceptable level of risk effectively guides the limits the EPA sets on exposure to various carcinogens, such as the establishment of maximum contaminant levels for drinking water.

Precautionary Principle

At times, those managing risk may desire to act with utmost caution and avoid an undesired outcome completely. This concern may arise even when the probability of an undesired outcome is not quantifiable, when all of the undesired outcomes are not foreseeable, or when one of the possible outcomes is an unacceptably large or irreversible loss. When this is the case, the safer action with a known outcome will be taken in an effort to avert all risk. In some cases, the safest action is not to undertake a proposed action when the risk is uncertain.

There have been many definitions of the precautionary principle, but all include the concept that if a proposed action may harm humans or the environment, and there is not scientific consensus regarding the risks of the proposed activity, no activity should be taken until those proposing the activity can prove it safe. This principle is founded on the concept of social responsibility: that those proposing new actions must ensure their actions will not cause harm before they undertake them. As a contrast, the precautionary principle is the opposite of giving the "benefit of the doubt," where a favorable judgment is given in the absence of certainty.

Proponents of Precaution

Another famous call for the use of the precautionary principle was in the 1998 Wingspread Statement. This statement further defined the precautionary principle and was the result of a three-day conference in Racine, Wisconsin, at the headquarters of the Johnson Foundation, called Wingspread. The statement cites numerous current environmental and health problems and asserts that current environmental regulations have failed to protect human and ecological health. The statement goes beyond an appeal to governments in implementing the precautionary principle, and states that "Corporations, government entities, organizations, communities, scientists and other individuals must adopt a precautionary approach to all human endeavors." The Wingspread Statement defines the precautionary principle as follows:

> When an activity raises threats of harm to human health or the environment, precautionary measures should be taken even if some cause and effect relationships are not fully established scientifically. In this context the proponent of an activity, rather than the public, should bear the burden of proof.

It goes on to provide the following advice for the application of the precautionary principle:

> The process of applying the Precautionary Principle must be open, informed and democratic and must include potentially affected parties. It must also involve an examination of the full range of alternatives, including no action.

The Wingspread Statement went beyond the Rio declaration in that it called not only for governments to use the precautionary principle but also for corporations, organizations, and scientists to use it. Since 1998, the Science and Environmental Health Network, one of the host organizations of the Wingspread conference, has been working toward promoting the application of the precautionary principle within government, businesses, and other organizations of the United States.

Implementing Precaution

On the 10th anniversary of the Wingspread conference and statement, the Science and Environmental Health Network performed a study to determine the implementation of the precautionary principle across various disciplines. The study included interviews with individuals representing agriculture, agribusiness, arts/culture, construction, healthcare, environmental health, environmental justice, food service, forestry, labor, law, medicine, peace/social justice, and public health. Government, nongovernmental organizations, business, nonprofits, academics, and labor were all represented. The study resulted in four main concepts for advancing the precautionary principle:

- *Institutionalizing precaution*: Recommendations with regard to institutionalizing precaution suggested that advocates of the precautionary principle "find common purpose with efforts to construct easy-to-navigate governmental and nongovernmental institutions that reflect and are accountable to the complex underlying systems they impact."
- *Measuring precaution*: The study also recommended the development of metrics to measure precaution that account for what is known, promote health, and result in a better understanding of system-wide impacts.
- *Economic drivers of precaution*: The study results "indicated that advocates (of the precautionary principle) should work to redress the distortions in society's allocation of resources and harness the largely untapped expertise of economists."
- *Envisioning precaution*: The study offered many suggestions for better envisioning the precautionary principle. Overall, they all concluded that advocates of the principle should start taking more systematic approaches to implementing the principle.

Within the United States, efforts to promote the implementation of the precautionary principle have been gaining momentum. Among the most famous implementations of the precautionary principle, in 2003, the San Francisco Board of Supervisors adopted the precautionary principle for the county and city, stating the following within its policy:

> Where threats of serious or irreversible damage to people or nature exist, lack of full scientific certainty about cause and effect shall not be viewed as sufficient reason for the City to postpone measures to prevent the degradation of the environment or protect the health of its citizens.

As the movement to implement the precautionary principle grows, more of our social institutions will take proactive measures to protect environmental and human health in the absence of scientific certainty regarding possible effects of our proposed actions.

See Also: Genetically Engineered Crops; Organic Trend; Pesticides; Precautionary Principle (Ethics and Philosophy).

Further Readings

Environmental Research Foundation. "San Francisco Adopts the Precautionary Principle." *Precaution Reporter*, 675 (March 19, 2003).

Science and Environmental Health Network, Johnson Foundation, W. Alton Jones Foundation, C. S. Fund, et al. "The Wingspread Consensus Statement on the Precautionary Principle." Wingspread Conference on the Precautionary Principle, January 1998.

Sutton, Patrice. *Advancing the Precautionary Agenda*. Science and Environmental Health Network, February 2009.

United Nations Environment Programme. "Rio Declaration on Environment and
    Development." Report of the United Nations Conference on the Human Environment,
    Stockholm, June 5–16, 1972. http://www.unep.org/Documents.multilingual/Default.asp?
    DocumentID=78&ArticleID=1163 (Accessed August 2010).
U.S. Environmental Protection Agency. *Science Policy Council Handbook: United States
    Environmental Protection Agency Risk Characterization Handbook*. EPA 100-B-00-002.
    December 2000.

Michelle E. Jarvie
Independent Scholar

R

RESOURCE CURSE

Some countries endowed with valuable natural resources experience worse economic growth than countries not equally resource rich due to having one sole export. Sierra Leone and its diamonds, being mined here, is an example.

Source: Laura Lartigue/USAID

The term *resource curse* refers to the situation in which a country endowed with valuable natural resources (especially oil, natural gas, or extractable minerals but also timber) experiences worse economic growth than countries not equally resource rich. About four dozen countries, mostly in the developing world, get more than 30 percent of their GDP from extraction industries, and many of them are considered resource curse countries. Their number is increasing as demand for hydrocarbons, minerals, and lumber increases.

The presence of valuable natural resources does not guarantee that a country will be affected by the resource curse. The plus side of developing such resources is that they offer a highly profitable, guaranteed source of export income, making them extremely attractive to poorer countries that have few alternatives. Demand for natural resources, especially in rapidly growing economies like India and China, has increased in the early 21st century and is expected to continue. Several countries have successfully avoided the problems the resource curse can bring through careful planning, management, governance, and good luck.

However, for many developing countries, such resources may also contribute to political problems, including instability, violence, corruption, human rights abuses, ethnic conflict, a more divorced rather than more responsive government, and economic problems.

One such economic problem is known as Dutch disease, wherein a resource country's currency becomes so strong that it makes its other exports too expensive to sell; a country's economic problems may include jobless growth, economic stagnation, budgetary fluctuations, economic predation, corruption, and severe environmental degradation.

A Way Out of Poverty?

Although most of the world seeks to transition to greener technologies, extractable resources are still essential, and the demand for them is increasing. In 2009, the International Energy Agency predicted that by 2030, global electricity demand will grow by 76 percent; oil demand will increase by 24 percent to 105 million barrels a day; natural gas demand will rise by 41 percent; and coal demand by 53 percent. These increases are fueled mostly by exploding growth in China, India, and other emerging economies. For example, *Mineweb*, an industry publication, estimates that new Chinese mining deals totaled $13 billion in 2009, 100 times what they were five years ago. The Heritage Foundation, a think tank in Washington, D.C., estimates that Chinese foreign investment will exceed $100 billion by 2014.

Such resources, especially oil, do make a country wealthier, bringing in billions of dollars annually. For example, it is estimated that in 2008, Nigeria earned more than $45 billion in export revenue, of which 95 percent came from petroleum. Most economists agree there are few other ways for extremely poor countries, and those that lack a developed educational, manufacturing, or technocratic sector, to otherwise earn such money.

And several countries have managed their natural resource–based wealth well. Chile, Norway (which is industrial), and Saudi Arabia all have Sovereign Wealth Funds, investment policies designed to create a positive cash flow for generations to come. Other countries with good governance like Botswana invest in infrastructure development to improve the country's future. Many resource-rich countries in Latin America maintain a healthy agricultural sector, exporting produce. For example, Brazil is also a leading exporter of meat and ethanol.

A Number of Hurdles

Although extractive resources do bring in money, they can also introduce the resource curse and its accompanying problems. Most industrialized countries tend to have well-developed political and technocratic institutions and a diversified economy that leaves them better equipped to avoid the problems that accompany the discovery and development of a profitable natural resource.

Because in most of the world (except in the United States and parts of Canada) national governments own the rights to subsurface minerals, the negotiations for and the income from these resources are managed by these governments. However, many governments in the developing world do not have the technical capacity or capital to extract the resources themselves and so turn to private companies to do so. Reliance on external sources for capital and technological expertise often leaves the resource country in a vulnerable position.

For one thing, despite improvements in environmental protections, oil, gas, and mining extraction projects offer some difficult environmental challenges. Direct environmental damage can range from deforestation to oil spills (from tankers, pipelines, storage, or other accidents). Regional water supplies can be overused to develop the resource or become contaminated. Most extraction projects yield toxic by-products (including heavy metals like lead and mercury or highly saline waters) that can leach into the land and drinking water and pose a threat to humans and wildlife. And there is the physical destruction of the land that holds the resource, especially from mining. Another indirect problem is the contribution of oil, gas, and coal to climate change—current energy use accounts for 65 percent of greenhouse gas emissions.

Site-specific environmental protections are rarely put in place because less developed countries or those without a history of resource development may simply not have the technological capacity to do environmental impact assessments. Also, environmental protections may cut into the amount of money an extraction company is willing to pay a country for its resource, and that may not be a trade-off a government is willing to make.

The resource curse also leads to widespread economic problems including Dutch disease, where, because of the ongoing presence of a reliably profitable resource, a country's currency is disproportionately strong. This makes a country's other economic sectors (like agriculture or manufacturing) uncompetitive as exports, weakening them, and leading in turn to increased unemployment and jobless economic "growth." The resource country becomes increasingly dependent on its one export: Venezuela, for example, received 64 percent of its export revenues from oil in 1998; in 2009, those revenues amounted to 92 percent.

Due to volatility in market price, resource countries have a difficult time accurately predicting their future income—and therefore their national budgets—years in advance. A downturn in prices can lead to increased debt, inability to service current debt, or reduced social services and infrastructure.

Countries may lack absorptive capacity. Infrastructure projects can be pushed too fast, without careful planning or with insufficient bureaucratic and technical support to be effective, leading to waste. And of course such valuable resources leave the door open to widespread corruption on all levels. In oil-rich Angola, more than $4.2 billion of the government's money went unaccounted for between 1997 and 2002, according to Human Rights Watch, a nongovernmental organization (NGO).

Resource curse countries can also face social threats. Perhaps the most notorious example was the vicious decade-long conflict in Sierra Leone, which was largely funded (on both sides) through its diamond resources. Ethnic tensions have been exacerbated in Nigeria, Iraq, and other places. Mining often causes large-scale displacement as communities are moved off the mining site, sometimes without adequate compensation or viable alternative destinations. Indigenous groups in Peru, Australia, and elsewhere have had their traditional practices and environment threatened as their countries sought to develop resources on their land. Finally, a country's security forces, in a quest to either increase their own wealth and/or "protect" the resource project, may harass, attack, even murder those who oppose them.

Recognizing the risks and threats of the resource curse, a number of international initiatives have been established to try to prevent it. At the instigation of then Prime Minister Tony Blair of England, the anticorruption Extractive Industries' Transparency Initiative was developed to encourage and aid countries to make their resource income and deals transparent and inclusive. Countries without legal negotiation experts can now turn to international NGOs and the United Nations for assistance. Environmental and governance groups, such as the Natural Resources Defense Council and Global Witness, are studying, opposing, and publicizing environmental threats and human rights violations. And there is general recognition that there needs to be better coordination of best practices, as in the Natural Resources Charter.

See Also: Capitalism; Ecocapitalism; Ecological Imperialism; North–South Debate; Sustainable Development (Politics).

Further Readings

Chadwick, John. "China's Enormous Appetite for Mining Assets." *Mineweb* (July 30, 2010).

Global Witness. "Paying for Protection: The Freeport Mine and the Indonesian Security Forces (2005). http://www.globalwitness.org/media_library_detail.php/139/en/paying_for_protection (Accessed September 2010).

Human Rights Watch. "Some Transparency, No Accountability: The Use of Oil Revenue in Angola and Its Impact on Human Rights," 16:1(A) (January 2004). http://www.hrw.org/en/reports/2004/01/12/some-transparency-no-accountability-0 (Accessed September 2010).

Humphreys, Macartan, Jeffrey Sachs, and Joseph E. Stiglitz. *Escaping the Resource Curse.* New York: Columbia University Press, 2007.

"IEA World Energy Outlook 2009." http://www.worldenergyoutlook.org/docs/weo2009/WEO2009_es_english.pdf (Accessed September 2010).

Karl, Terry Lynn. *The Paradox of Plenty: Oil Booms and Petrostates.* Berkeley: University of California Press, 1997.

Natural Resource Charter. "Preamble." http://www.naturalresourcecharter.org/content/precepts/preamble (Accessed March 2011).

Reno, William. *Warlord Politics and African States.* Boulder, CO: Lynne Rienner, 1999.

Ross, Michael. "Blood Barrels." *Foreign Affairs* (May/June 2008).

Shaxson, Nicholas. *Poisoned Wells: The Dirty Politics of African Oil.* New York: Palgrave Macmillan, 2007.

"Socialism With Cheap Oil." *The Economist* (December 30, 2008). http://www.economist.com/node/12853975?story_id=12853975 (Accessed December 2010).

Tsalik, Svetlana and Anya Shifrin, eds. *Covering Oil: A Reporter's Guide to Energy and Development.* New York: Open Society Institute, 2005.

Jodi Liss
Independent Scholar

ROUNDUP READY CROPS

The term *Roundup Ready crops* was developed through biotechnology by Monsanto in attempts to conserve resources such as soil and water during the farming process and provide farmers with a higher crop yield to promote sustainable agriculture. Along with no-till technology, it is one of many solutions for farmers to reduce the environmental impacts of farming and thus, there are both positive and negative views on the use and implementation of biotechnology for genetic modification of crops. Some of the pros and cons include the possible environmental impact of genetically modified (GM) crops, the debate over increased yield, monopolizing the farming industry, how the infrastructure of crop export has changed, and the human effects of consuming GM food.

Roundup, a glyphosate herbicide, was commercialized in the United States by Monsanto in 1975. The active ingredient in Roundup is the isopropylamine salt of glyphosate, and it contains a surfactant, polyethoxylated tallow amine (POEA), which enables the herbicide penetration into the plant, thus killing anything unwanted by farmers and gardeners alike. The Roundup Ready technology uses glyphosate-tolerant Roundup Ready seeds and markets a different glyphosate formulation named Roundup Ready Herbicide—meaning farmers using the GM seeds can spray the Roundup Ready Herbicide over the entire crop, and only the weeds will be affected.

Since weed control is necessary, Monsanto allows farmers to choose one herbicide to control their crop as opposed to several different herbicides, thus streamlining the farming industry. The major breakthrough was when Monsanto discovered the enzyme 5-enolpyruvylshikimate-3-phosphate synthase (EPSPS), which is naturally found in all plants, including microorganisms such as algae, bacteria, and fungi. However, this enzyme is not present in higher organisms in the animal kingdom. Therefore, the Roundup Ready formula uses the glyphosate active ingredient, which controls weeds by impeding the EPSPS enzyme. Roundup Ready crops also contain another enzyme that is a glyphosate-tolerant EPSPS, which comes from a soil bacteria, Agrobacterium sp. strain CP4, thus the enzyme is named CP4 EPSPS.

According to Monsanto, the first field trials of GM crops began in 1987; however, the first Roundup Ready crop was not introduced in the United States until 1996 with the onset of Roundup Ready soybeans. Then, in 1998, with the purchase of the DeKalb Genetics Corp., Roundup Ready corn was introduced. In addition to Roundup Ready corn and soybeans, there are also Roundup Ready cotton, alfalfa, and sugar beets.

Benefits of Roundup Ready Crops

With the introduction of many different Roundup Ready crops, farmers began looking at the benefits of using the GM seeds on their own farms. Monsanto reports several benefits for farmers, including broad-spectrum weed control and control of perennial weeds, which is essential in farming today. Because Roundup Ready crops are resistant to the active ingredient glyphosate, farmers are able to apply herbicides only on an "as-needed" basis, which is cost-effective and is said to increase yield in some crops.

In addition to the cost-effectiveness of Roundup Ready crops, there are also environmental benefits, including the no-till process created directly by the Roundup Ready technology. When the farmer no longer has to till the soil, it gains organic matter content, less compaction, and better use of water and fertilizer. There is also the benefit of decreased erosion as well as saved fuel, time, and production costs. When tillage is reduced, the soil structure is improved, leading to beneficial activity of the natural soil fauna and flora as well as preserving the fertility because organic matter and mineral availability to the crop are improved. Because soil erosion is reduced up to 90 percent, soil adsorption and permeability is improved, and with the reduction of tilling the soil, less energy is used, resulting in conserving fossil fuels.

Herbicide reduction is also a likely result of using Roundup Ready crops. For example, in the Roundup Ready sugar beet, a reduction in herbicide can reach nearly 50 percent, and is similar in soybeans. This is apparent because of the use of glyphosate and its efficiency to control annual and perennial weeds. Because of the environmental profile of glyphosate, it is in the soil for less than 40 days (50 percent dissipation) and studies show it does not accumulate in the soil over repeated applications because it is strongly absorbed into the soil, thus resulting in a low propensity to reach groundwater and surface water.

The active ingredient also has a low chance of affecting water and land animals. Because it has low vapor pressure and high water solubility, if it were to reach the ground or surface water, it would dissolve; and on land, the herbicide is said to be degraded by microorganisms. Studies also show that the herbicide is harmless to earthworms (essential to aeration of soil), which were tested at higher concentrations than normal projected application.

The benefits are also extended to the farmer in the form of lower maintenance, higher yield, and a higher net profit. When Roundup is applied to the field early, there is a delay in weed growth, and this results in fewer weed problems later in the season, requiring less treatments. This can result in higher yields in some crops such as alfalfa, according to Penn State College of Agricultural Sciences. Therefore, net profit is higher—not only does the grower save money on herbicides and application, but crop yields are higher, resulting in a higher net profit even after the initial cost of the Roundup Ready crop.

Potentially Negative Results of Continued Use

The effects of Roundup have not all been positive, and many farmers are reluctant to opt in to using the GM crop because many of their sales are exported. For example, Japan and countries in the European Union approve the import of only some GM crops but not all. Many farmers, too, are afraid of cross-pollination from a GM crop to an organic crop; however, Monsanto states that crops such as alfalfa are harvested before pollination occurs.

But this is not the only concern of many farmers. The herbicide is causing an increase in parasitic colonization at the roots of Roundup Ready soybeans and corn, and fungal growth has been causing sudden death syndrome (SDS) in the plants. Because of this, new growth of harmful bacterial colonies is destroying the beneficial ones, and the most apparent has been observed in the rhizobia that fix nitrogen in the soil—and without this, plants are not able to receive the nitrogen they need to survive.

In addition, many farmers are reporting an increase in "superweeds" that are resistant to glyphosate. Because of this, many third-party engineers continue to create stronger herbicides, and the weeds grow more virulent every year and continue to affect the overall composition of the soil. There are as many as 63 different species of glyphosate-resistant weeds encompassing 7 to 10 million acres of soil in the United States.

Reports also claim that the Roundup is actually penetrating the soil more deeply than previously thought and can threaten groundwater supplies. Similarly, depending on the condition and composition of the soil, the glyphosate can react differently and either leach deep into the soil or have a higher chance for runoff contamination in nearby streams and rivers around the GM farm, which is more likely to happen when the soil composition is very high in phosphorous. For example, the half-lives vary from 3 days at a site in Texas to 141 days at a site in Iowa, and in a site in Sweden, the glyphosate persisted up to two years. When the half-life is this variable, the chances of contaminating runoff increase and the chance of harm to wild life also increases. Amphibians are especially susceptible to glyphosate use, and even at concentrations one-third the maximum in expected use, glyphosate killed up to 71 percent of tadpoles in outdoor tanks.

While many farmers are leaning toward starting to grow organic crops, they are finding difficulty making the transition. It must be noted that continued use of Roundup Ready GM crops is unsustainable and is changing the agricultural system. Not only are the herbicides altering and destroying soil nutrients and microorganisms, but they are also creating a new class of "superweeds" that cannot be controlled. However, many farmers are going to continue using the GM seeds to gain a higher net profit while also shortening the amount of time and maintenance that was previously required.

See Also: Agriculture; Genetically Engineered Crops; Genetically Modified Organisms (GMOs); Organic Farming; Pesticides.

Further Readings

Allison, Melissa. "Deadline to Comment on Sale of Genetically Altered Alfalfa Nears." *Seattle Times* (March 2, 2010).

"Even Small Doses of Popular Weed Killer Fatal to Frogs, Scientist Finds." *ScienceDaily* (August 4, 2005). http://www.sciencedaily.com/releases/2005/08/050804053212.htm (Accessed January 2011).

Huff, E. "Scientists Reveal Negative Impact of Roundup Ready GM Crops." Natural News (March 11, 2010). http://www.naturalnews.com/028347_GM_crops_Roundup.html (Accessed January 2011).

Monsanto. "The Benefits of Roundup Ready Crops." Official European Fact Sheet (September 2001). http://www.teachingscience.org/BIOTECH/TeachScience.nsf/cf750919409ece6086256b27005a85cc/e61833b56dcef34086256b260062204e/$FILE/FactSheetSept.pdf (Accessed January 2011).

Sidhu, R. S., et al. "Glyphosate-Tolerant Corn: The Composition and Feeding Value of Grain From Glyphosate-Tolerant Corn Is Equivalent to That of Conventional Corn." *Agriculture Food Chemistry*, 48 (2000).

Amy Pajewski
Independent Scholar

S

SKEPTICAL ENVIRONMENTALISM

Skepticism is a political tool for opponents of environmental regulation. Exxon Mobil, seen here on the horizon, gave around $16 million between 1998 and 2005 to cast doubt on the concept of global warming.

Source: Wikimedia

Skeptical environmentalism and, more accurately, *environmental skepticism* are terms to describe the view that environmental problems are overstated or fabricated. Skeptics claim that they are open to persuasion by the scientific evidence while their critics claim that their skepticism is often driven by ideological opposition to environmental concerns or funded by those with vested interests in the status quo.

The term *skeptical environmentalism* is derived from the title of a 1998 book by Danish author Bjørn Lomborg, *The Skeptical Environmentalist: Measuring the Real State of the World*, which was published in English in 2001. At the time, Lomborg was an associate professor of statistics. His book cast doubt on the severity of many environmental problems, including deforestation, pollution, species loss, water shortages, and global warming. He argued that many of these problems could be solved with economic and social development and that the focus on these problems was diverting attention from more important problems such as poverty and AIDS. The book was heavily promoted in mainstream media outlets around the world and became an international best seller.

Not surprisingly, Lomborg's book was controversial and prompted many rebuttals, including a critical review in the scientific journal *Nature* and another in a 2002 issue of

Scientific American by several scientists who argued that Lomborg had misrepresented the scientific evidence. Experts commissioned by the Union of Concerned Scientists found that "Lomborg's assertions and analyses are marred by flawed logic, inappropriate use of statistics and hidden value judgments. He uncritically and selectively cites literature—often not peer-reviewed—that supports his assertions, while ignoring or misinterpreting scientific evidence that does not."

Lomborg was investigated by the Danish Committees on Scientific Dishonesty (DCSD), which are overseen by the Danish Ministry of Science, Technology, and Innovation. In 2003, the DCSD found that the book contained fabricated data, misleading use of statistics, plagiarism, deliberate misinterpretation of the results of scientific studies, and distorted interpretations of conclusions. However, it did not find Lomborg himself to be intentionally dishonest. On appeal, the ministry criticized the DCSD for treating the book as a scientific work, applying unclear definitions and standards, and not providing specific examples of errors.

In 2010, Lomborg backtracked on the issue of global warming, admitting that it was "undoubtedly one of the chief concerns facing the world today" and published a book calling for tens of billions of dollars to be spent to address the problem. It remains to be seen whether Lomborg's reversal of position will undermine the concept of skeptical environmentalism.

Skepticism Versus Denial

A skeptic is someone who needs to examine the evidence before accepting an argument or theory and may not be convinced by that evidence. Michael Shermer, publisher of *Skeptic* magazine, has stated the following:

> Scepticism is integral to the scientific process, because most claims turn out to be false. Weeding out the few kernels of wheat from the large pile of chaff requires extensive observation, careful experimentation and cautious inference. Science is scepticism and good scientists are sceptical.

In contrast, a denier is someone who has decided he or she opposes an argument or theory in advance and examines the evidence in order to find confirmation of his or her opposition, ignoring all the evidence that supports it. The denial is not based on examination of the evidence but rather on ideology, religious belief, self-interest, or some other predisposition.

Pascal Diethelm and Martin McKee describe denialism as the following:

> There is an overwhelming consensus on the evidence among scientists yet there are also vocal commentators who reject this consensus, convincing many of the public, and often the media too, that the consensus is not based on "sound science" or denying that there is a consensus by exhibiting individual dissenting voices as the ultimate authorities on the topic in question. Their goal is to convince that there are sufficient grounds to reject the case for taking action. . . .

Environmental science is full of scientific uncertainties, and opponents of environmental regulations tend to emphasize those uncertainties, so that the uncertainties seem to increase with the increasing relevance of the science to policy decisions. However, regulators are forced to make a decision even though there is scientific uncertainty and debate. A regulator generally does not have the luxury of waiting around until more compelling evidence comes in. Not acting on the given information is just as much a decision as acting.

One way of dealing with this uncertainty is to apply the precautionary principle, which says that scientific uncertainty cannot be used as a reason not to act to prevent or reduce the consequences of an action that may be serious. However, opponents of environmental regulation argue that no action should be taken until the environmental threats can be scientifically proved. Their skepticism becomes a political tool, and their strategy is to exaggerate the uncertainties. In other words, they become environmental deniers.

Diethelm and McKee identify five tactics common to denial movements:

- They claim there is a conspiracy, often aided by the scientific peer-review system that enables participating scientists to reject dissenting papers.
- They use fake experts (whose expertise is not central to the debate) to bolster their own position. At the same time, established experts may be denigrated and efforts made to discredit them.
- They selectively cite evidence that bolsters their own position even after it has been discredited, ignore evidence that contradicts their position, and seek to demonstrate weaknesses in papers that support the mainstream consensus.
- They demand unrealistic research findings.
- They use logical fallacies, false analogies, and misrepresentations to discredit the existing scientific consensus.

Vested Interests

Corporations and business interests have utilized think tanks and a few dissident scientists to cast doubt on the existence and magnitude of various environmental problems, including global warming, declining biodiversity, ozone depletion, the hazards of genetically modified organisms (GMOs), and the threat of industrial chemicals. This strategy is aimed at crippling the impetus for government action to solve these problems, action that would involve regulating business activities and might adversely affect corporate profits. The advantage to attacking the science rather than measures to avoid environmental problems is that the corporations are not directly opposing environmental measures, merely calling for them to be more soundly based in scientific certainty.

Phil Lesly, author of a handbook on public relations and communications, advised the following to corporations:

> People generally do not favor action on a non-alarming situation when arguments seem to be balanced on both sides and there is a clear doubt.
>
> The weight of impressions on the public must be balanced so people will have doubts and lack motivation to take action. Accordingly, means are needed to get balancing information into the stream from sources that the public will find credible. There is no need for a clear-cut "victory." . . . Nurturing public doubts by demonstrating that this is not a clear-cut situation in support of the opponents usually is all that is necessary.

Think tanks have played a key role in providing credible "experts" who dispute scientific claims of existing or impending environmental degradation and therefore provide enough doubts to ensure governments "lack motivation" to act. For example, most conservative corporate-funded think tanks have argued that global warming is not happening and that any possible future warming will be slight and may have beneficial effects.

A study by Peter J. Jacques and his colleagues of 141 English-language books published between 1972 and 2005 (most published since 1992) that denied environmental problems found that more than 90 percent of them (130) were either published by a conservative think tank or written by someone with a conservative think tank affiliation. Moreover, 90 percent of conservative think tanks that focus on environmental issues express skepticism in their publications and websites.

Jacques and his colleagues conclude that environmental skepticism has been a successful way to weaken environmental protection policies in the United States, which went from being a leader in environmental policies in the 1970s and 1980s to being a laggard since the 1990s, despite strong and majority public support for environmental protection and an increasingly strong environmental movement.

Global Warming

The high level of consensus among the world's climate scientists is not widely known because the fossil fuel industry and its allies that would be affected by measures to reduce greenhouse gas emissions have waged a deceptive campaign to confuse the public and policymakers on the issue. They have used corporate front groups, public relations firms, and conservative think tanks to cast doubt on predictions of global warming and its impacts, to imply that we do not know enough to act, and to argue that the cost of reducing greenhouse gases is prohibitively expensive.

In 2003, Republican Party political consultant Frank Luntz advised the following:

> Voters believe that there is no consensus about global warming within the scientific community. Should the public come to believe that the scientific issues are settled, their views about global warming will change accordingly. Therefore, you need to continue to make the lack of scientific certainty a primary issue in the debate. . . . The scientific debate is closing [against us] but not yet closed. There is still a window of opportunity to challenge the science. [emphasis in original]

It is in this way that corporate influence goes far beyond the millions of dollars in campaign donations made by the fossil fuel industry to politicians and political parties. In 2007, the Union of Concerned Scientists (UCS) reported that Exxon Mobil had given around $16 million to some 43 advocacy organizations between 1998 and 2005 "to deceive the public about the reality of global warming" and cast doubt on the idea.

By promoting global warming deniers, business takes advantage of the media's need for balance, and consequently, the debate appears to be evenly divided even though the vast majority of climate scientists are supporting the global warming consensus and there are only a handful of scientists disputing it. Similarly, the U.S. congressional hearings on climate change tend to feature the testimony of similar numbers of deniers as mainstream scientists representing the Intergovernmental Panel on Climate Change (IPCC).

The global warming denial movement, particularly think tanks and front groups, has played an active role in disseminating every tidbit of information, whether verified or not, that discredits the idea of global warming. For example, in 2006, Australian columnist Piers Akerman wrote an article labeling global warming warnings as alarmist in the *Daily Telegraph* newspaper. In it, he quoted former head of the IPCC John Houghton as having said, "Unless we announce disasters no one will listen." Although this quotation appears to have been fabricated, it was subsequently quoted in more than three books, more than 100 blog posts, and some 24,000 web pages, often attributed to a 1994 book by Houghton in which it does not appear.

New Scientist correspondent Jim Giles describes this phenomenon as an "informational cascade" that is amplified by the "echo chamber" of the Internet. He suggests that the more a person hears a piece of information, the more he or she is likely to believe it, particularly if it accords with existing beliefs ("confirmation bias").

See Also: Environmental Education Debate; Mitigation; Organic Trend; Precautionary Principle (Ethics and Philosophy).

Further Readings

Beder, Sharon. *Global Spin: The Corporate Assault on Environmentalism*, 2nd ed. Devon, UK: Green Books, 2002.

Diethelm, Pascal and Martin McKee. "Denialism: What Is It and How Should Scientists Respond?" *European Journal of Public Health*, 19:1 (2009).

Giles, Jim. "Giving Life to a Lie." *New Scientist* (May 15, 2010).

Jacques, Peter J., Riley E. Dunlap, and Mark Freeman. "The Organisation of Denial: Conservative Think Tanks and Environmental Scepticism." *Environmental Politics*, 17:3 (2008).

Lesly, Philip. "Coping With Opposition Groups." *Public Relations Review*, 18:4 (1992).

Lomborg, Bjørn. *The Skeptical Environmentalist: Measuring the Real State of the World*. New York: Cambridge University Press, 2001.

Michaels, David and Celeste Monforton. "Manufacturing Uncertainty: Contested Science and the Protection of the Public's Health and Environment." *American Journal of Public Health*, 95:S1 (2005).

Shermer, Michael. "I Am a Sceptic, but I'm Not a Denier." *New Scientist* (May 15, 2010).

Shulman, Seth. *Smoke, Mirrors & Hot Air: How ExxonMobil Uses Big Tobacco's Tactics to Manufacture Uncertainty on Climate Science*. Cambridge, MA: Union of Concerned Scientists, 2007.

Sharon Beder
University of Wollongong

SOCIAL ACTION

Within the political realm, social actions recruit segments of the public to join together for change. Once joined, the common voices focus on a preferred action and convince decision makers to act. In sociology, this concept refers more basically to an action that takes into account the responses of others. From a more classical definition, any action for advocacy—the typical end goal of social actions—is a form of rhetoric, the effort to engage the populace to support and act on the speaker's claims.

As a key strategy of the environmental and green movements, social actions are often triggered by the perceived failure of governance, mediation, or collaboration to respond to environmental concerns and problems, both ongoing and impending. As the actions are often initiated after other methods have failed, this approach may tend to spawn or exaggerate social conflict. Such conflict may support the end-goals of the action. But success is far more likely if social action advocates are able to create the "Trinity of Voice" effect. As identified by Susan Senecah, the Trinity of Voice occurs when all three key elements—access, standing, and influence—combine to ensure that a community's collaborative message succeeds with governing bodies.

Social actions serve to magnify the many voices into a common voice. United, these many voices gain access to pursue a common goal. They may do so with standing, based on the massing of voters, students, consumers, or other community blocs. These social-action advocates will likely gain influence, though if their gathered voices stray too far into an indecorous voice, they risk being ostracized by the decision makers they seek to influence.

Social actions are often prompted by threats posed from government or business activities; by inaction or inattention on evolving issues; and if access, standing, or influence are restrained for a group that seeks to address an environmental or social justice issue. As effective as such campaigning can be, the overreliance on extreme social actions may alienate the communities the advocates seek to engage, which may in turn limit access or standing. For instance, groups working along the edge of legal actions such as the more extreme actions of Earth First! and animal rights supporters loosely connected as the Earth Liberation Front may ostracize more opponents than the supporters they engage.

Within the environmental movement, these direct action strategies stir debate and concern. Does the need for action justify property damage, as practiced at times by the Earth and Animal Liberation Fronts? If international treaties and enforcement fail to protect whales, are the interceptions by Sea Shepherd therefore justified in order to prevent unethical actions?

Direct action as a strategy is shared by both civic justice and environmental movements, partly because their ethical precepts share a common rootstock. In fact, direct action is a key element of the United States' founding narratives. American culture is framed by revolutionaries whose social actions launched the Declaration of Independence. And witness the role of Henry David Thoreau, whose "Civil Disobedience," a protest of war taxes and of laws supporting slavery, defined the rationale and methodology of social actions. Add the transcendental vision of *Walden, or, Life in the Woods* and one can easily trace Thoreau's role in inspiring social action movements that advocate conservation, civil rights, environmental justice, and environment and green issues.

Thoreau's concept—that justice and "natural law" can guide us when protesting a government's unjust law—expanded into mass social action in Mohandas Gandhi's concept of *satyagraha*, or "firmness in adhering to truth." Gandhi paired nonviolent protest against unjust laws with a requirement that protesters also support organized social work to promote positive civic actions—insisting that one must act with a group to improve society even as one acts against the formal yet unjust government.

Thoreau and Gandhi inspired Martin Luther King Jr. and the civil rights movement, which also inspired the modern environmental movement that was triggered by Rachel Carson, whose *Silent Spring* identified toxic pesticide use that silenced birds and poisoned children and neighborhoods. Within the decade, we witnessed the social actions of Earth Day and of parents at Love Canal, who observed the toxic impact of dioxin on their children and gathered to demand relocation and treatment. Today, this movement to protect the future inspires college students who unite their communities around 350.org, a social-action group named for and organized in support for the 350 parts-per-million level of carbon dioxide that will slow or mitigate global warming. And animal rights advocates populate a range of claims and inspirations to act, incorporating undressed models and guerrilla videotapes to engage social-action protests of factory farming and animal testing.

From Conservation Movements to the New Environmental Paradigm

Initial conservation movements were triggered by external threats to wildlife and wild habitats (which were core to our cultural identity) but propelled as much by public opinion and opinion makers than by direct social action. In the United States, the parks and conservation movements of the 1870s to 1920s followed the closing of the frontier and the removal of Native American tribes from traditional homelands. During this era, environmentally motivated social actions focused more on the impacts of urbanization and industrialization on urban residents, including the work of muckraking journalists and social change advocates. These earlier protests for social equity and safe living and working conditions presaged the environmental justice movement, which has developed social actions in the past few decades to give voice to communities and community issues that are ignored as "outsider" or outlier issues by governments and even by mainstream environmental groups. For instance, "toxic tours" of marginalized communities serve to make the unseen toxics visible to the public and to decision makers.

Similar transitions from unvoiced to voiced concerns have relied on social action movements. The 1960s protests against DDT transitioned in the 1970s and 1980s to campaigns of mothers opposed to pesticides on apples and fresh food; today, groups seek to ban Bisphenol-A (BPA) and other estrogen-mimicking chemicals in our food systems. Likewise, the civil rights movement added support to the environmental justice movement and to the initial Earth Day, which enrolled more supporters in part because the indecorous voices of civil rights and antiwar protesters were given authority and some acceptance by the amount of television coverage the protesters received. Likewise, protests against oil spills in California and Alaska—which helped launch the New Environmental Paradigm of the 1970s—have been repurposed to protest the damage of Hurricane Katrina, the BP Deepwater Horizon Gulf oil spill, and that ultimate carbon spill—called "climate change," "global warming," or "climate hoax," depending on your interpretation of the science and politics of our carbon-fueled era.

The issues faced by proponents of social-action advocacy are immense. Many claim that the concerns have risen from a "should" stance to a "must." We must manage an overpopulated world and a warming globe. We must curtail our carbon dioxide output and sequester climate change gases if we are to prevent our human-caused thermal blanket from thickening, with impacts nearly beyond our imagination.

This a far cry from the "This Land Is Your Land" sentiment of our childhood. Yet within the realm of social action practitioners there are also those, typified by the Tea Party movement, who have gained that "Trinity of Voice"—access, standing, and influence—because they joined to protest against scientists whose analysis they prefer not to believe.

They succeed in part because fear is a poor motivator if it is not tied to a specific positive act that is likely to succeed. Against the unseen gas, one version of the "fight-or-flight" response is to fly from the predicted future while simultaneously fighting against the anti-American "green" advocates.

In contrast to the Tea Party response, an intertwined blend of advocates and analysts have chosen to face these challenges with specific actions. Paul Hawken, a leading chronicler of these actions, has observed and inspired the "movement of movements" in his book, *Blessed Unrest*, which combines a review of the history of such movements with a compilation of the many groups. In tandem, his organization gathered 40,000 of these activists on www.wiserearth.org, an online community dedicated to being "the social network for sustainability."

To mitigate global warming, 350.org organized online in order to physically gather at "7,347 events in 188 countries" for a global "10/10/10" work party, sending a message to global leaders that "If we can get to work, so can you!" Other groups—from Greenpeace's protest-based campaigns to the Sierra Club's feet-on-the-ground organizing—support species and landscape conservation, local food systems, intergenerational and social equity, sustainable development, and renewable energy.

Thousands of these groups seek to organize for change while at the same time collaborating with government, nonprofits, and corporations to foster "green" and technical solutions to local and global environmental issues. For example, the Worldwatch Institute supports these organizations with priorities to build a low-carbon energy system and create healthy agricultural systems that in turn support a sustainable global economy.

Some who oppose these advocacy campaigns align with a traditional conservative/ religious alliance as typified by the Tea Party movement. Others opposed to green advocacy include skeptics like Bjørn Lomborg, self-appointed as "The Skeptical Environmentalist," who claims that environmentalism takes the form of a "litany" of pessimistic claims. He admits the problems are authentic, but the typical green analysis may not take into account the shared roots of the problems—poverty—or the unequal scale of how and where the problems occur—often local and again tied to poverty. To presume a "global" nature of such local instances, he claims, leads to overly broad and unfocused solutions. This approach is shared by climate change skeptics. One often hears concerns against conspiracies of elite scientists tied to a fear of global governmental policies and "world government," which outweighs any concern for the modeled (and observed) effects of climate change on local social and natural systems.

To counter the skeptics, advocates argue for small steps, taken locally, to make cumulative changes, globally. One example is the "stabilization wedges," a series of seven steps of global carbon reduction that might counter dramatic climate change. In the midst of these debates, advocates can take heart that new media and social structures support the sharing of their voices, which are also protected by the Aarhus Convention, the shorthand title for the United Nations Economic Commission for Europe (UNECE) Convention on Access to Information, Public Participation in Decision-making and Access to Justice in Environmental Matters. While limited in its formal application to European countries within the UNECE, its scope and precedent has offered support for freedom of information and free speech guidelines in other countries. It specifically outlines citizen participation in sustainable development and management of environmental issues, which will in turn appear to ensure continued citizen participation in social action campaigns.

This guarantee of standing and access, though, does not guarantee influence. One may presume that social actions by those seeking influence will continue to be required as we seek to balance prosperity and civil-society priorities amid immense environmental challenges.

See Also: Earth Liberation Front (ELF)/Animal Liberation Front (ALF); Green Altruism; Green Anarchism; Individual Action Versus Collective Action.

Further Readings

Hawken, Paul. *Blessed Unrest: How the Largest Movement in the World Came Into Being and Why No One Saw It Coming.* New York: Viking, 2007.
"The Seeds of a Revolution: Earth Days." *American Experience.* Public Broadcasting Service. http://www.pbs.org/wgbh/americanexperience/films/earthdays (Accessed November 2010).
Senecah, S., S. P. Depoe, J. W. Delicath, and M. A. Elsenbeer, eds. "The Trinity of Voice: The Role of Practical Theory in Planning and Evaluating the Effectiveness of Environmental Participatory Processes." In *Communication and Public Participation in Environmental Decision Making.* Albany: State University of New York Press, 2004.

Ron Steffens
Green Mountain College

Socially Responsible Investing

Socially responsible investing (SRI), which is also known as sustainable, socially conscious, or ethical investing, describes an investment strategy that seeks to maximize social benefits and financial return. Ethical investors thus differentiate themselves from traditional investors who are concerned only with financial return. In general, ethical investors favor small business and corporate practices that promote environmental stewardship, consumer protection, and human rights. The areas of concern recognized by SRI can be summarized as environment, social justice, and corporate governance (ESG). There are three main strategies of SRI, which are utilized by lower-, upper-, and middle-class investors for competitive rates of return on their savings while promoting ESG goals: (1) social screening (both avoidance and affirmative), (2) community investing, and (3) shareholder advocacy.

Social screening can be either negative (avoidance screening) or positive (affirmative screening). Avoidance screening entails investors ensuring that their money is not used by corporations whose practices are counter to their moral or ethical values. For example, many ethical investors avoid tobacco companies (responsible for 400,000+ deaths each year in the United States alone in addition to a much higher rate of agricultural-related environmental damage than most food crops), land mine producers (land mines maim or kill approximately 26,000 civilians every year, including 8,000 to 10,000 children), or companies that are implicated in using slave labor. When such investors already own mutual funds or stocks in companies that do not agree with their ethical/moral values, they often divest and move their funds elsewhere. When this approach is applied to areas revolving around environmental stewardship, it is also known as "light green" investment strategy (e.g., light green funds or a light green screen).

A more aggressive or "dark green" strategic approach to SRI is the use of affirmative screening, which identifies and preferentially invests in companies based on positive social or environmental practices. For example, ethical investors interested in environmental stewardship might invest in the Ecodeposits program of ShoreBank Pacific, which lends money to small businesses and homeowners to strengthen environmental conservation-based practices. The majority of ethical investors that focus on the environment invest in mutual funds that target environmentally benign firms like those offered by Green Century or Portfolio 21. Portfolio 21 is a mutual fund that invests specifically in corporations that design "ecologically superior products," use renewable energy, and consistently improve their production efforts so that they reduce environmental impact. There is a very long and growing list of green mutual funds and other socially responsible funds maintained by Social Funds online. Finally, ethical investors may invest directly in stock of companies they want to support. For example, an ethical investor wishing to concentrate funds in the green sector might support investment in photovoltaic and wind turbine manufacturers, recycling firms, environmental remediation companies, ecotourism enterprises, and organic farms.

Community investing is the second SRI strategy and involves directing assets toward a particular locale by way of a community lending institution. Examples include purchasing certificates of deposit and opening savings/checking accounts at community development banks (e.g., Community Capital Bank) or community development credit unions (e.g., the Self-Help Credit Union). Ethical investors support community investing because they believe that the ethically questionable practices of some multinational corporations are counteracted by local community control. They contend that local communities tend to try to avoid harmful environmental practices that acutely pollute the local community. For example, a community might try to prevent a waste incinerator from operating near it for environmental reasons and have a greater chance of success dealing with local businesses than with a large multinational corporation, which may be less attuned to local concerns. This thinking has led to a movement in the United States to encourage people to move their money out of bigger banks and into smaller, community-oriented financial institutions that generally avoid what they believe are the reckless investments and schemes that helped cause the 2008 Great Recession. This work is centered at the organization called Move Your Money.

Microcredit is a relatively new community lending option. Microcredit banking offers business training, peer support, and small loans (even only a few dollars) to the world's poorest people. Proponents contend that such investments offer returns with high "social dividends." This approach was started by the Grameen Bank but has expanded to offer programs throughout the world at many different institutions. Overwhelming evidence from the Grameen Bank and many other similar institutions shows that microcredit creates personal and economic success for the poor and their communities as it gives people the little initial economic assistance necessary for them to pull themselves out of poverty. The initial microcredit organizations, such as the Grameen Bank, operated more as charities than as investments. An individual would donate money, and then the Grameen Bank would lend it out. When the money was repaid, the institution would lend it out again. Today, there are several ways for ethical investors to invest in microcredit rather than simply donating to a charity that supports microlending. One way that is gaining popularity is through the online organization Kiva.org, which is the world's first person-to-person microlending website. Kiva enables individuals to lend directly to unique entrepreneurs in the developing world. At Kiva, investors do not earn a financial return (e.g., they charge no interest), but socially responsible investors argue they can gain a large social return. After the entrepreneurs in the developing world repay their loans, Kiva investors can withdraw the funds or choose to reinvest in additional entrepreneurs. Some ethical investors believe this type of lending follows the zero-interest tenets of money lending described in the Bible and is consistent with early Christian teaching. A newer microloan website called

Microplace is an eBay affiliate, and it allows one to make direct loans to individuals or groups to help with development while earning a modest return that the investors choose for their money. This enables ethical investors to gain both a social and a financial return while participating directly in microlending. There are other groups such as Wokai.org, which operates in a model similar to Kiva with a special focus in rural China—one of the countries in which Kiva does not operate.

Shareholder activism is the third SRI strategy and encompasses several tactics used by ethical investors to improve social outcomes while using investments as leverage: (1) dialogue with the management of a company, (2) shareholder resolutions, and (3) divestment (or threats of divestment). For example, the Aquinas Fund, a group of mutual funds that draws from the U.S. Catholic Bishops' social teaching, has influenced six major drug companies' decisions to decline producing unsafe abortion drugs.

A recent twist to shareholder activism is to use the market to both profit from and punish "bad" companies using Karma Banque. Wealthy British scion Zak Goldsmith and investment activist Max Keiser want to financially ruin the Coca-Cola Co. because of social concerns, and they have developed an interesting improvement to the traditional tactic of the consumer boycott. They have opened a hedge fund designed to profit from any decline in the soft drink conglomerate's stock price. Karma Banque supporters believe it is at the center of a new activist movement that combines the civil disobedience of Gandhi with the financial savvy of George Soros to help change the economic landscape. Karma Banque endorses boycotts of companies to drive their stock price down, and at the same time it short sells the companies' stock. Short selling is an investment strategy that allows an investor to make money when a stock goes down in price. Individual investors can of course do this themselves for any company they wish to target, and Karma Banque's information on both the methodology and vulnerable companies is free and available to the public.

SRI Advantages

SRI can have a positive benefit on society through several mechanisms. First, in negative screening, the divestment of ethical investors effectively lowers "bad" companies' value and makes their access to capital more costly. This restricts their growth and retards their operations. As SRI expands beyond its current approximately 11 percent of investments and continues to have more stringent guidelines, this provides a major economic incentive for companies to avoid being socially and environmentally destructive. Similarly, positive screening raises the value of firms deemed "socially responsible" by ethical investors and thus enables them to have access to less costly capital. This allows these companies, which are deemed to be socially responsible by ethical investors, to grow faster. The positive effects of community lending in both the developed and developing world are clear for the communities that benefit from them. In some cases, this is not only about access to capital at a lower rate, but actual access to capital at all. For those in some communities without access to capital, SRI through microlending can have an enormous impact—according to proponents, it literally saves lives. Finally, there have been numerous examples of various forms of shareholder activism changing the behavior of companies, removing harmful products from the market, or expanding the product selection of socially or environmentally beneficial products.

There is an intuitive concern by most investors that, because SRI limits the range of investments, it would lower an investor's economic return. In the case of zero-interest microlending this is certainly true, but in this special case, ethical investors have knowingly chosen to forgo a financial return for a higher social return. However, in the more mainstream socially responsible investments such as mutual funds, there is substantial evidence that socially screened investments may have equivalent performance to or even outperform their nonscreened counterparts. A study was recently completed that compared portfolios comprising top socially responsible funds and top vice funds (or "sin funds" that are essentially anti-SRI funds targeted directly at investment in businesses focusing on alcohol, tobacco, gambling, weapons, etc.). This analysis found that in the short term, differences between the two portfolios' performance was negligible. Only the SRI funds had any significant positive risk-adjusted performance, and this was for the long term. For the 5- and 10-year periods of investment, the SRI portfolio did better, suggesting that the market valued the features of the companies in the SRI portfolio. These findings indicate that the market valued the features of the SRI portfolio over the longer time horizons, while not valuing the vice funds for any period, and it was the SRI portfolio that had superior risk-adjusted performance. There are several likely mechanisms for what at first appears to be a nonintuitive result. For example, companies that are included in SRI indices are better protected from long-term risks such as corporate scandals (better governance), employee lawsuits (better social practices), or environmentally related fines or the likely increases in the price of carbon emissions (better environmental performance).

SRI Challenges

SRI is slightly more challenging than conventional investing. Depending on the strictness of the ethical investor's criteria for investment, finding ethical investments can take extensive additional research on the part of the investor compared to standard investing. For this reason, most SRI is currently done through instruments such as SRI mutual funds where professionals do the research. The companies that offer these funds or consulting firms they employ are in a better position to do this research. Whether they do so properly varies between funds, and public scrutiny of their decisions is valuable. Socially responsible indices also can provide a guide, though their analysis is also open to question.

Another major challenge of SRI is that restricting investment opportunities makes a balanced portfolio more challenging. For example, for those concerned about environmental issues to be truly socially responsible in investing, it is necessary to seriously face questions of fossil fuel investment. In the case of the energy sector, a choice must be made between these approaches: (1) remove the entire segment of fossil fuel–related investment from the portfolio, favoring (for example) renewable energy technologies and sources, and (2) acknowledge that fossil fuel use may not disappear immediately and invest in the least damaging form of fossil fuel as a transitional form. Thus, the investor must recognize that there are very large differences in, for example, climate forcing impact between activities within the same sector, from natural gas (lowest climate change impact) to petroleum and coal (much higher) through to shale oil and tar sands (far more damaging). Finally, (3) ethical investors can reward certain fossil fuel–based companies that engage in the greenest behavior relative to their peers. In the last case, researchers and investors must be extremely wary of "greenwash" and consider the seriousness and commitment to the positive behavior of the company. For example, Royal Dutch Shell has a much-publicized corporate social responsibility (CSR) policy and was a pioneer in triple bottom line reporting, but this did not prevent the 2004 scandal concerning its misreporting of oil reserves, which seriously damaged its reputation and led to charges of hypocrisy. Similarly, BP's "Beyond Petroleum" campaign has been questioned as a classic "greenwash" by numerous commentators both before and after the Gulf of Mexico oil disaster.

Criticisms

Critics of SRI as well as proponents debate a number of concerns related to it. In addition to the questions raised by the efficacy of SRI-related companies' doing a good job and actually delivering the social returns they promise, SRI has been criticized from both the left and the right sides of the political spectrum.

On the left, it is pointed out that some SRI-related investment firms have made an almost arbitrary preference for some companies over others (e.g., BP over Exxon, and Reebok and Timberland over Nike) and are not strict enough about allowing companies onto a given SRI index. These critics argue that it is not the case that the SRI favorite is more ethical because the analysis used is shallow. For example, a vocal critic claimed that Nike was allowed back into KLD's SRI index on the basis of some favorable news clippings KLD found rather than on any substantial change in the company. KLD Research & Analytics, Inc., is major research firm in SRI, which many SRI mutual fund managers use for information. Even more radical commentators question whether bringing people (particularly in the developing world) into an exploitive capitalist system is a social good. These critics claim that the type of capitalism practiced in many developing countries is a form of economic and cultural imperialism, noting that these countries usually have fewer and weaker labor and environmental protections, and thus their citizens are at a higher risk of exploitation by multinational corporations.

As mentioned earlier, some critics believe that CSR programs are undertaken by companies such as British American Tobacco, the petroleum giant BP (well known for its high-profile advertising campaigns about environmental aspects of its operations), and McDonald's to distract the public from ethical questions posed by their core operations (e.g., the health and mortality threats from the use of tobacco, climate change/oil spills, and working conditions of fast food employees and the health impacts of fatty foods, respectively). These critics argue that some corporations start CSR programs for the commercial benefit they enjoy through raising their reputation with the public or with government and to attract SRI. Such critics suggest that corporations, which exist solely to maximize profits, are unable to advance the interests of society as a whole. Critics concerned with corporate hypocrisy and insincerity generally suggest that better governmental and international regulation and enforcement, rather than voluntary measures or meager economic incentives currently offered by SRI, are necessary to ensure that companies behave in a socially responsible manner.

It is noteworthy that such critics of SRI would agree with views of the more mainstream critics of SRI from the right discussed below. SRI proponents would counter these arguments with numerous demonstrated examples of SRI playing a pivotal role in encouraging socially responsible business practices. For example, in 2010, Domini (a social investment firm) and Nucor (the largest steel producer in the United States) entered into a written agreement in exchange for the withdrawal of a Domini shareholder proposal. Domini had been working for years to reduce the use of slave labor in the Brazilian pig iron industry. Nucor will require its top-tier Brazilian pig iron suppliers to either join the Citizens Charcoal Institute (ICC) or sign and adhere to the National Pact for the Eradication of Slave Labor.

On the far right, there is concern that SRI and the broader CSR is some form of anti-capitalist plot to push a socialist or similar agenda. These concerns include both SRI's and CSR's relationship to the fundamental purpose and nature of business and questionable motives for engaging in CSR including concerns about insincerity and hypocrisy discussed above.

Less radically, Milton Friedman and others on the right of the political spectrum have argued that a corporation's purpose is to maximize returns to its shareholders, and that since only people can have social responsibilities, corporations are only responsible to their shareholders and not to society as a whole. Although they accept that corporations should obey the laws of the countries within which they work, they assert that corporations have no other obligation to society. Those holding this view would generally not agree with the goals of SRI—as they are explicitly to improve society. It should be pointed out here that the stated (although perhaps not the real) original goal for the creation of corporations was to explicitly improve society by using the profit motive as a tool for social good. There are also many organizations that claim to support these goals today. In addition, many religious and cultural traditions hold that the economy exists to serve human beings, so all economic entities have an obligation to society.

SRI critics agreeing with Friedman (and the more politically left critics discussed above) argue that governments should set the agenda for social responsibility by way of laws and regulation that will allow a business to conduct itself responsibly. Such critics point out that organizations and individuals pay taxes to government to ensure that society and the environment are not adversely affected by business and corporate activities. SRI proponents and others argue, however, that the issues resulting from government regulation to impose social responsibility on corporations pose several challenges. Regulation in itself is unable to cover every aspect in detail of a corporation's operations. When, however, regulation tries to do this, it clearly can lead to burdensome legal processes bogged down in interpretations of the law and debatable gray areas. For example, General Electric failed to clean up the Hudson River after contaminating it with organic pollutants. Critics have pointed out that GE argued via the legal process on assignment of liability, while the cleanup remained stagnant. The second issue resulting from overregulation is the financial burden that regulation can place on a nation's economy. This view is supported by such examples as the Australian federal government's actions to avoid compliance with the Kyoto Protocol in 1997, based on concerns of economic loss and national interest. The Australian government took the position that signing the Kyoto pact would have caused more significant economic losses for Australia than for any other Organisation for Economic Co-operation and Development (OECD) nation. Several authors, however, have pointed out that for most businesses, acting responsibly and reducing greenhouse gas emissions (and thus surpassing the Kyoto targets) by treating energy conservation measures as investments would actually be profitable. They argue that environmentally beneficial direct investments (e.g., high-efficiency lighting) provide a better return than standard investments with less risk. Thus, such improvements to the physical infrastructure of a home or business could be considered green SRI.

Those who believe that a business's only purpose is to maximize profit for shareholders perceive SRI as opposed to the nature and purpose of business and, indeed, a hindrance to free trade. Those who assert that SRI is anticapitalist and are generally in favor of neoliberalism argue that improvements in health, longevity, and/or infant mortality have been created by the recent trend of economic growth attributed to free enterprise and that anything that opposes the maximization of economic growth indirectly hurts those social goals. Critics of this argument perceive neoliberalism as opposed to the well-being of society and a hindrance to human dignity and freedom. There are also several difficulties with some of the neoliberal argument against SRI. First, regardless of the social factors, SRI still represents an investment in businesses so most of those who comment on the topic would argue it is incorrect to label SRI as anti-capitalist. Second, there is some evidence that over the long term there is a financial benefit to corporate social responsibility because it reduces risks and inefficiencies while offering a host of potential benefits such as enhanced brand reputation and employee engagement. Thus, some argue that this would make SRI a pro-capitalist method of investing that does generate the maximum profit over the long term.

In addition to those who criticize SRI as a whole, critics of microfinance have additional concerns with the current process. Microlending still charges developing-world entrepreneurs relatively enormous interest rates when compared to the developed world. In recent years, there has been increasing attention paid to the problem of interest rate disclosure, as many suppliers of microcredit quote their rates to clients using the flat calculation method, which significantly understates the true annual percentage rate. Kiva provides the real annual percentage rates, and its numbers are instructive. For example, the average annual percentage interest rate charged to Kiva borrowers is about 36 percent, which would appear high to most borrowers in the developed world. However, it is clear that the nature of microcredit (e.g., small loans) is such that interest rates need to be high to return the cost of the loan, which comes from types of costs. The cost of the money lent and the cost of loan defaults are proportional to the amount lent.

For instance, if the cost paid by the microfinance institution (MFI) for the money it lends is 10 percent, and it experiences defaults of 1 percent of the amount lent, then these two costs will total $11 for a loan of $100, and $55 for a loan of $500. An interest rate of 11 percent of the loan amount thus covers both these costs for either loan. Transaction costs, however, are not proportional to the amount lent (e.g., the transaction cost of the $500 loan is approximately the same as for a $100 loan). Both loans require roughly the same amount of staff time for meeting with the borrower to appraise the loan, processing the loan disbursement and repayments, and follow-up monitoring. If the transaction cost is $25 per loan and the loans are for one year, to break even on the $500 loan, the MFI would need to collect interest of $80, which represents an annual interest rate of 16 percent. To break even on the $100 loan, the MFI would need to collect interest of $36, which is an interest rate of 36 percent. At first glance, a rate this high looks abusive to many people, especially when the clients are poor and living in the developing world. However, microfinance advocates argue this interest rate simply reflects the basic reality that when loan sizes get very small, transaction costs dominate because these costs cannot be cut below certain minimums without some sort of charity being involved.

Microcredit has also been criticized by some critics as a privatization of public safety-net programs. There is concern that enthusiasm for microcredit among government officials as an antipoverty program can have an unintended consequence of cuts in public health, welfare, and education spending. These critics coming from the left also argue that the success of the microcredit model has been judged disproportionately from a lender's perspective (repayment rates, financial viability) and not from that of the borrowers. For example, the Grameen Bank's high repayment rate does not reflect the number of women who are repeat borrowers that have become dependent on loans for household expenditures rather than for capital investments. Studies of microcredit programs have found that women often act merely as collection agents for their husbands and sons, such that the men spend the money themselves while women are saddled with the credit risk and as a result, borrowers are kept out of waged work and pushed into the informal economy. Many studies in recent years have shown that risks like sickness, natural disaster, and over-indebtedness are a critical dimension of poverty and that very poor people rely heavily on informal savings to manage these risks. It might be expected that microfinance institutions would provide safe, flexible savings services to this population, but—with notable exceptions like Grameen II—they have been very slow to do so. Some critics argue that most microcredit institutions are overly dependent on external capital as they are very slow to deliver quality microsavings services because of easy access to cheaper forms of external capital. In this way, SRI may actually be hindering microsavings development by providing inexpensive (or even free) capital to MFIs.

Critics have also pointed out that there can be other unintended consequences of microfinance. As field officers are in a position of power locally and are judged on repayment rates as the primary metric of their success, they sometimes use coercive and even violent tactics to collect installments on the microcredit loans. Some loan recipients sink into a cycle of debt, using a microcredit loan from one organization to meet interest obligations from another. Also, counter to the original intention of the microcredit system to empower women, one of the effects of an infusion of cash into local economies has been to increase dowries, with women forced at times to take microcredit loans as the only means to pay these increased dowries for their daughters. Microcredit proponents contend that such problems are not widespread and that the good that microcredit does for those in poverty greatly outweighs these unintended consequences seen only in a small minority of cases.

Current Trends

The popularity of SRI is increasing and becoming mainstream among investors. The aggregate SRI assets under management in the United States have grown steadily in recent years. According to the Social Investment Forum, at the end of 2007, $2.7 trillion in assets used one or more of the strategies that commonly define SRI. That $2.7 trillion is about 11 percent of all U.S. assets under professional management. Microfinance is also expanding rapidly. For example, Kiva has been operating for less than five years, but has raised more than $153 million dollars from more than 700,000 users for nearly 400,000 entrepreneurs in the developing world. These entrepreneurs have a 99 percent repayment rate and most of the money when repaid is recycled to fund the next wave of entrepreneurs.

See Also: Corporate Social Responsibility; Ecosocialism; Green Jobs (Cities); Individual Action Versus Collective Action.

Further Readings

Brill, H., J. A. Brill, and C. Feigenbaum. *Investing With Your Values*. Gabriola Island, British Columbia, Canada: New Society Publishers, 2000.

Domini, A. *Socially Responsible Investing*. Chicago, IL: Dearborn Trade, 2001.

Friedman, M. "The Social Responsibility of Business Is to Increase Its Profits." *New York Times Magazine* (September 13, 1970).

Kiva. "About Microfinance." http://www.kiva.org/about/microfinance (Accessed August 2010).

McKibben, B. "Hope vs. Hype." *Mother Jones* (November/December 2006).

Pearce, J. M., D. Denkenberger, and H. Zielonka. "Energy Conservation Measures as Investments." In *Energy Conservation: New Research,* Giacomo Spadoni, ed. New York: Nova Science, 2009.

Robinson, M. *The Microfinance Revolution: Sustainable Finance for the Poor.* Washington, DC: World Bank, 2001.

Shank, T. M., D. K. Manullang, and R. P. Hill. "Is It Better to Be Naughty or Nice?" *Journal of Investing* (Fall 2005).

Social Funds. http://www.socialfunds.com (Accessed August 2010).

Social Investment Forum. http://www.socialinvest.org (Accessed August 2010).

Joshua M. Pearce
Queen's University

SOLAR ENERGY (CITIES)

The U.S. Department of Energy chose 25 cities to serve as models in the area of solar technology and implementation. One of these is Denver, Colorado, which installed this photovoltaic system at its Colorado Convention Center.

Source: Solar America Communities/U.S. Department of Energy

Conversations about our future frequently contain phrases like "sustainability," "peak oil," and "energy independence." With growing concern about the consequences of extracting, refining, and burning finite fossil fuels, many have turned to alternative energy sources such as solar energy as the solution for meeting our energy needs while causing the least harm to both environmental and human health. Solar energy has been harnessed to provide power to both buildings and vehicles, yet it has happened in piecemeal fashion. There is a strong movement desiring to expand upon this framework and to embrace solar energy more fully, like the movement to use solar energy to power cities. Proponents see it as a viable option that can meet community needs while causing little or no harm to the environment; opponents, on the other hand, note that solar energy can be unreliable except in sunny climates, that it does not work at all once the sun sets each day, and that it is expensive to install. Further, critics note that the manufacture of solar panels is not as "green" as many assume.

The Technology

The sun generates more than 10,000 times the amount of energy that the entire world consumes annually. While schemes to harness that energy have been around for many years, recent technological advances have helped bring us closer to doing so efficiently and effectively. Solar thermal and photovoltaic are the two methods most commonly thought of when thinking of how to convert solar energy into power that we can use. Solar thermal methods involve using the sun's energy directly to generate heat; by using solar panels in conjunction with "solar thermal collectors," or tubing through which water circulates, the panels collect the sun's heat and transfer it to the solar thermal collectors, which can then be used to heat water and buildings.

The photovoltaic process converts the sun's power into electricity with the use of photovoltaic or solar cells. These cells are often silicon based, and they absorb the sun's light as opposed to its heat. The electrons in the cells are excited by the solar energy, and this creates direct current (DC) electricity. This DC energy is transformed to alternating current (AC) energy with the use of an inverter and can then be used to provide energy to power households and businesses. With the photovoltaic method, excess energy can be fed back into the conventional power grid.

Some of the criticisms of solar energy of the past are no longer relevant, as the technology has advanced to the point that solar panels are now being manufactured that work not only on cloudy days and in climates that do not receive as much sunlight, but that also work at night. Further, the use of battery banks that store the energy absorbed by conventional solar panels throughout the day enables that energy to be utilized at night or on cloudy days. Thus, arguments about solar panels not being viable in cloudy climates or at night have been weakened by technology such as battery banks attached to conventional solar panel grids and the new solar panel technology that works with both ultraviolet and infrared light. These technological advances stand to deeply affect movements such as the one to bring solar energy to cities.

Harnessing the Sun

The U.S. Department of Energy (DOE) has chosen the following as the 25 Solar American Cities (SAC) to serve as a model for cities around the nation:

- Ann Arbor, Michigan
- Austin, Texas
- Berkeley, California
- Boston, Massachusetts
- Denver, Colorado
- Houston, Texas
- Knoxville, Tennessee
- Madison, Wisconsin
- Milwaukee, Wisconsin
- Minneapolis–Saint Paul, Minnesota
- New Orleans, Louisiana
- New York City, New York
- Orlando, Florida
- Philadelphia, Pennsylvania
- Pittsburgh, Pennsylvania
- Portland, Oregon
- Sacramento, California
- Salt Lake City, Utah
- San Antonio, Texas

- San Diego, California
- San Francisco, California
- San Jose, California
- Santa Rosa, California
- Seattle, Washington
- Tucson, Arizona

The SAC program works with more than 180 municipal, county, and state agencies as well as with solar companies, universities, utilities, and nonprofit organizations, all committed to powering their cities with solar energy. The DOE provides financial and technical assistance to these cities and helps the cities to identify and overcome barriers to embracing solar energy, such as zoning regulations, codes, covenants, and other urban planning practices that can discourage the installation of solar energy systems; these partnerships also help cities increase citizen and business awareness of solar energy technologies. It is believed that by reducing barriers and increasing awareness, the growth of solar markets in these cities will accelerate, which will help bring prices down and make solar energy cost competitive with conventional utility grid electricity.

DOE chose to focus on cities for applications of solar energy because cities are the site of the most intense energy concentrations in the United States. This places cities in a unique position to challenge global climate change, foster U.S. energy independence, and support the transition to a clean energy economy. The cities that have been chosen by DOE are undertaking various projects in order to meet the goals of removing market barriers to solar energy, increasing consumer awareness and demand for solar energy, and helping solar energy manufacturers and suppliers to succeed. The projects include solar panel installation at farmers markets, airports, convention centers and other public buildings, solar-powered parking meters, solar-powered bus stops that offer Wi-Fi connectivity, and many other projects. Further, many of the cities host workshops to educate residents, business owners, property assessors, and architects about solar energy, and many have integrated solar technology into affordable housing projects.

Municipal financing has changed the face of solar energy, making it affordable for many who could not have considered it before. By lending money to consumers who wish to install solar energy systems and then allowing those loans and interest to be paid back over a period of 20 years as part of their property taxes, municipal financing has dramatically increased the spread of solar energy across the private market; in some places, these loans are also available to businesses that wish to "go solar." California and a dozen other states now offer municipal financing for solar energy projects; any homeowner is eligible, not only those with good credit, and the loan is attached to the house, so the obligation to pay it off passes to future buyers, should the original homeowner wish to sell. Other government supports such as rebates and income tax credits also decrease the financial burden of going solar, and there are efforts to change the federal tax code so that cities can borrow the money to lend to homeowners tax-free.

The Dark Side of the Sun

While there is some overlap between those who are critical of solar power in general and those who are critical of the movement to bring solar power to cities, there are a few concerns that are related strictly to the notion of using solar energy to power our cities. For example, many are critical of municipal financing for solar energy projects and argue that government is subsidizing and boosting a form of energy production that would otherwise not be cost-effective. Further, some of the schemes used in certain cities to encourage home-owners to install solar systems result in squeezing out smaller solar installation companies. For example, the Solarize Portland movement, run by several Portland neighborhood associations, works by purchasing photovoltaic systems in bulk and then relying on a single solar contractor for installation to keep costs down. Smaller solar contractors are not able to make competitive bids and lack the manpower to take on large installation projects.

One of the strongest criticisms of solar energy relates to solar energy in general and certainly has profound implications for cities embracing solar energy: the manufacture of solar panels not only relies on toxic chemicals but also generates hazardous waste. Manufacture of solar panels requires the mining of raw materials such as quartz and metal ore, and the actual creation of solar panels results in air pollution, heavy metals emissions, and greenhouse gases. Solar panels do not last forever, which means the creation of more waste once the solar panels have reached the end of their life span. Still, supporters of solar energy point out that while there is certainly a downside to solar panels, the consequences of mining, refining, and burning fossils fuels still outpace the consequences of using solar panels.

Embracing solar energy to power cities, then, faces many challenges. Supporters insist that embracing solar energy is a necessary step toward creating a sustainable future and that cities are perfectly positioned to help nurture this movement toward a green future. Opponents, however, remain skeptical and cry foul at some of the methods used to nurture this movement.

See Also: Alternative Energy; Ecocapitalism; Green Jobs (Cities); Sustainable Development (Cities).

Further Readings

Bradford, Travis. *Solar Revolution: The Economic Transformation of the Global Energy Industry.* Cambridge, MA: MIT Press, 2008.
"The Ugly Side of Solar Panels." *Low-Tech Magazine* (March 3, 2008). http://www .lowtechmagazine.com/2008/03/the-ugly-side-o.html (Accessed September 2010).
U.S. Department of Energy. "Solar American Cities: Frequently Asked Questions." http:// www.solaramericacities.energy.gov/about/frequently_asked_questions/#9 (Accessed September 2010).

Tani E. Bellestri
Independent Scholar

Sustainability in TV Shows

Is it possible that a medium as pervasive as television, and so exquisitely tuned to market a consumer culture, can itself become a medium leading us to sustainability?

The easy answer is "no," if only due to our near-addictive relationship with this and other visually based media technologies. In 1990, writer and environmental advocate Bill

McKibben ran an experiment that became the book *The Age of Missing Information*. For 24 hours, McKibben took notes of his time on a mountain near his home in the Adirondacks, making his own record of experience. Meanwhile, his friends videotaped the same 24 hours of cable television. He then contrasted his day of experience with some 2,400 hours of television from more than 90 stations, the contents of a single potential viewing day in the media-saturated lives of Americans, circa 1990.

The physical experience and lessons of the natural world offer cyclical time; the televised world is linear and time coded, honoring the visual and auditory sensory channels over more reflective media experiences. In nature, we participate in growing, gathering, and hunting food; on television, the hunter prowls the shopping channels. Walk five miles into the woods, McKibben notes, and there is nothing to buy. Watch TV for the same period and the mediated narratives will emphasize the status quo of a merchandized world.

Fast-forward from 1990. If you were a typical child born in that year, your first 18 years of life would include 350,000 commercials, 100,000 of those for beer. You would witness 200,000 acts of televised violence (and a vast majority of studies on this matter indicate that watching violence on TV is a causal factor correlated with real-life violence). The only thing you would do more than watch TV is sleep.

TV would involve 20 percent of your waking life, 9 percent of your hobbies, 3.5 percent of your homework. You would, on average, spend roughly a third of that TV time consuming food or drink; on your other primary screen, the Internet, you would consume calories as little as 2–4 percent of the time. Altogether, with TV and computers joined with smart phones, you would spend 53 hours a week connected with media.

In part because your life has become so mediated, you spend less time outside than your parents did. And recent studies find that the less time outside, the greater the incidence of nearsightedness and obesity. And more time outside may reduce the symptoms of attention deficit disorder.

So ask yourself: What will I choose, a TV or a mountain? Research provides one answer: as a culture, Americans have chosen the screen. Viewership of television is up; attendance at national parks and time spent camping is down.

And if we choose the TV, then (to roughly paraphrase the Talking Heads), you may ask yourself: Is this my sustainable house? Is this my sustainable TV?

To recreate McKibben's experiment today, the available stations could easily number more than 150 (not including the near-infinite number of pay-per-view and Internet video channels). Yet the average number of television channels per U.S. household has gained only slightly since his experiment, reaching a recent high of 118.6 channels.

And how do these TV channels engage us today? We remain creatures of its domain:

- We watch more than eight hours of TV per day per U.S. household.
- Our average household has 2.5 people and 2.8 TVs.
- And in the intervening years, our habits have become more singular—more of us are likely to be watching alone and making our own channel selections; we are more likely to switch channels to multitask during the show.

This typical household also averages 2.28 automobiles, a number that might be intuitively correlated with the penetration of TVs, where car ads are a dominant marketing presence. And commercial TV is intricately tied to marketing, with 40 percent of U.S. advertising focused on TV. Recent studies indicate that TV accounts for roughly 50 percent of the media impact on our consumer selection process, from awareness to purchase.

Can one hope, when operating within this economic machine, that a concept like sustainability might surface and survive as it questions the consumer process of a medium dedicated to consumerism? Perhaps, if you use the visual engagement of the medium itself. Witness the shift to define environmentalism as simply "green"—the color is the meaning—which uses the medium's visual properties to question some of the very messages its delivers.

Sustainability Definitions and Broadcasting

What would sustainable TV shows and channels contain? In its essence, a "sustainable TV show" might anchor its roots in the Brundtland Commission report (World Commission on Environment and Development, 1987), which defines sustainability as "meeting the needs of the present without compromising the ability of future generations to meet their own needs." The Brundtland Commission clarified this process by applying Paul Ehrlich's IPAT formula, which defines the human role in environmental degradation as:

$$\text{Impact} = \text{Population} \times \text{Affluence} \times \text{Technology}$$

Watching TV in America, one witnesses a force-multiplier, with an advancing Technology (from tubes to LCD screens) multiplied by Affluence (ever-larger cars) and a sizable American Population = a disproportionate Impact by Americans on the environment. Using this formula, Ehrlich claims that Americans on average have 50 times the impact as Bangladeshis (though Bangladesh faces the highest risk from global warming).

The medium is the message, as Marshall McLuhan famously observed. The TV glows, flashes, speaks, sings—the medium itself communicates the value of increased technology and perpetually increasing demands for technological innovation. To switch from this consumptive-messaging machine to deliver a sustainable message, communicators must consciously redefine messages so that they convey social, economic, and environmental meaning. This is the advice to advertisers offered by the United Nations Environment Programme (UNEP), which offers tools to advertisers and producers for communicating sustainable and human right issues. UNEP recruits advertisers by speaking the language of the profession—noting that advertising professionals may miss an expanding market for sustainable consumption if they do not become leaders in their own right.

In television, though, the environment (and our impact on it) is secondary to the story that propels the show (the social), which in turn serves the economic marketing role of the channel. TV shows that break this consumption model often do so by embedding sustainability messages into the story. This embedding approach was pioneered by Miguel Sabido, a Mexican television executive who pioneered an "Entertainment-Education (E-E)" formula from the 1970s on, in which wellness and health-related messages were embedded in soap opera story lines.

This concept—framed by labels such as behavior modification, social marketing, identity campaigning, and socially responsible behavior—steers away from brainwashing and greenwashing, relying instead on the use of story to support authentic changes within individual lifestyles. Facts remain a component of the environmental message, but facts alone will not engage us, the social marketing proponents claim. A leading production team of online campaigns, Jonah Sachs and Susan Finkelpearl of Free Range Studios, make this argument in a chapter in the 2010 "State of the World" report by Worldwatch Institute, which focuses specifically on "Transforming Cultures: From Consumerism to

Sustainability." In their report, they explain their motivation when producing "The Story of Stuff," an Internet animation with a viral audience. The rationale: to engage change today, producers must draw upon archetypal stories of heroes vanquishing a nemesis, of broken worlds that require healing. These stories will engage the emotions and the actions of a society that still operates at a tribal scale, though our tribes today connect by Facebook and text messaging.

An Affluent Population watching Technology = Impact. Combine the scale of our population impact with the rising impact of our current sustainability and climate problems, and one might wonder if something as seemingly simple as a mediated story, even one that leads the audience to act sustainably, may seem too little, too late. The response by many is to embrace the existential challenge and simply act (a message familiar to our TV heroes). But from the stereotypes of TV heroes, we now meet the new, hyperlinked hero, who may be an ordinary person, a neighbor who acts for our neighborhood. While television features eco-heroes and "survivor" experiences, some argue that the use of the computer and smart-phone screens offers greater avenues for our heroes and tribes to act for sustainability, such as the organizers who gathered social-media networks and meet-ups to promote www.350.org. A television show is not enough to enact the change that Sachs and Finkelpearl argue for in their hypertexted media stories. Nor is a televised show enough for environmentalists like McKibben and his 350.org movement, or for someone like writer and social-change advocate Paul Hawken, who observes and organizes the "movement of movements," a new extragovernmental process in which a dispersed yet mass movement is evolving from the work of everyday heroes and Internet communities.

Channels and Shows Focused on Sustainability Issues, Themed Channels, and Green Weeks

Two significant changes in contemporary television support its movement to support sustainability messages: the increase in TV bandwidth (more voices equal more choices) and convergent integration with new media (which offers more marketing bandwidth, yet also the potential for more "social choice" branding).

Examples abound of channels and TV shows that are evolving to a converged media ecosystem and simultaneously offering a new approach to environmental broadcasting. A good place to start is The Weather Channel. Today, one can point and click with a finger on the Weather Channel's iPad app and select the weather for home and for the homes of your friends—sweeping through the globe with a certain godlike authority, as if we control the weather with our finger. Compare that to the broadcast version, where our weather streams below in text, unstoppable, and we wait for the "Local on the 8s," a remarkably regular commitment to our own locale—but it seems so archaic, so 1982 (the year the channel was founded) to wait that long today. But as we click through our personal weather, one is struck by a note below a sunny day's icon, claiming that this future sunny day would be a "Perfect day for a dream drive." Click through the link to discover the brand of car you should be driving on this dreamy sunny day.

While The Weather Channel is uniquely suited to inform its audience about climate change, it often contrasts versions of "how is my commute today" (or my weekend activities) with extreme weather trauma. The everyday remains simple: umbrella or shorts. And the traumas are often far off. Yet when a channel devotes this much focus to weather, day after day, an audience might soon begin to picture our own climate. We might wonder if these many hours of 24-hour weather broadcasts might prime its audience to be hyper-alert for global warming. Or do we simply become inured to the travesties of storms and droughts? And does the fear of mass disaster and lack of any ability to change the day's weather convert us from actors into mere audience?

If The Weather Channel offers one model of covering an issue that is key to our sustainability—the very weather and climate we live in—then the "Green Is Universal" campaign of NBCUniversal offers another approach: to focus on green and sustainability issues for one key week, with the week's worth of stories serving to highlight all the green activities it does year round. Yet this one-week version of behavior placement (similar to product placement) has been criticized, oddly enough, as too overt, since sometimes the messages are not core to the shows or characters. Or simply because it appears as a crass approach to marketing a sales pitch while also trying to balance a beleaguered corporate budget. Which brings up the fundamental irony and internal conflict of sustainable television: to reach a mass market, the networks and shows require advertisements proposing consumptive behavior. Yet the sustainability movement, while not shy when highlighting green products, is anchored in frugality and a "locavore's" diet, a call to right-size our appetite for "stuff," reducing the fossil fuel carbon we exhaust to support our energy comforts, automobile commutes, and jet-setting winter fruits—goals that undercut the messages of traditional TV advertisers.

Despite the Green Week focus, some critics observe that The Weather Channel, which was acquired by NBCUniversal in 2008, has edged toward more entertainment and less science. But green efforts across all channels managed by NBCUniversal (even the company's film studio) at least beg the question: Can a sustainably focused Green Week serve the public good and the corporation's profits? Green Week may reach a general audience that is only experimenting with "green" and after this week, year after year, one might imagine this mainstream audience edging greener, toward such "mediated" messages as "locavore" food, life-cycle analysis, and other sustainability concepts.

Why Broadcast Sustainability? To Reach for the Green Demographic

The Environmental Media Association (EMA) estimates that 70 percent of children gain their primary knowledge of nature from television. This is one reason why the EMA borrows the common practice of product placement when it works to place environmental messages in television and film. This form of "behavior placement"—to embed "change stories" in everyday broadcasts—echoes NBC's Green Week and Sabido's Entertainment-Education formula.

Within the past five years (2005–10), though, organizations like the Nielsen Company and the Natural Media Institute have begun to define, quantify, and market to a new consumer demographic, one that is guided by green consumer purchasing. Segmentation of green versus typical consumers led to creation of the LOHAS model (Lifestyles of Health and Sustainability). Whereas typical consumers such as Conventionals, Drifters, and Unconcerned rarely select a product for their green values, these new ecoconsumers drive a green marketplace estimated at more than $400 billion in 2010.

In this model, the LOHAS consumer offers the predominant green demographic. This consumer self-identifies as an active environmental steward and is twice as likely as other consumers to buy a green-branded product. When the LOHAS consumers are combined with the next tier in this demographic, the Naturalites (who are defined by a pursuit of personal health), these two segments compose a third of the current consumer market. This is why Nielsen offers a variety of ways to track and market to the green consumer, who is moving from niche into mass market. And even the Drifters are trending green, in part because green goods are seen to be more durable in troubled economic times. As this consumer shift occurs, however, the risk of greenwashing rises. Recent commercials by McDonald's, for instance, imply that an isolated farmer delivers food to consumers, when food sourcing by McDonald's rarely features such an "Old McDonald Had a Farm" farmer. In response to concerns about false-light adverting, in October 2010, the Federal Trade Commission proposed stricter guidelines in their "Green Guides" to address the perceived sense that some ecofriendly claims resort to greenwashing.

Will the Sustainability Revolution Be Convergently Televised?

Convergent media—typified by the blending of video, audio, text, and interactivity delivered via a "smart" device—is leading this shift to reach a green audience on a smaller scale, as typified by "The Story of Stuff." Another approach is the mostly online "Planet Forward" series hosted by Public Broadcasting Service. Here the innovations and messages and stories are primarily driven by viewer suggestions and submissions, similar to the "Movement of Movements" concept proclaimed by Paul Hawken in his book *Blessed Unrest*. Yet the question remains whether "Planet Forward" will gather a larger demographic (or funding source); in its current incarnation, it feels more like a video version of a call-in talk show than a threat to the momentum of mainstream media. Yet "Planet Forward" directly emphasizes the forward-looking focus—to choose hope against decline—that mirrors Hawken's definition of sustainability, which focuses on "stabilizing the currently disruptive relationship between Earth's two most complex systems—human culture and the living world."

Online television shows join the viewing shift to Internet delivery of video—a key shift from the top-down conglomerated messages of the networks—yet the "viral" experience of multimedia shows like "Stuff" or "Planet Forward" remain mere blips compared to the audience for dancing cats and lip-synching in the living room offered on YouTube. Yet in both new and traditional delivery, we can witness two key media processes at work: the concept of framing—that the issues we are aware of and the frames within which we act are shaped in part by our media—and the concept of symbolic legitimacy boundaries—that we may not select to go beyond the bounds if a new route is not first made legitimate by an engaging symbology.

So we seek to follow the symbolic hero on his quest, and both Internet video and television are opening the boundaries of legitimate behaviors: we await that knight-errant, Ed Begley, the seeker who fights for the lowest carbon footprint against his neighbor, Bill Nye (formerly the Science Guy). Ed's efforts add symbolic legitimacy to our own sustainability competitions, in our neighborhoods and our schools and our workplaces. By watching *Living With Ed* or the *Treehugger* shows on Planet Green, we join a movement of green messages that are expanding into multiple platforms. And in an odd mark of cultural valuation, Planet Green has just recently been added to the Nielsen-rated family of television channels; green is paying off with green profits. And Planet Green joins other sustainability focused shows in The Discovery Channel family, which claims to the most distributed network worldwide. Other channels offer shorter frames of programming, such as the coalition of "The Green" series on the Sundance Channel, including *Green Porno*, *The Lazy Environmentalist*, and *Big Ideas for a Small Planet*.

We argue climate change and bemoan the spilled oil being televised from the Gulf of Mexico, in our daily conversations and in our televised media. Green shows and green channels proliferate; the green demographic grows. All of which makes one realize, these days, that we do have big ideas on the small screen, whether we watch via an orbiting satellite or a smart phone.

The medium may yet deliver the sustainability message.

See Also: Ecocapitalism; Ecosocialism; Sustainable Development (Business); Sustainable Development (Politics).

Further Readings

Hawken, Paul. *Blessed Unrest: How the Largest Movement in the World Came Into Being and Why No One Saw It Coming.* New York: Viking, 2007.

McKibben, Bill. *The Age of Missing Information*, 1st ed. New York: Random House, 1992.

Planet Green. "Sustainable Living, Energy Conservation, Earth Day." http://planetgreen .discovery.com (Accessed December 2010).

Sachs, Jonah and Susan Finkelpearl. "From Selling Soap to Selling Sustainability: Social Marketing." *State of the World 2010: Transforming Cultures*. New York: Worldwatch Institute/W. W. Norton, 2010.

World Commission on Environment and Development [Brundtland Commission]. *Our Common Future*. Oxford, UK: Oxford University Press, 1987.

Ron Steffens
Green Mountain College

Sustainable Development (Business)

The modern sustainability movement traces its roots back to the United Nations World Commission on Economic Development, which is often cited as first defining "sustainable development." The work of John Elkington expanded traditional business accounting to include what are now known as the three pillars of sustainability: society, environment, and economics. Today, investors are increasingly demanding that businesses act in an ethical and sustainable manner. This has led to the rise of sustainability-based investment indexes. Additionally, investors are increasingly requesting that companies publicly disclose their performance with regard to sustainability. Many large companies today are producing sustainability reports in an effort to be unceasingly transparent and accountable to the public. Companies truly committed to sustainable development also use this data to drive internal initiatives to manage their social and environmental impacts. Life-cycle assessment is another tool that companies use to manage sustainability by comparing the various impacts of products and designs, enabling a selection of more sustainable product options.

United Nations World Commission on Economic Development

At the end of the 20th century, the United Nations became increasingly aware of global pollution issues such as pesticide bioaccumulation and ozone depletion. The international community was also facing increasing industrialization among the developing nations of the world. It was generally acknowledged that the path many nations had taken in their industrial development was not the best for the health of humans or the environment, and a more careful path to development would be prudent.

To answer the global community's questions on how to best guide development, the United Nations convened the World Commission on Environment and Development (WCED) in 1983, chaired by Gro Harlem Brundtland, the prime minister of Norway. Due to her pioneering work to address the issues surrounding continuing global development, the WCED came to be known as the Brundtland Commission. The Brundtland Commission was created to address the growing concern "about the accelerating deterioration of the human environment and natural resources and the consequences of that deterioration for economic and social development." The Brundtland Commission's 1987 report, *Our Common Future*, defined "sustainable development" as "development that meets the needs of the present without compromising the ability of future generations to meet their own needs." This definition is considered the original and probably the most quoted definition of "sustainable development."

The Triple Bottom Line

Another major force in the integration of sustainability concerns into the business world is the work of John Elkington. Traditionally, business has only concerned itself with the economic bottom line as a measure of success. In his 1998 book *Cannibals With Forks: The Triple Bottom Line of 21st Century Business*, John Elkington coined the phrase *triple bottom line* (TBL) as a new means of accounting. Elkington advised an expansion of the traditional accounting framework to consider environmental and social performance in addition to the traditional business metric of financial performance. The TBL includes the three elements (or pillars) of sustainability: environment, economy, and society, which must all be considered for truly sustainable development to occur, as represented in the Venn diagram in Figure 1.

In this scenario, a product or design that balances social and environmental impacts but does not account for economic impacts may not be profitable in the long run. Products

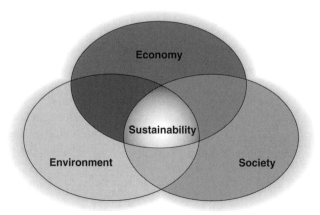

Figure 1   Sustainable Development Business

that balance environmental and economic impacts may fail due to lack of social acceptance or undesired social impacts. And products that balance economic and social impacts, but do not account for the environmental impacts, may cause unacceptable environmental damage. Sustainable solutions and products exist only at the intersection of economic, social, and environmental concerns. When all three of these impacts are accounted for and the TBL is considered, the most sustainable solution can be selected. Commitment to this TBL is considered a step beyond environmental awareness of corporations into the realm of corporate social responsibility.

When accounting for social impacts in the business context, the emphasis is often on the role of human capital within the TBL. Human capital can mean the employees who work for a company, the surrounding community, or even society at large. Accounting for human capital in the TBL means acting in a fair and responsible manner in an attempt to provide a positive impact on society. Taking voluntary measures to positively affect society may come at an economic cost for a business. But the cost of ignoring the social context of a business can be greater. For example, in the United States, when a local community does not have faith in a company's ability to protect the environment, its operating permit and applications and/or renewals may be challenged and held up in the courts. In this scenario, the company has lost its human (or social) capital necessary to operate.

In the context of sustainable business, the environment is considered natural capital. Similar to managing their social impacts, sustainable businesses must also manage their environmental impacts, many of which are governed by environmental laws in the industrialized world. Within industrializing nations, however, efforts to safeguard the environment may be voluntary but are still an essential part of sustainable development. In this case, the business is managing its environmental impacts because it is the sustainable thing to do, not because of economic drivers in the short term. However, a sustainable business will have the financial gain of economic longevity. Businesses with a commitment to the TBL make a commitment to conduct their business in the most environmentally sustainable manner as a standard course of doing business.

There are many critiques regarding TBL as a business concept. Environmental concerns are often viewed as a luxury of the rich, or at least the economically comfortable. In countries where many are starving, dwindling forests may be cut down to make room for agriculture. The need for food may take immediate importance over long-term environmental needs such as biodiversity and carbon. Similarly, businesses that are profitable can better afford to examine their TBL, while struggling and developing businesses may be more likely to examine only the economic bottom line.

Proponents of the division of labor that occurs in most capitalistic societies argue that businesses and other organizations contribute the most to society when they focus on what they do best. An automobile manufacturer, for example, can create the most profits when it focuses on just that, rather than worrying about social and environmental impacts beyond those regulated by government. Nonprofit organizations meant to benefit society or the environment such as religious organizations, Goodwill, and the Sierra Club can achieve their best results when focusing on their activities, rather than trying to make a profit for owners or shareholders.

Dow Jones Sustainability Index

As a business approach, sustainability programs manage risks and develop opportunities from the economic, environmental, and social impacts of development, resulting in long-term shareholder value. As the investment community's education and awareness of sustainable development practices grew, so did demand for the ability to invest in companies committed to sustainable practices. Investors desired an index to benchmark the performance of sustainability-driven investments.

In 1999, the first global index tracking the financial performance of the leading companies committed to sustainability—the Dow Jones Sustainability Index (DJSI)—was launched by Dow Jones Indexes, STOXX Limited, and SAM Group. The leading businesses with regard to sustainability are identified through an assessment of social, economic, and environmental performance. For a company's social performance, the SAM Group considers philanthropy, labor practice indicators, human capital development, social reporting, and talent attachment and retention. The economic factors considered are codes of conduct, compliance, corruption/bribery, corporate governance, and risk and crisis management. Environmental reporting is also considered when assessing the sustainability performance of a company. Each company submits an annual sustainability questionnaire. The companies are then ranked within industry groups and selected for the DJSI. At the end of December 2009, the assets managed based on the DJSI portfolios were worth approximately US$8 billion.

However, inclusion in the DJSI does not guarantee a business is acting in the most sustainable manner. For example, in 2010, after the BP oil spill in the Gulf of Mexico, the DJSI removed BP from its sustainability indexes, citing the extent of the spill and long-term damages. But none of the index's ranking methods prevented BP's original inclusion in the index. Critics of the index raise this company's previous inclusion in the index as evidence that the index's screening and ranking methods may not be effective at measuring a company's actual commitment to sustainable development and integration of sustainability concepts into its business practices. In fact, the question of how to appropriately measure sustainability is one of the most significant issues within the field.

Measuring Sustainability

For a business to be able to manage and balance its TBL, there must be some metric by which the impacts on each of the three pillars can be measured. Economic impacts can be quantified using currency, such as U.S. dollars. But social and environmental impacts do not have a similar common unit by which to measure all impacts. A business may affect both the environment and society in a myriad of ways, and some form of measurement is needed for each of these.

Indicators are measurable effects that can be tracked over time. Common indicators have been developed for each of the three pillar areas, which are usually selected and tailored to serve as meaningful metrics for a unique business. For example, common environmental indicators may include annual emissions of hazardous air pollutants and annual freshwater withdrawn from surface water, while social indicators may include impacts on native cultures and lifestyles. For an urban dry cleaning operation, the annual emissions of hazardous air pollutants is a more meaningful metric than the annual freshwater withdrawn from surface water or the impacts on native cultures and lifestyles. But for a rural trout farm, the more meaningful metrics may be the annual freshwater withdrawn from surface water and the impacts on native culture and lifestyles. This example demonstrates the importance of selecting meaningful indicators for a specific business. However, it is also important that indicators be standardized so that they are comparable across businesses.

Sustainability Reporting

To accurately report on sustainability, the global business community needed a common set of agreed-to metrics on which they could report. Started by the nonprofit CERES in 1997, the Global Reporting Initiative (GRI) is an international organization with the goal of standardizing sustainability indicators and reporting internationally. In 2000, the Global Reporting Initiative launched its first sustainability reporting framework, including general sustainability reporting guidelines and special guidance for different industry sectors. Also in 2000, the Global Reporting Initiative was inaugurated into the United Nations as a collaborating center of the United Nations Environment Programme.

Since the issuance of the initial guidance, Global Reporting Initiative has worked to continually update the reporting guidelines and expand them to include new industries. Sector supplements are developed for new sectors as needed and include the following sectors: airports, apparel and footwear, automotive, construction and real estate, electric utilities, events, financial services, food processing, logistics and transportation, media, mining and metals, nongovernmental organizations (NGOs), oil and gas, public agency, and telecommunications.

Both the sector supplements and the general sustainability reporting guidelines have been created through consensus-seeking processes involving representatives from business, academia, governments, labor, professional societies, and others worldwide. The sustainability reporting guidelines include guidance on appropriate indicators and are meant to be reviewed and updated as needed. The third version of the guidance, commonly referred to as "G3 Guidelines," was published in 2003 and is available for free use globally. Guidelines include suggested sustainability indicators for use in reporting. Sector supplements include additional indicators that may be unique to a given sector of business. Protocols for each indicator provide its definition, reporting units, and the methodology for data gathering and calculation.

To encourage beginners to venture into sustainability reporting, various reporting levels exist. Reports do not need to address all existing indicators, but reporting on more indicators results in a higher reporting level. Reports are also identified as to whether they were validated (or audited) internally or externally. The report ranking system allow readers to identify how much confidence to place in reported data based on the degree of scrutiny it received prior to publication.

Critics of sustainability reporting argue that it only requires companies to gather data and report it to the public. The reports, without any follow-up action based on the data or company initiatives to address areas of greater impact, are viewed by some as little more than greenwashing, an effort to appear green on the part of businesses. For sustainability reporting to be an effective business tool, the data gathered must not only be reported but also analyzed internally. Follow-up sustainability initiatives must be implemented to minimize negative social and environmental impacts and to provide increased opportunities for positive impacts. Companies accomplish this by establishing sustainability departments to address sustainability year-round, not just at the end of each reporting cycle. However, all of these efforts are voluntary, and GRI does not verify that companies that use their protocols for sustainability reporting actually create goals and objectives based on that information.

Critics of voluntary sustainability reporting within business argue that through these reports, businesses focus on the "win-win" situations good for both business and the environment, neglecting the situations where environmental and social concerns are dismissed for the sake of profit. As the content of sustainability reports is determined by each reporting organization, situations where negative social and environmental impacts occur are often not highlighted within these reports. In essence, a reporting organization is only as transparent as it chooses to be. These reports are viewed by skeptics as little more than greenwashing.

Life-Cycle Assessment

A common business tool used to assist companies in making more sustainable designs and products is life-cycle assessment (LCA). This cradle-to-grave approach (from raw materials to manufacture to use to reuse/remanufacture to disposal) considers the environmental impacts of all aspects of a product's manufacture, use, and ultimate disposal. Sometimes so-called LCAs are performed not on the product's complete life cycle but only for the manufacturing phase of the product. In this case, only the production portion of the LCA has been performed, addressing cradle-to-gate; the full product life cycle has not been addressed.

When used to comprehensively address a product's life cycle, LCA results in designed products that account for ease of disassembly and recycling throughout a product's life cycle. Some of these components may cost more but ultimately cause less impact on the environment. LCA also allows comparisons of several products or designs, providing the information necessary for businesses to select the option that minimizes environmental impacts, while still being profitable. The International Organization for Standardization (ISO) has standards for LCA, which designate four steps as follows:

- *Goal and scope*: The goal and scope of an LCA must be established, including defining the functional unit or object of study.
- *Life-cycle inventory*: Collecting data, validating, and modeling the various inputs (such as raw materials and energy) and outputs (such as solid waste and air emissions) within the bounds of the LCA. Validating data is often viewed as one of the most critical steps in LCA, as data quality determines the reliability of the predicted impacts.
- *Life-cycle impact assessment*: Evaluating the contribution to various impact categories, such as global warming, acid rain, and so forth. Each category is then weighted based on relative importance.
- *Interpretation*: Includes analysis of the major contributions, sensitivity analysis, uncertainty analysis, and a review of whether the goal and scope were met.

Although LCA can be performed using internal company data, gathering that data can be costly and time consuming. Most LCAs performed in the business world utilize LCA software packages such as SimaPro and GaBi Software with associated databases of various raw products and manufacturing methods. Critics of LCA often address the issue the issue of data quality when questioning the results of an LCA. For this reason, it is important to verify data or use an established, already verified data set when performing LCAs.

In fact, one of the most common critiques of LCA is that it can only consider quantifiable impacts. Some impacts, such as social impacts, are more qualitative than quantitative and cannot be accounted for quantitatively. Additionally, as each LCA requires the establishment of unique system boundaries, LCAs for various products or processes may not always be comparable. By drawing different system boundaries around the LCA, impacts may increase or decrease. Similarly, the weight attributed to various factors that contribute to an overall LCA score can be changed. In this manner, the results can be manipulated to show the desired results. Critics of integrating concepts of sustainability with economic development may see it as a threat to the traditional goal of maximization of profits.

See Also: Corporate Social Responsibility; Precautionary Principle (Ethics and Philosophy); Socially Responsible Investing.

Further Readings

Elkington, J. *Cannibals With Forks: The Triple Bottom Line of 21st Century Business.* Gabriola Island, British Columbia, Canada: New Society Publishers, 1998.

Elkington, J. "Towards the Sustainable Corporation: Win-Win-Win Business Strategies for Sustainable Development." *California Management Review*, 36/2 (1994).

Global Reporting Initiative. "Sustainability Reporting Guidelines Version 3.0." 2006.

International Organization for Standardization (ISO). *ISO 14040: Environmental Management—Life Cycle Assessment—Principles and Framework.* Geneva: ISO, 2006.

International Organization for Standardization (ISO). *ISO 14044: Environmental Management—Life Cycle Assessment—Requirements and Guidelines.* Geneva: ISO, 2006.

SAM and Dow Jones Indexes. "BP Removed From the Dow Jones Sustainability Indexes" (June 1, 2010). http://www.sustainability-index.com/djsi_pdf/news/PressReleases/20100531_Statement%20BP%20Exclusion_Final.pdf (Accessed August 2010).

Schmidheiny, S. *Changing the Course: A Global Business Perspective on Development and the Environment.* Cambridge, MA: MIT Press, 1992.

Schmidheiny, S., R. Chase, and L. De Simone. *Signals of Change: Business Progress Towards Sustainable Development.* World Business Council for Sustainable Development, 1997.

World Commission on Environment and Development. *Our Common Future.* Oxford, UK: Oxford University Press, 1987.

Michelle E. Jarvie
Independent Scholar

Sustainable Development (Cities)

Cities are the most important geographic unit, having a strong impact on economic, social, and environmental activities. In 2010, there were more people living in cities than in rural areas. Throughout the world, population growth is very rapid within cities. In the future, the greatest challenge to sustainable development in cities will be managing urban growth. This is because although cities consume less land per person than rural areas, they consume very high levels of natural resources in the form of manufactured goods. Therefore, in the context of cities, the concept of sustainable development needs careful attention.

The most widely known definition of sustainable development is given by the Brundtland Commission, which defined it as "development that meets the needs of the present without compromising the ability of future generations to meet their own needs." From this definition, it is clear that "development" is compatible with sustainability. However, this is an optimistic view of development that should not harm the environment. For example, if

Urban sprawl and low-density housing, like this area in Tokyo, causes congestion, consumes energy, and destroys the urban environment.

Source: Wikimedia

cities draw particular resources from nature such as water from a lake, it should be done in such a way as to not overdraw water from the lake, which would damage the ecology of lake. In such a scenario, development and environmental protection are not incompatible but instead are fundamental pillars of sustainable development in cities.

Sustainability in cities could be understood by applying the Metabolism Approach to human settlements. This approach looks at the resource inputs and waste outputs as well as the livability of a city. A city would be sustainable if it were using less resource inputs (land, water, food, energy, and building material) and reducing outputs (e.g., solid waste, liquid waste, toxics, sewages, air pollutants, heat, and noise) at the same time as improving its livability for its people. This approach provides a simple way of applying sustainable development to the complexity of a city. However, the result of this approach will vary from city to city due to its urban form.

Sustainable Urban Form

Sustainable development in cities is closely bound up with urban growth due to the worldwide phenomena of urbanization. However, urban growth is not necessarily a bad phenomenon. The important consideration is where in the city population growth is occurring. After World War II, most cities in the developed world managed population growth using low-density single-family housing built at the edges of cities on greenfield sites. Population growth at the edge of a city consumes large amounts of potentially rich agricultural and forest land and demands heavy investment in infrastructure, including new roads. Such development makes these areas viable only for cars. Accordingly, car ownership and distance and miles traveled in cities have steadily increased. This large share of private transport usage in cities can be observed around the world. This causes congestion, consumes energy, and destroys the urban environment. This low-density housing development makes cities edgeless and is known as "urban sprawl." Urban sprawl has been allowed to happen because of the assumption that we have unlimited natural capital, including land. In addition, natural capital and man-made capital have been seen to substitute for one another, which is not the case as they instead complement each other. Urban sprawl is not a substitute for a forest or agricultural land. In short, urban sprawl makes our cities unsustainable and is completely undesirable.

However, the question is: What does a sustainable pattern of urban growth look like? The "compact city" concept emphasizes increasing the population density of a city in order to address the negative impacts of urban sprawl. Different scales of high population density have been proposed in the literature. Some argue in favor of higher population density at the metropolitan level, some favor high density in the central business district (CBD), and some favor high densities at the neighborhood or community level. However, everyone agrees that we should stop urban sprawl. Some cities in the world use a "green belt" or "urban growth boundary" as a tool to control future urban sprawl.

The opponents of compact cities argue that high density is not the desirable living arrangement for most of the people living in developed countries. Additionally, the revolution in telecommunications makes location of future population and economic activities impossible to organize. It is more likely that people and jobs will move wherever they want to live and work.

The change in the economies of developed countries from manufacturing to knowledge/information-based economies provides opportunities to reorganize cities based on mixed land use for employment and residential uses, which can reduce the negative impacts of urban sprawl. Following industrialization, "undesirable" mixtures of land use, for example, industrial and residential areas, were separated by the use of zoning regulations in cities. However, such land use increased commute distances and affected the overall sustainability of cities. The sustainable cities concept encourages mixed land use development supported by public and nonmotorized transport.

Employment densities play an important role in making cities sustainable. Historically, employment opportunities in world cities have been concentrated in the CBD, which provides opportunities for the development and use of public transport. London, Paris, and New York are example of such cities. However, since World War II, like population, growth in employment has been occurred in the suburbs. The location and nature of employment in the suburbs makes it hard to serve by public transport, which ultimately affects the sustainability of cities. Therefore, to achieve sustainable development in cities, compact development might include the concentration of employment and housing in mixed land use patterns. Due to economic restructuring, jobs are increasing in the CBD and suburban centers. It is more likely that high-density mixed land use cities could achieve new agglomeration economies of firms, lower automobile dependency for households, reduced environmental destruction, lower pollution levels, and higher-quality services.

Economic Sustainability

The economic sustainability of cities could be improved by applying energy-efficient land use, investing in sustainable modes of transport, and adopting appropriate pricing mechanisms for promoting sustainable development.

Energy is one of many scarce resources in the world and is a key indicator associated with sustainability of the city. It appears that energy demand is now higher than energy supply, which in the future will lead to world energy crises. Historically, cheap oil/energy has encouraged low-density suburban development in our cities accessible by the private car. Therefore, transport per capita energy use has increased over time. However, increasing energy demands raise questions about such developments, and it is more likely that such cities will suffer from global energy crises in the future. Compact cities could reduce vehicle kilometers traveled and hence reduce energy consumption. For example, low-density U.S. cities consumed five times more in total per capita transport energy than wealthy and compact Asian cities, and 2.5 times and 1.6 times more than European and Australian cities, respectively.

Low-density suburban development is only accessible by car. Therefore, car ownership and usage has increased sharply over time, causing congestion. Congestion puts pressure on the economic sustainability of cities. Initially, further investment in low-density suburban development and highway networks was presented as a successful mechanism for reducing congestion. However, heavy investment in roads has not reduced traffic congestion. Increased road traffic has contributed to greenhouse gas emissions and vehicle accidents and raises questions about the economic viability of private transport. Low-density cities make public transport systems unattractive due to the difficulty of designing services that can fulfill community needs. Over time, public transport operational costs increase, ultimately leading to increased fares that further reduce public transport patronage. As a result, in most of the cities in the developed world, travel to work is heavily biased toward cars (more than 95 percent) rather than to public transport.

In short, low-density suburban development and investment in roads transformed our cities into car-dependent cities. Reversing this trend into a more sustainable one is not easy. However, some cities are setting examples in reversing these trends by investing in public transport infrastructure and services. Perth, Curitiba, Portland, Seoul, Hong Kong, and Singapore are examples. In European and wealthy Asian cities, sustainable patterns of transport occur because the cities are better structured and investment in transport infrastructure for sustainable modes has been made.

Cycling and walking are the most economical and environmentally friendly modes of transport. These modes are successful in compact urban forms when accompanied by investment in their infrastructure. Most cities in the world ignore these modes of transport. However, cities such as Copenhagen and Amsterdam provide good examples, where more than 30 percent of work trips are made by bicycle. These cities achieved these figures by investment in these modes of transport and giving them clear priority in their transport policies. Many cities are developing walking-friendly infrastructure and formulating policies that give pedestrians priority over other modes of transport. In short, cities can improve their sustainability by providing opportunities to travel by sustainable modes of transport.

Currently, policies and price mechanisms such as the taxation system on suburban homes, insurance, subsidies for car usage, investment in high-speed road systems, parking subsidies, lack of congestion pricing, and ignorance of social and environmental externalities favor urban sprawl. The cost of infrastructure and utility service provision in low-density areas are 10 times higher than in high-density areas. This is understandable given that cost of infrastructure increases when distance increases from the central core. Moreover, the huge investment in highway systems subsidizes the cost of motorization, which includes parking, air pollution, road crashes, and cheap oil. Therefore, the cost of low-density urban development jeopardizes the economic sustainability of cities. It is now widely accepted that market failure and public policies and investment are the major factors in urban sprawl. It is also widely argued that people show a strong preference for suburban living, especially in developed-world cities, and that market forces therefore cause developers to provide houses that reflect consumer choice. This phenomenon can be observed in many cities in the United States, Canada, Australia, and New Zealand.

However, this argument remains questionable, given the lack of full understanding of the above mentioned government policies and prices that shape consumer preferences for suburban living. In Singapore and Hong Kong, policies and investment clearly give preferences to high-quality public transport and land-use development around stations. As a result of these government policies, individual preferences favor high-density living near public transport hubs. It can be concluded that individual preferences are shaped by government economic policies that may favor sustainable or unsustainable development.

Social Sustainability

Social sustainability in cities could be increased by improving the affordability of houses, providing green or public spaces for social interaction, and designing a city that could promote healthy lifestyles.

The biggest challenge in cities is providing houses that are both affordable and inclusive of the whole community. Ideally, housing should be of different sizes, designs, and prices in order to address the needs of a broad range of people, including the poor, disabled, older and young people, and immigrants. Patterns of housing could be managed by designing compact cities. The opponents of compact cities argue that low-density housing is more affordable then high-density housing. This is true in cases where few high-density housing projects are implemented. Wherever a holistic approach to providing diverse and compact housing development is implemented, housing affordability will be improved.

Currently, low-density standard housing is the result of market processes that produce socially inequitable results. Outer suburban housing patterns are segregated in terms of both income and ethnicity. Therefore, the divisions between different population groups living in a city increase over time and need to be addressed. Additionally, such housing development is only accessible by cars. Therefore, those who cannot drive or cannot afford to drive such as the elderly, young, and poor have limited access to community facilities, causing further social exclusion. Changing housing patterns in compact cities both increases social sustainability and reduces social exclusion. Compact cities are more humane or at least human in scale.

The sustainable development agenda in cities urges an increase in social interaction in and among local communities. This raises the question: How can social interaction be improved in cities?

Public and green spaces provide a social and communal place for recreation and interaction among different community groups. Throughout history, nature and green open places have remained central to the planning of cities. Present-day planners intuitively try to incorporate green spaces to enhance the livability of the city and to moderate the stress of urban life. For example, Central Park in New York City was specifically designed to provide healthy contact between people and nature and to help alleviate the stress of urban economic life. This is because people think of natural places in the city differently from other places like the office, home, and markets. The natural places within the city help people to escape from their personal concerns and obligations. At the same time, these natural places can also be a habitat for plants and birds, which are central to biodiversity in the city. Therefore, there is a need for communal green space in cities. It is regarded as the "nature" within the city space, and as nature matters to all of us, everyone living in a city needs access to and to be afforded the benefits of contact with nature. Besides the direct physical and psychological benefits that people derive from exposure to green space/nature in the city, these spaces can provide space for social interaction.

In addition to parks serving as communal places, a sustainable city should also have some communal buildings so that people can arrange social gatherings. In some cultures, places of worship can be used as meeting places, so the sustainable city should provide places of worship for people of different faiths.

The pre-modern settlements in the world speak about the livability and sustainability inherent in design of buildings and the social life around these buildings. These buildings were nonuniform but were part of a natural landscape. Nature was not lost in those cities. Water and trees were central to their public space, waste was recycled, and rural productive land use was strongly integrated with the urban centers. Streets were filled with people, who could interact with other community members on a regular basis, and major destinations like schools, shops, religious places, and workplaces were accessible by walking. These urban places of the past were embedded in nature and provided ideal livability for the people. However, the physical uniformity resulting from urban sprawl discourages social interaction, and there is a loss of a sense of community where neighbors interact regularly and feel a sense of belonging and a responsibility for one another. Public or open space, local shops, and community centers accessible by walking are central to greater social interaction in any community. However, in low-density suburbs, open or public spaces are surrendered to private space or gardens. Therefore, it is widely quoted that residents of mixed-use areas have a greater sense of community because they interact more. In short, the sustainable development agenda in cities supports an increase in social interaction and a healthy lifestyle.

The natural vegetation in cities provides physical health benefits to people as well. Though little research has been done on this issue, what has been done highlights that merely being exposed to nature in the city has some real, measurable benefits. For example, some research finds that viewing natural and vegetated landscapes reduces healing time after surgery, and patients with exposure to nature needed shorter hospital stays and fewer painkillers and anti-anxiety pills compared to patients who had very little exposure to natural vegetation.

Environmental Sustainability

Cities provide an extreme separation of human or man-made life from the natural world on which city life depends. Our cities draw in environmental raw materials and spew out waste. The problem lies both in consumption and in production. The globalization of production puts ever-greater distance between the people who consume and the land and sea that ultimately produce. Japanese whale meat ends up in city restaurants. Rainforest timber adorns city boardrooms. Vegetables on supermarket shelves may come from Brazil, timber from Malaysia, and most manufactured goods from China. Unlike the situation in early industrial cities, the factories are thousands of miles away from the consumers of their products. Everyday life in cities depends on the long chain of production and waste of the natural world. Therefore, it is fair to say that modern cities are wonderfully efficient machines for the consumption of nature. Unless cities focus on sustainable consumption and production, environmental damage will escalate. Environmental sustainability in cities can be divided into two parts: built environment and natural environment.

A vast area of the city is made up of houses. Every house consumes the environment by swallowing materials and energy, but the issue is how cities can be designed to become less burdensome on the environment and more energy efficient. In the sustainable city model, housing is designed to conserve energy and provide comfortable living by using renewable building materials for the construction of houses. The buildings are designed to trap the sun, and thick internal walls prevent overheating in summer and store heat in winter. Kitchens are designed incorporating energy-saving appliances and low-energy lighting. After the 1973 oil crisis, the focus turned toward energy efficiency and renewable energy in buildings. Efforts were put into finding solutions that use less energy. One of the key solutions that emerged was that compact cities not only save transportation fuel but also save energy use in buildings. Further, the shared insulating effect of joining buildings together saves a lot more energy. This provides the hint that planners should consider the bigger picture and collective actions in order to achieve sustainability in buildings in our cities.

Urbanization is predominantly responsible for the world's loss of biodiversity. Biodiversity includes a range of different species, from bacteria to gorillas, and the ecological systems of relationships that these species have evolved over time. Cities, through their very construction, replace the habitat of plants and animals with the human habitat. Similarly, cities indirectly affect biodiversity through a city's agricultural "footprint," where natural ecological habitats are forced to the margins and small remnant pockets. Having said this, sustainable development urges that cities must coexist with nature and that they must become resource pools by reusing their own energy and resources—both natural and human. Sustainable development in cities marks a larger agenda of improvement and hope, not only for improving the local environment, but also for improving the global commons such as biodiversity.

Cities are continuously consuming rich agricultural land to accommodate urban growth. On the other hand, the increasing worldwide population increases demands on agricultural land and productivity. Though increased productivity of land is possible with new and more productive crops, these crops need nutrients and water that in turn will face depletion. Urban sprawl consumes more agricultural land. The consumption of land for housing not only causes the loss of prime agricultural land but also harms environmentally sensitive areas. Compact cities usually save prime agricultural land from suburban development; however, opponents argue that it is not important to preserve agricultural land but instead to make the "best" use of land. Although world food production per capita has increased over time, it still falls short in the face of increasing global demand. Additionally, the declining productivity of the world's rangelands and fisheries places pressure on agricultural land at the same time that biotechnology has failed to respond to increasing demand of food.

Sustainable development in cities needs environmentally responsible water management systems. Some of the goals for sustainable water management in cities include removing unneeded ocean and river outfalls, recycling water for various urban uses, recycling nutrients, and increasing soft surfaces for storm water retention and reduced requirement for large pipes. To fulfill these goals, new urban water management processes should be developed by integrating urban planning and water planning. This would help to link urban growth with the management of water supply, wastewater collection and treatment, and storm water drainage services while keeping in view the environmental constraints. Water-sensitive design could improve both the quantity and quality of water in our cities, the alternative being increased water pollution, especially river pollution and drainage.

Urban forestry through trees and vegetation is an important element of sustainable development in cities. It is not a cosmetic beautification of city life; rather, it is a basic infrastructure that not only serves the city ecologically but also contributes to the psychological well-being of people. Currently, cities give low priority to urban forestry due to high demands on land for the built environment. Sustainable development in cities also requires the establishment of urban forestry and vegetable gardens. Both these techniques not only provide a more soothing and greener look to the city and enhance air quality, but in certain cases also provide economic benefits to urban dwellers, especially in developing countries. European cities have been actively involved in urban agriculture for many decades through allotment programs and the establishment of urban farms. These urban farms remain focal points for unemployed youth to learn practical skills and to provide locally grown food to the cities. Zurich is a good example of urban agriculture in a European country. It has maintained its green character through the use of allotment gardens and, as a result, has preserved large forested areas on the ridges of the city while retaining common garden areas around apartments and other housing units. The allotment gardens contain areas of land divided into series of small parcels, which are allotted to different families to grow plants, vegetables, and flowers. These kinds of areas not only provide benefits related to good air quality and a better environment, but also provide a welcoming communal space for the people.

Sustainable development is a global agenda, and efforts are being made at the global level. It is not possible to achieve global results if efforts are not made at the local level. For urban change toward sustainable development, a smooth working relationship between the market, government, and civil society is necessary. The market provides the necessary wealth and resources for development to happen, government regulates and creates the rules and laws for this development, and civil society serves as the custodian of ethics and culture in a particular society. Above all, experts in the form of engineers, economists, architects, designers, planners, environmental scientists, administrators, and political advisers have influenced cities in the past, and they will be instrumental in the change toward sustainability. It is now widely agreed that integrated, interdisciplinary solutions are the best way to tackle difficult and complex problems in cities. It is time to create space that allows decision makers and other stakeholders to go beyond their specializations in order to consider the deeper issues of sustainable development in cities. The change to sustainable development in cities may come when local-level community-based approaches are taken into account. Local communities should take the leading role in determining how to solve their economic, social, and environmental problems, which could lead to sustainable solutions.

See Also: Agriculture; Brownfield Redevelopment; Sustainable Development (Business); Sustainable Development (Politics).

Further Readings

Ewing, Reid. "Is Los Angeles–Style Sprawl Desirable?" *Journal of the American Planning Association*, 63 (1997).

Gordon, Peter and Harry Richardson. "Are Compact Cities a Desirable Planning Goal?" *Journal of the American Planning Association*, 63 (1997).

Low, Nicholas, Brendan Glesson, Ray Green, and Darko Radovic. *The Green City, Sustainable Homes, Sustainable Suburbs*. Sydney: UNSW Press, 2005.

Newman, Peter and Jeffery Kenworthy. *Sustainability and Cities: Overcoming Automobile Dependence*. Washington, DC: Island Press, 2000.

Yencken, David and Debra Wilkinson. "The Sustainability of Settlements." *Resetting the Compass, Australia's Journey Towards Sustainability*. Victoria, Australia: CSIRO Publishing, 2001.

Muhammad Imran
Massey University

SUSTAINABLE DEVELOPMENT (POLITICS)

Sustainable development sets out to make necessary modifications that will enable economic development to be sustainable into the future. It recognizes that serious and irreversible environmental degradation should be prevented because it could diminish the ability of the planet to sustain human development. It requires that economic activity that is carried out now to meet current requirements should not degrade or deplete the environment so much that people will not be able to meet their needs in the future. Critics of sustainable development often fear that so long as the concepts of sustainability and development are linked, economic developmental goals will predominate.

In October 1987, the goal of sustainable development was endorsed by the governments of 100 nations in the United Nations General Assembly. The UN endorsement followed the completion of a report, *Our Common Future*, by the World Commission on Environment and Development (the Brundtland Commission), which defined sustainable development as "development that meets the needs of the present without compromising the ability of future generations to meet their own needs."

The concept of sustainable development succeeded in gaining widespread support among the world's decision makers and power brokers because it aims to protect the environment without the need for radical social or political change. Sustainable development offers the promise that economic activities can be harmonized with environmental protection and that technologies can be found and implemented that will ensure economic growth does not harm the environment.

Governments, industry groups, and business associations have produced numerous documents and policy statements on sustainable development, outlining how the environment can be protected in a context of economic growth, freed-up markets, and industrial self-regulation.

The support of environmentalists for the concept of sustainable development has been less universal. While some welcome the newfound attention being paid to environmental protection and the opportunity to negotiate with governments and developers on these issues, others are more wary because of the minimalist approach that seems to be inherent in the sustainable development approach. They argue that more fundamental institutional and social changes need to take place, including a shift toward steady-state economies.

Accommodating Economic Growth

Sustainable development is part of a second wave of modern environmentalism and heralds a newer approach to tackling environmental problems. The first wave of environmentalists argued that the exponential growth of populations and industrial activity could not be sustained without seriously depleting the planet's resources and overloading the ability of the planet to deal with pollution and waste materials. Some argued that new technologies and industrial products such as pesticides and plastics also threatened the environment.

First-wave environmentalists, following the protest mood of the times, did not hesitate to blame industry, Western culture, economic growth, and technology for environmental problems. The environmental movement was easily characterized as being antidevelopment. Nevertheless, their warnings captured the popular attention, resonating with the experiences of communities facing obvious pollution in their neighborhoods.

Although many governments did not recognize the importance of global environmental problems, they were forced by community pressure to respond to local pollution problems. During the 1970s, many countries introduced new environmental legislation to cope with the gross sources of pollution. Governments around the world introduced clean air acts, clean water acts, and legislation establishing regulatory agencies to control pollution and manage waste disposal.

The decade that followed saw a backlash against the early environmentalists. Various writers argued that global catastrophe was the fantasy of doomsday forecasters and that scientific discoveries and technological innovations would easily cope with any problems that might arise. Government departments and agencies found it extremely difficult, in this new climate of opinion, to administer properly the legislation that had been put in place at the height of the first wave of environmentalism and businesses did their best to ignore the laws or get around them.

The second wave of environmentalism, which began in the late 1980s, has had much broader support and has involved governments, businesspeople, and economists in the promotion of sustainable development. Scientific evidence about the buildup of greenhouse gases in the atmosphere and the depletion of the ozone layer made it difficult to deny the threat of global environmental problems. Many of the concerns of environmentalists were taken up by senior politicians (including prime ministers and ministers of foreign affairs, of finance, and of agriculture) from countries around the world, as well as eminent scientists, jurists, and international bureaucrats.

Sustainable development succeeded in arenas of influence denied to first-wave environmentalism because of its central idea that environmental protection is not necessarily opposed to development. The concept of sustainable development accommodates economic growth, business interests, and the free market and therefore does not threaten the power structure of modern industrial societies.

Many of the ideas associated with sustainable development were articulated in the 1980 World Conservation Strategy produced by the International Union for Conservation of Nature and Natural Resources (IUCN) in collaboration with the United Nations Environment Programme (UNEP) and the World Wildlife Fund (WWF, now the World Wide Fund for Nature). This document was circulated to all governments, and national conservation strategies based on the World Conservation Strategy were adopted in 50 countries. The World Conservation Strategy argued that while development aimed to achieve human goals through the use of the biosphere, conservation aimed to achieve those same goals by ensuring that use of the biosphere could continue indefinitely.

The renewed interest in sustainability in the 1980s marked a shift from first- to second-wave environmentalism. Earlier environmentalists had used the term to refer to systems in equilibrium: they argued that exponential growth was not sustainable, in the sense that it could not be continued forever because the planet was finite and there were limits to growth. Sustainable development, however, seeks to make economic growth sustainable, mainly through technological change. In 1982, the British government began using the term *sustainability* to refer to sustainable economic expansion rather than the sustainable use of resources.

This new formulation recognizes that economic growth can harm the environment but argues that it does not need to. The conflict between economic growth and environmental protection was denied. The foreword to *Our Common Future* said, "What is needed now is a new era of economic growth—growth that is forceful and at the same time socially and environmentally sustainable." Nearly all definitions of sustainable development are premised on the assumed compatibility of economic growth and environmental protection.

Sustainable development seeks to change the nature of economic growth rather than limit it. It aims to achieve economic growth by increasing productivity without increasing natural resource use too much. Instead of being the villains, as they were in the 1970s, technology and industry are now expected to provide the solutions to environmental problems.

Because so many environmentalists are ready to give way on the issue of economic growth and deny there is a conflict, environmental protection must be argued in terms of its contribution to economic growth. When the inevitable conflicts come up in particular instances, the environment will only be protected where the economic costs are not perceived to be too high.

More radical environmentalists argue that while it is theoretically possible that economic growth could be achieved without additional impacts on the environment, this would mean many activities that might otherwise provide economic growth would have to be forgone, and this will not happen while priority is given to achieving economic growth. Whether they believe economic growth and environmental protection are compatible, almost everyone agrees that there will inevitably be situations in which the goals of economic growth and environmental protection are irreconcilable, and choices will have to be made.

From Conflict to Consensus

The language of sustainable development is clearly aimed at replacing protest and conflict with consensus by asserting that economic and environmental goals are compatible. The concept of sustainable development enabled a new breed of professional environmentalists to partner with economists, politicians, businesspeople, and others to achieve common goals rather than confront each other over whether economic growth should be encouraged or discouraged. By avoiding the debate over limits to growth, sustainable development provided a compromise that, on the face of it, suited everyone.

Sustainable development spawned a number of consensus decision-making processes, including Round Tables on Environment and Economy in Canada, the President's Council on Sustainable Development in the United States, and the Ecologically Sustainable Development (ESD) process in Australia. These processes involved bringing together various interest groups or stakeholders to reach a consensus about how sustainable development should be achieved and how business interests, economic interests, and environmental protection could be reconciled.

During the mid-1980s, Canada developed a "multistakeholder" consultation process for environmental policy. The National Task Force on Environment and Economy was established in 1986 to facilitate dialogue between the environment ministers, senior business executives, leaders of environmental groups, and academics. These groups sought consensus among the different interests represented about how environmental and economic concerns could be integrated into government policies and in private-sector strategies to achieve sustainable development. There were literally hundreds of municipal or regional roundtables in Canada. One of the key aims of these roundtables was to resolve conflicts and disputes over environmental issues and to foster a cooperative rather than a confrontational approach. However, surprisingly, the Canadian government did not use the roundtables to develop its national sustainable development strategy, the Canadian Green Plan.

In June 1990, the Australian government set up working groups to study how sustainable development would be applied to nine different industry sectors that were thought to use or have a significant impact on natural resources. Membership of the working groups was composed of representatives from government, industry, unions, consumer/social welfare organizations, and conservation groups. In addition, there were a few academics and government scientists. The groups were, however, dominated by bureaucrats from the federal government, particularly those from development-oriented departments such as the Department of Primary Industries and Energy (DPIE), which also provided the secretariat for the working groups.

Membership of the working groups was limited to representatives of recognized interest groups who had faith in the process. Dissenters outside the process were marginalized. The working group discussions were not open, and their very existence tended to remove discussion about sustainable development from the public arena to closed meetings of select people. The final ESD working group reports were passed on to dozens of government committees, which produced a National Strategy on Sustainable Development. Environmental groups, business groups, and others were disappointed with the watered-down document produced by the government, which, like the Canadian Green Plan, concentrated on information generation and dissemination rather than on action and change.

Talks and negotiations were held between national governments in the lead-up to the Earth Summit in Rio de Janeiro in June 1992. At the summit a number of agreements and conventions were finalized, including the Rio Declaration; Agenda 21, an action plan for the 21st century; a nonbinding statement of Forest Principles; a Climate Change Convention; and a Convention on Biological Diversity. A Commission for Sustainable Development was also formed to review progress in implementing Agenda 21 and to rationalize intergovernmental decisions related to sustainable development.

For many, however, the Earth Summit agreements were a disappointment. In the negotiations leading up to the Earth Summit in Rio de Janeiro in June 1992, the United States insisted on removing all references to consumption from the Agenda 21 document. The George H. W. Bush administration would not brook suggestions that lifestyles would need to change in affluent nations. "The American lifestyle is not negotiable," said President Bush. Low-income nations retaliated by removing references to the urgent need to slow population growth. They wanted to shift responsibility for environmental problems onto industrialized nations.

Agenda 21 was therefore criticized as being weak and lacking any strong statements on some of the more important yet contentious issues such as population control and resource consumption by people in affluent countries. It was also suggested that transnational corporations had managed to assert an undue amount of influence on the summit. A coalition of third-world activists and international environmental organizations condemned the summit for failing to address issues such as climate change, consumption, the power of transnational corporations, export of hazardous wastes, the causes of deforestation, and the conflict between free trade and environmental protection.

The roundtables, working groups, and the Earth Summit were all early manifestations of an unprecedented degree of consensus, at least among the powerful and influential, achieved by the concept of sustainable development. This consensus approach favored the status quo because change had to be agreed to by all parties to meet the consensus criterion. Traditional frameworks of business activity, the priority of economic goals over environmental goals, and existing social or political structures, institutions, and goals were all taken as given. Radical change rarely emerges from such a process. The consensus approach also favors the status quo by taking the discussion behind closed doors.

The reduction in public participation in environmental decision making combined with the renewed emphasis on economic growth and the incorporation of the environment into the economic system has ensured that business can go on as usual, but critics argue that this means the environment will continue to deteriorate. The imperative that environmental deterioration might once have had for social and political change has been dissipated by the consensus-forging notion of sustainable development.

Dependence on Technology

At the heart of the debate over the potential effectiveness of sustainable development is the question of whether technological change can reduce the impact of economic growth sufficiently to ensure that political and social change will not be necessary. The inability to reach a consensus on the issues of population and consumption and the political need for the concept of sustainable development to accommodate economic growth mean that the achievement of sustainable development will depend on our ability to reduce the environmental impact of resource use through technological change.

Many interest groups accept this political reality. They see continual growth in a finite world as possible through the powers of technology, which will always be there to help us find new sources or provide alternatives if a particular resource appears to be running out. Otherwise, technology will help us use and reuse what we have left in the most efficient manner.

But can technology give us both environmental protection and economic growth? Can it ensure equity between and within generations so that everyone, now and in the future, our far neighbors and our great-grandchildren, can enjoy the standard of living we do? Such an accomplishment would require more than just a few adjustments to existing technological systems. It would require a radically different technology. Yet if technology is socially shaped, as modern scholars of technology studies argue, can we achieve radical technological change without equally radical social change taking place?

Technological change can be encouraged through government funding of research and development and/or private firms can be motivated to change their technologies through legislative requirements or economic incentives. Many sustainable development policies assume that if the environment were incorporated into private economic decisions, this would be enough to encourage the desired technological change.

The Economic View

Sustainable development policies call for the integration of environmental, social, and economic concerns into policies and activities. For economists, this means incorporating the environment into the economic system. David Pearce and his colleagues, in their report on sustainable development to Margaret Thatcher, the British prime minister at the time, said that the principles of sustainable development meant recognizing that "resources and environments serve economic functions and have positive economic value." Considered as a component of the economic system, the environment is seen to provide raw materials for production and to be a receptacle for wastes from production.

Consequently, the discussion surrounding sustainable development borrows heavily from the language and concepts of economics. Nature and the environment are described in economic terms—as natural resources or natural capital, and as part of the community's stock of assets. Thatcher described environmental protection in these terms: "No generation has a freehold on the Earth. All that we have is a life tenancy—with full repairing lease."

Sustainable development policies also call for environmental assets to be appropriately valued. For economists, this entails putting a price on the environment and creating artificial markets and price mechanisms through economic instruments and tradable rights to pollute (emissions trading). The idea is that "the power of the market can be harnessed" to environmental goals. Many governments have taken up this view, which encompasses the idea that the loss of environmental amenity can be substituted for by wealth creation; that putting a price on the environment will help us protect it unless degrading it is more profitable; that the "free" market is the best way of allocating environmental resources; that businesses should base their decisions about polluting behavior on economic considerations and the quest for profit.

Advocates of "weak sustainability" claim that so long as loss of natural capital (environmental degradation) is compensated for by increased man-made capital (wealth and infrastructure), that is, the total amount of capital is constant, then future generations will be taken care of. Economist David Pearce says that this means that the Amazon rainforest can be removed so long as the proceeds from removing it "are reinvested to build up some other form of capital." Pearce points out that this principle requires that "environmental assets be valued in the same way as man-made assets, otherwise we cannot know if we are on a 'sustainable development path.'"

Weak sustainability provides a rationale for continuing to use nonrenewable resources at ever-increasing rates. Economists claim that although there may be temporary shortages, rising prices will ensure that new reserves will be found, substitutes discovered, and more efficient use encouraged. Governments also often view the idea of avoiding consumption of nonrenewable resources as extreme and unrealistic.

While the economic value of natural resources can be easily replaced, their functions are less easily replaced. Most people, even economists, agree that there are limits on the extent to which natural resources can be replaced without changing some biological processes and putting ecological sustainability at risk. Pearce, for example, recognizes that some environmental assets could not be "traded off" because they are essential for life-support systems and as yet they cannot be replaced.

Most environmentalists do not agree that human and natural capital are interchangeable and believe that a loss of environmental quality cannot be substituted with a gain in human or human-made capital without loss of welfare. Therefore, they argue that future generations should not inherit a degraded environment, no matter how many extra sources of wealth are available to them. This is referred to as "strong sustainability."

Even if the production and consumption values and absorption capacity provided by "natural capital" can be replaced or extended, particularly through technological innovation, critics argue this is not the case with other environmental values. To maintain recreational, entertainment, and aesthetic values, the environment must not be spoiled.

Environmental Critics

Understandably, environmentalists generally reject the concept of weak sustainability even if it incorporates the idea of maintaining essential environmental functions. They claim that human welfare can only be maintained over generations if the environment is not degraded; in economists' terms, if natural capital is not declining. They point out that we do not know what the safe limits of environmental degradation are; yet if those safe limits are crossed, the options for future generations would be severely limited.

The purpose of giving environmental "resources" a price is that they will be valued in the same way that other resources are valued. Some environmentalists may be reassured that this means more account will be taken of the environment. On the other hand, this can be seen as a devaluation of the environment because it brings it down to the same level as other commodities that can be bought and sold. Its value is reduced to an economic value, and it is then treated as a substitutable part of the economic system.

For many environmentalists, the real problem is not that the environment is not privately owned or valued on the market but rather that economic considerations take priority in most countries around the world. The tragedy of the commons is not that there are commons but rather the freedom of the commons. The lack of legal sanctions combined with a value system that promotes the raising of individual economic interest to a primary decision-making principle is what destroys the commons.

Seen in this light, pricing the environment, economic instruments such as emissions trading, and privatization are all mechanisms for perpetuating the central problems that caused environmental degradation in the first place. They ensure priority is still given to economic goals and they enable individuals and firms to make decisions that affect others on the basis of their own economic interests. The primacy of "free" markets in environmental decision making ensures that power remains in the hands of those who direct and control financial resources: the wealthy, the corporations, and the economists they employ.

Vandana Shiva, an Indian activist, points out that sustainability should require that markets and production process be reshaped to fit nature's logic rather than "the logic of profits and capital accumulation, and returns on investment" determining nature's fate. Instead, she says, sustainable development "protects the primacy of capital. It is still assumed that capital is the basis of all activity."

For those environmentalists who see businesspeople and economists as part of the problem, sustainable development is an oxymoron—a contradiction in terms. They despair at the lack of discussion of the role played by economic growth and business corporations in environmental degradation, and they abhor the reduction of environmental values to monetary values and the commodification of nature.

Sustainable development is human centered; that is, it is primarily concerned with maintaining human welfare through meeting human needs and ensuring the quality of human life. The Rio Declaration on Environment and Development, agreed to by more than 170 countries at Rio de Janeiro in June 1992, has the following first principle: "Human beings are at the centre of concerns for sustainable development. They are entitled to a healthy and productive life in harmony with nature."

Sustainable development does not guarantee the needs or quality of life of animals or other living organisms, except inasmuch as this will benefit humans. For many environmentalists, this concept does not go far enough. They argue that all living creatures have an inherent right to exist that is separate from their usefulness or value to humans.

The international environmental organizations that originally put together the World Conservation Strategy put together a set of principles for sustainable development that were published in 1992 in a report titled "Caring for the Earth: A Strategy for Sustainable Living." These principles place more emphasis on maintenance of biodiversity and Earth's natural limits. Its first principle is "Respect and care for the community of life":

All life on Earth is part of one great interdependent system, which influences and depends on the nonliving components of the planet—rocks, soils, waters and air. Disturbing one part of this biosphere can affect the whole. Just as human societies are interdependent and future generations are affected by our present actions, so the world of nature is increasingly dominated by our behaviour. It is a matter of ethics as well as practicality to manage development so that it does not threaten the survival of other species or eliminate their habitats. While our survival depends on the use of other species, we need not and should not use them cruelly or wastefully.

See Also: Anthropocentrism Versus Biocentrism; Ethical Sustainability and Development; Intergenerational Justice; Sustainable Development (Business); Sustainable Development (Cities).

Further Readings

Beder, Sharon. *The Nature of Sustainable Development*, 2nd ed. Melbourne, Australia: Scribe Publications, 2003.

Blewitt, John. *Understanding Sustainable Development*. London: Earthscan, 2008.

"Caring for the Earth: A Strategy for Sustainable Living." Gland, Switzerland: International Union for Conservation of Nature/United Nations Environment Programme/World Wildlife Fund, 1991.

Pearce, David, ed. *Blueprint2: Greening the World Economy*. London: Earthscan, 1991.

Rogers, Peter P., Kazi F. Jalal, and John A. Boyd. *An Introduction to Sustainable Development*. London: Earthscan, 2007.

Shiva, Vandana. "Ecologically Sustainable: What It Really Means." *Third World Resurgence*, 5 (January 1991).

World Commission on Environment and Development. *Our Common Future*. Melbourne, Australia: Oxford University Press, 1990.

Sharon Beder
University of Wollongong

T

TRAGEDY OF THE COMMONS

Garrett Hardin's theory described how herdsmen grazing cattle on common ground would increase the number of cattle to maximize individual profits, thereby destroying the common environment.

Source: Massachusetts National Resources Conservation Service

Garrett Hardin's article titled "The Tragedy of the Commons," published by *Science* in 1968, examines the fate of a common pasture shared among rational, utility-maximizing herdsmen. The parable, first developed by William Forster Lloyd (1794–1852), was used by Hardin to demonstrate the importance of collective action through cooperation and/or coercion to protect finite resources. The basis for Hardin's story is that unlimited population growth coupled with the overconsumption of finite materials ultimately causes environmental, social, and economic destruction. Using widespread population growth as the example, Hardin demonstrates the implications for resources and the collective and seeks to highlight the need to regulate and control the commons for the good of the whole. While he uses increase in population as the root cause for destruction, the concept has wider and growing significance in terms of understanding how we can manage and prevent overarching global challenges we face, including global warming, environmental pollution, and declining natural resources.

Hardin's "Tragedy of the Commons" involves a pasture of land shared by all. Individuals graze their herds on the land, and each increases his herd numbers in order to maximize his profits. Although the commons is damaged with every additional animal brought onto the commons, the economic gain to be had by the individual herdsman in increasing the herd proportionately outweighs the degradation to the commons. The story assumes that each individual operates on a rational, self-interest-based value system and pursues his/her rights over the commons to increase his/her herd. Eventually, this cumulative and steady increase in the herd numbers results in the destruction of the commons. Hardin summarizes: "Each man is locked into a system which compels him to increase his herd without limit—in a world that is limited. Ruin is the destination toward all men rush, each pursuing his own best interest in the freedom of the commons. Freedom in a commons brings ruin to all."

This cautionary tale clearly illustrates the limits of the environment and natural resources, above which we are unable to sustain ourselves—even taking into account beneficial advances in technological innovation presupposed by Boserup's theory that necessity is the mother of invention. In fact, this scenario is not only a hypothesis, but it has been played out more than once throughout history and thus serves as a reasoning for how we have come to experience or be at the brink of environmental catastrophes in the past and how we can prevent them in the future. According to "The Tragedy of the Commons," the problem of the commons arises when individuals are required through necessity (in this case, to prevent the degradation of the commons) to cooperate to achieve a goal that is in both their collective and individual interests (in this case, the keeping of cattle) but when the costs to individuals of cooperating exceed the short-term benefits of cooperating. In such a case, the rational, self-interested herdsman performs a basic cost-benefit calculation that suggests that they should each withhold their cooperation for personal reward in the short term even though this will be to the detriment of the collective in the long term.

The parable challenges the belief held by theorists including the Scottish moral philosopher and political economist Adam Smith that decisions reached individually are the best decisions for an entire society. Smith's popularized theory is predicated on an "invisible hand" that guides decision makers toward decisions that benefit the collective. He maintained that "by pursuing his own interest, [the individual] frequently promotes that of the society more effectually than when he intends to promote it." In contrast, Hardin argues that common property rights are more likely to lead to exploitation on the basis of individual interest resulting in destruction for all. Consequently, he advocates the transformation of common property into private property as a means of securing its long-term protection and utility for the collective.

One of the key insights in the tragedy of the commons is that of human behavior. The premise is that in a situation without a system of control and in which an individual has the freedom to act independently, individuals will adopt a self-interest-based approach to decision making. In Hardin's story, each herdsman chooses to exploit the shared commons by increasing his herd number for short-term economic gain. Hardin's message here is twofold. First, individual decision making is founded on short-term cost-benefit analyses that take little if any account of long-term impacts likely to be felt either individually or by others. Second, the user alone is not capable of changing the system of the commons and his values but regulatory controls are required to shape how the commons is used and to protect it for the benefit of all.

Consequently, Hardin concluded that in order to avoid the tragedy, the commons should be privatized or kept as public property to which rights of entry could be granted—institution of government regulation of activities and operators (uses and users). The notion of private property rights is that whoever holds these rights (individuals, businesses, or groups) will be more likely to protect their resources and less inclined to exploitation, ruination, and pollution. In this way the long-standing and well-established market-based incentives schemes to control environmental problems—introduced by Arthur Cecil Pigou and developed into transferable pollution rights such as the U.S. Emissions Trading Program and its European counterpart emissions trading allowance scheme—provide an incentive for rights holders to consider the wider, common interest on the basis of the transferability of these rights and the market price to be gained.

Hardin's tragedy, although unarguably revolutionary and indeed prophetic, was foreshadowed by Harold Demetz's paper also discussing the problem of the commons. Demetz explained the underlying problem of the commons to be a lack of coordination. He argued that the consequences of self-interest on the whole could have been prevented if the individuals of the commons had come together and agreed to reduce their consumption through a coordinated approach. The obstacle to such an agreement, while benefiting all, is that coordination of such an agreement involves high transaction costs, negotiations, and compromise in a situation where each individual member has a common right to exploit, which to some extent would be surrendered under an agreement on how to treat the commons.

James Krier considers Hardin's argument to be flawed on the basis that his parable assumes there is no coordination between the individuals and their utility of the commons. They argue that in emphasizing the need for "mutual coercion, mutually agreed upon," Hardin ignores the likelihood that if members can agree to a program of mutual coercion, then they are likely to be able to restrain themselves by other means, resulting in a system for the collective benefit of all. Krier goes on to acknowledge the three defenses to Hardin's contradiction in "The Tragedy of the Commons." First, the core purpose of his story was a political piece to demonstrate that population growth is a problem of the commons that can only be controlled through coercion rather than through voluntary programs. Second, while Hardin pointed to the corruptibility of government agents, he emphasized the important role they have to play in driving public coercion. Krier went on to point to the relationship between the public and government in which political agendas, at least in democratic countries, are significantly shaped and influenced by the society as the electorate. Last, the perspective from which Hardin developed his argument was grounded firmly in ecology rather than economics or political theory.

"The Tragedy of the Commons" has dire implications for any commonly held resource being utilized near capacity, not just pastures. It is not only consumable resources that endanger the commons but also the pollution of a resource. This problem occurs when the cost of degrading the resource, for example, through emissions to air and water or the dumping of wastes, is lower than the cost of conserving and protecting the resource through treatment and control.

Sustaining survival requires a fundamental step change in the ways in which we understand, manage, and use our shared (common) resources and the environment upon which we depend. The enduring nature of the tragedy of the commons arises from its pervasive and continuing, if not mounting, relevance in the way we utilize and interact with our environment. In this way, Hardin's concept can be applied much more broadly to environmental problems such as emissions to air, pollution of waters, overfishing, and deforestation.

See Also: Carbon Tax; Carbon Trading/Emissions Trading; Common Property Theory.

Further Readings

The Bundeena Maianbar Water Cycle Management Study, the Environment Industry Development Network, and the Hacking River Catchment Management Committee. "Avoiding Another Tragedy of the Commons." Port Hacking, Australia: Port Hacking Protection Society, 1997.

Demetz, Harold. "Toward a Theory of Property Rights." *American Economic Review* (1967).

Hardin, Garrett. "The Tragedy of the Commons." *Science*, 162 (1968).

Krier, James. E. "The Tragedy of the Commons, Part Two." *Harvard Journal of Law and Public Policy* (1992).

Pandey, Bindhy. *Wasini Natural Resource Management*. New Delhi, India: Mittal Publications, 2005.

Smith, Adam [1759]. "The Theory of Moral Sentiments." In *Glasgow Edition of the Works and Correspondence of Adam Smith*, Vol. 1, D. D. Raphael and A. L. Macfie, eds. Indianapolis, IN: Liberty Fund, 1982.

Hazel Nash
Cardiff University

UTILITARIANISM

Utilitarianism, in its simplest definition, is the notion that the moral worth of an action can be determined by assessing its utility in providing happiness or pleasure to the greatest number of people. It is related to consequentialism, the belief that the consequences of a particular action provide the criteria for making valid moral judgments about that action. Notions of utilitarianism go back to the Greek philosopher Epicurus, though it is in the works of 18th-century philosopher and social reformer Jeremy Bentham that utilitarianism found its niche as a specific school of thought. Nineteenth-century philosopher John Stuart Mill is probably the most famous proponent of utilitarianism in recent history, and he is credited with improving the structure, meaning, and application of Bentham's ideas regarding utilitarianism. Today, Australian philosopher Peter Singer and Swedish philosopher Torbjörn Tännsjö are two of the best-known contemporary utilitarians, though many associate utilitarianism in general with socialism and communism. Since Bentham's early work in quantifying morality based on the consequences of human actions, many branches of utilitarianism have emerged, and criticism of the basic concept of utilitarianism and of these subdivisions of the philosophy have arisen as well.

Bentham proposed the notion of a "felicific calculus"; that is, he believed that we could assign a net value to each human action by considering, for all those affected by the action, how strongly its pleasure is felt, how long the pleasure lasts, how quickly the pleasure follows the action, and how likely the action is to lead to collateral benefits while avoiding collateral harms. For Bentham, the happiness of a community as a whole is composed of the sum of individual interests, and so moral obligation is intimately tied to providing the greatest amount of happiness for the greatest number of people. This is known as "the greatest happiness principle" or the notion that actions should be judged by their ability to produce the greatest balance of pleasure over pain for the greatest number of people.

Mill also accepted the notion of the greatest happiness principle, though he believed that it was not possible to quantify all differences of various pleasures, because pleasures differ from each other in qualitative ways. While Bentham valued all forms of happiness as equal, Mill argued that intellectual and moral pleasures are superior to more physical forms of pleasure. Mill contended that simple pleasures tend to be preferred by people who have not experienced more complex pleasures such as exposure to "high art" and that those who have experienced both high and low forms of pleasure are in the best position to judge what is best for a society. Applying his theories to the world around him, Mill contended that governmental infringement upon the freedom of individuals is rarely defensible and that the tyranny of the majority is especially dangerous to individual liberty. He believed that the only purpose for which power can be rightfully exercised over any citizen, against his will, is to prevent harm to others, arguing that while society has a clear obligation for protecting its citizens from each other, there is no justification for interfering in anything else a person does. Mill also tackled the issue of women's place in society with *The Subjection of Women,* and many of his notions regarding women mark him as an early feminist. He argued that men's dominance over women was strictly related to the brute strength and force of men and was not in any way evidence of men's superiority in intellectual or moral capacities. He also contended that women deserved compensation for their contributions to their families and to society in general and railed against the patriarchal culture that trapped women in a web of social expectations that rarely, if ever, reflected women's true desires.

One of the main distinctions in utilitarianism is that between "act utilitarianism" and "rule utilitarianism." Act utilitarianism dictates that, when faced with a choice, one must contemplate the consequences associated with each action associated with the choice and choose the action that will produce the most pleasure. Conversely, rule utilitarianism is rooted in evaluating the potential rules of an action. When deciding whether a rule should be followed, one must look at the consequences of always following the rule; if following the rule creates more happiness than flouting it, then morally, the rule must be followed at all times. And so, the distinction between act and rule utilitarianism is in whether consequences should be considered as specific to each case or generalized to rules. In addition to act and rule utilitarianism, other types of utilitarianism include motive utilitarianism, negative utilitarianism, total utilitarianism, and average utilitarianism.

Those who embrace utilitarianism see it as an effective, rational, and moral way to nurture the general welfare of a population. By weighing the consequences of our actions on the world around us, we will make decisions that lead to happiness and prevent suffering. Utilitarians believe that its principles can be applied to politics and economics as well as to individual actions. Utilitarian notions can be found in animal rights movements, environmental movements, and, as mentioned, in aspects of socialism and communism. Critics, however, find many flaws in utilitarian concepts, classifying it as distasteful, impractical, and impossible. More specifically, they note the following:

- All political arguments about society claim that their proposed solution is the one that most increases general human happiness.
- Utilitarianism can lead to conclusions contrary to "commonsense" morality.
- Utilitarianism focuses too much on the results of actions and not on intentions.
- Utilitarianism's reliance on felicific calculus not only assumes that happiness can somehow be quantified, but that the happiness of different people can be compared and measured quantitatively.
- Utilitarianism has not successfully answered why individual actions must be weighed in relation to their benefit to society as a whole; that is, why is it wrong for someone to act in her own self-interest? Thus, utilitarianism threatens individual rights.
- Utilitarianism can be incompatible with human rights; for example, if slavery is beneficial for the population as a whole, it could theoretically be justified by utilitarianism.

Proponents of utilitarianism argue that by tailoring our individual actions to "promote the greatest good for the greatest number," individual and societal happiness will increase while suffering will decrease. Critics, however, find that defining what is meant by "the greatest good" is difficult, as is quantifying pleasure and happiness. Further, they are disturbed by what they see as the potential for oppression of individual rights. While general notions of utilitarianism are most likely acceptable to most people, the difficulties associated with strictly following utilitarian principles suggest that it would require a major shift in societal and cultural ideas in order for it to be fully embraced.

See Also: Animal Ethics; Individual Action Versus Collective Action; Mitigation; Precautionary Principle (Ethics and Philosophy); Utilitarianism Versus Anthropocentrism.

Further Readings

Mulgan, Tim. *Understanding Utilitarianism.* Stocksfield, UK: Acumen Publishing, 2007.
Sen, Amartya and Bernard Williams, eds. *Utilitarianism and Beyond.* Cambridge, UK: Cambridge University Press, 1982.
Smart, J. J. C. and Bernard Williams. *Utilitarianism: For and Against.* Cambridge, UK: Cambridge University Press, 1973.

Tani E. Bellestri
Independent Scholar

Utilitarianism Versus Anthropocentrism

Environmental philosophers debate the moral responsibilities that humans bear to the natural environment. The rise of the environmental movement in the late 20th century occurred as humans realized that their interconnectedness with the natural world meant that its preservation was essential to human survival. Environmentalists did not agree, however, on the form of human responsibilities to the natural world and how they should be weighed against human needs and interests. Environmental utilitarians believe that human interaction with the natural world should be based on which rules and actions will achieve the greater good. Alternative environmental philosophies instead argue that such views are anthropocentric, resulting in environmental degradation that ultimately threatens the human survival it is meant to ensure.

Utilitarianism

Utilitarianism is an ethical or philosophical theory first fully developed by English philosopher Jeremy Bentham in the late 18th and early 19th centuries. Bentham stated that humans were naturally governed by pleasure and pain, which are key determinants of their actions. Bentham's utilitarian ideal stated that an ethical person should base his/her actions on the promotion of the greatest good (pleasure) for the greatest number of people. There are also further distinctions between act and rule utilitarianism, based on the consideration of either individual actions or categories of actions, respectively. Utilitarianism is frequently applied to modern environmental philosophy.

Environmental philosophies such as utilitarianism rose to prominence along with the global environmental movement in the late 20th century, as society became increasingly concerned with the negative environmental impacts of human civilization and technological development, including pollution, the growth of endangered species, and declining fossil fuels and other natural resources. An early example of this concern was the 1972 Stockholm Declaration on Human Environment, which noted the threat of environmental degradation to humans and their interconnectedness with the natural world. Some environmentalists and environmental philosophers have adopted utilitarianism to moral human decision making with regard to their interaction with the natural world. Conservationist Gifford Pinchot, for example, stated that decisions on the conservation of natural resources should be based on utilitarian principles of the greatest good.

Environmental utilitarianism, like other environmental philosophies, is often used to judge particular actions or rules as either right (moral or ethical) or wrong (immoral or unethical), in this case based on their utility in service of the greater good and their consequences in this regard. An ethical person would examine all of the possible actions given a set of circumstances and choose the action calculated to achieve this greatest good. This reasoning, however, can reduce the philosophy of utilitarianism to an all-or-nothing scenario: either all humans will engage in a certain action when it meets the universal greatest good or no humans will engage in a certain action when it fails to meet the universal greatest good.

The greatest universal good based on rule utilitarianism, however, is not always the best immediate choice based on action utilitarianism. For example, a rule stating that one should never hunt and eat an endangered animal would conflict with the best utilitarian action if a man were lost in the wilderness without any other obtainable food source. In such an instance, would the man's decision to eat the animal be considered morally wrong, and by whom? Critics of the utilitarian philosophy as a whole might further question whether the man's goal of survival, giving utilitarian value to the animal as a food source, should take precedence over the long-term survival of an endangered species based on its own intrinsic value outside its human utility.

Problems also arise based on who is determining the greater good and how it is being determined. English philosopher John Stuart Mill elaborated on Bentham's early view of utilitarianism by stating that some goods or pleasures are inherently of more value than others, further introducing the role of human judgment in determining the greatest good. Some environmental philosophers argue that utilitarianism could be used to justify actions to which they are morally opposed. For example, some environmentalists feel that medical benefits are a justification for animal testing while others disagree.

The philosophy of utilitarianism has other inherent problems besides its capacity to pass judgment. One is the impossibility of measuring the utility of actions or rules; there is no scientific determination of the greatest good. There is also the question of whether considerations of the good of the majority should always take preference over that of the minority, such as in cases where the minority hold a much stronger belief in their position. This argument is similar to political arguments over whether a minority should be subjected to the tyranny of the majority within a governmental system.

Environmental ethicists also debate whether utilitarian concerns for the greater good extend beyond humans to the rest of the natural world, considering both sentient beings such as animals and inanimate natural objects such as trees, rivers, or the soil. Should animals as sentient beings be subjected to negative consequences such as pain or confinement in the interests of the greater human good such as the production of food sources or medical advances? Those who answer in the negative advocate such issues as vegetarianism, a ban on fur coats, and animal liberation, with some even arguing that humans should not own domestic pets.

Yet another consideration is the impossibility of determining all of the future outcomes (consequences) of a present decision and whether they will meet future definitions of the greatest good or future moral approbations. Other environmentalists, however, point out the fact that determinations of utility often change over time, resulting in present criticisms of past actions that were justified as utilitarian at the time. Environmental impacts of past decisions have included deforestation and desertification, endangered and extinct species of plants and animals, and air, water, and soil pollution.

Other utilitarians have argued that the promotion of total utility calls for an increasing greater good, which can only be attained through continual population growth. They argue that average utility is a better goal that also allows for population control to avoid the overcrowding that can decrease average utility and result in environmental degradation. The most recent official estimation of the world population stood at 6,790,062,216 in mid-2009. Questions of future considerations also encompass future generations of humans who will inherit the planet: Should their good be considered too, and if so, should it be equally weighted with the present good? Already, populations in many parts of the world are facing shortages of critical natural resources such as potable water and fossil fuels. Many scientists argue that a critical impasse will arrive at some point in the future if changes in philosophy and action do not occur in the present. Some environmentalists posit the question: What kind of world do we want our children to inherit?

Supporters of environmental utilitarianism can struggle with ethical dilemmas, or competing ethical claims, which occur when two courses of action are ethically valid, pose harm, or are incompatible with each other. Environmental philosopher Paul Taylor used a series of ethical guidelines to overcome some of the difficulty posed by such ethical dilemmas. These included the right of self-defense, the precedence of basic survival, the determination of the lesser harm in competing ethical claims, equal distribution of the negative consequences of competing ethical claims, and the importance of compensating on an equal order for negative consequences.

Anthropocentrism

Some environmentalists argue that utilitarianism is anthropocentric because most proponents focus on a rule or action's human benefits and see the value of inanimate objects in their usefulness to humanity (use or instrumental value) rather than their own intrinsic value. Anthropocentrism is a human-centered ethical system. Anthropocentrism is the view that humans are the most important or valuable species in the world as determined by their place at the top of the evolutionary ladder, their technological and cultural development, and their advanced civilizations. Anthropocentrists argue that ethics apply to human interaction with the natural world because nature has important human utilitarian value.

Anthropocentrism posits that the greater good should emphasize human utility and maximization of the natural world's resources based on its contributions to this utility. Other environmental philosophies note, however, that the same technological and cultural developments that produced modern civilization also threaten its long-term survival, or what environmentalists term "sustainability." Technological developments in business, home, transportation, communication, and other areas combined with population growth have required ever-increasing amounts of natural resources and have resulted in environmental exploitation, degradation, and possible global climate changes.

Thus, a strictly utilitarian environmental philosophy based on anthropocentric beliefs is not sustainable in the evolutionary long run if the natural resources needed for basic survival are depleted.

Some environmental philosophers counter that anthropocentrism is counter to modern biology, which posits that there are no superior species. They also feel that utilitarianism can lead to the unethical treatment of animals or other sentient beings when they are considered primarily as food sources or other human uses without considering their intrinsic value as part of the natural world. Opponents also argue that strict anthropocentrism does not account for the fact that animals as sentient beings suffer pain from human activities such as hunting, fishing, or scientific animal research experiments. Environmentalists blame such anthropocentric thinking as a leading cause of the environmental degradation and threats affecting the Earth.

Other Philosophies

Environmentalists who oppose utilitarianism because of the belief in its anthropocentric basis have proposed other philosophies to replace it, such as environmentalist Paul Taylor's development of biocentrism. Biocentrism is based on the tenet that the entire animate natural world is an interdependent system of which humans are an important part but not of sole importance, that it has intrinsic value independently of its human usefulness, and that humans as a species are equal to all other species rather than superior, as anthropocentrism posits. Taylor also posited a series of ethical principles to guide human conduct. The first principle stated that any natural entity with its own intrinsic value should not be harmed, although he excluded inanimate natural objects. The second principle stated that humans should not interfere with the normal operation of a natural entity or ecosystem, and the third principle stated that humans should not deceive any animal capable of such deception, such as a fish baited to bite a worm on a hook.

Other environmental philosophers posit what is called the greater value assumption, which states that humans have more value than animals but that animals do have intrinsic value that must be considered. Ecocentrism is an environmental philosophy that recognizes humans as a component of Earth's living and inanimate ecosystems and their dependence on the interactivity between these ecosystems for survival. Ecocentrism argues that ethics apply to human interaction with the natural world because of nature's intrinsic value, not merely its usefulness to humans. They also argue that humans are not the only beings or objects that possess intrinsic value and that their intrinsic value is not greater than that of other natural beings or objects.

Ecocentrism is a nature-centered ethical system. The founder of ecocentrism was Aldo Leopold, who wrote on what he termed the "land ethic." Other proponents include Arne Naess and George Sessions, founders of deep ecology, and Ted Mosquin and Stan Rowe, authors of *A Manifesto for Earth*. The manifesto's core principles included realization of the central value of the ecosphere to humanity, the need for integrity to maintain the creativity and productivity of its ecosystems, the evidence found in natural history to support an ecocentric worldview, the centrality of awareness of the human place in the natural world to ecocentrism, the valuation of the diversity of ecosystems and cultures, and the link between ecocentrism and social justice. The manifesto's action principles include the defense of Earth's creative potential, the reduction of the global population and human consumption of natural resources, and the promotion of ecocentrism in politics and public awareness of the ecocentric philosophy.

The environmental movement known as ecofeminism offers yet another approach, claiming that utilitarianism and its counter-environmental philosophies are based on the patriarchal domination inherent in Western culture, which includes man's dominion over nature and historical views of nature as a wilderness to be tamed according to human prescriptions. Ecofeminism emphasizes human connectedness with the environment, basing ethical interactions with the natural world on a more nurturing footing.

Utilitarians counter such arguments with the knowledge that humans rely on the usefulness of the natural world for survival, arguing that some environmental philosophies do not account for this inherent need. For example, hunting and fishing for food sources, the control of plants and animals that threaten human life, and the use of plants and animals to develop medicines such as insulin and penicillin are essential activities for human survival. Others counter on the basis that human intelligence and knowledge does place them apart and gives them the added obligation of responsible stewardship of the natural world, of which they are an essential component. They call for humans to be good stewards of the natural world through the adoption of actions and philosophies that promote environmental sustainability.

Anthropocentrism can lead to the view that natural resources with human uses are vulnerable to human exploitation, which has led to environmental destruction, possible climate change, and the endangerment and extinction of a number of plants and animals. These results ultimately threaten human survival. What some environmental philosophers term "speciesism" can also affect how humans make ethical judgments pertaining to human interactions with animals. For example, much of the medical and consumer-product testing that is allowed on animal subjects is not permitted on human subjects, and medical trials involving humans are more strictly regulated. Another common example is what some environmentalists feel is the mistreatment of animals in the modern transnational agribusiness systems, such as beef cattle confined to tight feedlots in an effort to produce more tender meat.

Environmental utilitarians counter that most environmental philosophies are necessarily anthropocentric to a certain degree, as humans must utilize the world's natural resources in order to survive. Such arguments hark back to Charles Darwin's theory of evolution and the ultimate utilitarian goal of survival of the species. They feel that many modern environmental philosophies either ignore or minimize this key need and must be modified to incorporate it. The ultimate goal of environmentalism is survival of the natural planet so humans can maintain their place within it.

See Also: Anthropocentrism Versus Biocentrism; Intrinsic Value Versus Use Value; Utilitarianism.

Further Readings

Attfield, Robin and Andrew Belsey. *Philosophy and the Natural Environment*. New York: Cambridge University Press, 1994.

Baron, Jonathan. *Against Bioethics*. Cambridge, MA: MIT Press, 2006.

Beckerman, Wilfred and Joanna Pasek. *Justice, Posterity, and the Environment*. New York: Oxford University Press, 2001.

Beckmann, Suzanne C., William E. Kilbourne, et al. *Anthropocentrism, Value Systems, and Environmental Attitudes: A Multi-National Comparison*. (1997). http://e-archivo.uc3m.es/bitstream/10016/8451/1/anthropocentrism_pardo_1997.pdf (Accessed January 2011).

Branch, Michael P. and Scott Slovic. *The ISLE Reader: Ecocriticism, 1993–2003*. Athens: University of Georgia Press, 2003.

Callicott, J. Baird. *Beyond the Land Ethic: More Essays in Environmental Philosophy*. Albany: State University of New York Press, 1999.

Callicott, J. Baird. *In Defense of the Land Ethic: Essays in Environmental Philosophy*. Albany: State University of New York Press, 1989.

Dasgupta, Partha. *Human Well-Being and the Natural Environment*. New York: Oxford University Press, 2001.

Dobson, Andrew. *Green Political Thought*. London: Routledge, 2007.

Elliot, Robert. *Environmental Ethics*. New York: Oxford University Press, 1995.

Faber, Daniel. *The Struggle for Ecological Democracy: Environmental Justice Movements in the United States*. New York: Guilford Press, 1998.

Gillroy, John Martin and Joe Bowersox. *The Moral Austerity of Environmental Decision Making: Sustainability, Democracy, and Normative Argument in Policy and Law*. Durham, NC: Duke University Press, 2002.

Goldfarb, Theodore D. *Taking Sides: Clashing Views on Controversial Environmental Issues*. Guilford, CT: McGraw-Hill/Duskin, 2001.

Gruen, Lori and Dale Jamieson. *Reflecting on Nature: Readings in Environmental Philosophy*. Oxford, UK: Oxford University Press, 1994.

Jamieson, Dale. *Ethics and the Environment: An Introduction*. New York: Cambridge University Press, 2008.

Johnson, Lawrence E. *A Morally Deep World: An Essay on Moral Significance and Environmental Ethics*. New York: Cambridge University Press, 1991.

Lindahl Elliot, Nils. *Mediating Nature*. London: Routledge, 2006.

O'Neill, John, Alan Holland, and Andrew Light. *Environmental Values*. New York: Routledge, 2008.

Postma, Dirk Willem. *Why Care for Nature? In Search of an Ethical Framework for Environmental Responsibility and Education*. Dordrecht, Netherlands: Springer, 2006.

Sarkar, Sahotra. *Biodiversity and Environmental Philosophy: An Introduction*. New York: Cambridge University Press, 2005.

Scriven, Tal. *Wrongness, Wisdom, and Wilderness: Toward a Libertarian Theory of Ethics and the Environment*. Albany: State University of New York Press, 1997.

Squatriti, Paolo. *Nature's Past: The Environment and Human History*. Ann Arbor: University of Michigan Press, 2007.

Stevis, Dimitris and Valerie J. Assetto. *The International Political Economy of the Environment: Critical Perspectives*. Boulder, CO: Lynne Rienner, 2001.

Taylor, Paul W. *Respect for Nature: A Theory of Environmental Ethics*. Princeton, NJ: Princeton University Press, 1986.

Traer, Robert. *Doing Environmental Ethics*. Boulder, CO: Westview, 2009.

Warren, Karen and Nisvan Erkal. *Ecofeminism: Women, Culture, Nature*. Bloomington: Indiana University Press, 1997.

Wenz, Peter S. *Environmental Ethics Today*. New York: Oxford University Press, 2001.

Marcella Bush Trevino
Barry University

V

Veganism/Vegetarianism as Social Action

In India, avoiding meat is associated with the Brahmin class. This vegetarian Indian meal consists of fruit and vegetable salad sprinkled with chaat masala, and poppadoms and chutney.

Source: iStockphoto

Meat consumption has increased steadily throughout the years. For example, the United Nations Food and Agriculture Organization (FAO) has reported that in 1970, meat consumption in East Asia was 100 kilocalories per person per year; this figure rose to about 400 kilocalories in 2000 and is projected to rise to 650 kilocalories in the year 2050.

Yet there has been a perennial countertrend to meat consumption. As early as 1908, the International Vegetarian Union was formed to specifically promote vegetarianism worldwide. Indeed, the first known vegetarian society—the British and Foreign Society for the Promotion of Humanity and Abstinence from Animal Food (the precursor for the United Kingdom's Vegetarian Society)—was formed even earlier, in 1843. The two main reasons for establishing vegetarian societies and promoting vegetarianism are related to health and to the promotion of a more humane relationship with animals. Beyond health and animal welfare reasons, historically, vegetarianism in many societies is a marker of, among other things, culture, ethnicity, religion, and even class status. For example, followers of Jainism are all vegetarians, and a significant proportion of them are vegans (who not only avoid meat but all animal by-products as well). A significant proportion of high-caste Hindus are vegetarians as well.

Contemporary vegetarian societies have to constantly navigate these markers as they advance their goal of reducing meat consumption among the population. Overcoming these cultural-historical markers of vegetarianism is imperative if activists want their message to reach more people. Put simply, vegetarianism advocacy is essentially a form of social action whose goal is to realize a considerable lifestyle change in the mainstream population. This is by no means easy because a person's dietary choice is both personal and grounded in particular social and cultural structures.

Vegetarianism in Context

India offers an apt context to discuss the idea of vegetarianism as a form of socio-political action. Vegetarianism in India is rooted in the religious concept of *ahimsā* (nonviolence); contrary to popular belief, not all Hindus are professed vegetarians. Nonetheless, meat avoidance has enjoyed a privileged status in Indian society because of its association with the purity of the Brahmin class. In that sense, the lower castes are said to have turned vegetarian in part to enhance their social status. Moreover, at the broadest level, vegetarianism has been appropriated by Hindu nationalists to distinguish themselves from other meat-eating minorities in India (most notably, the Muslims). As evidence of such dietary-based divisive lines being drawn, there have been incidents reported in the Indian press of localized politics where vegetarians pressured supermarkets to stop selling meat.

In Western societies, vegetarianism is much less associated with a particular religion or class. The consumption of meat, however, has long been linked to masculinity and patriarchal power. In that sense, vegetarianism has been consistently portrayed as a countercultural practice that is aligned with the animal rights and feminist movements through a common discourse of "oppression." Indeed, some feminists have argued that meat is a symbol and celebration of male dominance and hence meat consumption simultaneously reinforces patriarchal values. Therefore, all feminists must necessarily be vegetarians.

The proliferation of vegetarian societies around the world attests to the fact that activism and education on meat reduction and avoidance is ever present. The specific politics of vegetarianism activism are, however, played out differently in different places and at different scales.

Vegetarianism and Social Action

Large-scale events that are held to promote vegetarianism for health reasons are generally collegial and nonconfrontational. For example, Meatout, a result of grassroots social activism, is held the first day of spring in the United States to educate communities, friends, and families to reduce meat consumption. First organized in 1985, the nationwide event aims also to persuade people that a vegetarian diet is more wholesome.

Vegetarianism advocacy is somewhat unique compared with other socio-environmental movements in that its members are often expected to make visible, material changes in their lifestyles (to a meat-free diet). Individual vegetarians and vegans also negotiate a very personal politics at a day-to-day level, for example, when they are in social gatherings where food is served. In their frequent interactions with families, friends, and colleagues about their dietary choices and explaining to them their reasons for adopting a meat-free lifestyle, vegetarians and vegans are essentially politicizing vegetarianism in subtle ways. Put another way, they are exemplifying the fact that "the personal is political" and that being vegetarian is arguably a form of embodied social action that affects the immediate social sphere of vegetarians. Indeed, research has shown that people generally change their diets through personal social interaction with already committed vegetarians.

There have also been city-level efforts to promote vegetarianism through the symbolic declaration of a Meatout Day (or Veggie Day) each week. The earliest city to adopt this was the Belgian city of Ghent, where each Thursday is designated as a Veggie Day. Other cities that have adopted a meat-free day each week include Cape Town in South Africa, Bremen in Germany, and São Paulo in Brazil. While it is impossible to legally enforce such a gesture, it is to the credit of vegetarianism activists to have successfully launched such high-profile socio-political campaigns amid skepticism from various quarters (e.g., the restaurant industry). Indeed, there are signs that such Meatout initiatives are increasingly replicated in universities and in private companies.

An increasingly important realm for vegetarianism activism is the meat commodity chain where livestock producers have marketed meat consumption as a viable and enviable dietary choice. A good example is the pork producers in the United States, who have for years advertised pork as the "other white meat." Similarly, the attempts by these producers to sell leaner cuts of meat are aimed at people who are wont to reduce meat consumption for health reasons. Such attempts, however, are unlikely to sway people who object to meat consumption for environmental and ethical grounds. For example, the FAO's 2006 landmark study "Livestock's Long Shadow," which revealed that the livestock sector releases 18 percent of greenhouse gas emissions (measured in carbon dioxide equivalent), has provided good fodder to revalorize vegetarianism advocates' environment-based objections to meat consumption. Animal advocacy groups like People for the Ethical Treatment of Animals (PETA) have continually exposed the cruelty prevalent in the meat production process through undercover tactics. For advocates, such social-political exposés highlight the fact that animal cruelty is collateral to the modern production of cheap meat; it compels ordinary consumers to think of their culpability in the meat commodity chain.

Conclusion

In a recent study based on 29 countries representing 54 percent of the world's population, it is extrapolated that there are 75 million vegetarians by choice in the world and many more million vegetarians by necessity. The latter are defined as consumers who are too poor to afford meat but would likely consume meat once their income level improves. There is evidently much more work to be done for vegetarianism advocates. Through their social actions that are appropriate to specific countries and regions, the number of vegetarians will likely increase in the years to come, thereby tempering the projected increases in meat consumption around the world.

See Also: Animal Ethics; Animal Welfare; Green Anarchism; Organic Consumerism.

Further Readings

Beardsworth, Alan and Teresa Keil. "The Vegetarian Option: Varieties, Conversions, Motives and Options." *The Sociological Review*, 40:2 (1992).

Gaard, Greta. "Vegetarian Ecofeminism: A Review Essay." *Frontiers: A Journal of Women Studies*, 23:3 (2002).

Jabs, Jennifer, Carol Devine, and Jeffery Sobal. "Model of the Process of Adopting Vegetarian Diets: Health Vegetarians and Ethical Vegetarians." *Journal of Nutrition Education*, 30:4 (1998).

Klein, Jacob. "Afterword: Comparing Vegetarianisms." *South Asia: Journal of South Asian Studies*, 31:1 (2008).

Leahy, Eimear, Sean Lyons, and Richard Tol. "An Estimate of the Number of Vegetarians in the World." Working Paper No. 340. Economic and Social Research Institute, Dublin, Ireland, March 2010.

Maurer, Donna. "Meat as a Social Problem: Rhetorical Strategies in the Contemporary Vegetarian Literature." In *Eating Agendas: Food and Nutrition as Social Problems*, Donna Maurer and Jeffery Sobal, eds. Berlin, Germany: Aldine De Gruyter, 1995.

Steinfeld, Henning, et al. *Livestock's Long Shadow: Environmental Issues and Options*. Rome: Food and Agricultural Organization of the United Nations, 2007.

Harvey Neo
National University of Singapore

Vertical Policy Integration

Policy integration implies the incorporation of "new" objectives into existing sectoral policies. It is seen as a strategy for "greening" other policy domains by including environmental concerns. However, relatively autonomous policy sectors obstruct such integration attempts. Hence, policy integration, including different strategies to it, is a frequently underestimated challenge.

Policy integration may be referred to as the incorporation of specific policy objectives that are extrinsic to a policy domain into existing sectoral policies. For example, might gender aspect be integrated into research policy or might agricultural policy be greened by integrating environmental objectives, such as biodiversity conservation and reductions of pesticides? Policy integration is frequently requested for making public policy more coherent. This claim is not new, and the idea of policy integration dates back to the 1970s' discussions on cross-cutting policies. The issue was largely discussed in 2010 under the term *environmental policy integration*. Still other objectives also are being discussed in the context of policy integration, for example, rural development, gender, food safety, and freshwater conservation, which should be included into strong sectors' policies such as, for example, agricultural, energy, transport, or innovation policy.

Existing interpretations of policy integration are diverse and partly describe the phenomenon either as a process, a political result, or a mixture of both. The former comprises the management of cross-cutting issues that surpass established policy sectors and organizational responsibilities. The output of policy may be considered integrated where the policy elements are in accord with each other. Viewed as a conflicting organizational process, policy integration can be accomplished when various public organizations work together and do not produce either redundancy or gaps in services.

Debates in Vertical Policy Integration

In order for policy integration to be effective, the extrinsic objectives need to be formally included into other sectors' policies, and they need to be actually implemented. Both aspects must be seen as serious challenges for policy integration. The literature distinguishes between two major political strategies for realizing integration: vertical and horizontal policy integration (→ *horizontal policy integration*).

The term *vertical policy integration* is used in two ways. First, a traditional notion refers to attempts of an advocate of extrinsic policy objectives to incorporate these into a specific sectoral policy. Emphasis lies on procedures and mechanisms for effective implementation at all political levels within the respective sector. Here, a single department takes an intrasectoral approach to the integration of policy objectives. For example, after realizing that horizontal policy integration indeed leads to the formal uptake of environmental objectives into agricultural and transport policies, but still its implementation has not been achieved, the environmental agencies pay due attention to the vertical process of policy implementation within the sectors. Ensuring proper procedures and instruments within the sector's policy is a main focus under this notion, trying to secure implementation at all tiers of government and subsequent on-the-ground effects.

A second interpretation of the term refers to the assignment for incorporating extrinsic policy objectives into multiple sectors' policies by a high-ranking governmental body, such as the cabinet or the parliament. Here "the government" or parliament in a strategic manner instructs relevant departments to include new objectives into their policies. Implementation is monitored through mandatory mechanisms for reporting. Examples include the recent strategy processes in Western countries such as national biodiversity or sustainability strategies, which are commissioned and monitored by the cabinet and implemented by sectors such as agriculture and nature conservation.

Pros and Cons

The general idea of policy integration recently gained momentum due to two reasons. On the one hand, political issues are of a cross-cutting character. Environmental, energy, and growth policies have multiple dimensions and cut across different sectors of the economy. On the other hand, this change in the problem structure of political issues goes along with changes in policy approaches to them. Traditional sectoral and uncoordinated approaches such as agricultural policy have been criticized as being ineffective or inconsistent with the holistic idea of sustainable development. In these cases, policy integration is expected to increase the efficacy and coherence of policies.

Although the concept of policy integration enjoys high political acceptance, realization in political practice encounters significant resistance due to two major obstructs. First, the integration of political objectives such as environmental goals, which are extrinsic to a given policy field, or such as agricultural or energy policy, disputes with the logic of a sectorally designed government, which follows the concept of rational and effective public administration. In this conception, a number of separate government departments or ministries and other public agencies are charged with delivering public policy. Overlaps of responsibilities are avoided, and consequently each department develops a relatively autonomous policy sector, which delivers highly specialized, functional, and vertically organized policy through separate administrations. Second, policy integration goes against the economic interests of actors surrounding these relatively autonomous administrations. The extrinsic policy objectives often conflict with vested interests of their industrial clientele. For example, will agricultural stakeholders, such as farmers' associations, influence the ministry of agriculture and its agricultural policy much more effectively than would, for example, environmental groups, because they share specific material interests that are touched upon by environmental claims?

Both interpretations of the vertical approach criticize the mere formal integration and stress the need to also effectively implement the new objectives for creating impact on the ground. In doing so, they go beyond the technocratic notion of horizontal integration and make visible additional bottlenecks obstructing integration impact. In particular, the second interpretation recognizes the need for high-level backing in the course of any attempt at integration. The examples of high-level national sustainability and biodiversity strategies illustrate this backing, which seems to be crucial for changing, for example, the yield-maximizing practices of farmers that are supported by agricultural policy.

According to the former and rather technocratic understanding, however, integration seems to be mainly a question of proper procedures, which—following a rather linear model—merely need to be developed and applied. Yet it acknowledges that the process of policy implementation as such is crucial for the success of integration. This notion is of an implementation-oriented character that takes for granted the formal integration of a supposed rational objective into sectoral policies (→ *horizontal policy integration*) and assumes that this could be reached without conflict.

The latter and more realistic interpretation acknowledges the discretion among existing policy sectors and addresses the integration task at both the political as well as the administrative branch of departments and government. It comprises elements of the traditional notion of vertical integration as well as aspects of horizontal integration.

Merging these two understandings leads to comprehensive vertical integration, which refers to attempts of an advocate of extrinsic policy objectives to incorporate these into other sectors' policies, which are backed and further legitimized by the plea of a high-ranking political body and which pay due attention to procedures and mechanisms for effective implementation at all political levels.

Vertical environmental integration may currently be observed in many Organisation for Economic Co-operation and Development (OECD) countries. Still, environmental departments are facing the challenge of coordinating diverse sectoral policies. But they now enjoy the political support of their governments or parliaments as high-ranking political bodies, instructing the integration. Examples are the national biodiversity and sustainability strategies and comprehensive rural development plans at national and subnational levels of European Union member states. The degree to which this may be considered comprehensive vertical integration, however, varies and must in general be seen skeptically because the fundamental conflicts of interests have not changed yet, as did the prevailing policy paradigm that currently favors integration.

See Also: Horizontal Policy Integration; Individual Action Versus Collective Action; Sustainable Development (Politics).

Further Readings

Briassoulis, Helen. *Policy Integration for Complex Environmental Problems—The Example of Mediterranean Desertification.* Burlington, VT: Ashgate, 2005.

Giessen, Lukas, et al. "Politikintegration für ländliche Räume? Die (Nicht-) Koordination der Förderung [Rural Policy Integration? The (Non-) Coordination of Funding Programs]." In *Land–Stadt Kooperation und Politikintegration für ländliche Räume* [Rural–Urban Cooperation and Rural Policy Integration], Sebastian Elbe, ed. Aachen, Germany: Shaker, 2008.

Jänicke, Martin and Helge Jörgens. "Neue Steuerungskonzepte in der Umweltpolitik [New Concepts for Political Steering in Environmental Policy]." *Zeitschrift für Umweltpolitik und Umweltrecht* [Journal of Environmental Policy and Environmental Law], 3 (2004).

Peters, B. Guy. "Managing Horizontal Government—The Politics of Coordination." *Public Administration*, 76:2 (1998).

Lukas Giessen
University of Göttingen

WIND POWER

Wind, created from uneven energy and temperatures, powers turbines on wind farms, which are connected to a power collection system and communications network.

Source: Joshua Winchell/U.S. Fish and Wildlife Service

Wind power, the conversion of wind into a form of energy, is an alternative power source. As concern has increased regarding the effects of fossil fuels and other energy sources on the environment, interest in alternative fuels and energy options has grown. Many alternative forms of energy have been explored, including use of ethanol as a fuel, solar panels, and biofuels. Wind power, however, has attracted a great deal of interest and investment due to its reliability, efficiency, and the promise of minimal effects on the environment. Wind power has also been championed for its ability to create jobs for those who construct and maintain windmills and other related infrastructure. Financial, ecological, and other reasons have caused wind power to become an area of interest for those who are interested in green technology and who favor sustainable development initiatives. Some observers, however, have questioned the cost-effectiveness of wind power as well as the consequences on open land held by farmers and other landowners located near wind farms.

Wind Energy

Wind energy is derived from multiple sources. The uneven distribution of heat from the sun creates an atmospheric convection system. "Convection" refers to the movement of liquids and gases as the major mode of heat transfer. The sun distributes heat in such a way that both the north and south poles receive less energy than does the equator. In addition to this uneven distribution of energy, dry land heats up and cools down in a process quicker than do oceans and other large bodies of water. This differential in heating, in addition to the rotation of the planet, causes wind. Wind velocity varies based on location. Some locations experience more wind velocity than others located nearby. Because of this variation, the power of wind in a given area is not the sole indicator of the amount of energy a wind turbine could produce. Placement of the turbine is also an important factor. Because the majority of wind power is generated by higher wind speeds, the greater proportion of the energy is created by short bursts of wind.

Wind Farms

Wind farms use turbines—large rotary devices—to extract energy from the wind. Specifically, wind farms are composed of individual turbines that are interconnected to a power collection system and communications network. Substations increase the electric current voltage using a transformer to connect to the greater electric power transmission system. Generators used for wind power are generally induction generators, which generate electrical power when the shaft rotates faster than the normal frequency of the equivalent induction motor. Induction motors are popular for generating wind power due to their ability to produce power and varying rotor speeds. These generators are not self-exciting, meaning that the energy produced stems from an external force. In the construction of wind farms, this makes the substations highly important because they tend to include substation capacitor banks for power factor correction. All of these mechanisms assist in the penetration that occurs with wind energy. Wind energy penetration refers to the portion of energy produced by wind compared with the total available generation capacity. Although the potential for wind power as a production of power has been examined, there is no widely accepted maximum level of wind penetration.

Approximately half of the energy generated by wind farms is generated as the result of only 15 percent of the operating time of the turbines. Despite the efficiency of a single turbine or wind farm, wind power does not have the same consistent output as fuel-fired plants. As a result, utilities that use wind power must allocate power from initiating energy generation to compensate for times of weak wind power. Because of this, wind power is currently used primarily as a fuel saver rather than as a capacity saver. Proposed innovations that might assist in increasing the amount of energy derived from wind power include use of strong transmission lines to link wind farms, which would ensure that spare capacity is available in case of a failure or shortage on any other part of the network.

Power Generation Variability

Electricity generated from wind power varies by location, but also over time. Wind power's lack of consistency measured over hours and days is even more significant when examined over different seasons of the year. In determining the output of wind farms, conventional methods of scheduling such as used in other electricity sources are used. For wind power in particular, forecasting methods are used, but there remains a lack of predictability in the output of wind plants. Because of this unpredictability, concerns have arisen about the incorporation of wind power into a grid system. Electrical generation and consumption must remain in balance to assist the current grid system's stability. Wind power's lack of consistency has the potential to increase the costs of operation. For example, utilities are concerned about incremental operation reserve, which is the capacity available to the system within a short period of time in the event a generator goes down or other problems arise. Incremental operation reserve also assists in dealing with unexpected increases in energy demand. Increasing wind power capacity results in a corresponding need for increased incremental operation reserve. This is because although wind power can be replaced during periods of low wind, electric transmission networks must also account for outages at generation plants and daily shifts in electrical demands. Consequently, systems that rely on a larger percentage of wind power may need more conventional plants operating at less than full load to deal with this uncertainty.

Environmental Concerns

By traditional standards, wind power has a relatively minor effect on its surroundings. Over time, a wind turbine offsets energy used in its initial creation. Energy used for installation of turbines is also offset over time, with most estimates concluding this occurs within two and a half years of a turbine's operation. Although wind power does not produce emissions or use fossil fuels, there are certain dangers to the environment from the use of wind turbines. For many, the danger turbines cause to birds and bats is of great concern. It is estimated that between 10,000 and 40,000 birds die annually as a result of collisions with wind turbines in the United States. Fears persist that increased wind capacity will cause that number to rise. Proponents of wind power contest these estimates, however. Wind power advocates assert that the number of birds killed by wind turbines is low compared to death rates caused by other electrical generating sources. Birds and bats are also placed at risk by other environmental impacts caused by the use of non–clean power sources, such as air pollution and other toxic emissions.

Future Growth

As of 2009, the normal maximum output of wind power generating sources was 159.2 gigawatt-hours (GWh), while the total energy production was 340 terawatt-hours (TWh), which represents approximately 2 percent of worldwide electricity usage. This production has doubled since 2006. Currently, nearly 80 countries use wind power on a commercial basis. Large-scale wind farms are connected to the electric power transmission network, commonly referred to as high-voltage electric transmission. High-voltage electric transmission is the bulk transfer of energy from generating power plants to substations near population centers. As an economic incentive, utility companies increasingly are buying back surplus electricity produced by small domestic turbines, as these can supply energy to the electric grid, offsetting the consumption of others. The total amount of extractable wind power is substantially greater than present use from all sources. Estimates have stated that 72 TWh of wind power on the entire planet might become commercially viable. As a result, interest in wind power can be expected to increase in the future.

See Also: Biofuels (Business); Carbon Emissions (Personal Carbon Footprint); Hydroelectric Power; Nuclear Power (Energy); Solar Energy (Cities).

Further Readings

Burton, T., D. Sharpe, N. Jenkins, and E. Bossanyi. *Wind Power Handbook*. Hoboken, NJ: John Wiley & Sons, 2001.

Gipe, P. *Wind Power: Renewable Energy for Home, Farm, and Business*, 2nd ed. White River Junction, VT: Chelsea Green, 2004.

Tester, J. W., E. M. Drake, M. J. Driscoll, M. W. Golay, et al. *Sustainable Energy: Choosing Among Options*. Cambridge, MA: MIT Press, 2005.

Stephen T. Schroth
Jordan K. Lanfair
Knox College

Y

YUCCA MOUNTAIN

Located in Nevada, Yucca Mountain was chosen in 1987 as a nuclear waste repository, but no waste has been stored there yet as of 2010.

Source: Wikimedia

Eight years after the 1945 atomic bombings of Hiroshima and Nagasaki, in his "Atoms for Peace" speech, President Dwight Eisenhower lauded the "miraculous inventiveness" of humans and pledged that the United States would develop peaceful applications of nuclear power. In 1955, Arco, Idaho, became the first town in the United States to be powered by nuclear energy; today, there are more than 100 nuclear power plants in the United States, and intense debates over issues such as nuclear waste disposal are widespread. Yucca Mountain, located in Nevada, was chosen by Congress in 1987 as a suitable site for a nuclear waste repository—and debate over the site has raged ever since. Originally scheduled to begin accepting waste in 1998, as of December 2010, no waste was yet stored at Yucca Mountain.

Politics of Yucca Mountain

The 1982 Nuclear Waste Policy Act (NWPA) made the Department of Energy (DOE) responsible for siting, building, and operating a permanent underground nuclear waste repository; in 1987, Congress amended the NWPA and directed DOE to study only Yucca Mountain. Yucca Mountain was chosen, in part, because of its location—a fairly remote, sparsely populated, federally owned tract of land. While this dismayed residents of Nevada, other states welcomed the choice, particularly South Carolina, Washington, and several other states, all of which were storing millions of gallons of liquid nuclear waste and tons of spent fuel rods, the result of nuclear weapons production. From the beginning, then, the Yucca Mountain project has had a political dimension, as states that built atom bombs during the Cold War have been promised since 1987 that the massive amounts of nuclear waste they store would be transported to Yucca Mountain. In preparation for this, the federal government has spent more than $10 billion to develop the Yucca Mountain repository; the funding has come from a surcharge paid by nuclear power customers, as required by the NWPA. In 2002, President George W. Bush signed the Yucca Mountain Development Act, allowing the DOE to move forward on the Yucca Mountain project. Still, fierce debate has continued over the site and, during the 2008 election, Barack Obama stated his opposition to the project.

In March 2010, the DOE filed a motion with the Nuclear Regulatory Commission (NRC) to withdraw the license application for the Yucca Mountain repository. Although this move delighted those opposed to the repository, the celebration was short-lived because in June 2010, the NRC ruled that the DOE could not withdraw its application. The NRC's three-member panel of judges ruled that because Congress had designated Yucca Mountain to be a nuclear waste repository, the president and the DOE lacked the power to unilaterally close the repository. Ending the Yucca Mountain project, the judges ruled, would require another act of Congress. As of December 2010, the DOE had appealed the NRC's decision and was awaiting review by the NRC's five-member board, which has the authority to overturn rulings made by NRC judges.

South Carolina legislators and others are bitter about the halting process of the project and about the Obama administration's moves to shut it down, arguing that their states will be saddled with nuclear waste indefinitely if the Yucca Mountain project does not move forward. These legislators are particularly concerned about the response of their constituents should the project be discontinued because they have been paying the surcharge designated by the NWPA, which has generated more than $32 billion. In response to the Obama administration's declared intent to shut down the project, some legislators introduced bills into Congress that would give nuclear utility customers rebates from the Yucca charges. Further, the president's budget for 2011 proposes no money at all for the project; thus, Congress would have to appropriate hundreds of millions of dollars a year for the DOE to continue to pursue the repository.

The Nuclear Energy Institute, which is the nuclear industry's trade association, several Nevada counties, and the National Association of Regulatory Utility Commissioners all oppose termination of the project. In response to concerns about the reactions of electricity consumers whose surcharges have funded the Yucca Mountain project, state commissioners have asked that payments to the fund be suspended until a final decision is made regarding the project. Despite his opposition to the repository, President Obama has established a commission to study recycling and reusing nuclear waste. Although recycling and reusing the nuclear waste would reduce the number of repositories needed, at least one would still be required because national policy dictates that nuclear waste be buried.

Yucca Mountain Repository Makes Sense

Those who support building the repository cite the following points in favor of continuing the project:

- On-site storage such as in South Carolina and other states is a temporary measure, and a permanent solution such as Yucca Mountain must be embraced.
- Yucca Mountain is remote and sparsely populated.

- Yucca Mountain is federally owned.
- Yucca Mountain is located in the southern Great Basin, so if there were leakage of nuclear waste, it would be contained within the basin.
- The planned safety measures ensure that the nuclear waste will be contained for thousands of years.
- The United States has already spent more than $10 billion on the project; halting it would waste those billions of dollars and years of research spent on environmental, engineering, and safety plans for the repository.

In addition to holding waste generated by nuclear weapon manufacture, Yucca Mountain would also take in waste from every nuclear reactor in the United States. Champions of the project argue that consolidating the nation's nuclear waste makes security much easier; now, nuclear waste is distributed among 120 temporary storage facilities in 39 states. Finally, supporters point out that the DOE has determined that seismic and tectonic activity at Yucca Mountain will not compromise the repository's performance.

Yucca Mountain Repository Is Unsafe and Unfair

Opponents of the Yucca Mountain repository cite these broad reasons for their resistance:

- It is unfair to Nevada, which does not have any nuclear power plants.
- Located only about 100 miles from Las Vegas, Yucca Mountain poses the potential of drastic harm to that city's tourism industry.
- Transporting nuclear waste is dangerous; it would have to be shipped from across the United States via trucks and trains, and this constant transportation dramatically increases the chances of a radioactive spill or of terrorist attacks on the moving waste.
- Its location holds the potential for earthquakes and volcanoes; there is less concern over a major seismic event than there is over the long-term effects of small events that could lead to increased fractures and, eventually, water leaking into the containment areas, which would cause corrosion and leaking of the nuclear waste.
- It is impossible to predict what might happen over the next 100, 1,000, or 1 million years; models, at best, offer only guesses as to how safe the Yucca Mountain repository would be.
- Yucca Mountain is designed to hold 77,000 tons of nuclear waste materials, and there are already nearly 50,000 tons in temporary storage; thus, the site would reach capacity around 2036, at which time the country would again have to deal with the issue of creating another repository.

Since the site was chosen in 1987, citizens of Nevada as well as other concerned parties from across the United States have fought the Yucca Mountain project bitterly, both in court and in Congress. Despite the DOE's assurances, scientific concerns about the site have arisen since then, once it was understood that water flows through Yucca Mountain faster than was initially supposed, which raises concerns about nuclear waste seeping into the water table. The waste containers are predicted to break down in tens of thousands of years, at which time waste could feasibly infiltrate the water table through the many fractures present within Yucca Mountain. Critics believe they have reason to question whether Yucca Mountain has been rigorously and carefully investigated, citing the 2007 discovery of a major fault line that runs beneath the facility, hundreds of feet from where it was thought to be located. The fault line is directly beneath a storage pad where canisters of nuclear waste would be cooled before being sealed into tunnels. Further, Holtec International, a nuclear equipment supplier, has criticized the DOE's safety plans for handling containers of waste prior to their being buried; the canisters would be unanchored and subject to movement should they be disturbed in any way.

The debate over the Yucca Mountain repository does not promise to be settled anytime soon. Questions regarding the safety of the site remain pertinent, and the anger of those states that have been paying to fund the project and that were promised they were storing nuclear waste only temporarily is justifiable. Nuclear energy, in general, remains a polarizing topic, and the Yucca Mountain project illustrates the complexity and volatility of these debates.

See Also: Not in My Backyard (NIMBY); Nuclear Power (Energy); Nuclear Power (Health); Social Action.

Further Readings

Eureka County (Nevada) Nuclear Waste Office. "FAQs." http://www.yuccamountain.org/faq.htm#character (Accessed September 2010).

Main Justice: Politics, Policy and the Law. "Liability for Scrapping Yucca Mountain Could Run in the Billions." http://www.mainjustice.com/2010/07/27/liability-for-scrapping-yucca-mountain-could-run-in-the-billions (Accessed September 2010).

McClatchy. "Judges Rule Obama Can't Close Yucca Mountain Nuclear Dump." http://www.mcclatchydc.com/2010/07/04/96995/judges-rule-obama-cant-close-yucca.html (Accessed September 2010).

Tani E. Bellestri
Independent Scholar

Green Issues and Debates Glossary

A

Air Pollution: Contaminants or substances in the air that interfere with human health or produce other harmful environmental effects.

Alternative Energy: Energy from uncommon sources such as wind power or solar energy, not fossil fuels. Alternative energy is usually environmentally friendly.

Anarchism: A nonhierarchical form of social relations that seeks to abolish the authority of the state and the power of large property owners.

Anthropocentric: Literally, "human-centered." Anthropocentrism is the belief, taken for granted in most cultures through most of human history and argued more explicitly by some schools of philosophy today, that humans are the figurative "center of the universe" and that ethical systems should thus be principally concerned with human benefit.

Anthropogenic: Man-made; used especially to underscore the human origins of a substance or phenomenon, as in "anthropogenic climate change" or "anthropogenic toxic compounds."

B

Behavioral Change: As it affects energy efficiency, behavioral change is a change in energy-consuming activity originated by, and under the control of, a person or organization. An example of behavioral change is adjusting a thermostat setting or changing driving habits.

Biodiversity: The total variety of life on Earth. Modern science considers biodiversity to be an inherently good thing for the ecosystem and for the loss of species and of species diversity to be an alarming consequence of environmental damage. From an evolutionary standpoint, genetic diversity—the diversity of genes within a species—is also especially important.

Biomass: Any organic matter that is available on a renewable basis, including agricultural crops and agricultural wastes and residues, wood and wood wastes and residues, animal wastes, municipal wastes, and aquatic plants.

Brownfields: Abandoned, idled, or underused industrial and commercial facilities/sites where expansion or redevelopment is complicated by real or perceived environmental contamination. They can be in urban, suburban, or rural areas. The U.S. Environmental Protection Agency's Brownfields Initiative helps communities mitigate potential health risks and restore the economic viability of such areas or properties.

C

Capitalism: A form of social relations in which markets are presumed to be the most efficient way of distributing goods or services.

Carbon Footprint: A popular term describing the impact a particular activity has on the environment in terms of the amount of climate-changing carbon dioxide and other greenhouse gases it produces. An individual's carbon footprint is the amount of greenhouse gases that his or her way of life produces overall. It is also a colloquialism for the sum total of all environmental harm an individual or group causes over their lifetime. People, families, communities, nations, companies, and other organizations all leave a carbon footprint.

Carbon Offsets: Financial instruments, expressed in metric tons of carbon dioxide equivalent, which represent the reduction of carbon dioxide or an equivalent greenhouse gas. Carbon offsets allow corporations and other entities to comply with caps on their emissions by purchasing offsets to bring their totals down to acceptable levels. The smaller voluntary market for carbon offsets exists for individuals and companies that purchase offsets in order to mitigate their emissions by choice. There is a great deal of controversy over the efficacy and truthfulness of the offsets market, which is new enough that, in a best-case scenario, the kinks have not yet been worked out, while in the worst-case scenario, it will turn out to be a dead end in the history of environmental reform.

Carbon Sequestration: The process by which carbon is taken from the atmosphere and stored in a carbon sink using underground reservoirs or biomass.

Certified Organic: Food products that meet or exceed standards set forth by the U.S. Department of Agriculture's National Organic Program (NOP). Products "made with organic ingredients" include 70 percent organic ingredients and cannot contain the organic label. "Organic" products must have at least 95 percent organic ingredients and may feature the USDA organic seal. "100% Organic" is the most stringent, but does not count water or salt.

Clean Coal: This refers to coal burning that attempts to capture some of the resulting carbon emissions. Typically, clean coal power plants require 30 percent more energy to operate the carbon capture equipment.

Climate Change: A term used to describe short- and long-term effects on Earth's climate as a result of human activities such as fossil fuel combustion and vegetation clearing and burning.

Common Property: A property or resource whose use can exclude others from using it, though it is held in common by the public.

Compost: A process whereby organic wastes, including food wastes, paper, and yard wastes decompose naturally, resulting in a product rich in minerals and ideal for gardening and farming as a soil conditioner, mulch, resurfacing material, or landfill cover. Consumers can make their own compost by collecting yard trimmings and vegetable scraps.

Conservation: Preserving and renewing, when possible, human and natural resources.

D

Dioxin: Any of a family of compounds known chemically as dibenzo-p-dioxins. Concern about them arises from their potential toxicity as contaminants in commercial products. Tests on laboratory animals indicate that it is one of the more toxic anthropogenic (man-made) compounds.

E

Energy: The capability of doing work; different forms of energy can be converted to other forms, but the total amount of energy remains the same.

Energy Star: A joint program formed between the U.S. Environmental Protection Agency and the U.S. Department of Energy to identify and label high-efficiency building products.

Entropy: A measure of the unavailable or unusable energy in a system; energy that cannot be converted to another form.

Environmental Equity/Justice: Equal protection from environmental hazards for individuals, groups, or communities regardless of race, ethnicity, or economic status. This applies to the development, implementation, and enforcement of environmental laws, regulations, and policies and implies that no population of people should be forced to shoulder a disproportionate share of negative environmental impacts of pollution or environmental hazard due to a lack of political or economic strength levels.

Epidemiology: Study of the distribution of disease, or other health-related states and events in human populations, as related to age, sex, occupation, ethnicity, and economic status in order to identify and alleviate health problems and promote better health.

Ethics: The study of moral questions. Ethics can refer to specific types of ethics (such as applied ethics or medical ethics) or to specific systems of ethics (such as Catholic ethics or Marxist ethics). Though the religions of the world always include an ethical dimension to their belief systems, ethics and religion are not coequal, and the term *secular ethics* is sometimes used to describe systems of ethics that derive their conclusions from logic or moral intuition rather than from religious teachings or revealed truths. Secular ethics and religious ethics can and often do reach the same conclusions, and may do so by the same means; there are both secular and religious articulations of utilitarianism, for instance. Major types of ethics include descriptive ethics (which describes the values people live by in practice), moral psychology (the study of how moral thinking develops in the human species), and applied ethics (addressing the ethical concerns of specific real-life situations and putting ethics into practice).

Exposure: The amount of radiation or pollutant present in a given environment that represents a potential health threat to living organisms.

F

Fair Trade: A certification scheme that evaluates the economic, social, and environmental impacts of the production and trade of agricultural products, in particular coffee, sugar, tea, chocolate, and others. Fair trade principles include fair prices, fair labor conditions, direct trade, democratic and transparent organizations, community development, and environmental sustainability.

Farmers Market: Farmers markets are where local farmers gather to sell their produce or specialty goods in a specific place at a designated time. All food bought at a farmers' market is probably not produced using green or organic practices, but in general, the selection of organic food is broader than at a supermarket.

FSC Certification: The Forest Stewardship Council (FSC) is an international nonprofit organization promoting responsible stewardship of the global forests. FSC certifies forests and forest products that fulfill its requirements for responsible forest stewardship.

Fugitive Emissions: Emissions not caught by a capture system.

G

Genetic Engineering: The manipulation of an organism's genome by recombining rDNA from one or more organisms.

Geothermal Energy: Any and all energy produced by the internal heat of the Earth.

Greenhouse Effect: The warming of Earth's atmosphere attributed to a buildup of carbon dioxide or other gases. Some scientists think that this buildup allows the sun's rays to heat Earth, while making the infrared radiation atmosphere opaque to infrared radiation, thereby preventing a counterbalancing loss of heat.

Greenhouse Gas Emissions: Any emissions that are released by humans (though naturally occurring in the environment), mainly through the combustion of fossil fuels. These emissions have a warming potential as they persist in the atmosphere, contributing to the greenhouse effect.

Green Job: Broadly defined as a job that contributes to improving environmental quality.

Green Purchasing: The practice of selecting products and services that minimize the ecological impact of an individual's or organization's day-to-day activities. Many organizations implement a green purchasing policy with guidelines for purchasing agents to select the "greenest" products and services available.

Greenwashing: A marketing ploy for businesses to jump onto the green movement bandwagon. They are not genuinely interested in sustainability, but are simply trying to improve their standing with the public by paying lip service. A company interested in "going green" for public relation reasons is greenwashing.

H

Hybrid Vehicle: Vehicles that use both a combustible form of fuel (gasoline, ethanol, and so forth) and an electric motor to power them. Hybrid vehicles use less gasoline than traditional combustion engines, and some even have an electric plug-in to charge the battery.

I

Irradiation: Exposure to radiation of wavelengths shorter than those of visible light (gamma, x-ray, or ultraviolet), for medical purposes, to sterilize milk or other foodstuffs, or to induce polymerization of monomers or vulcanization of rubber.

L

LD 50/Lethal Dose: The dose of a toxicant or microbe that will kill 50 percent of the test organisms within a designated period. The lower the LD 50, the more toxic the compound.

LEED (Leadership in Energy and Environmental Design): Responsible for creating the Green Building Rating System that encourages and accelerates global adoption of sustainable green building and development practices through the creation and implementation of universally understood and accepted tools and performance criteria.

Life Cycle of a Product: All stages of a product's development, from extraction of fuel for power to production, marketing, use, and disposal.

Lifetime Exposure: Total amount of exposure to a substance that a human receives in a lifetime (usually assumed to be 70 years).

Lowest Acceptable Daily Dose: The largest quantity of a chemical that will not cause a toxic effect, as determined by animal studies.

M

Megawatt: One thousand kilowatts, or 1 million watts; standard measure of electric power plant generating capacity.

Moral Relativism: The acknowledgment that different cultures have different moral standards. There are various levels of moral relativism, from the weak descriptivist articulation that simply acknowledges and describes those differences, to the normative position that says that there is no universal moral standard, only culturally derived morals. Moderate positions often propose that there are certain key moral standards that form a universal ethical core, such as taboos on murder, incest, or parental neglect. The question of which moral standards are universal becomes important when cultures deal with one another, and when international bodies mediate between them. Most differences are not about matters as obvious or seemingly clear-cut as murder, but may instead bear on matters of justice, or on the distribution of responsibility. The questions of who has the responsibility to do something about climate change, or of the ethical importance of avoiding polluting behaviors, vary widely around the world. The opposite of relativism is universalism.

N

Nanotechnology: The manipulation of matter at the nano-scale, typically 100 nanometers or less. Some nanoparticles have been suggested to have adverse impacts on human or animal health.

Net Metering: A method of crediting customers for electricity that they generate on-site in excess of their purchased electricity consumption. Customers with their own generation offset the electricity they would have purchased from their utility. If such customers generate more than they use in a billing period, their electric meter turns backward to indicate their net excess generation. Depending on individual state or utility rules, the net excess generation may be credited to the customer's account (in many cases at the retail price), carried over to a future billing period, or ignored.

Net-Zero Energy: Characteristic of a building that produces as much energy as it consumes on an annual basis, usually through incorporation of energy production from renewable sources such as wind or solar.

NGO: A nongovernmental organization is an institution operating independently from government that does not function as a private business. Also known as civil society organizations, these groups typically act in the public interest or toward some broader political, cultural, or social goals.

NIMBY: An acronym for "not in my backyard" that identifies the tendency for individuals and communities to oppose the placing of noxious or hazardous materials and activities in their vicinity. It implies a limited or parochial political vision of environmental justice.

North–South: A model of the world that contrasts the industrialized, developed, wealthy countries of the global north with the developing, poorer countries of the global south. Geography here is partially figurative, with Australia and New Zealand included in the global north, and a number of African, Middle Eastern, and Asian nations in the northern hemisphere included in the global south. The term became popular in the wake of the Cold War, when a new way of distinguishing between the developed (first and second worlds) and developing (third world) was desired. However, while there are many political and cultural ties between the nations of the global north, the global south—much like the third world—is varied enough to invite criticism of the model's accuracy and usefulness.

P

Pacific Gyre: Otherwise known as the Great Pacific Garbage Patch, it is a gyre of small bits of marine garbage, including chemical sludge and pelagic plastic, thought to be larger than the state of Texas.

Persistent Toxic Chemicals, Persistent Pollutants: Detrimental materials, like styrofoam or DDT, that remain active for a long time after their application and can be found in the environment years, and sometimes decades, after they were used.

Photochemical Smog: Air pollution caused by chemical reactions of various pollutants emitted from different sources.

Planned Obsolescence: The art of making a product break or fail after a certain amount of time. The failure of the product does not occur in a period of time that you will blame the manufacturer, but soon enough for you to buy another one and make more profit for the manufacturer.

Political Ecology: A field of research concerned with the relationship of systems of social and economic power to environmental conditions, natural resources, and conservation.

Pollution: Generally, the presence of a substance in the environment that, because of its chemical composition or quantity, prevents the functioning of natural processes and produces undesirable environmental and health effects. Under the Clean Water Act, for example, the term has been defined as the man-made or man-induced alteration of the physical, biological, chemical, and radiological integrity of water and other media.

Pollution Prevention: Identifying areas, processes, and activities that create excessive waste products or pollutants in order to reduce or prevent them through alteration or eliminating a process. Such activities, consistent with the Pollution Prevention Act of 1990, are conducted across all U.S. Environmental Protection Agency programs and can involve cooperative efforts with such agencies as the Departments of Agriculture and Energy.

Polychlorinated Biphenyls: A group of toxic, persistent chemicals used in electrical transformers and capacitors for insulating purposes, and in gas pipeline systems as a lubricant. The sale and new use of these chemicals, also known as PCBs, were banned by law in 1979.

Power: Energy that is capable or available for doing work; the time rate at which work is performed.

Precautionary Principle: A philosophy that states that policymakers should not wait for scientific proof of harmful effects before taking steps to limit harmful environmental and human health impacts from new products or activities. Specific areas of application include genetically modified food products and chemicals that may have harmful developmental effects in low doses.

R

Radioactive Waste: Any waste that emits energy as rays, waves, streams, or energetic particles. Radioactive materials are often mixed with hazardous waste, from nuclear reactors, research institutions, or hospitals.

Recycling: The process by which materials that would otherwise become solid waste are collected, separated or processed, and reused in the form of raw materials or finished goods.

Risk: A measure of the probability that damage to life, health, property, and/or the environment will occur as a result of a given hazard.

Risk Assessment: Qualitative and quantitative evaluation of the risk posed to human health and/or the environment by the actual or potential presence and/or use of specific pollutants.

S

Semiconductor: Any material that has a limited capacity for conducting an electric current.

Smog: Air pollution typically associated with oxidants. The word is a portmanteau of "smoke" and "fog."

Sustainability: To give support or relief to, to carry, to withstand, or to meet the needs of the present without compromising the ability of the future generations to meet their needs.

Sustainable Seafood: The act of not overfishing, which causes the possibility of extinction or adverse effects on a habitat.

T

Toxicity: The degree to which a substance or mixture of substances can harm humans or animals.

Turbine: A device for converting the flow of a fluid (air, steam, water, or hot gases) into mechanical motion.

U

U.S. Department of Agriculture (USDA): Established by President Abraham Lincoln in 1862, the USDA is an umbrella organization encompassing all aspects of farming production that has executive and legislative authority to ensure food safety and protect national resources. Active operating units include the National Organic Program, Agricultural Resource Service, Food Safety and Inspection Service, Risk Management Agency, and Animal and Plant Health Inspection Service.

V

VOCs (Volatile Organic Compounds): Gases emitted from liquid or solid substances that may cause short-term and long-term harmful health effects. Examples of products containing VOCs include paints and lacquers, paint strippers, cleaning supplies, pesticides, building materials and furnishings, office equipment such as copiers and printers, correction fluids and carbonless copy paper, graphics and craft materials including glues and adhesives, permanent markers, and photographic solutions.

W

Water Pollution: Includes chemicals and debris that render water unusable for natural habitat, human consumption, and recreation.

Watershed Approach: A coordinated framework for environmental management that focuses public and private efforts on the highest-priority problems within hydrologically defined geographic areas, taking into consideration both ground and surface water flow.

Wildlife Refuge: An area designated for the protection of wild animals, within which hunting and fishing are either prohibited or strictly controlled.

Dustin Mulvaney
University of California, Berkeley

Sources: U.S. Environmental Protection Agency (http://www.epa.gov/OCEPAterms), U.S. Energy Information Administration (http://www.eia.doe.gov/tools/glossary)

Green Issues and Debates Resource Guide

Books

Aldy, J. E., and R. N. Stavins, eds. *Architectures for Agreement: Addressing Global Climate Change in the Post-Kyoto World.* New York: Cambridge University Press, 2007.

Allen, Gary and Ken Abala. *The Business of Food: Encyclopedia of the Food and Drink Industries.* Westport, CT: Greenwood Press, 2007.

Allin, Craig W. *The Politics of Wilderness Preservation.* Westport, CT: Greenwood Press, 1982.

Beauchamp, T. and J. Childress, eds. *Principles of Biomedical Ethics.* Oxford, UK: Oxford University Press, 2008.

Beder, Sharon. *Environmental Principles and Policies.* Sydney, Australia: UNSW Press, 2006.

Berry, R. *The Ethics of Genetic Engineering.* London: Routledge, 2007.

Berry, Wendell. *The Unsettling of America: Culture and Agriculture.* New York: Avon, 1977.

Brohé, Arnaud. *Carbon Markets. An International Business Guide.* London: Earthscan, 2009.

Burley, J., ed. *The Genetic Revolution and Human Rights.* Oxford, UK: Oxford University Press, 1999.

Caldicott, Helen. *Nuclear Power Is Not the Answer.* New York: New Press, 2006.

Callicott, J. Baird and Michael P. Nelson, eds. *The Great New Wilderness Debate.* Athens: University of Georgia Press, 1998.

Chen, Joseph S., Philip Sloan, and Willy Legrand. *Sustainability in the Hospital Industry.* Oxford, UK: Butterworth-Heinemann, 2009.

Clifton, Sarah-Jayne. *A Dangerous Obsession: The Evidence Against Carbon Trading and for Real Solutions to Avoid a Climate Crunch.* London: Friends of the Earth, 2009.

Cravens, Gwyneth. *Power to Save the World: The Truth About Nuclear Energy.* New York: Alfred A. Knopf, 2007.

de Sadeleer, Nicolas. *Environmental Principles: From Political Slogans to Legal Rules.* Oxford, UK: Oxford University Press, 2002.

Domini, A. *Socially Responsible Investing.* Chicago, IL: Dearborn Trade, 2001.

Donaldson, Scott. *The Suburban Myth.* New York: Columbia University Press, 1969.

Durham, Leslie A. 2005. *Good Growing: Why Organic Farming Works.* Lincoln: University of Nebraska Press, 2005.

Dyson, A. and J. Harris, eds. *Ethics and Biotechnology*. London: Routledge, 1994.

Elliott, David, ed. *Nuclear or Not? Does Nuclear Power Have a Place in a Sustainable Energy Future?* New York: Palgrave Macmillan, 2007.

Escohotado, Antonio. *A Brief History of Drugs: From the Stone Age to the Stoned Age*. Rochester, VT: Park Street Press, 1999.

Esty, Daniel and Andrew S. Winston. *Green to Gold: How Smart Companies Use Environmental Strategy to Innovate, Create Value, and Build Competitive Advantage*. Hoboken, NJ: John Wiley & Sons, 2009.

Feenberg, Andrew. *Questioning Technology*. London: Routledge, 1999.

Finegold, D., et al. *Bioindustry Ethics*. Oxford, UK: Elsevier, 2005.

Fitzpatrick, Kevin and Mark LaGory. *Unhealthy Places: The Ecology of Risk in the Urban Landscape*. London: Routledge, 2000.

Friedheim, Robert L., ed. *Toward a Sustainable Whaling Regime*. Seattle: University of Washington Press, 2001.

Friedman, Laura S., ed. *Organic Food and Farming*. Farmington Hills, MI: Greenhaven Press, 2010.

Gipe, P. *Wind Power: Renewable Energy for Home, Farm, and Business*. White River Junction, VT: Chelsea Green Publishing Co, 2004.

Glover, J. *What Sort of People Should There Be?* New York: Penguin, 1984.

Goodman, Paul. *New Reformation: Notes of a Neolithic Conservative*. Oakland, CA: PM Press, 2010.

Graedel, T. E. and Braden R. Allenby. *Industrial Ecology and Sustainable Engineering*. Upper Saddle River, NJ: Prentice Hall, 2009.

Green, J. D. and J. K. Hartwell, eds. *The Greening of Industry: A Risk Management Approach*. Cambridge, MA: Harvard University Press, 1997.

Guglar, Josef, ed. *The Urban Transformation of the Developing World*. New York: Oxford University Press, 1996.

Guha, Ramachandra. *How Much Should a Person Consume? Environmentalism in India and the United States*. Berkeley: University of California Press, 2006.

Halweil, Brian. *Home Grown: The Case for Local Food in a Global Market*. Washington, DC: Worldwatch Institute, 2002.

Harris, J. *Clones, Genes and Immortality*. Oxford, UK: Oxford University Press, 1998.

Harris, J. *Enhancing Evolution*. Princeton, NJ: Princeton University Press, 2007.

Herbst, Alan M. and George W. Hopley. *Nuclear Energy Now: Why the Time Has Come for the World's Most Misunderstood Energy Source*. Hoboken, NJ: John Wiley & Sons, 2007.

Hillman, Mayer and Tina Fawcett. *How We Can Save the Planet*. New York: Penguin, 2004.

Hinrichs, C. C. and T. A. Lyson. *Remaking the North American Food System: Strategies for Sustainability*. Lincoln: University of Nebraska Press, 2007.

Hodge, R. *Genetic Engineering*. New York: Facts on File, 2009.

Jacobson, Mark Z. *Atmospheric Pollution: History, Science and Regulation*. New York: Cambridge University Press, 2002.

Jamuna, Carroll, ed. *The Pharmaceutical Industry*. Farmington Hills, MI: Greenhaven, 2008.

Jansen, Kees and Sietze Vellema. *Agribusiness and Society: Corporate Responses to Environmentalism, Market Opportunities and Public Regulation*. New York: Zed Books, 2004.

Kincheloe, J. L. *The Sign of the Burger: McDonald's and the Culture of Power*. Philadelphia, PA: Temple University Press, 2002.

Kovel, Joel. *The Enemy of Nature: The End of Capitalism or the End of the World?* Zed Books: London, 2007.

Magdoff, Fred, John Bellamy Foster, and Frederick H. Buttel, eds. *Hungry for Profit: The Agribusiness Threat to Farmers, Food, and the Environment.* New York: Monthly Review Press, 2000.

Milani, Brian. *Designing the Green Economy: The Postindustrial Alternative to Corporate Globalization.* Lanham, MD: Rowman & Littlefield, 2000.

Nelson, Michael P. and J. Baird Callicott, eds. *The Wilderness Debate Rages On: Continuing the Great New Wilderness Debate.* Athens: University of Georgia Press, 2008.

Newkirk, Ingrid. *Free the Animals: The Untold Story of the U.S Animal Liberation Front and Its Founder, "Valerie."* Chicago, IL: Noble Press, 1992.

Norberg-Hodge, Helena, Todd Merrifield, and Steven Gorelick. *Bringing the Food Economy Home: Local Alternatives to Global Agribusiness.* West Hartford, CT: Kumarian Press, 2002.

Nussbaum, M. and C. Sunstein, eds. *Clones and Clones.* New York: W. W. Norton, 1998.

Perrow, Charles. *Normal Accidents: Living With High-Risk Technologies.* Princeton, NJ: Princeton University Press, 1999.

Regan, Tom. *The Case for Animal Rights.* Berkeley: University of California Press, 1983.

Ruse, M. and C. Pynes, eds. *The Stem Cell Controversy.* New York: Prometheus Books, 2003.

Sarkar, Saral. *Eco-Socialism or Eco-Capitalism? A Critical Analysis of Humanity's Fundamental Choices.* London: Zed Books, 1999.

Sarni, William. *Greening Brownfields: Remediation Through Sustainable Development.* New York: McGraw-Hill, 2009.

Shrader-Frechette, Kristin. *Environmental Justice: Creating Equality, Reclaiming Democracy.* Oxford, UK: Oxford University Press, 2002.

Singer, Peter. *Animal Liberation.* New York: Random House, 1990.

Singer, P. and H. Kuhse. *Should the Baby Live?* Oxford, UK: Oxford University Press, 1985.

Siracusa, Joseph M. *Nuclear Weapons: A Very Short Introduction.* New York: Oxford University Press, 2008.

Sorrell, S. and J. Skea, eds. *Pollution for Sale: Emissions Trading and Joint Implementation.* Cheltenham, UK: Edward Elgar, 1999.

Stewart, R. B. and J. B. Wiener. *Reconstructing Climate Policy: Beyond Kyoto.* La Vergne, TN: AEI Press, 2003.

Tietenberg, T. H. *Emissions Trading: Principles and Practice.* Washington, DC: RFF Press, 2006.

Victor, D. G. *The Collapse of the Kyoto Protocol and the Struggle to Slow Global Warming.* Princeton, NJ: Princeton University Press, 2004.

Wisnioski, Matthew. *Engineers for Change: America's Culture Wars and the Making of New Meaning in Technology.* Cambridge, MA: MIT Press, 2010.

Journals

Advances in Energy Research
Animal Science Paper and Review
Annual Review of Phytopathology

Biocontrol Science and Technology
Biological Conservation

Biological Control
Biomass and Bioenergy

Energy Policy
Environmental Economics
Environmental Science & Technology
Environment and Behavior
Ethical Corporation
The European Journal of Public Health
European Physical Journal Special Topics

International Journal of Life Cycle Assessment
Issues in Legal Scholarship

Journal of Consumer Behavior
Journal of Ecology and Natural Environment
Journal of Environmental Economics and Management
Journal of Environmental Engineering
Journal of International Biotechnology Law
Journal of International Wildlife Law & Policy
Journal of Invertebrate Pathology
Journal of Personality and Social Psychology
Journal of Phycology

Nature Geoscience
New Phytologist
New Scientist

The Plant Cell

Renewable Agriculture and Food Systems

Science

Trends in Ecology and Evolution

Vermont Journal of Environmental Law

Websites

Biomass Energy Resource Center
   www.biomasscenter.org

Carbon Tax Center
   www.carbontax.org

Carbon Trade Watch
   www.carbontradewatch.org

Clean Energy
   www.eap.gov/clearnrgy/index.html

Dirty Oil Sands
   www.dirtyoilsands.org

Environmental Protection Agency
www.epa.gov

European Sustainable Investment Forum
www.eurosif.org

Industry and Technology: EU Ecolabel
http://ec.europa.eu/environment/ecolabel/index_en.htm

Monterey Bay Aquarium's Seafood Watch
www.montereybayaquarium.org/cr/cr_seafoodwatch/sfw_aboutsfw.aspx

United Nations Environment Programme
www.unep.org

United Nations 2015 Millennium Development Goals
www.un.org/millenniumgoals

Yale Environment 360: Opinion, Analysis, Reporting & Debate
http://e360.yale.edu

Green Issues and Debates Appendix

Carbon Tax Center

www.carbontax.org

This website was created to publicize and support the work of the Carbon Tax Center, a nonprofit nongovernmental organization founded in 2007 by Charles Komanoff and Daniel Rosenblum and headquartered in New York City. The website collects information about carbon taxation and its alternatives (e.g., a cap-and-trade system for carbon emissions or government subsidy of alternative fuels) and advocates for adoption of a carbon tax (or carbon fee: basically a tax imposed on fuels according to their carbon content) in the United States. The website explains in basic terms what a carbon tax is and why some feel it is the best alternative to achieve the goals of reducing U.S. carbon emissions and thus the U.S. contribution to global warming. The website includes numerous documents and multimedia materials supporting this point of view, information about current carbon taxes in the United States and globally, information about local initiatives, and a digest of links to news items related to global warming and carbon taxation.

Clean Energy

www.epa.gov/cleanenergy/index.html

This website is run by the U.S. Environmental Protection Agency (EPA) that provides information about Clean Energy Programs run by the EPA as well as basic information such as the impact of different methods of energy generation on the environment. EPA Clean Energy Programs covered on this website include the Combined Heat and Power Partnership, the Green Power Partnership, the State and Local Climate and Energy Program, and the Energy Star program. Tools available on the website include the Waste Energy Recovery Registry, a database of clean energy resources, the searchable eGRID database that has information about greenhouse gas emissions by ZIP code, a calculator to move between different measures of greenhouse gases (for instance, 1 million metric tons of carbon dioxide emissions is the same as the greenhouse gas emissions produced annually by 183,000 passenger vehicles or the electricity use of 110,095 homes), the Rapid Deployment Energy Efficiency toolkit that contains design and implementation guides for 10 energy efficiency programs, and downloadable information comparing historical and projected costs of various types of wind, solar photovoltaic, solar thermal, and geothermal energy technologies.

European Sustainable Investment Forum

www.eurosif.org

This is the website of Eurosif, a pan-European not-for-profit organization whose mission is to address sustainability through financial markets. Eurosif was founded by national social investment forums from five European nations (France, Germany, Italy, the Netherlands, and the United Kingdom) and includes affiliates such as pension funds, financial service provides, academic institutes, nongovernmental organizations, and research associations. The organization's activities include lobbying, research, initiatives such as trustee education and guideline development and organization, and participation in European and international events. The website includes basic information about socially responsible investment, resources for socially responsible investment in 13 European countries, press releases and press contact information, news relating to socially responsible investment, and information about upcoming events. Many publications and multimedia resources relating to socially responsible investment are available on the website that are organized by content area (e.g., biodiversity, green real estate, safety in the workplace) and investment sector (e.g., banking, real estate, hotel, and tourism).

Industry and Technology: EU Ecolabel

http://ec.europa.eu/environment/ecolabel/index_en.htm

This website is available in all European Union (EU) languages and is run by the European Commission on the Environment about the EU licensing Ecolabel scheme, begun in 1992, which is intended to recognize environmentally friendly products across all product classes (as opposed to schemes devoted to one class of product such as footwear or laundry detergents) or that are specific to a single country or small group of countries (e.g., the Nordic Swan, the Blue Angel). The website explains how the Ecolabel application and licensing process works, what the Ecolabel means, and its relationship to similar schemes. The website includes a searchable list of companies that produce products that carry the Ecolabel and a searchable catalog of Ecolabel products by country, product or service category, manufacturer or service provider, and retail source. The website also includes statistical information about the EU Ecolabel (e.g., the number of licenses awarded annually and the distribution of licenses among product groups, information about the Green Public Procurement process, information about carbon footprinting in general and with regard to the Ecolabel, and links to relevant legal documents and reports and studies).

United Nations 2015 Millennium Development Goals

www.un.org/millenniumgoals

This website of the United Nations (UN) is dedicated to the eight Millennium Development Goals (MDGs) established in September 2000 with the goal of reducing extreme global poverty by 2015. There is a separate section for each of the eight goals (End Poverty and Hunger, Universal Education, Gender Equality, Child Health, Maternal Health, Combat HIV/AIDS, Environmental Sustainability, Global Partnership), and each includes details on specific measurable targets for each goal and news about the goal, including progress toward meeting the goal. For instance, under Environmental Sustainability, there are four targets that specifically address sustainable development and environmental protection, reduction in biodiversity loss, increasing access to safe drinking water and sanitation, and achieving significant improvements in the lives of slum dwellers. The website includes links to media contacts and UN Partners on the MDGs (e.g., the World Bank, the World Health Organization, and the United Nations Children's Fund), annual reports on the MDGs, links to statistics relevant to the MDGs, and press releases.

Yale Environment 360: Opinion, Analysis, Reporting, and Debate

http://e360.yale.edu

This website is an online magazine published by the Yale School of Forestry and Environmental Studies that aggregates information about global environmental issues. The website includes news, analysis, and feature and opinion articles written specifically for the magazine as well as a daily digest of environmental news with links to the sources (newspapers and magazine, press releases, etc.); multimedia content (e.g., a 20-minute video on mountaintop removal mining in Appalachia); and links to other sources of information about the global environment. Information is organized by department (opinion, reports, analysis, interviews, e360 digest, and video reports), topics (biodiversity, business and innovation, climate, energy, forests, oceans, policy and politics, pollution and health, science and technology, sustainability, and water), and regions (Antarctica and the Arctic, Africa, Asia, Australia, Central and South America, Europe, Middle East, and North America). Website content is also available for mobile devices and as an RSS feed.

Sarah Boslaugh
Washington University in St. Louis

Index

Articles titles and their page numbers are in **bold**.